The Comprehensive Valve Dictionary

This book is a pioneering reference work designed to address the complex terminology, technical specifications, and diverse applications of valves in industrial systems.

The Comprehensive Valve Dictionary: Terminology, Applications, and Insights for Oil, Gas, and Renewable Energy is an authoritative and practical reference that demystifies the technical language surrounding industrial valves. The dictionary provides detailed definitions for hundreds of valve-related terms, including valve types, components, materials, standards, testing methods, and failure mechanisms. In addition to traditional applications, this dictionary places significant emphasis on valves used in renewable energy systems, including those for hydrogen production, hydropower, carbon capture and storage (CCS), geothermal energy, and offshore wind power. Moreover, it explores the definitions and real-world applications of cutting-edge technologies such as smart valves, condition monitoring, digital twin technology, and additive manufacturing. It also includes cross-references to relevant international standards (e.g., API, ASME, ISO), troubleshooting guides, and operational insights. By bridging gaps in understanding, this resource will support professionals in making informed decisions, improving system reliability, and ensuring compliance with industry best practices.

This work is specifically tailored for engineers, designers, manufacturers, and researchers working in oil and gas, petrochemicals, power generation, and renewable energy industries.

The Comprehensive Valve Dictionary

Terminology, Applications, and Insights for Oil, Gas, and Renewable Energy

Karan Sotoodeh

CRC Press
Taylor & Francis Group
Boca Raton London New York

CRC Press is an imprint of the
Taylor & Francis Group, an informa business

Designed cover image: ©Shutterstock Images

First edition published 2026
by CRC Press
2385 NW Executive Center Drive, Suite 320, Boca Raton FL 33431

and by CRC Press
4 Park Square, Milton Park, Abingdon, Oxon, OX14 4RN

CRC Press is an imprint of Taylor & Francis Group, LLC

ISBN: 9781041073789 (hbk)
ISBN: 9781041075318 (pbk)
ISBN: 9781003640967 (ebk)

DOI: 10.1201/9781003640967

Typeset in Times
by KnowledgeWorks Global Ltd.

Contents

Preface ix
About the Author xi

Chapter 1 Introduction to Valves 1

Introduction 1
Part 1: Overview of Valve Functions and Applications 2
Part 2: Historical Development of Industrial Valves 15
Part 3: Importance of Valves 19
Questions and Answers 22

Chapter 2 Types of Valves 26

Introduction 26
Part 1: Classification of the Valves 27
Function 27
Material Design 28
Operational Considerations 29
Part 2: Detail Description of Industrial Valves 29
Combination Valve 39
Part 3: Valve Comparative Analysis 64
Part 4: Valve Selection Criteria 67
Questions and Answers 68

Chapter 3 Valve Components 71

Introduction 71
Questions and Answers 96

Chapter 4 Materials of Construction 99

Questions and Answers 154

Chapter 5 Valve Design and Standards 159

Questions and Answers 205

Chapter 6 Valve Manufacturing, Testing, and Quality Assurance 210

Introduction 210
Part 1: Valve Manufacturing 210
Part 2: Valve Testing 232

Part 3: Quality Assurance 249
Questions and Answers 257

Chapter 7 Valve Installation and Commissioning 262

Introduction 262
Part 1: Preparation and Preinstallation 263
Part 2: Valve Installation 267
Part 3: System Integration 275
Part 4: Testing and Verification 278
Part 5: Documentations and Records 282
Part 6: Packing and Preservation 285
Questions and Answers 290

Chapter 8 Valve Operation and Performance 293

Questions and Answers 331

Chapter 9 Valve Inspection, Maintenance, and Troubleshooting 335

Introduction 335
Part 1: Common Valve Failure Modes 336
Part 2: Inspection 341
Part 3: Maintenance Practices 342
Part 4: Troubleshooting 348
Part 5: Reliability Practices 355
Part 6: Operational Challenges 358
Part 7: Best Maintenance Practices 361
Questions and Answers 363

Chapter 10 Valve Safety, Reliability, and Compliance (SIL, SAR) 366

Questions and Answers 417

Chapter 11 Actuators and Valve Automation 421

Questions and Answers 453

Chapter 12 Valves in the Energy Industry: From Conventional to Emerging Applications: Hydrogen and Carbon Capture 456

Questions and Answers 507

Chapter 13 Valve Life Cycle and Cost of Ownership 513

Introduction to Valve Life Cycle and Cost of Ownership 513
Part 1: Valve Life Cycle Stages 513
Part 2: Cost Analysis 517
Part 3: Maintenance and Reliability 523
Part 4: Materials Technology and Valve Life Cycle 527
Part 5: Energy-Related Terms and Valve Cost 529
Questions and Answers 536

Chapter 14 Environmental Impact and Sustainability of Valves 541

Questions and Answers 565

Index 569

Preface

Valves are among the most essential components in industrial systems, playing a critical role in controlling the flow of liquids, gases, and slurries across a wide range of applications. From oil and gas production facilities to chemical plants, refineries, and power stations, valves are the essential elements ensuring the safety, reliability, and efficiency of these operations. Despite their widespread use, the vast diversity of valve types, materials, and functions makes them one of the most complex and challenging elements of engineering design and specification.

This book, *The Comprehensive Valve Dictionary: Terminology, Applications, and Insights for Oil, Gas, and Renewable Energy*, was born out of a desire to fill a critical gap in the industry's literature. As an engineer with nearly two decades of experience working on valves, actuators, piping, and material engineering, I realized there was a need for a definitive resource that could serve both professionals and newcomers alike. This dictionary is designed to provide a clear, concise, and comprehensive guide to the terminology, concepts, and standards associated with industrial valves.

The book is structured as a reference tool, offering definitions, illustrations, and explanations of more than thousands of terms relevant to valve technology. It covers a wide range of topics, from valve design and material selection to maintenance and troubleshooting. Each term is explained in a straightforward manner, with practical insights drawn from real-world applications in the oil and gas industry.

To enhance the learning experience, I have included multiple-choice questions at the end of each section, encouraging readers to test their knowledge and apply the concepts in practical scenarios. I have also provided a bibliography section for those who wish to delve deeper into specific topics.

This book is not only for valve engineers but also for piping, mechanical, process, and reliability engineers, as well as students, procurement professionals, and anyone involved in the energy industries. It is my hope that this dictionary will become an indispensable resource, equipping readers with the knowledge they need to make informed decisions and excel in their respective fields.

Finally, I would like to extend my gratitude to all the colleagues, mentors, and valve suppliers I have worked with over the years. Their expertise, guidance, and shared experiences have greatly enriched my understanding of valves and inspired me to compile this book.

I sincerely hope you find this book valuable and that it serves as a reliable companion in your engineering endeavors.

Karan Sotoodeh
November 2024

About the Author

Karan Sotoodeh is an Iranian author and engineer with nearly two decades of experience in the oil and gas industry, specializing in valves, actuators, piping, and material engineering. In his most recent role, he served as Senior/Lead Valve and Actuator Engineer at Baker Hughes, focusing on subsea oil and gas projects. Karan Sotoodeh earned his PhD in Safety and Reliability in Mechanical Engineering from the University of Stavanger in 2021. Over the years, he has established himself as a prolific contributor to the field, authoring 15 technical books on topics such as valves, piping, actuators, corrosion, material selection, and safety engineering. He has also published more than 50 papers in peer-reviewed journals and has been invited to present his work at international conferences in the United States, Germany, and China. Karan has collaborated with numerous valve suppliers across Europe, including those in the UK, Italy, Spain, France, Germany, and Norway, gaining extensive insight into global industry practices. Beyond his professional accomplishments, he is an avid traveler, runner, and swimmer who enjoys spending time in nature.

1 Introduction to Valves

INTRODUCTION

This chapter gives a comprehensive overview of valves, highlighting their significance as fundamental components across various industries for regulating and directing fluid flow within piping systems. It begins with an introduction section highlighting principles of valve functionality, their historical development, and their pivotal role in the energy sector. Through understanding essential historical milestones, technological advancement, and the diverse applications of valves, this chapter gives a deeper understanding of valve technology, which can help readers to understand the following chapters.

For greater clarity and ease of navigation, the chapter is organized into three distinct sections:

1. **Overview of Valve Functions and Applications**
   This section focuses on the functions of valves, including flow control, pressure regulation, and system isolation. It also highlights their applications across various industries. Furthermore, the importance of valves in the piping system is highlighted.
2. **Historical Development of Industrial Valves**
   In this section, readers and students will trace the evolution of valve technology, from ancient designs used in irrigation systems to the advanced valves employed in modern industrial operations. Key historical milestones and innovations are explained to provide context for the development of this essential technology.
3. **Importance of Valves in the Energy Industry: Oil and Gas Industry as well as Renewable Energy**
   The final section emphasizes on the critical role the valves play in the upstream, midstream, and downstream oil and gas industry as well as renewable energy. This part highlights their importance in maintaining safety, efficiency, and environmental protection within this demanding industry.

Each section is organized alphabetically, presenting key terms and concepts related to the topic. This structure makes the content easy to navigate as well as enabling readers to quickly locate specific terms or concepts of interest.

By the end of this chapter, readers will have a foundational understanding of valve technology, its historical significance, and its indispensable role in industrial systems, particularly in the oil and gas sector. This knowledge serves as a gateway to the more detailed discussions and technical insights covered in later chapters.

DOI: 10.1201/9781003640967-1

PART 1: OVERVIEW OF VALVE FUNCTIONS AND APPLICATIONS

This part focuses on terms related to what valves do, their general functions, and how they are used in various industries.

A

Actuator

A device that automates valve operation using electric, pneumatic, or hydraulic power. For a detailed discussion, refer to Chapter 11. Figure 1.1 illustrates a ball valve equipped with a pneumatic actuator mounted on top of the valve.

API

American Petroleum Institute (API), providing guidelines for valve design and testing. For a detailed discussion, refer to Chapter 5.

ASME

American Society of Mechanical Engineers (ASME), offering valve design, manufacturing, inspection, and testing codes and standards. For a detailed discussion, refer to Chapter 5.

Automated Valve

A valve that operates without manual intervention, using an actuator powered by electric, pneumatic, or hydraulic energy. Figure 1.1 shows an automated ball valve.

FIGURE 1.1 Ball valve with a pneumatic actuator mounted on top.

Automated valves are commonly used in industrial processes to improve precision, efficiency, and safety.

B

Body

The main casing of a valve that contains the internal components. See Chapter 3 for a detailed discussion on body.

Bonnet

The *bonnet* is an essential component of a valve that acts as a cover for the valve body. It serves multiple purposes, including enclosing and protecting the valve's internal parts such as the stem and packing. See Chapter 3 for a detailed information about bonnet. Body and bonnet of the ball valve are highlighted in Figure 1.1.

C

Cavitation

Cavitation in valves occurs when a liquid flowing through the valve has a rapid pressure drop below its vapor pressure. As a result of pressure drop, vapor bubbles form. As the fluid pass the pressure drop area, the pressure starts to recover. Due to this, vapor bubbles collapse severely, creating shock waves. This phenomenon can lead to significant damage to the valve's internal surfaces. Cavitation can create issues such as erosion, pitting, and noise, reducing the valve's performance and lifespan. Cavitation is most common in high-pressure drop applications. It can be mitigated by selecting appropriate valve designs, such as multistage or anti-cavitation trims, and optimizing operating conditions to minimize pressure fluctuations. Figure 1.2 shows the erosion of a valve stem in a globe valve due to cavitation.

Chemical Industry

The *chemical industry* is a sector that produces chemicals and chemical products. In these industries, raw materials such as oil and natural gas are transformed into a several substances used in various applications. These include industrial chemicals, pharmaceuticals, plastics, fertilizers, paints, and specialty chemicals. These products serve essential roles in industries like agriculture, healthcare, construction, and manufacturing.

Coating

Valve coating refers to the application of protective or functional layers on the surface of a valve. Coating is done to enhance valve material resistance to

FIGURE 1.2 Control valve plug erosion due to cavitation. (With permission: Shutterstock.)

corrosion, wear, erosion, and extreme environmental conditions. These coatings may include materials like epoxy, polyurethane, ceramic, or specialized metallic alloys. Coatings improve the valve's durability and extend its service life. Furthermore, they ensure reliable performance in challenging applications such as offshore environments and underground piping systems. See Chapter 4 for a detailed discussion on coating.

Control Valve

A *control valve* is a device used to regulate the flow, pressure, or temperature of a fluid. Control valves in process systems are changing the size of the flow passage in response to signals from a controller. Figure 1.3 shows a control valve and its internal components as well as the flow path. The flow enters the valve from the left and exit from the right. "Backseat" is incorrectly labeled in the image and should be "Seat."

Corrosion

Corrosion is a natural process through which metals and other materials deteriorate over time. This deterioration can occur for several reasons, including chemical or electrochemical reactions with the environment, as well as exposure to oxygen. Additionally, moisture, acids, and salts are other factors that contribute to corrosion. This degradation weakens the structure of the materials.

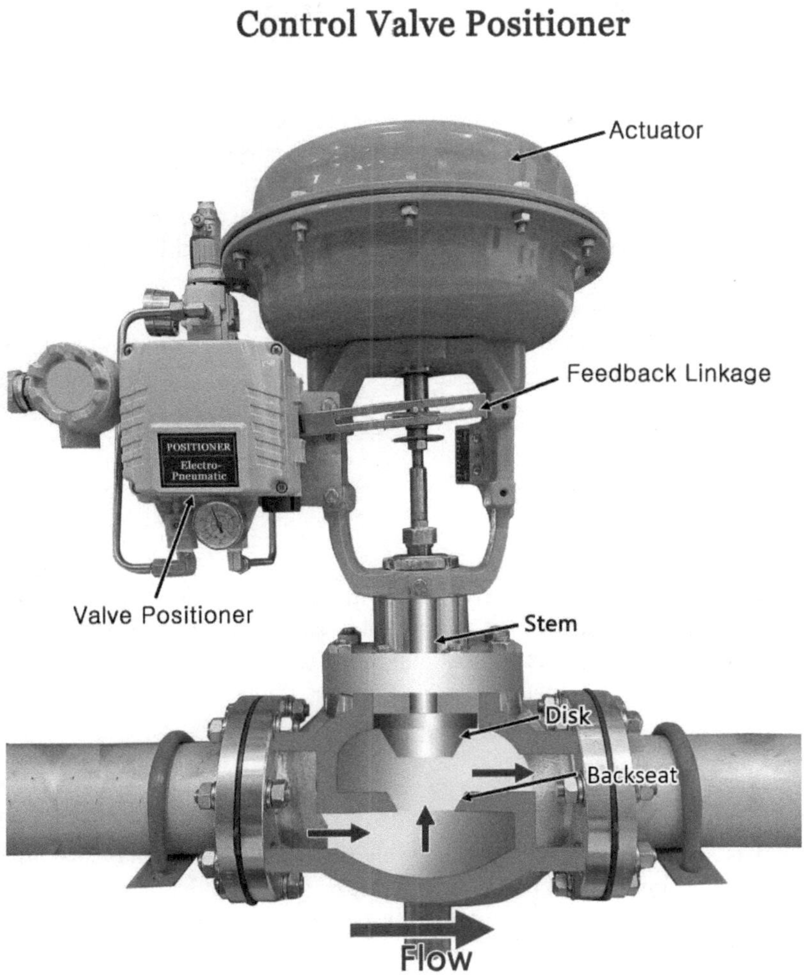

FIGURE 1.3 A detailed diagram of a control valve showing internal components and flow path.

Corrosion makes materials unsafe and unreliable for use in industries or everyday applications. Chapter 4 provides more detailed information about the different types of valve corrosion and the mechanisms involved.

Cryogenic Applications

Valves designed for extremely low-temperature conditions in hydrogen or liquified natural gas (LNG) plants. See Chapter 12 for a detailed discussion on cryogenic application.

D

Downstream

In the oil and gas industry, downstream refers to the final stage of the petroleum value chain. Downstream sectors include refining, distributing, and marketing petroleum products. This process begins after the extraction and production of crude oil and natural gas (upstream) as well as their initial processing (midstream). In oil and gas or other industrial systems, downstream refers to the direction of flow after a particular point in the process. Units or valves that are located or operate after the upstream component or unit are referred to as downstream components or units. The upstream is the opposite of the downstream. If, for example, a valve is located after a pump, we can say it is downstream of the pump. On the other hand, if the valve is located before the pump, it is called upstream valve.

E

Energy Industry

The energy industry includes various activities associated with the production, distribution, and consumption of energy. This encompasses fossil fuels such as oil, natural gas, and coal, as well as renewable sources like solar, wind, hydro, geothermal energy, and nuclear energy. Chapter 12 provides information about renewable energy plants and valves.

Erosion

Erosion is the process in which materials degrade or wear away due to the mechanical action of solid particles, liquids, or gases moving across a surface. In industrial systems, erosion commonly occurs when high-velocity fluids or abrasive particles strike components, resulting in surface damage, thinning, or gradual loss of material over time.

F

Flow Control

The process of managing the rate and direction of fluid movement.

Flow Path

The designed route for fluid movement through a valve.

Fluid

Any substance, liquid, gas, or slurry, controlled by valves in industrial processes.

Fugitive Emission

Fugitive emissions refer to the unintended release of gases or vapors from pressurized equipment, including valves, flanges, and seals, into the atmosphere. These emissions typically consist of greenhouse gases or volatile organic compounds (VOCs) and can occur due to leaks, wear, or improper sealing. They pose environmental, health, and safety concerns, particularly in industries such as oil and gas and chemical processing. More information about greenhouse gasses and VOCs is provided in Chapter 14.

H

High-Pressure Systems

High-pressure systems are specially engineered setups designed to operate at elevated pressures, typically exceeding 1,000 psi. These high pressures are commonly found in industries such as oil and gas, as well as power generation. Several types of valves are used in these industries to manage fluids or gases under extreme pressure conditions.

In high-pressure systems, valves play a crucial role, particularly those classified as Pressure Class 600 (Pressure Nominal [PN] 50 bar) and above, according to the ASME B16.34 valve standard. These valves regulate, isolate, and control fluid flow while being able to withstand high pressures. The safety, efficiency, and reliable performance of high-pressure class valves are essential in these demanding applications, as they contain high-pressure fluids. High-pressure valves are utilized in various systems, including hydraulic systems, steam pipelines, and high-pressure reactors.

HVAC

HVAC (heating, ventilation, and air-conditioning) refers to systems designed to regulate indoor temperature, humidity, and air quality. HVAC systems are found in residential, commercial, and industrial buildings. HVAC system valves control water, refrigerants, and airflow. These valves ensure precise temperature regulation, energy efficiency, and proper system operation by modulating or isolating fluid movement as needed.

Hydraulic System Valves

Hydraulic system valves are components within hydraulic systems that control the flow, pressure, and direction of hydraulic fluid (typically oil). These valves are essential for transmitting power and controlling motion in hydraulic machinery and components such as hydraulic actuators. Hydraulic system valves are used in industries like oil and gas, manufacturing, aerospace, and automotive.

I

Industrial Revolution

The Industrial Revolution was a period of significant transformation during the 18th and 19th centuries. This era is characterized by the shift from manual labor to mechanized production, driven by advancements in technology and machinery. It began in Great Britain and gradually spread around the world. The revolution had a profound impact on various industries, including textiles, manufacturing, and transportation. Additionally, advancements in valve technology played a crucial role in enabling the safe and efficient control of steam, water, and other fluids during this time.

Isolation Valves

A kind of valve (e.g., ball, gate, butterfly) used to completely shut off fluid flow for maintenance or safety purposes. See Chapter 2 for a detailed discussion on isolation valves.

L

Leakage

Unintended fluid leakage from a valve can be minimized through effective sealing. Figure 1.4 illustrates a control valve and its potential leakage points. The most critical area for leakage is the stem sealing, followed by the welding or flange connections on both the right and left sides where the valve interfaces with the piping. Another possible leakage point is the connection between the valve body and bonnet.

FIGURE 1.4 Potential leakage areas in a control valve. (With permission: Shutterstock.)

M

Manual Valve

A valve that is operated manually by means of a handle or handwheel in order to control the flow of fluid.

O

On/Off Valve

A valve type (e.g., ball, gate, butterfly) designed to start or stop fluid flow. For a detailed discussion, refer to Chapter 2.

P

Packing

Sealing material used to prevent leakage from the area between the valve stem and the valve body/bonnet. See Chapter 3 for more information about packing.

Piping/Pipeline

A system of connected pipes controlled by valves for fluid transportation.

The terms *piping* and *pipeline* are often used interchangeably. However, they refer to distinct systems with unique characteristics, applications, and design considerations. The main difference between piping and pipeline lies in their application and scope. *Piping* refers to the network of pipes within industrial facilities, such as refineries or power plants, used to transport fluids like steam, chemicals, or gas within a confined area. It is complex, with numerous fittings, valves, and branches, designed to handle dynamic conditions and varying pressures and temperatures. *Pipelines*, on the other hand, are long-distance systems primarily used to transport fluids like oil, gas, or water between distant locations, such as cities or production sites. Pipelines are simpler in design, primarily linear, and operate under stable conditions, with minimal fittings. Figure 1.5 shows a pipeline running through a desert landscape, supported by elevated structural supports to maintain stability.

Pneumatic Piping Valves

Pneumatic piping valves are devices designed to control the flow, pressure, and direction of compressed air in pneumatic systems. These valves are essential for operating various pneumatic devices, actuators, and equipment, including compressors. They are responsible for directing and regulating the compressed air needed to perform mechanical tasks. Pneumatic valves play a critical role in systems used in manufacturing, automation, and transportation.

FIGURE 1.5 A long-distance pipeline installed in a desert environment, supported by elevated structural supports. (With permission: Shutterstock.)

Pressure Control Valve

A pressure control valve is a type of valve used in fluid systems to regulate or maintain pressure within a specified range. Examples of pressure control valves include pressure relief valves (PRVs) and pressure safety valves (PSVs). Additionally, some control valves can adjust pressure by modifying the flow within the piping system. In certain cases, a valve may direct the flow to a specific location in the plant to increase pressure; these valves are also classified as pressure control valves.

Process Systems

Process systems are interconnected networks of equipment, piping, and controls that transport, manipulate, and process fluids or gases to achieve specific industrial objectives. These goals are refining, chemical production, or power generation. Valves are essential components in these systems since they regulate the flow, pressure, temperature, and direction of fluids. Industrial valves ensure safe, efficient, and precise operations under various process conditions.

R

Regulating Valve

A valve that controls the flow rate or pressure in a system. For a detailed discussion, refer to Chapter 2.

S

Safety Valve

A safety valve protects systems by releasing excess pressure to prevent failures.

Figure 1.6 illustrates a typical offshore piping system configuration with *three safety valves installed in parallel.* This setup allows for switching between the safety valves when one is removed for maintenance or calibration. The parallel arrangement ensures continuous overpressure protection, even when one safety valve is offline. In the picture, *three ball valves* are positioned upstream of the safety valves. These ball valves serve as isolation devices, enabling safe removal of an individual safety valve from the line without interrupting the operation of the remaining two safety valves or compromising system protection. This arrangement highlights the importance of maintaining system integrity while facilitating regular maintenance and calibration of critical safety equipment.

Seat

The part of the valve where the closure element seals to stop flow. See Chapter 3 for a detailed discussion on seat.

Smart Valve

A modern valve integrated with sensors and control systems for digital monitoring. See Chapter 12 for a detailed discussion on smart valve.

FIGURE 1.6 Offshore piping system with three parallel safety valves and upstream ball valves for isolation, allowing maintenance and calibration without disrupting system operations. (With permission: Shutterstock.)

Stem

Stem connects the actuator or handle to the valve closure mechanism. See Chapter 3 for a detailed discussion on stem.

Subsea Valves

Subsea valves are specialized components designed to control the flow of fluids (oil, gas, or water) in offshore subsea production systems. Subsea valves are operating at extreme depths and pressures.

T

Trim

A *valve trim* refers to the internal components such as disk and seat of a valve that come into contact with the fluid being controlled.

V

Valve

A mechanical device that regulates, directs, or controls fluid flow within a system.

Valve Automation

Valve automation refers to the process of using actuators and control systems to operate valves remotely or automatically, eliminating the need for manual intervention. It enhances efficiency, precision, and safety in industrial systems by enabling real-time control, monitoring, and integration with new technologies.

Valve Design

Valve design refers to the engineering process of creating valves to meet specific functional, safety, and performance requirements for controlling the flow, pressure, or direction of fluids in a system. It involves selecting appropriate materials, configurations (e.g., globe, gate, ball), sealing mechanisms, and actuation methods.

Figure 1.7 is a visualization of a valve design using finite element analysis (FEA), showing stress distribution across the valve components. The high-stress areas are highlighted in red, and low-stress areas are in blue, with a mesh overlay representing the analysis. FEA is a numerical simulation technique used to predict the physical behavior of structures and components under various conditions, such as mechanical stress, thermal loads, or fluid dynamics. FEA divides a complex structure into smaller, manageable finite elements (a process called meshing), allowing precise analysis of how forces and conditions affect each part of the structure.

FIGURE 1.7 Stress distribution in valve using finite element analysis (FEA). In the original color version, no red regions are present, indicating that no high-stress zones were detected. Most areas appear in blue, corresponding to low-stress levels, while bolt holes in all flange connections are shown in yellow, representing medium stress. In the black-and-white print version, these stress levels may appear with similar grey tones.

Valve Life Cycle

The *valve life cycle* refers to the stages a valve goes through from its initial design and manufacturing to installation, operation, maintenance, and eventual decommissioning or replacement.

Valve Maintenance

Valve maintenance involves the regular inspection, servicing, and repair of valves to ensure their optimal performance, safety, and duration in industrial systems. Key tasks include cleaning, lubricating moving parts, inspecting seals and gaskets for wear, and replacing damaged components such as stems, seats, or packing.

Valve Sizing

Determining the correct valve size based on flow and pressure needs.

Valve Standards

Valve standards are documented specifications established by recognized organizations that define the design, material, manufacturing, testing, and performance requirements for valves used in industrial applications. These standards, such as ASME, API, ISO, and DIN, ensure consistency, safety, compatibility, and quality across valve products. In addition, they enable engineers to select and use valves that meet specific operational and regulatory requirements. Chapter 5 provides more detailed information about the valve standards mentioned above.

Valve Testing

Valve testing is the process of evaluating a valve's performance, safety, and reliability under specified conditions by valve standards to ensure it meets operational and industry standards. Tests typically include pressure tests on the valve body (e.g., hydrostatic, pneumatic), leak tests, and functional tests to verify the valve's sealing. Specialized tests like fire testing, cryogenic testing, or fugitive emissions testing are performed for valves used in extreme (e.g., subsea) or safety-critical applications. Figure 1.8 shows a pressure test on hydraulically actuated through conduit gate valve.

Valve Types

A valve type can be categorized into various categories such as gate, globe, ball, or butterfly valves, which are used in a variety of applications.

FIGURE 1.8 Operator performing a pressure test on a hydraulically operated through conduit gate valve.

PART 2: HISTORICAL DEVELOPMENT OF INDUSTRIAL VALVES

Ancient Valve Design

Early civilizations like Romans used rudimentary valves such as wooden plugs. These valves were employed to manage water flow in irrigation systems and aqueducts. These basic mechanisms marked the first steps in fluid control engineering.

API Standards Establishment

The American Petroleum Institute established its first standards in 1924, formalizing valve design and testing requirements for the oil and gas industry.

Automation Revolution

See valve automation in this section.

Ball Valve Evolution

Popularized in the 1950s, ball valves became a versatile and reliable option for on/off control in various industries. First widely used in the 20th century, ball valves offered reliable, tight shutoff and became an essential component of modern fluid control technology.

Bronze Valves

Early valves made of bronze, used in ancient plumbing systems for durability and corrosion resistance. Bronze is an alloy primarily composed of copper and tin.

Butterfly Valve Invention

It was developed in the mid-20th century. The butterfly valve introduced a lightweight and compact design for controlling fluid flow in industrial pipelines.

Cast Iron Valves

Introduced during the Industrial Revolution, cast iron valves became popular due to their strength and cost-effectiveness. *Cast iron* is an iron–carbon alloy that typically contains *more than 2% carbon*. This high-carbon content differentiates cast iron from *carbon steel*, which usually has less than 2% carbon.

Control Valve Invention

It was introduced in the 20th century. The control valves were used to modulate flow, pressure, and temperature precisely.

Cryogenic Valve Innovation

Cryogenic valves were developed in the mid-20th century to handle extremely low-temperature fluids. An example of cryogenic fluid is LNG that is critical for modern energy applications. Figure 1.9 illustrates a cryogenic butterfly valve for an LNG plant.

Diaphragm Valves Development

These valves were originally developed in the 1920s for use in chemical applications due to their ability to handle corrosive fluids.

Digitalization of the Valves

The recent incorporation of digital technologies and smart systems into valve designs is aimed at improving monitoring and predictive maintenance. Chapter 9 offers more information about monitoring as a best practice for maintenance and explores various maintenance types, including predictive maintenance. A modern smart valve equipped with integrated sensors and a digital monitoring system, representing a cutting-edge solution for real-time industrial automation and precision. For a detailed discussion about digitalization, please refer to Chapter 12.

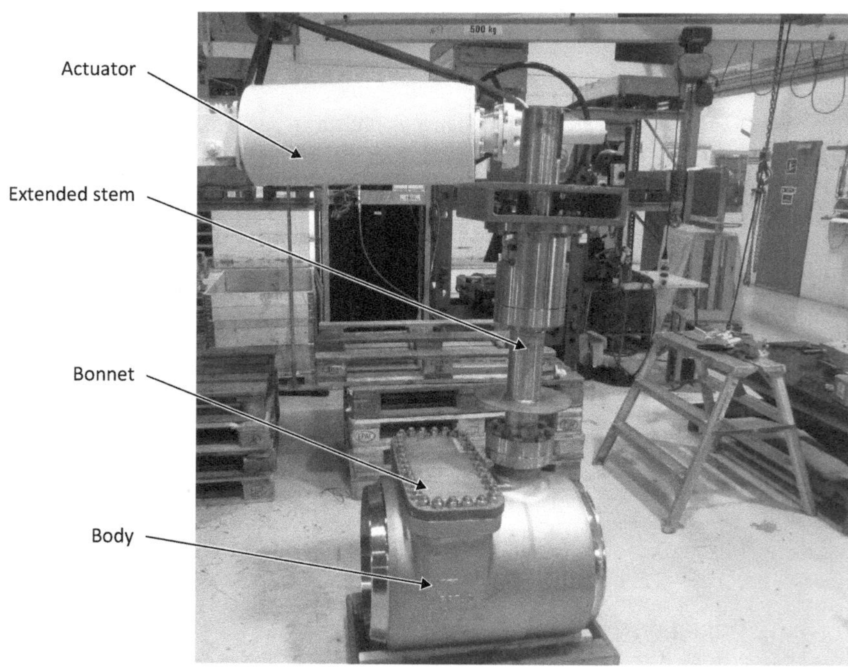

FIGURE 1.9 A cryogenic butterfly valve with manual operation and stem extension for an LNG plant to be used in cryogenic (extremely cold) service.

Industrial Revolution

This period in the late 18th and early 19th centuries saw significant advancements in valve technology. That was mainly driven by the need to control steam in engines and industrial machinery.

Industrial Standardization

The emergence of organizations like API and ASME in the early 20th century to standardize valve design and manufacturing. For a detailed discussion about valve standards, refer to Chapter 5.

Modern Valve Standards

As a result of the focus on environmental and safety standards of the 21st century, low-emission and fire-safe valve designs have been developed. Chapter 5 provides a detailed explanation of valve standards addressing fugitive emissions and fire-safe design.

Pipeline Expansion

The rise of large-scale pipelines in the early 20th century, which drove innovations in valve design to handle higher pressures and flow rates.

Plug Valves

One of the earliest valve designs is the plug valve, which features a cylindrical or conical plug to regulate flow. The plug valve was used extensively in ancient times. For a detailed discussion about the plug valve, refer to Chapter 2.

Pneumatic Actuation Invention

Introduced in the early 1900s, pneumatic actuators allowed valves to be operated using compressed air, enhancing efficiency and reducing manual labor.

Roman Aqueducts

The Roman Empire pioneered large-scale water distribution systems. They used simple valves to regulate water flow and pressure within their advanced aqueduct networks.

Safety Valve Patents

Steam-powered equipment was transformed into a safer environment through the invention of safety valves during the 19th century. Figure 1.10 shows couple of safety valves installed on the piping system to prevent overpressure scenarios in the piping systems, protecting assets, onsite personnel, and environment.

Steam Engine Valves

As steam engines began to power locomotives, factories, and ships during the Industrial Revolution, valves to manage steam pressure became essential components.

FIGURE 1.10 Installation of two safety valves on the piping system to prevent overpressure scenarios. (With permission: Shutterstock.)

Steam Trap Valves

Invented in the 19th century, steam trap valves remove condensate from steam systems while preventing the loss of live steam.

Valve Automation

The transition from manual to automated valves in the mid-20th century revolutionized process control, allowing for precise and remote operation of industrial systems. Figure 1.11 shows an industrial control room where an operator

FIGURE 1.11 An industrial control room with multiple screens displaying real-time data analysis and advanced automation of valves and pipeline systems. (With permission: Shutterstock.)

TABLE 1.1
Key Historical Valve Developments

Time Period	Key Development	Description
Ancient era	Basic water valves	Early civilizations, like the Egyptians and Romans, used rudimentary valves in aqueducts and irrigation systems
17th century	Invention of stopcock valves	Stopcock valves were introduced to control the flow of liquids in early plumbing systems
18th century	Industrial Revolution: steam valves	Steam valves became critical components in steam engines, powering factories, trains, and ships
19th century	Known materials used for valves commonly	Cast iron and bronze were commonly used for valve construction, improving durability and pressure-handling capabilities
Early 20th century	Introduction of standards (ASME/API)	Industry standards for valve design and performance were established, ensuring uniformity and safety
Early 20th century	Advancement in materials	New materials such as nickel alloys, stainless steels, and plastics were developed and used for piping and valves
Mid-20th century	Specialized valves for oil and gas	The oil and gas boom led to the development of high-pressure, high-temperature valves designed for harsh environments
Late 20th century	Automation and actuators	Pneumatic, hydraulic, and electric actuators became widely used, allowing remote and automated valve operation
21st century	Smart valves and IoT integration	Digital technologies and sensors enabled real-time monitoring, predictive maintenance, and integration into automated systems

monitors actuated valves and pipeline systems on multiple screens. The picture shows advanced automation and real-time data analysis.

Valve Material Advancement

Valve materials have evolved historically, such as stainless steel being introduced in the 20th century for better corrosion resistance.

To understand the innovations in modern valve technology, it is essential to recognize the key developments that have shaped their evolution, as summarized in Table 1.1.

PART 3: IMPORTANCE OF VALVES

Downstream Refining/Petrochemical Plants

Refineries and petrochemical plants rely on valves to regulate complex chemical reactions, control product flows, and ensure safety in high-pressure systems. Figure 1.12 shows a refinery and petrochemical plant, both integral components of the

FIGURE 1.12 A refinery and petrochemical plant representing the downstream sector of the oil and gas industry. (With permission: Shutterstock.)

downstream sector in the oil and gas industry. The facility includes large storage tanks, an extensive piping network, and various processing equipment used for refining crude oil and producing petrochemical products. The industrial infrastructure is designed to handle hydrocarbon processing, separation, and transportation.

Environment Protection

It is possible for poorly designed and tested valves to leak and cause pollution in the environment. As a result, valves are designed to minimize emissions and leaks, thereby contributing significantly to the protection of the environment. Chapter 14 provides more information about environment protection by valves.

Equipment Protection

Valves safeguard sensitive equipment, such as pumps and compressors. This is done by isolating them during maintenance or preventing overpressure conditions.

Flow Regulation

Valves are fundamental in managing the flow of liquids, gases, and slurries in industrial systems. This is done to ensure processes run smoothly and efficiently.

Midstream Application

Valves are vital in the transportation and storage of oil and gas in midstream oil and gas industry. Valves maintain safe flow control through pipelines and storage facilities.

Piping/Pipeline Integrity

Valves help maintain the structural integrity of pipelines by regulating pressure and preventing leaks that could cause damage or system failure.

Pressure Control

A valve maintains stable pressure within a system in order to protect the equipment, provide the reliability of the process, and prevent catastrophic failures.

Process Safety

Valves play a critical role in industrial safety by controlling pressure, isolating sections of a system, and preventing hazardous fluid releases.

System Efficiency

Properly designed and installed valves optimize fluid dynamics, reducing energy consumption and improving overall process efficiency.

Upstream Operation

In the exploration and production phases of oil and gas, valves control the flow of hydrocarbons, manage pressure, and protect equipment from harsh environments. An example is a subsurface safety valve that shuts down a well to prevent blowout in case of high pressure in the well. Figure 1.13 shows two interconnected

FIGURE 1.13 Two offshore platforms interconnected, with one dedicated to drilling and the other for oil and gas production in the upstream sector. Numerous valves are installed on both platforms, playing a crucial role in controlling flow and pressure during extraction and processing. (With permission: Shutterstock.)

offshore platforms—one designated for drilling and the other for oil and gas production in the upstream sector. Various valves are utilized on both platforms as essential components of the oil and gas extraction process.

QUESTIONS AND ANSWERS

1. What is the function of valve body?
   A. To seal the stem and prevent leakage
   B. To measure flow rates in pipelines
   C. To automate the control of fluid flow
   D. To house the internal components of the valve

   Answer: Option D is the correct answer.

2. Which term refers to the material used to seal the stem of a valve and prevent leakage?
   A. Packing
   B. Seat
   C. Trim
   D. Operator

   Answer: Option A is the correct answer.

3. What is the purpose of the valve seat in a valve system?
   A. To connect the valve to the actuator
   B. To reduce pressure fluctuations
   C. To provide a sealing surface for the closure element
   D. To reduce pressure fluctuations

   Answer: Option C is the correct answer.

4. Which part of a valve connects the actuator or handle to the closure mechanism?
   A. Bonnet
   B. Stem
   C. Trim
   D. Seat

   Answer: Option B is the correct answer.

5. What is the primary purpose of an actuator in a valve system?
   A. To measure fluid pressure
   B. To control the temperature of fluids
   C. To automate valve operation using power sources
   D. To reduce pipeline friction

   Answer: Option C is the correct answer.

6. Which ancient civilization is credited with using rudimentary valves in their water distribution systems?
   A. Greeks
   B. Egyptian
   C. Chinese
   D. Romans

   Answer: Option D is the correct answer.

7. What material was commonly used for valves in ancient plumbing systems due to its durability?
   A. Iron
   B. Wood
   C. Bronze
   D. Lead

   Answer: Option C is the correct answer.

8. During which historical period did valve technology see significant advancements due to the rise of steam engines?
   A. Industrial Revolution
   B. Ancient times
   C. Modern era
   D. Renaissance

   Answer: Option A is the correct answer.

9. Which valve type was widely used in early steam engines for pressure regulation?
   A. Globe valve
   B. Butterfly valve
   C. Safety valve
   D. Ball valve

   Answer: Option C is the correct answer.

10. The butterfly valve was first introduced in which century?
    A. 17th century
    B. 18th century
    C. 19th century
    D. 20th century

    Answer: Option D is the correct answer.

11. The first API standards for valves were introduced in which year?
    A. 1915
    B. 1924
    C. 1936
    D. 1948

    Answer: Option B is the correct answer.

12. What innovation in valve design became popular in the mid-20th century due to its versatility and reliable sealing?
    A. Plug valve
    B. Ball valve
    C. Gate valve
    D. Diaphragm valve

    Answer: Option B is the correct answer.

13. Which of the following contributed to the rise of digital and smart valve technologies?
    A. Hydraulic systems
    B. Industrial Revolution
    C. Advances in sensors and automation
    D. Development of cast iron valves

    Answer: Option C is the correct answer.

14. What is the primary function of a valve in an industrial system?
    A. Generating energy
    B. Regulating fluid flow
    C. Measuring fluid temperature
    D. Connecting pipelines

    Answer: Option B is the correct answer.

15. Valves play a critical role in which of the following safety functions?
    A. Preventing unauthorized access
    B. Regulating system pressure to prevent failure
    C. Controlling electrical flow
    D. Managing solid waste disposal

    Answer: Option B is the correct answer.

16. Which type of valve is used to completely shut off fluid flow for maintenance or safety purposes?
    A. Control valve
    B. Safety valve
    C. Needle valve
    D. Butterfly valve

    Answer: Option D is the correct answer.

17. How do valves contribute to environmental protection?
    A. By improving fluid velocity
    B. By preventing fluid leaks and emissions
    C. By controlling equipment vibrations
    D. By increasing pipeline pressure

    Answer: Option B is the correct answer.

18. In which part of the oil and gas process are valves crucial for transportation and storage?
    A. Upstream
    B. Midstream
    C. Downstream
    D. Refining

    Answer: Option B is the correct answer.

19. Valves protect sensitive equipment like pumps and compressors by:
    A. Releasing overpressurized liquids and gases from the equipment and isolating equipment during maintenance
    B. Enhancing fluid velocity
    C. Increasing system temperature
    D. Preventing fluid contamination

    Answer: Option A is the correct answer.

20. How do automated valves improve industrial systems?
    A. By eliminating the need for regular maintenance
    B. By reducing the need for system testing
    C. By increasing the size of pipelines
    D. By enabling precise and remote control of operations

    Answer: Option D is the correct answer.

BIBLIOGRAPHY

Crosby Valve Inc. (1997). Crosby Pressure Relief Valve Engineering Handbook. Technical document number TP-V300.

Hellemans, M. (2009). The Safety Relief Valve Handbook: Design and Use of Process Safety Valves to ASME and International Codes and Standards. 1st edition. Butterworth-Heinemann: Oxford, UK.

Nesbitt, B. (2007). Handbook of Valves and Actuators: Valves Manual International. 1st edition. Elsevier: Oxford, UK.

Petroleum Extension Service (PETEX). (2016). Introduction to Valves and Piping Systems: A Comprehensive Guide for the Oil and Gas Industry. 2nd edition. PETEX, University of Texas at Austin: Austin, TX.

Skousen, W. L. (2012). Valve Handbook. 3rd edition. McGraw-Hill Education: New York, NY.

Sotoodeh, K. (2021). A Practical Guide to Piping and Valves for the Oil and Gas Industry. 1st edition. Gulf Professional Publishing: Houston, TX.

Sotoodeh, K. (2021). Subsea Valves and Actuators for the Oil and Gas Industry. 1st edition. Gulf Professional Publishing: Houston, TX.

Sotoodeh, K. (2021). Prevention of Valve Fugitive Emissions in the Oil and Gas Industry. 1st edition. Gulf Professional Publishing: Houston, TX.

Sotoodeh, K. (2021). Prevention of Actuator Emissions in the Oil and Gas Industry. 1st edition. Gulf Professional Publishing: Houston, TX.

Spirax Sarco Ltd. (2007). Steam Trapping and Condensate Handling Handbook. 3rd edition. Spirax Sarco Ltd.: Cheltenham, UK.

Zappe, R. W., & Smith, P.. (2003). Valve Selection Handbook: Engineering Fundamentals for Selecting the Right Valve Design for Every Industrial Flow Application. 4th edition. Gulf Professional Publishing: Houston, TX.

2 Types of Valves

INTRODUCTION

Valves are important in industrial systems and plants because they control the flow of fluids, pressure, temperature, and direction in piping and process systems. Each type of valve is designed for a specific application, such as controlling the flow, stop the fluid flow by isolating parts of the system, or maintaining safety by releasing excess pressure in pressurized applications such as piping or pressure vessel. Various valves are available in energy industries (oil and gas, petrochemical, and renewable energy plants). From simple fluid flow on/off functions to complex fluid control systems in subsea or wind turbine applications, valves improve safety and reliability in many areas.

The purpose of this chapter is to provide an overview of the most commonly used valves by giving more details regarding valve design, operation, applications, and selection. Better understanding of the characteristics, working principles, and applications of each valve type can assist engineers in making decisions lead to improvement of the piping and process system efficiency, reliability, and safety. All valves are critical for industrial processes, whether gate valves for tight shutoff, ball valves for quick operation, or safety valves for overpressure protection. The following sections provide an overview of these valve types by giving more detail information about advantages, limitations, and practical considerations. This discussion aims to give readers the technical knowledge to choose and use the correct valve for their applications.

This chapter is organized to help readers to understand different aspects of valve types that is divided into four key sections:

1. **Classification of Valves**
   A description of the main types of valves is given in this section. This section provides a general overview of various valve categories. It covers flow control valves such as needle and globe valves, on/off valves like ball and butterfly valves, and more specific valves such as check and safety valves. The readers will have better understanding of valve classification based on their specific applications by reading this section.
2. **Detailed Description of Valves**
   Detailed description of each valve type enables readers to find more about valves with specific designs, operations, and applications in the industry. For example, gate valves are good at providing tight shutoff or isolation, whereas globe valves are used for accurate flow control. Advantages and limitations of each valve type are discussed in more detail in this section.

DOI: 10.1201/9781003640967-2

This discussion and analysis will help readers and engineers to make more suitable decision in the projects regarding the valve selection.

3. **Comparative Analysis**
   The most important valves are compared to each other in this section to understand better the advantages and disadvantages of each and every valve type. Essential parameters such as ability of valves for flow control or regulation, pressure drop created across the valve, and valve operation or opening and closing speed are considered in this comparison. By comparing the valves in a table, readers can choose the best valve for their applications.
4. **Valve Selection Criteria**
   The final section provides a practical guide for selecting the right valve based on factors like fluid properties, design, and operating pressure and temperature. It also addresses common problems and mistakes in valve selection process as well as how to avoid them.

   Readers and engineers will understand different types of valves, their characteristics, including working mechanism and operation, and how to select the most appropriate valve for various industrial applications. This foundational knowledge will be a valuable reference for engineers and professionals working with fluid control systems such as piping, valves, instruments, and mechanical equipment.

PART 1: CLASSIFICATION OF THE VALVES

Valves are classified according to their functionality (purpose of operation), design (e.g., type of body material), and operational features (e.g., manual or actuated). Knowing these classifications are fundamental knowledge for engineers and operators in the industry to choose the best valve for their specific applications. The following is an alphabetically organized overview of valve categories and their operational uses.

FUNCTION

Flow Control (Regulation) Valves

A flow control valve regulates (controls) the flow rate of a fluid (liquid or gas) in a piping system. These valves are essential components for accurate flow control in piping and process systems where achieving specific flow rates is critical.

- *Globe Valve:* Designed for throttling and regulating fluid flow with high accuracy.
- *Needle Valve:* Offers control of flow in small-diameter piping systems.
- *Control Valve:* An actuated globe valve that is used to adjust the flow rate based on signals from a control system.

Isolation Valves

Isolation valves are mainly used to completely stop or allow fluid flow in a piping system. These valves are vital for maintenance and safety of the operators during maintenance.

- *Gate Valve:* Provides a tight shutoff for a straight flow and minimizing the pressure drop.
- *Ball Valve:* A robust valve with a rotating ball, ideal for quick on/off applications.
- *Butterfly Valve:* It is a lightweight, compact, and economic valve with rotating disk. Due to its quarter-turn stem and closure member motion same as the ball valve, it is suitable for fast operation.
- *Plug Valve:* Uses a cylindrical or conical plug for on/off application and limited flow control. In addition, it is suitable for dirty (particle containing) services (fluids).

Non-Return Valves

Non-return valves allow fluid to flow in only one direction and prevent backflow by closing. Reverse flow could damage equipment like pumps and compressors.

- *Check Valve:* Prevents reverse flow by automatically closing when flow stops or reverses.

On/Off Valves

Same as isolation valves.

Safety Valves

Safety valves are critical pressure relief devices designed to automatically release excess pressure from a pressurized system when it exceeds a predefined limit. This action prevents damage to the equipment and piping systems and maintaining safety and reliability.

MATERIAL DESIGN

Valves are categorized by the materials used for their body and bonnet and internals, which determine their compatibility with different fluids.

- *Metallic Valves:* Materials such as stainless steel, nickel alloy, or carbon steel that can be used in high-pressure and high-temperature applications.
- *Non-Metallic Valves:* Constructed from plastic, ceramic, or composite materials for corrosion resistance, lightweight, and lower cost.

OPERATIONAL CONSIDERATIONS

Flow Direction

- *Unidirectional Valves:* Allow fluid to flow in one direction, such as globe and check valves.
- *Bidirectional Valves:* Permit flow in both directions, such as standard gate and ball valves.
- *Multi-Port Valves:* Diverting or mixing fluid flow between multiple piping, as seen in multiway ball valves.

Operation Mode

Valves are classified based on the type of operator that is also called operation mechanism.

- *Manual Valves:* They are operated by human and using a handwheel and gearbox, or lever.
- *Actuated Valves:* They are automated by pneumatic, hydraulic, electrical, or other type of actuators that are explained in more detail in Chapter 11.

PART 2: DETAIL DESCRIPTION OF INDUSTRIAL VALVES

Valve are components that play an essential role in piping systems such as flow control, equipment isolation, safety, and backflow prevention. Selecting the right valve to meet both design and operational requirements requires knowledge of the unique features, mechanisms, and applications of different valve types. This section provides more details about commonly used valves, including their designs, key components, working principles, advantages, limitations, and typical applications. Professionals and engineers can ensure optimal piping and valve performance and reliability in various industrial systems by understanding the details of each valve type.

A

Axial Flow Valve

- *Design and Working Mechanism:* An *axial flow valve* is designed to control fluid flow by using a disk that moves along the axis of the valve body. The disk or sleeve adjusts its position along the flow path, varying the flow area and, consequently, regulating the flow rate. The valve body typically features a straight-through design, minimizing turbulence and pressure drop. When the valve is operated, the disk or sleeve moves axially (in the direction of flow) to either open or close the valve, providing precise flow control. Figure 2.1 illustrates an axial flow valve, also

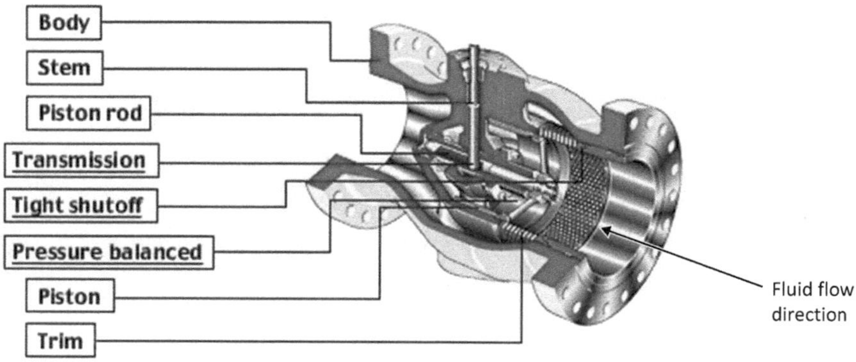

FIGURE 2.1 An axial flow (control) valve.

known as an axial flow control valve. The fluid enters the valve from the right side and exits through the left. The actuator adjusts the stem's position, and the linear movement of the stem is converted into horizontal motion of the rod, which moves the disk back and forth to regulate the flow.

- *Key Components:* Key components of an axial flow valve include the valve body, a disk or sleeve that moves along the axis, the seat, operator, spring, and rod (stem) behind the disk. All these components are designed to control the flow by adjusting the position of the flow-regulating element.
- *Applications (Advantages):* The axial movement is suitable for flow control (throttling) applications. This valve ensures smooth and stable flow passage which makes axial flow valves ideal for applications where smooth flow with low pressure drop is critical.
- *Limitations:* The limitations of an axial flow valve are its relatively complex design, which can lead to higher maintenance, especially in high-flow or high-pressure applications. Additionally, it may not be suitable for particle-containing or high-viscosity fluids due to the potential for increased erosion and friction and reduced performance over time.

B

Ball Valve

- *Design and Working Mechanism:* A ball valve is a quarter-turn valve that uses a spherical closure element, known as the ball, with a hole (bore) through its center to control (stop and start) fluid flow. The valve operates by rotating the ball 90° via a stem connected to a handle or actuator. When the bore aligns with the piping, the valve is fully open, allowing unrestricted flow with minimal pressure drop. Rotating the ball perpendicular to the piping blocks the flow, sealing it off completely.

During valve closure, the ball fits tightly against the valve seats, ensuring a secure seal even under high pressure. This simple and effective design makes ball valves ideal for quick on/off applications in the piping systems. Figure 2.2 illustrates a ball valve by highlighting essential components such as the ball, seat, stem, body, and bonnet. The hole inside the ball is illustrated in an approximately half-open position, meaning the valve is neither fully open nor closed. Since a ball valve is mainly designed and operated as an on/off valve for start and stop the fluid flow, operating it in this partially open position can cause damage to both the seats and the ball due to turbulence and erosion. It is recommended by valve experts to avoid such positions to ensure long life span and proper functioning of the valve.

- *Key Components:* The key components of a ball valve include the *ball*, which serves as the closure element with a central bore for passage of flow and controlling it; the *stem*, which connects the ball to the valve operator (handwheel or actuator); the *seats*, which provide a tight seal around the ball and fill the area between the body and ball; and the *body*, which houses and protects these internal parts. These components work together in a ball valve to ensure reliable and efficient operation in various applications.
- *Applications:* Ball valves are widely used in applications requiring quick and reliable flow stop/start. For example, blowdown valves require fast release of overpressure fluid in case of high pressure that require to be ball valves in many cases. In addition, ball valves are critical in

FIGURE 2.2 Ball valve in a partially open position showing key components and potential operational concerns. (With permission: Shutterstock.)

emergency shutdown systems to stop the fluid fast to ensure safety in operation by isolating sections of a piping system during emergencies. Moreover, ball valves are ideal for *isolation during maintenance*, so sections of piping, pipeline, or equipment need to be safely serviced. It should be noted that ball valves are extensively used in *piping and pipeline systems* for their tight shutoff and minimal pressure drop.
- *Advantages:* Reliable sealing, minimal pressure drop in case of full-bore design. More information about valve design features is given in Chapter 5.
- *Limitations:* Unsuitable for throttling.

Butterfly Valve

- *Design and Working Mechanism:* A butterfly valve is a quarter-turn valve designed to regulate or isolate flow in a system. Its primary component is a flat, circular disk connected to a rotating shaft that passes through the valve body. When the valve is fully open, the disk aligns parallel to the flow direction, allowing minimal resistance and the highest possible flow rate passing through the valve. When the valve is closed, the disk rotates perpendicular to the flow and creates a seal against the valve seat to block the passage.
- *Key Components:* The key components of a butterfly valve include the *disk*, which rotates to regulate flow; the *stem*, which connects the disk to the valve operator such as handwheel or actuator; and the *seat*, which provides a seal when the valve is closed. The *body* is the main pressure containing part that contains these components. Figure 2.3 shows a rubber lined (soft seat) butterfly valve in a closed position.
- *Applications:* Butterfly valves are widely used in applications requiring flow regulation or isolation such as large pipelines, liquified natural gas (LNG) plants, HVAC systems, and power plants, due to their lightweight design and quick operation. They are also suitable for industries like chemical processing and onshore and offshore oil and gas, where their compact size and cost-effectiveness make them ideal for handling large volumes of fluid with minimal pressure drop.
- *Advantages:* Butterfly valves are lightweight especially in wafer (flangeless) design, compact, and cost-effective, making them ideal for large-diameter pipelines and space-constrained applications. They offer quick operation, minimal maintenance due to simple design, and low-pressure drop when fully open.
- *Limitations:* A butterfly valve's primary limitation is its weaker sealing capacity compared to a ball valve. Due to this weakness, butterfly valves are less commonly used for high-pressure applications. In some systems, the disk of the valve remains in the flow path even when fully open, causing turbulence and pressure drops. In addition, ball valves are more robust than butterfly valves and they are more suitable for high-cycle (frequent opening and closing) applications.

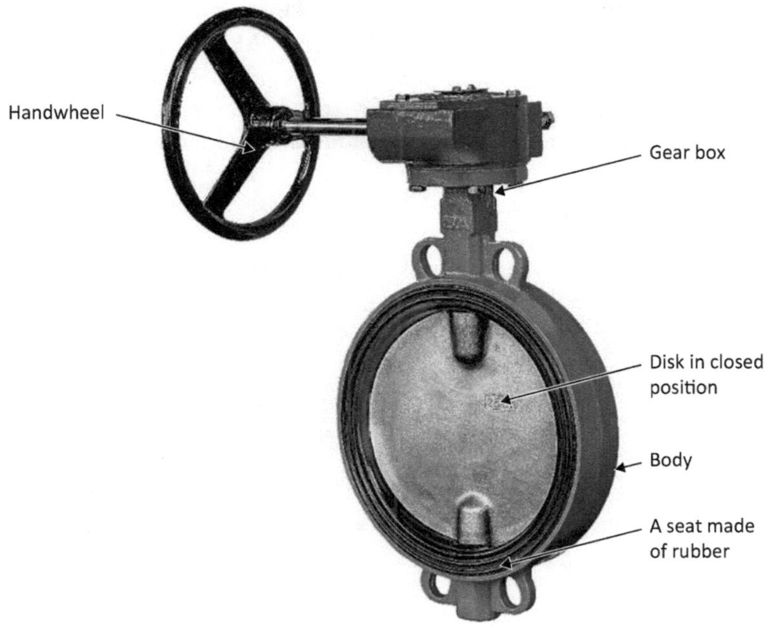

FIGURE 2.3 Rubber-lined butterfly valve with key components highlighted. (With permission: Shutterstock.)

C

Check Valve

- *Design and Working Mechanism:* A check valve is designed to allow fluid to flow in one direction while preventing reverse flow. It operates automatically by the fluid force using a movable disk, ball, or piston that opens under forward flow pressure and closes against a seat when the flow reverses. Thus, check valves maintain unidirectional flow and system protection.
- *Key Components:* Key components of a check valve include the body, closure member (disk or ball), seat, and spring (if applicable). For example, there is no spring in the swing check valve.
- *Applications:* Check valves are commonly used in piping systems to prevent reverse flow, keeping that fluid or gas flows in only one direction. They are critical in systems such as water distribution, oil and gas piping, and mechanical and HVAC systems to protect equipment such as pumps and compressors from damage caused by backflow.
- *Advantages:* Check valves provide the advantage of preventing backflow, which protects equipment from damage and ensures proper flow direction in systems. They are low maintenance, reliable, and help maintain system efficiency by automatically closing when flow reverses.

- *Limitations:* Check valves have limitations such as the potential for leakage if the valve seat is damaged or worn. Also, they are not suitable for particle containing fluids as they can be prone to clogging if debris accumulates. Additionally, they may require a minimum flow rate to function properly and can create pressure drops in the system.

Check Valve Types (CHVT)

This section covers five types of check valves – axial flow check valve, ball lift check valve, dual plate check valve, piston check valve, and swing check valve.

CHVT1: Axial Flow Check Valve

- *Design and Working Mechanism:* The valve is relatively simple in design, with an axially moving disk, allowing for a smooth, turbulent-free flow through the valve. Figure 2.4 illustrates an axial flow check valve where fluid enters from the right, flows through the valve, and exits to the left after moving the disk away from its seat. In the event of reverse flow, the returning fluid and the spring (installed behind the seat) push the disk back onto the seat, closing the valve and preventing backflow.
- *Applications:* Ideal for applications requiring low-pressure loss and quick closing such as downstream of the pumps and compressors
- *Advantages:* Quick opening and closing due to short distance between the seat and disk in open position, non-slamming, low-pressure drop, providing very smooth flow passage. Chapter 5 provides more information about non-slamming under the heading of "Non-Slam Check Valve."

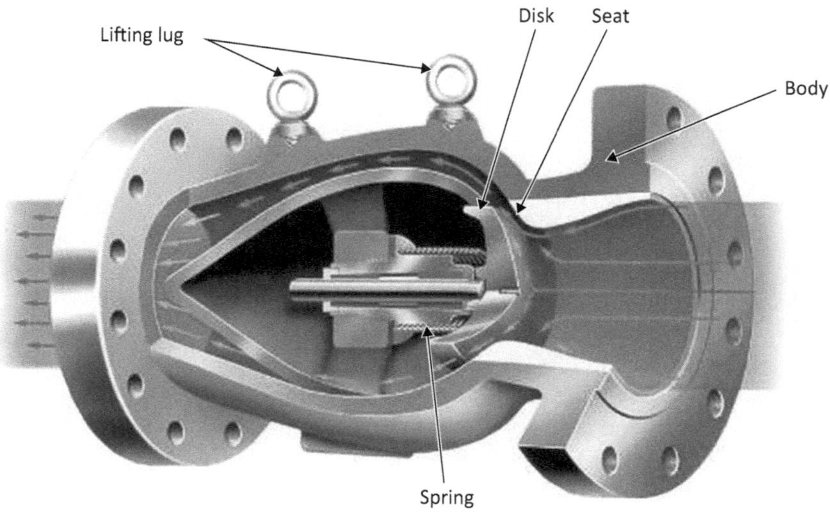

FIGURE 2.4 Axial flow check valve demonstrating flow direction and disk operation.

- *Limitations:* When the flow rate becomes low, the disk in an axial flow check valve can chatter between the open and closed positions. This occurs because the flow is not sufficient to keep the disk fully open, and not low enough for the spring to close it completely. This unstable intermediate position causes the disk to move back and forth rapidly, leading to wear and erosion of the disk and seat as well as vibrations and noise. Increased wear on the valve components can lead to leakage and failure over time.

CHVT2: Ball Lift Check Valve

- *Design and Working Mechanism:* A ball lift check valve operates similarly to a piston check valve but uses a ball as the movable element rather than a piston or plug. When fluid flows in the normal direction, the fluid pressure lifts the ball, allowing the flow to pass through the valve. If the flow reverses or pressure drops, the ball is pushed back on the seat, sealing the flow path inside the valve and prevent backflow. This type of the valve is very useful for low- to medium-pressure applications, where saving space is important.
- *Applications:* Ball lift check valves are commonly used in various industries, including water treatment, and oil and gas, to prevent backflow and ensure unidirectional flow. They are common for piping systems in 2" size and smaller.
- *Advantages:* Like piston check valves, ball lift check valves offer reliable backflow prevention with minimal maintenance requirements. Their design is simple, durable, and cost-effective.
- *Limitations:* The limitation of a ball lift check valve is that it can be prone to *ball sticking* or *seating issues*, particularly in applications involving viscous fluids, debris, or low-flow conditions. Over time, wear or contamination can also affect the sealing performance, potentially leading to leaks or inefficient operation. Additionally, ball lift check valves are not suitable for high-pressure piping systems unlike dual-plate and swing check valves.

CHVT3: Dual-Plate Check Valve

- *Design and Working Mechanism:* This valve has two half-disk or plate loaded by springs that open with the fluid flow and close with the spring force. Figure 2.5 illustrates a dual-plate check valve featuring two plates that remain in the open position due to the spring forces around the pin. When fluid pressure decreases and a backflow tendency arises, the spring force closes the disks, preventing reverse flow. The closing of the disk could be done by both reverse flow and the spring forces. The valve body is of a flangeless (wafer) design, offering advantages such as compactness, reduced weight, and cost-efficiency. In general, a dual

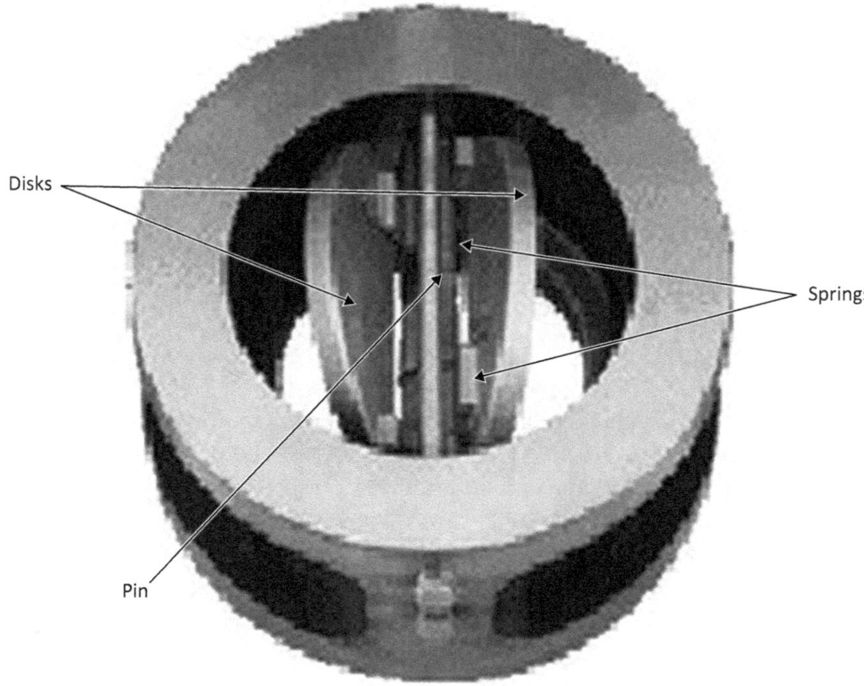

FIGURE 2.5 Dual plate check valve with flangeless (wafer) design, featuring spring-loaded plates that prevent backflow.

plate check valve has a lower pressure loss compared to swing check valves and ball lift check valves when fully open.

- *Applications:* It is suitable for various piping systems in onshore and offshore production, refineries, and petrochemical plants.
- *Advantages:* Lightweight and compact.
- *Limitations:* The limitation of a dual-plate check valve is that it may experience *fluttering* or *chattering* under low flow rates, which can lead to noise, vibration, and wear of the plates and hinges over time. Additionally, its design can cause higher *pressure drop* compared to axial check valve types due to half disks in the middle of the flow. In addition, the valve's internal components, such as the springs or hinge pins, may require regular maintenance due to wear and erosion.

CHVT4: Piston Check Valve

- *Design and Working Mechanism:* A piston check valve has a piston also called plug that is guided by a spring. The plug moves inside the valve body by either fluid or spring force. When fluid flows in normal flow direction, the piston (closure member) is pushed up and open the valve which allows fluid to pass through the valve. If the flow reverses because of any reason including pressure drop, the piston is forced to close, preventing backflow and protecting the upstream piping system from damage. Figure 2.6 illustrates a piston check valve, highlighting key components, including the body, bonnet, spring, and piston (or plug). In this valve, when fluid flows from left to right, the fluid pressure overcomes the spring force, pushing the piston upward and allowing the fluid to pass through the valve. In case of reversal flow or pressure drop, the spring forces the piston back down, sealing it against the valve seat and preventing backflow. The combination of the spring and piston provide possibility of automatic valve operation, as well as maintaining unidirectional flow and protecting the system from reverse flow damage.

- *Application:* Piston check valves are used in piping and process systems to prevent backflow. They are commonly found in oil and gas small piping systems, water treatment, and HVAC systems, as well as in pump protection systems to prevent reverse flow that could damage equipment or cause cavitation. These valves are typically employed in smaller sizes, 2″ and below, where their compact design offers reliable backflow prevention in low-to-medium pressure applications.
- *Advantages:* Piston check valves have several benefits such as excellent backflow prevention, even in systems with fluctuating pressures or low flow rates. Their tight sealing ability prevents leakage as much as

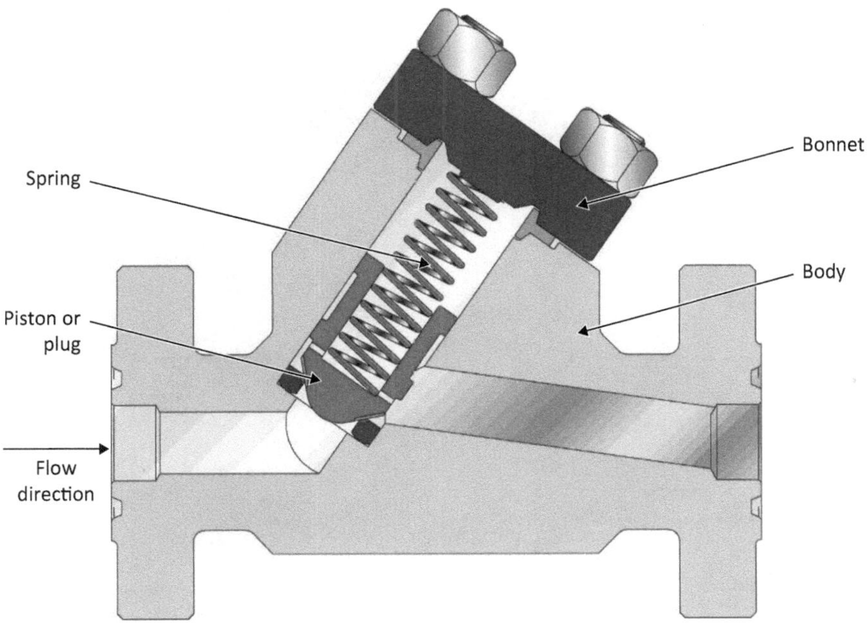

FIGURE 2.6 Illustration of a piston check valve showing the key components and the flow direction.

possible, enhances system efficiency, and protects downstream equipment from damage. Additionally, their compact design makes them ideal for use in smaller piping (2″ and below), where space and weight are concerns. Piston check valves also have low maintenance requirements due to their simple design and are durable, providing long-lasting performance in various industrial applications.

- *Limitations:* Same as limitations for ball lift check valve.

CHVT5: Swing Check Valve

- *Design and Working Mechanism:* A swing check valve has a hinged disk, also called flap, that pivots on a hinge pin and moves around it inside the valve body. When fluid flows in the normal or preferred direction, the disk is pushed open by the fluid, allowing the fluid to pass through the valve. If the flow reverses or pressure drops, the disk swings to the closed position by both force of the fluid and gravity to prevent backflow and protect the downstream piping systems and equipment.

  Figure 2.7 illustrates a swing check valve and its key components, including the valve body, bonnet, disk (or flap), hinge, hinge pin, and seat. The disk is attached to the hinge and hinge pin and moves inside the valve body. In the preferred flow direction, from right to left (refer to the picture), the fluid pressure pushes the disk upward to open the valve, enabling the flow to pass through the valve. If the flow reverses, the disk swings

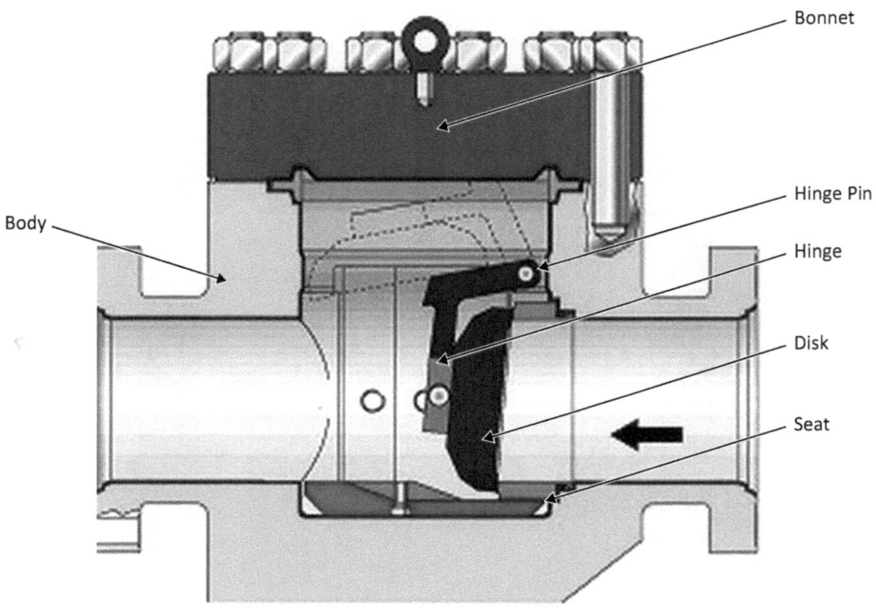

FIGURE 2.7 Picture of a swing check valve showing key components.

back to the closed position, utilizing the fluid pressure and gravity to seal against the valve seat and prevent backflow. The bonnet also called body housing covers the valve body and provides structural integrity. The seat ensures a tight seal with the disk in closed position of the valve. The hinge and hinge pin allow movement of the disk, allowing the valve to operate efficiently and reliably to protect the system from reverse flow.

- *Applications:* Swing check valves are typically used in applications where lower flow rates and larger diameter piping or pipeline are involved. These include water distribution systems, and oil and gas industrial applications. They are effective for preventing backflow in large pipelines or systems without the high backflow requirements simply because of closure of disk under gravity force. In the past, swing check valves were typically found in pump discharge lines, where axial check valves are now preferred for their efficiency and smooth operation without slamming effect. Swing check valves remain a good option for systems where space is less of a concern since they can have wafer design. Also, they are suitable for applications where the flow direction is less frequently reversed.
- *Advantages:* Swing check valves prevent backflow and protect downstream piping systems and equipment. They are suitable for piping and larger pipelines in the oil and gas industry flow. The valves are relatively cheap and simple to operate. Additionally, their robust construction and ease of maintenance make them a popular choice for various industries, including water treatment and oil and gas.
- *Limitations/Disadvantages:* It is possible for swing check valves to slam shut against the seat when flow velocity drops or changes direction rapidly, causing damage to the valve's internal components and causing noise. As a result of this "slamming" action, the valve can wear out quicker and lose efficiency. New plants and projects select dual plate check valves, which operate with less slamming, or axial check valves, which don't slam. In more critical applications such as offshore oil and gas, these alternatives minimize damage, reduce noise, and provide smoother, more reliable operation.

COMBINATION VALVE

A *combination valve* is a type of valve that integrates different valve functions into a single body. While it may seem similar to a modular valve that is often used for double block and bleed applications, a combination valve serves a little different purpose. For example, a combination valve can include a double ball valve with a needle valve and a check valve, as illustrated in Figure 2.8. This setup can be used in chemical injection lines, where the *double block and bleed* feature (two ball valves and the needle between) provides double isolation between the chemical injection line and the process line, while the *check valve* prevents the backflow of the injected chemical into the system.

(*Note:* Chemicals are injected into the process line for a variety of reasons, including breaking emulsions between water and oil to optimize separation

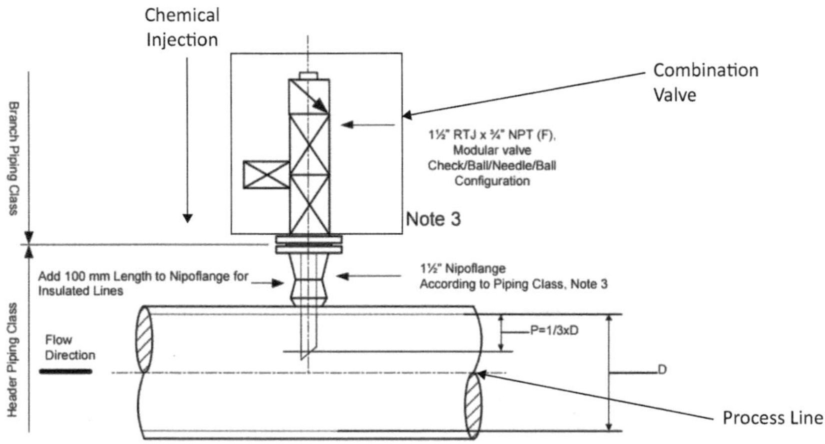

FIGURE 2.8 Combination valve in a chemical injection line, providing double isolation and preventing backflow to ensure safe injection of chemicals into the main process line.

between oil and water-in-oil production facilities, preventing corrosion, scale formation, and hydrate formation.)

Control Valve

Control valves are types of globe valves operated automatically by means of an actuator mounted on the top of the valve. An essential component of the process industry is the control valve, which regulates the flow, pressure, temperature, and level of fluid within a system. This is accomplished by altering the size of the flow passage, which alters fluid flow in response to signals from the controller. A number of industries, including oil and gas, chemical processing, water treatment, and power generation, rely heavily on control valves to maintain a safe and efficient operating environment. As mentioned before, control valves are controlled by actuators, positioners, and controllers. Controllers send signals to positioners and actuators (pneumatic, electric, or hydraulic), which adjust valve openings to maintain setpoints. There are a number of applications in which control valves are used, including temperature control in steam systems, pressure regulation in pipelines, and fluid level control in storage tanks where conditions fluctuate frequently. Chapter 3 provides more information about valve positioners.

G

Gate Valve

A gate valve is a type of valve used to start and stop the flow of fluid by raising or lowering a gate (or wedge) within the valve body. Thus, gate valve is used only for fully open and fully closed conditions.

Gate Valve Types (GTVT)

Four main gate valve types are reviewed in this section – *wedge gate* type, where the gate is tapered to achieve a tight seal, or a *through conduit gate valve*, which allows uninterrupted flow when open. Through conduit gate valves can be further divided into *slab type*, with a flat gate, or *expanding type*, which may be single or double, where the gate expands to seal tightly against the valve seat as it is lowered. Other types of gate valves include the *knife gate valve* and the *slide gate valve*.

GTVT1: Knife Gate Valve

A knife gate valve has a sharp-edged gate that is designed to cut through thick, viscous fluids, slurries, or solids. This type of valve provides reliable sealing and on/off flow in specific challenging applications. It is commonly used in industries such as mining, wastewater treatment, pulp and paper, and biomass where the media may contain suspended particles or abrasive materials. Figure 2.9

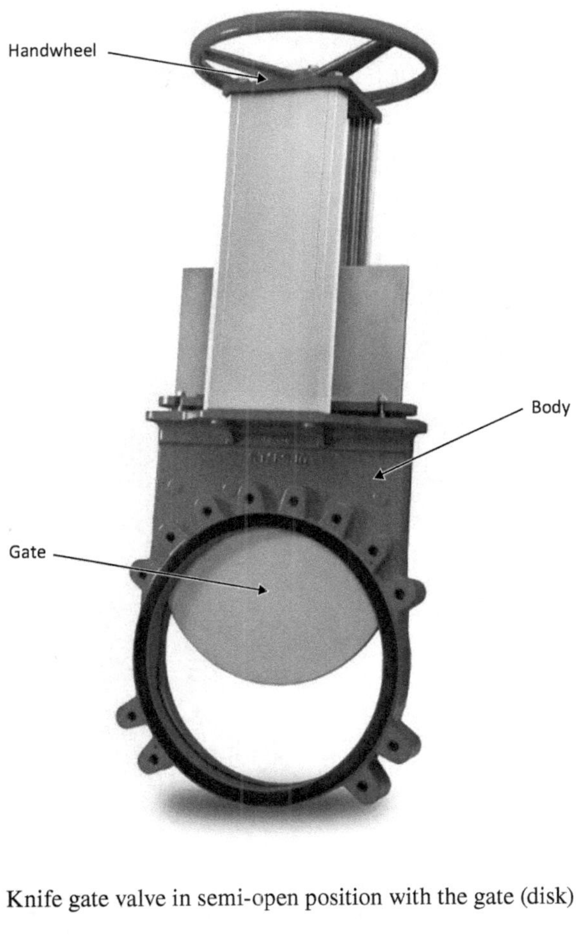

FIGURE 2.9 Knife gate valve in semi-open position with the gate (disk) descending.

illustrates a knife gate valve in a semi-open position, with the gate (or disk) moving downward.

GTVT2: Slide Gate Valve

A slide gate valve is a type of valve commonly used for stop and start the flow of bulk materials, such as powders or granular substances, in industries like mining and food processing. It has a sliding plate that moves perpendicular to the flow path to either open or close the valve. Unlike a knife gate valve, which uses a sharp-edged gate to cut through thick, viscous fluids or slurry, a slide gate valve is designed for bulk solids and typically provides a better seal to prevent leakage of powders and granular materials. While knife gate valves are suited for high-pressure liquid applications, slide gate valves are more appropriate for low-pressure, material-handling systems. Figure 2.10 shows a slide gate valve in the closed position.

A *slide gate valve* typically has a flat, rectangular gate that slides horizontally within a frame to stop and start the flow of bulk materials like powders or granules. Its body is often rectangular, and the gate does not have a sharp edge.

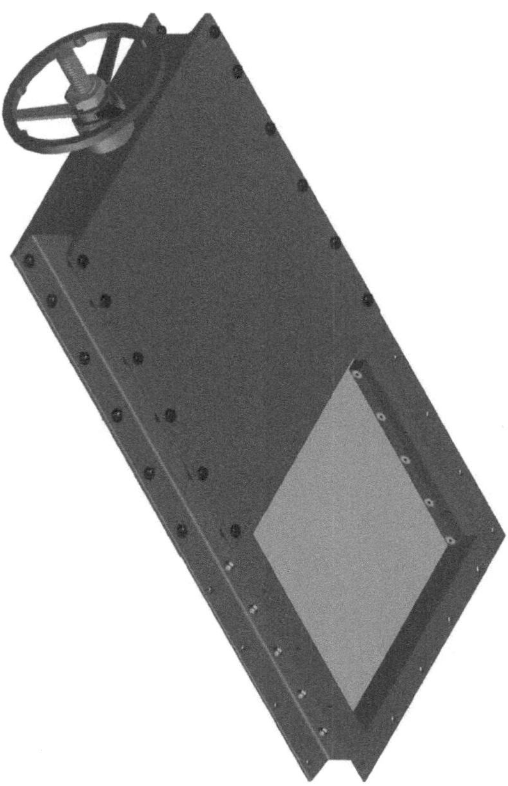

FIGURE 2.10 A slide gate valve in closed position. (With permission: Shutterstock.)

In contrast, a *knife gate valve* has a wedge-shaped, sharp-edged gate that moves vertically to cut through thick, viscous fluids or slurries.

GTVT3: Through Conduit Gate (TCG) Valve

The through conduit gate valve is designed to allow unobstructed flow when open, with its gate fully sliding into the valve body when closed. Typically, it has a straight-through flow path that reduces turbulence and provides a smoother flow, making it suitable for applications involving large particles or high viscosities.

GTVT3.1: Expanding (Single Expanding/Double Expanding)

Design and Working Mechanism: An expanding gate valve features a gate that expands to create a tight seal against the valve seats when closed. The disk of the gate consists of male and female components that interlock with a sealing element between them, ensuring a secure closure. When the valve is operated, the stem causes a wedging effect on the disk, which then expands radially, increasing the seal's tightness. In a *single expanding gate valve*, the gate expands only in the closed position, ensuring a reliable seal under pressure. In a *double expanding gate valve*, the gate or disk containing male and female half sections expands both when open and closed, providing an even tighter seal and enhanced reliability in high-pressure applications.

- *Key Components:* Key components of an expanding gate valve include the valve body, gate, or disk (with male and female components), sealing element between half male and female disks, stem, and seats.
- *Advantages:* It is advantageous to use an expanding gate valve due to its reliability, tight seal under extreme conditions such as high pressure and high temperature, suitability for dirty liquids or fluids containing particles, minimal pressure drop, and minimal leakage risk. Despite varying pressures and extreme harsh conditions, the expanding mechanism provides a strong, self-adjusting seal.
- *Limitations:* There are a number of limitations or disadvantages of expanding gate valves, including their relatively complex design with many internal components, which can lead to higher initial costs as well as more costly maintenance in comparison to simpler valve types. The expanding mechanism of these valves may result in a slower response time during opening and closing. Consequently, expanding gate valves are less suitable for applications requiring quick operation than ball valves. As a result of the complexity of the design of the expanding gate closure member, more weak points may be present, potentially reducing reliability. Last but not least, they are not suitable for regulating flow.

Figure 2.11 shows (on the right side) a through conduit gate valve (expanding type), and on the left side, the sealing mechanism between the valve's male and female disks, achieved through a stem force (wedging effect).

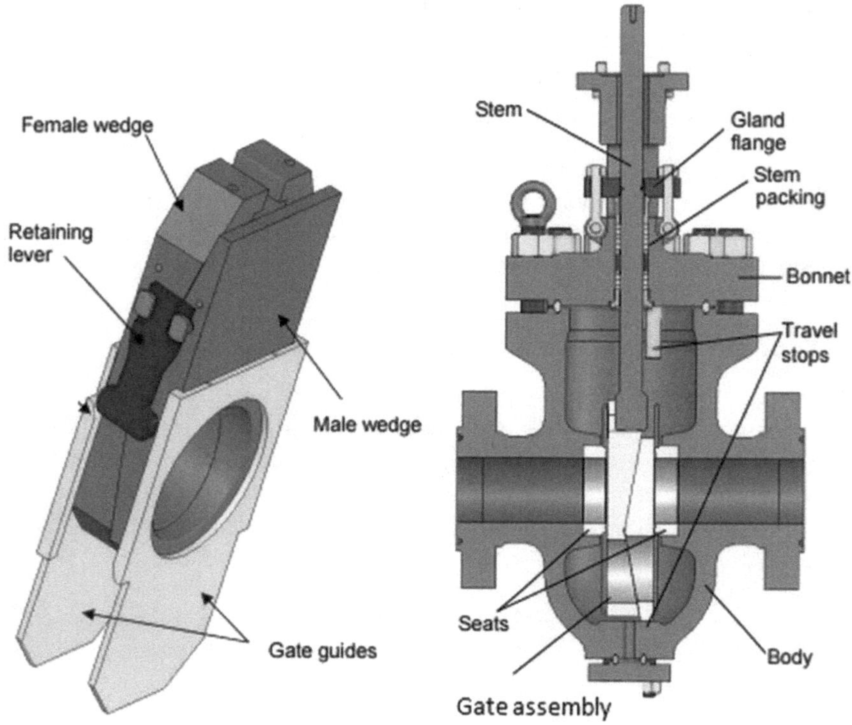

FIGURE 2.11 Through conduit expanding type gate valve (*right*) with male and female half disks in contact with seat rings (*right side*), and sealing between the male and female half disks (*left*).

GTVT3.2: Slab

- *Design and Working Mechanism:* A *slab gate valve* contains a single, flat gate (or slab) that moves perpendicular to the flow path to start/stop the fluid or gas flow. The gate is typically sealed by a metal-to-metal in most cases or elastomeric seal. When the valve is in closed position, the gate presses tightly against the seat to prevent leakage. Figure 2.12 illustrates a slab gate valve including the part list. In the picture, the disk is pushed against the downstream seat (shown on the right) by the fluid flow from the left side to provide seal during operation.
- *Key Components:* Key components of a slab gate valve include the gate (slab), body, seat rings, bonnet, stem, and operator (manual or actuator).
- *Applications:* Slab gate valves offer advantages such as reliable sealing, minimal maintenance, and efficient flow control in large pipeline and piping systems, particularly for on/off service.

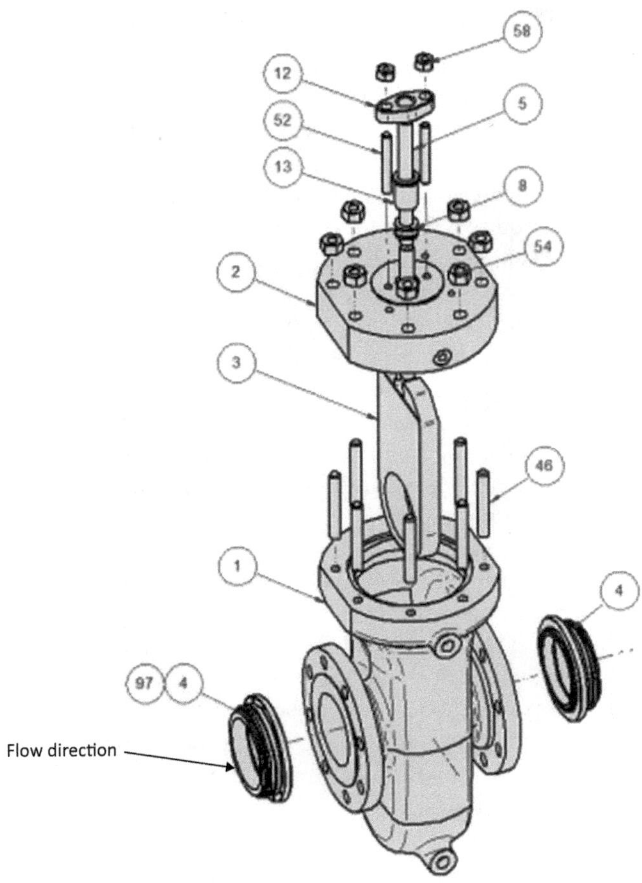

FIGURE 2.12 Slab gate valve with its essential components.

- *Limitations:* The limitations of slab gate valves are summarized in their relatively slow operation and potential for leakage if the gate becomes worn or misaligned over time.

POS.	Part name
1	Body
2	Bonnet
3	Gate
4	Seat
5	Stem
8	Lantern ring
12	Gland flange
13	Gland
46	Stud bolt
52	Stud bolt
54	Nut
58	Nut
97	Lip seal

GTVT4: Wedge Type Gate (TCG) Valve

- *Design and Working Mechanism:* A *wedge-type gate valve* uses a wedge-shaped gate that is pushed against the seat to provide a seal during the valve closure and ensure minimum leakage through the valve. The wedge design provides a strong and reliable seal even under high-pressure conditions. Thus, wedge gate valves are ideal for applications where isolation of the flow is required. Figure 2.13 shows a wedge-type gate valve, with key components highlighted. In the closed position, the gate is down, not allowing flow to pass through the valve. The valve is bidirectional, meaning flow can enter from either direction. The figure also illustrates the wedging effect of the stem force, which causes the disk to expand linearly and press tightly against the seat, providing a tight seal.
- *Key Components:* Key components of a wedge gate valve are valve body, wedge-shaped gate, seat rings, bonnet, stem, and operator.
- *Applications:* Wedge gate valves are commonly used in applications where a reliable, tight shutoff is required in oil and gas plants, including onshore and offshore.
- *Limitations:* The limitations of wedge gate valves include their relatively slow operation, as they require full opening or closing to control flow. In addition, these valves are not suitable for throttling (flow control) applications due to the risk of chattering, vibration, and disk and seat damage from partial opening. Additionally, they can be prone to seat wear and leakage over time, especially in high-cycle or high-pressure environments. Wedge gate valves unlike expanding gates are not suitable for particle-containing fluids.

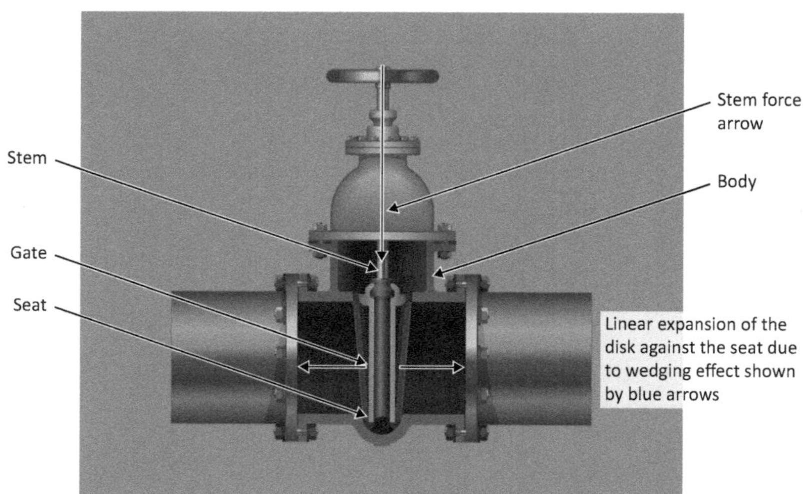

FIGURE 2.13 Wedge-type gate valve in closed position and illustration of the wedging effect created by stem on the disk that ensures tight sealing by pressing the disk against the seat. (With permission: Shutterstock.)

Globe Valve

A *globe valve* has linear motion (up and down stem and plug) that uses a movable disk and a stationary ring seat to regulate flow, making it ideal for throttling or flow control applications.

Globe Valve Types (GLVT)

Variants of the globe valve include the *T-type globe*, *Y-type globe*, and *angle globe*, each designed for specific flow control needs and installation configurations, with differences in body shape and internal flow paths.

GLTV1: Angle Globe Valve (Angle Valve)

- *Design and Working Mechanism:* An *angle globe valve* is designed with the inlet and outlet ports at a 90° angle, reducing the need for an elbow to change the direction of the flow. It is primarily used for flow control, providing precise regulation of fluid flow in systems where space constraints or directional changes are important, while maintaining efficient performance. Figure 2.14 illustrates an angle globe valve where fluid

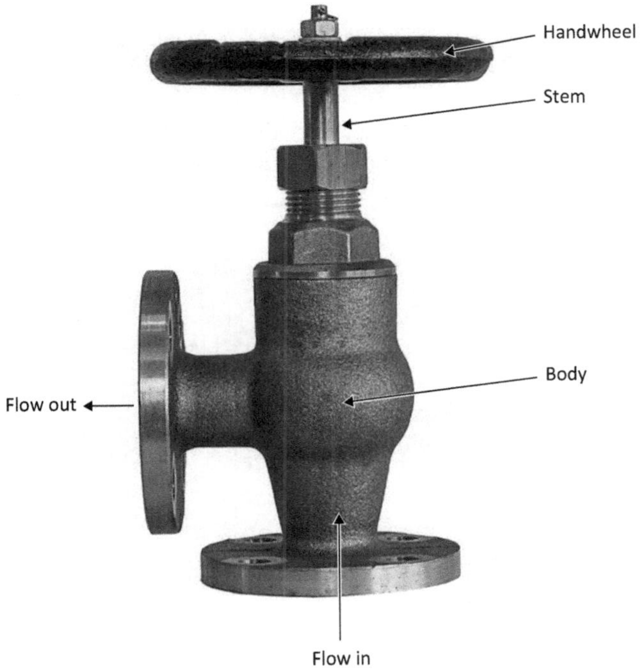

FIGURE 2.14 Angle globe valve with flange connections, allowing fluid to flow in from the bottom and exit through the left outlet, with flow control via handwheel adjustment of the stem and disk. (With permission: Shutterstock.)

enters from the bottom and exits through the outlet on the left after a 90° rotation. Both the inlet and outlet are equipped with flange connections for the piping. The operator can adjust the handwheel to raise or lower the stem and disk, thereby adjusting the position of the disk relative to the seat and controlling the flow.

- *Key Components:* Key components of an angle globe valve include the valve body, bonnet, stem, disk (or plug), seat, all designed to control the flow of fluid through the angled ports.
- *Applications:* Angle globe valves are commonly used in applications where space is limited, or where the flow direction needs to be changed, such as in heating systems, steam lines, and chemical processing plants for precise flow regulation.
- *Limitations:* The limitations of an angle globe valve include its relatively higher pressure drop compared to other valve types and the potential for increased maintenance due to the complexity of its design (fluid rotation and high-pressure drop) and moving parts.

GLTV2: T-Type Globe Valve

- *Design and Working Mechanism:* A *T-type globe valve* features a valve body shaped like a "T," with the inlet, outlet, and flow control components aligned in a straight path, allowing for efficient flow regulation. Figure 2.15 shows a globe valve operated manually using a handwheel.

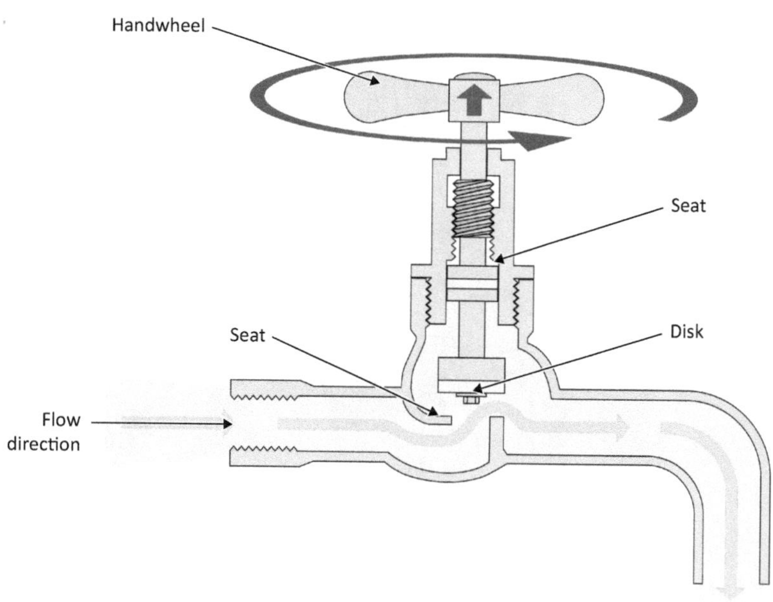

FIGURE 2.15 Manually operated globe valve with flow direction from left to right. (With permission: Shutterstock.)

The fluid flows from left to right, passing beneath the disk. The valve is in fully closed position. In the fully closed position, the disk is located on the seat and no flow is passing through the valve. However, in the fully open position, the disk is lifted off the seat ring, allowing unrestricted flow. Turning the handwheel moves the stem up or down in a linear motion, adjusting the position of the disk and thereby regulating the flow through the valve.

- *Key Components:* Key components of a T-type globe valve include the valve body, bonnet, stem, disk (or plug), seat, all designed to control the flow of fluid through the angled ports.
- *Applications:* A T-type globe valve is commonly used in applications where precise flow control is required in piping systems with limited space or where the flow needs to be regulated in a straight line. It is typically used in industries such as water treatment, oil and gas, chemical processing, where accurate throttling and minimal pressure loss are important.
- *Limitations:* The limitations of a T-type globe valve include the potential for *cavitation* due to pressure loss in liquid applications, especially when there is a significant pressure drop across the valve. Additionally, these valves can be *heavy* and difficult to handle in larger sizes, which can be problematic for high-pressure applications. The relatively *complex design* can lead to higher maintenance requirements, as the moving parts are more prone to wear. In some cases, the *pressure drop* across the valve may reduce efficiency, and the valve may not be suitable for high-velocity or pulsating flow conditions.

GLTV3: Y-Type Globe Valve

Y-type globe valves, also known as Y-pattern globe valves, are variations of standard globe valves designed to reduce flow resistance and pressure drop. Globe valves with a Y-type body have an angled body that aligns the stem and seat at approximately 45° to the flow path. As a result, the valve forms a "Y" shape in the flow path. When compared to a straight-through globe valve, this reduces the number of sharp turns the fluid must make, resulting in smoother flow and less pressure drop. As a result of its design, it minimizes the disruption of flow that is typically experienced with traditional (T-type) globe valves, making it ideal for applications requiring a lower pressure loss and higher flow efficiency. The Y-type globe valve has a low-pressure drop, which makes it less susceptible to operational problems such as erosion and cavitation. Consequently, the angled design of the Y-type globe valve makes it easier to have access inside the valve and easier maintenance than other globe valves, especially in applications where it is necessary to regularly inspect. In a Y-type globe valve, the bonnet is positioned at an angle (approximately 45°) to the flow path. This angled orientation reduces the valve body depth compared to a T-type, where the bonnet is positioned directly perpendicular to the flow path. This brings the internal components, such as the *seat*, *stem*, and *disk/plug*, closer to the top opening of the valve. This makes them easier to reach during disassembly.

M

Modular Valve

- *Design and Working Mechanism:* A *modular valve* is used for double block and bleed applications. This refers to a valve system that incorporates two ball valves installed in series within a single body for double isolation. In addition, a needle valve is positioned between two ball valves to drain or vent the trapped fluid or gas between two balls. The design allows for two independent sealing points, ensuring complete isolation of the process fluid when the two ball valves are closed. The needle valve in the middle provides a bleed or venting option, allowing for pressure monitoring and ensuring safety by enabling the release of trapped pressure between the two ball valves. Figure 2.16 shows a modular valve including double ball and a needle valve in between. It is important to note that a double block and bleed configuration can also be achieved using other valve combinations, such as butterfly or plug valves (see Figure 2.17), for double isolation.
- *Key Components:* Key components of a modular valve include two ball valves in series, a needle valve for pressure bleed, a body that houses the components, seats, and seals for isolation.
- *Applications:* Modular valves are commonly used in high-pressure piping systems, critical isolation processes like flammable and toxic process fluids, and applications requiring double block and bleed functionality for safety and pressure management.
- *Limitation:* A limitation of modular valves is their complexity and higher cost compared to single isolation valves, as well as the potential for increased maintenance due to multiple internal components.

FIGURE 2.16 Modular valve including two ball valves in half-open position and a needle in between. (With permission: Shutterstock.)

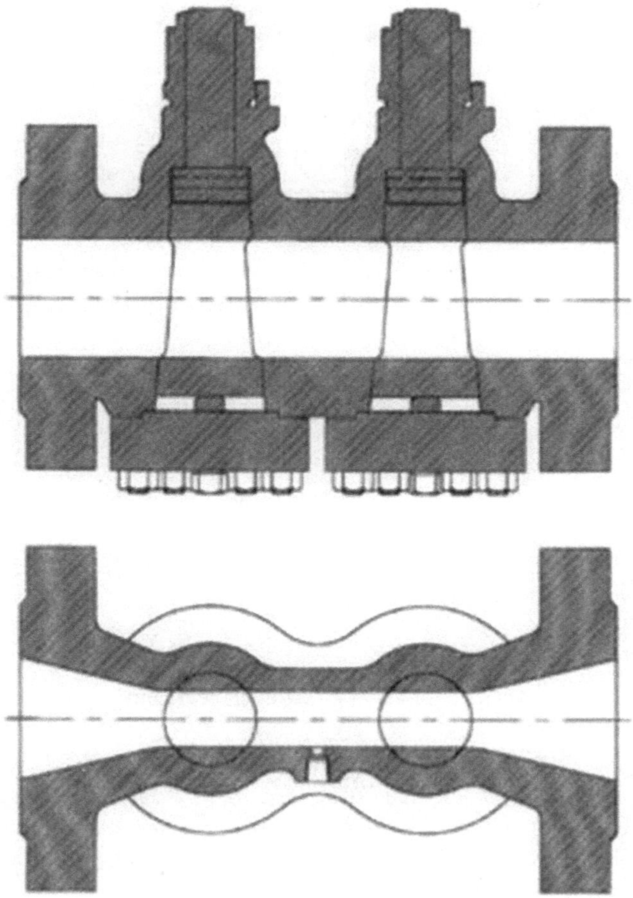

FIGURE 2.17 Double-plug valves with a bleeder in between.

N

Needle Valve

- *Design and Working Mechanism:* A *needle valve* is designed to precisely control the flow of fluid by adjusting the position of a tapered needle-like plunger within a corresponding seat. The valve operates by turning the handwheel, which moves the stem and needle either closer to or away from the seat, allowing fine adjustments to the flow rate. Due to its design, it provides excellent throttling capability, especially in low-flow or high-pressure applications, ensuring accurate flow control and minimal fluid turbulence. Figure 2.18 illustrates a needle valve in the closed position, with the needle or plunger seated firmly against the valve seat.

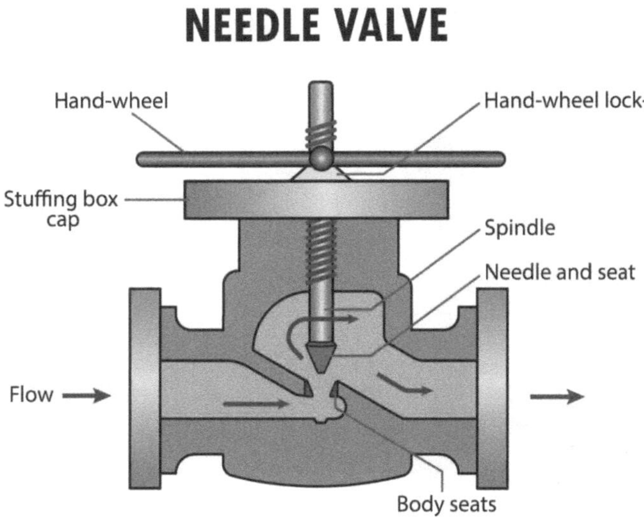

FIGURE 2.18 A needle valve with an integrated stem and plug. The essential components of the valve are highlighted. (With permission: Shutterstock.)

- *Key Components:* The key components of a needle valve include the body, bonnet, stem, needle-shaped plunger, seat, and handwheel or actuator for precise flow control.
- *Advantages and Applications:* Needle valves offer precise flow control, leak-tight performance, and durability, making them ideal for applications in instrumentation, sampling, hydraulic systems, and low-flow rate processes in small sizes typically 2″ and less and industries like oil and gas, chemical, and pharmaceuticals.
- *Limitations:* Needle valves are limited by their high-pressure drop, susceptibility to wear in abrasive or high-velocity flows, and unsuitability for large flow rates due to their small orifice size.

P

Plug Valve

- *Design and Working Mechanism:* A plug valve consists of a tapered or cylindrical plug with a hole or port through the center, which is rotated within the valve body to control flow. When the port of the plug aligns with the inlet and outlet, fluid can flow through the valve; when the plug is rotated 90°, the port is closed, stopping the flow. The design of the plug valve allows for quick, reliable shutoff with minimal leakage, and is commonly used in on/off applications in various industries, including oil and gas. Figure 2.19 illustrates a plug valve in the open position, where the hole in the plug is aligned with the flow path, allowing fluid to pass

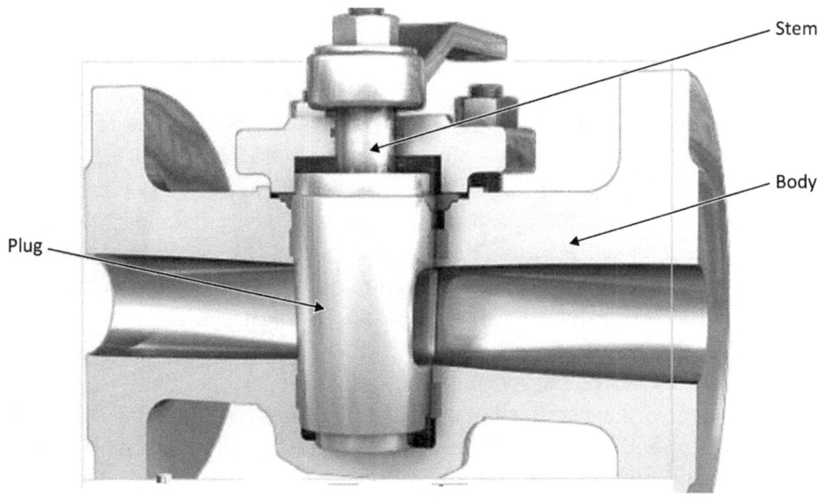

FIGURE 2.19 Picture of a manually operated plug valve in the open position.

through. The valve is manually operated using a lever, which transfers motion to the stem, rotating the plug to control the flow. When the lever is turned, the plug is rotated, and the flow path can be opened or closed, providing a simple and effective means of controlling fluid flow.

- *Key Components:* Key components of a plug valve include the valve body, the plug (which may be tapered or cylindrical), a stem to rotate the plug, and an actuator or a handle to control the plug's position.
- *Advantages:* Plug valves offer several advantages, including simple, reliable operation with quick on/off control, making them ideal for systems requiring frequent shutoff. They provide a tight seal when closed, reducing leakage and ensuring minimal maintenance. Additionally, their design allows for low-pressure drop, smooth flow control, and durability, especially in applications with viscous fluids or slurry or dirty (particle containing) services. Plug valves are also versatile, available in various configurations, and are effective in both low- and high-pressure systems. Plug valves are primarily designed for on/off control, but they can be used for limited flow regulation in certain applications. However, their flow control capability is not as precise as that of other valve types, such as globe or control valves.
- *Limitations:* Plug valves have limitations in flow regulation, as they are primarily designed for on/off control and cannot provide precise flow modulation (throttling). They can experience wear, especially in abrasive or high-viscosity applications, leading to leakage. Additionally, plug valves have a limited pressure and temperature range, making them less suitable for extreme conditions. Larger plug valves can also be bulky and require more force to operate due to increased friction, and their seals may need regular maintenance to prevent leakage.

S

Safety Valves

- *Design and Working Mechanism:* A safety valve is designed to protect pressure vessels, piping and pipelines, or equipment from overpressure by automatically releasing excess pressure when it exceeds a predetermined limit. In normal operation, the spring holds the disk tightly against the seat in closed position, sealing the system. When the pressure in the system exceeds the set limit, the force exerted by the pressure overcomes the spring force in the valve disk, lifting the disk and allowing the excess pressure to escape through the outlet. Once the pressure drops back to a safe level, the spring pushes the disk back onto the seat, resealing the valve. This simple but effective mechanism ensures the safety of equipment and personnel by preventing catastrophic pressure buildup. Figure 2.20 illustrates a safety valve with its key components. The inlet port is positioned at the bottom, and the outlet port is on the right side. The disk is lifted off the seat, indicating that the valve is open, allowing pressurized fluid or gas to flow through by overcoming the force of the spring. When the pressure of the fluid or gas decreases to a certain level, the spring forces the disk back onto the seat, closing the valve.
- *Key Components:* The valve consists of key components such as a valve body, inlet and outlet ports, a spring mechanism, a disk, and a seat.
- *Advantages:* Safety valves provide critical protection by automatically relieving excess pressure in systems, and preventing potential damage

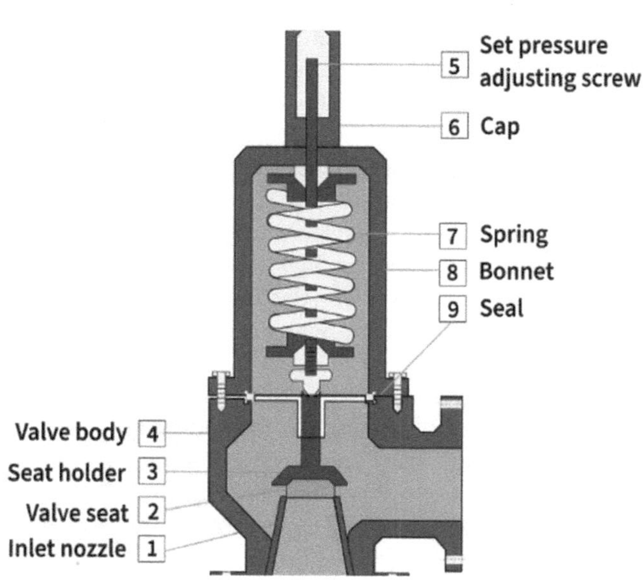

FIGURE 2.20 Safety valve and key components. (With permission: Shutterstock.)

or catastrophic failures due to overpressure. They are commonly used in industries such as oil and gas, chemical processing, and power generation, where maintaining pressure within safe limits is essential. Their advantages include reliability, simple operation, and the ability to respond quickly to overpressure situations, ensuring the safety of both equipment and personnel.

- *Limitations:* The limitations of safety valves include the potential for *frequent cycling* or *chattering* under fluctuating pressures, which can lead to wear and reduced lifespan. Additionally, they may not be effective in *low-pressure relief situations* or in systems with *high-viscosity fluids* that could impede valve operation.

Safety valves are known by many different names depending on their function and whether they are used for gas or liquid. It should be noted that the pressure relief valve (PRV) and the pressure safety valve (PSV) are two of the most important names that are explained as follows.

Pressure Relief Valve

PRVs are designed to relieve excess pressure in a controlled manner in order to protect equipment and systems. In the event that the pressure within the system rises above the set pressure, it gradually opens, releasing fluid and maintaining a safe pressure level. As pressure increases, PRVs open gradually and close gradually as pressure decreases, making them proportional in their operation. PRVs are commonly used in liquid systems (such as water or oil lines) where pressure increases slowly and a gradual pressure release is sufficient to protect the system.

Pressure Safety Valve

A PSV is also designed to release excess pressure, but it is typically used in emergency situations in high-pressure systems where a quick response is necessary. As soon as the set pressure is reached, the PSVs open fully and immediately, which is known as the pop action valve. As soon as the pressure drops below the set point, the valve shuts. PSVs are typically used in gas and vapor systems (such as steam, air, or compressed gas applications) where pressure increases rapidly and a full opening is required to protect the system from damage. While the phrases pressure relief valve and pressure safety valve are frequently used interchangeably, they have different functions and have different applications depending on the specific requirements of the system.

Special Valves

Special valves are designed for specific applications, offering unique features to meet particular process requirements.

Special Valve Types (SPVT)

This section provides explanations for choke valve, C-type ball valves, diaphragm valves, multi-port valves, orbit ball valves, pinch valves, solenoid valves, vacuum breaker, and air release valves.

SPVT1: Choke Valve

A *choke valve* is a specialized type of valve used to regulate the flow of fluids, especially in high-pressure applications like oil and gas production. Its primary function is to control the flow rate by restricting or "choking" the fluid to a desired level, often in situations where pressure needs to be reduced in a controlled manner. The valve typically consists of an *orifice plate* or a *plug* that can be adjusted to vary the flow restriction. The *working mechanism* involves the movement of the plug (or another controlling element) relative to the orifice, which controls the size of the opening and, thus, the flow rate. Choke valves are designed to handle high-pressure differentials, and they can be manually or automatically operated. They are commonly used in *oil and gas production*, *well testing*, and *pressure relief* systems, where precise control of fluid flow is critical to prevent equipment damage and optimize production. A choke valve can also be used to stop production in the event that something goes wrong downstream. The downstream part refers to the piping and separator located after the choke valve. A choke valve can be observed on the right side of some Christmas tree valves in Figure 2.21 (right). In Figure 2.21, the fluid enters the choke valve from the bottom part inside the vertical line; once it has rotated 90° within the valve, it exits to the horizontal line, as shown on the right-hand side of the choke valve. As a result of the rotation of the fluid inside the choke valve and the resulting pressure drop, choke valves are at high risk of erosion and cavitation. A choke valve can either be operated manually with a handwheel or automatically with an actuator.

FIGURE 2.21 Choke valve on a wellhead. (With permission: Shutterstock.)

SPVT2: C-Type Ball Valve

A *C-type ball valve* is a specialized ball valve featuring a C-shaped ball with a notch, designed for enhanced flow control and throttling capabilities. This design provides precise modulation of flow (unlike standard ball valve) and is particularly suited for applications requiring both shutoff and control functions. Key parts of a C-type ball valve include the C-shaped ball, body, seat, stem, and actuator or handle. The advantages of C-type ball valves include their ability to handle high-pressure and high-temperature conditions, smooth operation, and suitability for abrasive or viscous fluids due to reduced turbulence. However, their limitations include potential wear and erosion of the seat in highly abrasive environments and higher initial costs compared to standard ball valves. These valves are commonly used in industries such as oil and gas, chemical processing, and power generation where precise flow control is essential.

Figure 2.22 shows a C-type ball valve in the closed position, with the C-shaped ball securely seated against the valve seat. The valve can be throttled by rotating

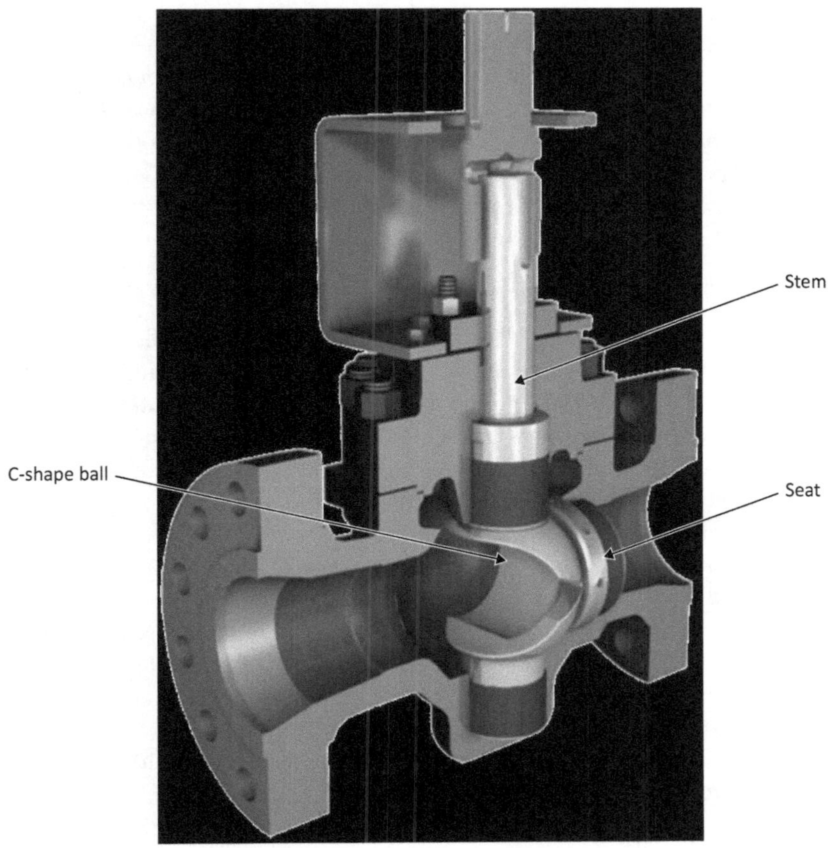

FIGURE 2.22 A C-type ball valve in closed position.

the handwheel, which moves the stem to adjust the position of the C-shaped ball. This allows precise modulation of the flow rate, as the notch in the ball gradually opens or closes the flow passage, providing controlled regulation of the fluid flow through the valve.

SPVT3: Diaphragm Valve

A diaphragm valve is a valve that regulates fluid flow using a flexible diaphragm. Flow control is achieved by the diaphragm pressing against a seat to seal the valve or lifting away from the seat to allow flow. Key components include the diaphragm (usually made from elastomers like rubber or thermoplastics like PTFE), valve body, bonnet, stem, and actuator. The main advantage of diaphragm valves is their ability to handle corrosive, abrasive, or viscous fluids. The fluid does not contact with internal metal parts due to presence of diaphragm, thus minimizing wear and contamination risks for valve internals. They also offer a reliable shutoff. However, their limitations include limited pressure and temperature tolerance due to presence of nonmetallic diaphragm compared to other valve types. The other operational risk involved with this valve is potential diaphragm wear over time, which may require maintenance or replacement. Diaphragm valves can be used for *flow throttling*, but they are generally more suited for *on/off* control or shutoff applications. Diaphragm valves are commonly used in industries like pharmaceuticals, food processing, water treatment, and chemical processing where clean, leak-tight, and corrosion-resistant fluid control is essential.

SPVT4: Multiport Valves (Two-Way and Three-Way Ball Valves)

A *multiport valve* is a valve designed to control the flow of fluid in systems requiring more than two ports, allowing for fluid diversion, mixing, or selection between multiple pathways. Multiport valves include *three-way ball valves*, which are among the most commonly used types. Key components of multiport valves include the valve body, the ball, or rotary plug (with multiple ports or channels), the stem, actuator, and seals. A *two-way valve* is *not* a multiport valve, but it is important to know this type of valve. In a two-way ball valve, flow is either directed through or blocked, whereas a three-way ball valve can divert or mix fluids depending on its configuration, often using a T- or L-shaped ball.

When comparing a *two-way ball valve* and *a standard ball valve with one inlet and one outlet*, they are essentially the same in terms of functionality. Two-way valves, as their name implies, have two ports: an inlet port “A” and an outlet port “AB.” These ports provide a straightforward flow path – fluid enters through one port and exits through the other. Moreover, it consists of a ball with a hole (bore) through its center for allowing or blocking the flow of fluid. As fluid enters the inlet (port A) of a two-way valve, the relative position of the plug determines the amount of fluid allowed to exit the outlet (port AB). The valve is fully closed from ports A to AB as soon as the ball is fully positioned so that its hole is

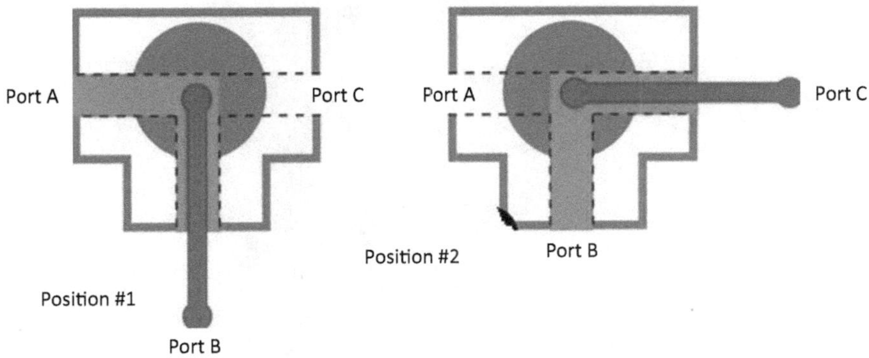

FIGURE 2.23 L-port ball valve.

perpendicular to the flow path in port A. On the other hand, when the ball hole allows fluid to pass through, the valve is fully open from ports A to AB. It is common to find two-way valves used in basic on/off applications, where they are often referred to as isolation valves.

The L-port three-way ball valve illustrated in Figure 2.23 has a ball with the shape of an "L." In fact, three-way ball valve with L port has one inlet and one outlet at a time. However, rotation of ball can change the positions of inlet and outlet ports. The figure on the left shows Position #1, in which the flow enters the valve from port B, moves 90° to the left, and exits the valve from port A. The ball is rotated 90° counterclockwise in position #2 to divert the flow from port A to port C. In this case, the flow enters the valve from port B, gets diverted to the right, and exits the valve through port C. L-port three-way ball valve cannot be used for mixing the flow. A *T-port three-way ball valve* is a type of ball valve with three ports (one inlet/two outlet combinations) and a ball that has a *T-shaped bore* (see Figure 2.24). The valve shown in the figure is connected to the piping system on three sides using three flanges. This design allows for multiple flow configurations, making it versatile for mixing, diverting, or isolating fluid flow in a system.

There are different flow patterns for a three-way ball valve in both L-port and T-port. In some rare cases, a Y-port may be used. The main advantages of multiport valves include their ability to control the flow in complex systems with fewer valves, saving space and reducing costs. They also offer high flow capacity and require less actuation (in case of automation requirement for valves). Applications of multiport valves are widespread, including in chemical processing, water treatment, and power generation, where multiple flow paths need to be managed. However, their limitations include potential complexity in maintenance due to multiple ports and seals, as well as a higher risk of wear and leakage in the case of three-way valves when used for throttling or in high-pressure systems. Additionally, multiport valves may not provide as fine a flow control as single-port valves in some cases.

FIGURE 2.24 T-port ball valve connected to piping on three sides via three flanges. (With permission: Shutterstock.)

SPVT5: Orbit Ball Valve

Orbit ball valve is same as the standard ball valve with one key difference. One of the key advantages of the Orbit ball valve is that it is designed for high-cycle applications. This is largely due to its unique design, where the ball does not touch the seat during movement. In traditional ball valves, the ball often makes contact with the seat when the valve is opened or closed, which can lead to wear over time, especially in high-cycle applications. In contrast, the Orbit ball valve uses a floating ball design where the ball moves within the valve without coming into direct contact with the seat during the opening and closing process. This feature significantly reduces friction, wear, and tear. This makes the valve to withstand frequent cycles and operate for a longer period with minimal maintenance. These may have limitations in handling high-viscosity fluids. In addition, this valve can be prone to wear from abrasive materials.

SPVT6: Pinch Valve

A *pinch valve* is a type of valve that uses a flexible elastomer sleeve or tubing as the primary flow control mechanism. It is designed to regulate or completely stop the flow of abrasive, corrosive, or particulate-laden fluids. The valve operates by mechanically or pneumatically compressing the sleeve, pinching it shut to block flow or partially compressing it to control flow. Pinch valves are highly resistant to wear and corrosion, as the fluid only contacts the elastomer sleeve and not the valve body and valve internals. They are commonly used in industries like mining, chemical processing, wastewater treatment, and food production. Pinch valves are valued for their simplicity, reliability, and low maintenance requirements. Figure 2.25 illustrates the operation of a pinch valve in both open and closed positions. On the left side, the image shows the pinch valve in the open position, where the stem is fully retracted, allowing fluid to flow freely through the tube. The left image depicts the valve in the closed position, where the stem has moved downward to compress the flexible tube, effectively stopping the fluid flow.

SPVT7: Solenoid Valve

A *solenoid valve* is an electromechanically operated valve used to control the flow of liquids or gases in a variety of applications. It consists of a solenoid (a coil of wire that acts as an electromagnet when energized) and a valve mechanism that opens or closes the fluid path. When electrical current flows through the solenoid, it generates a magnetic field, causing the valve's plunger or armature to move and change the valve state, either opening or closing the flow. Solenoid valves are available in two primary types: *normally closed* (NC), which remain closed until energized, and *normally open* (NO), which remain open until energized. Proportional solenoid valves are a specialized type of solenoid valve that

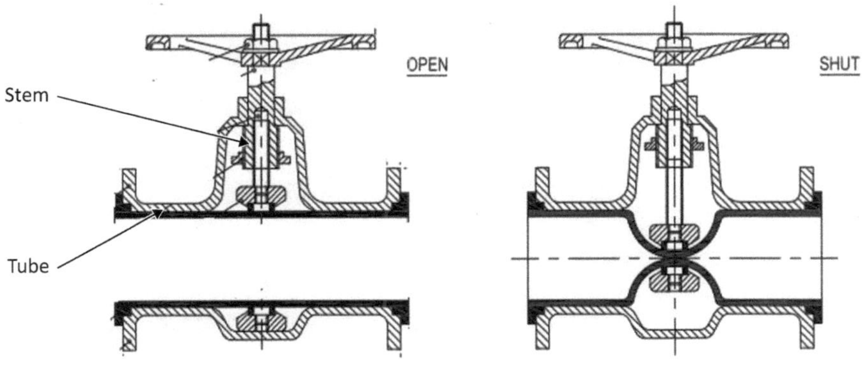

FIGURE 2.25 Pinch valve operation showing the open position (*left*) with the stem retracted, allowing fluid flow, and the closed position (*right*) with the stem compressed, blocking the flow.

allows for a smooth and continuous variation in flow or pressure based on the electrical input. Unlike standard solenoid valves, which operate in an on/off manner, proportional solenoid valves function as control valves. Key components of the solenoid valve are body, disk or diaphragm or plunger, coil, and spring. Figure 2.26 illustrates the operation of a solenoid valve, with the left side showing the de-energized (closed) state and the right side displaying the energized (open) state. In the left section, the coil (electromagnet) is not energized, so the spring pushes the plunger down, blocking the fluid flow and keeping the valve closed. In the right section, the coil is energized, generating a magnetic field that lifts the plunger (spring-loaded piston), allowing fluid to pass through the valve. The blue area at the bottom shows the fluid flowing through the open passage.

In renewable energy systems, solenoid valves are vital for controlling fluid or gas flow in a precise, efficient manner. For example, in *wind farms*, solenoid valves are often used in hydraulic or pneumatic systems to operate brakes, pitch control mechanisms, and lubrication systems for turbines. By enabling fast and reliable operation, solenoid valves contribute to the efficient management of energy generation and mechanical performance in wind turbines.

Solenoid valves offer several advantages, making them a popular choice in various industries. They provide fast and precise flow control, switching between open and closed states almost instantaneously, which is crucial for automated systems. Their compact design allows for easy integration into tight spaces, while their minimal moving parts ensure reliability and low maintenance requirements. Solenoid valves are energy-efficient, requiring low power to operate, and

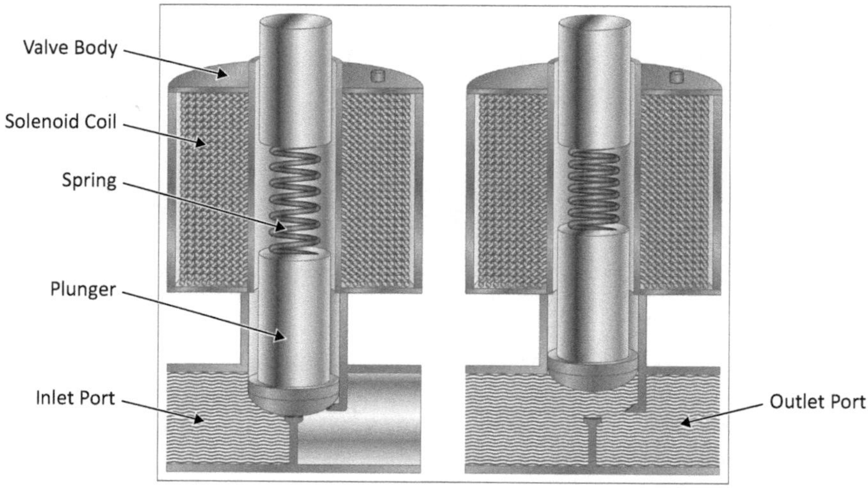

FIGURE 2.26 Operation of a solenoid valve, showing the plunger held down by spring force on the left to block fluid flow (closed position). The plunger is lifted up by electromagnetic force when the coil is energized on the right side, allowing fluid to pass (open position).

are available in a wide range of materials to handle different fluids and environmental conditions. Their versatility and ability to operate remotely make them ideal for applications in industries like renewable energy, water treatment, and manufacturing.

SPVT8: Vacuum Breaker and Air Release Valve

A *vacuum breaker* and an *air release valve* are devices used to maintain the integrity and functionality of fluid systems by managing air and pressure. A *vacuum breaker* prevents the formation of a vacuum within a system, which can cause structural collapse or disrupt fluid flow. It allows air to enter the system when internal pressure drops below atmospheric pressure, safeguarding equipment like pipelines and tanks. An *air release valve*, on the other hand, is designed to expel trapped air during normal system operation, preventing air pockets that can reduce flow efficiency, cause water hammer, or lead to uneven pressure. Together, these valves ensure the smooth operation and safety of fluid systems.

This paragraph explains how *vacuum breaker* works. In normal conditions, the system pressure is higher than atmospheric pressure, so the valve remains closed. The sealing element (like a spring-loaded disk, weighted disk, or ball) is held in place by the internal pressure and a spring mechanism. When the internal system pressure drops below atmospheric pressure (e.g., during draining, pump shutdown, or rapid cooling), a vacuum forms, which could damage the system. The force of the vacuum is greater than the resistance provided by the sealing mechanism (spring, weight, or atmospheric force). This pressure differential causes the sealing element to lift or unseat, opening the valve. The open valve allows atmospheric air to flow into the system, equalizing the pressure and preventing potential damage from the vacuum. Once the pressure is equalized, the atmospheric force or spring mechanism pushes the sealing element back into place, closing the valve and stopping further air entry.

This paragraph explains how *air release* works. During system operation, air can get trapped in high points of the system, forming pockets that disrupt fluid flow, reduce efficiency, or cause water hammer. The air release valve has a float inside, which rises or falls with the liquid level. When air accumulates in the valve, it displaces the liquid inside, causing the float to drop. This movement opens a small orifice or vent. The trapped air escapes through this vent until the liquid level rises again. As the liquid reenters the valve, it lifts the float, which seals the orifice, stopping any further air release.

Figure 2.27 illustrates a vacuum breaker (left) and an air release valve (right). In the air release valve, the float is shown in the lowered position, allowing trapped air to escape from the system into the atmosphere through the vent hole. Since air is present in the system, there is no vacuum, meaning the vacuum breaker valve on the left remains closed. As shown, its disk is seated securely on the valve seat.

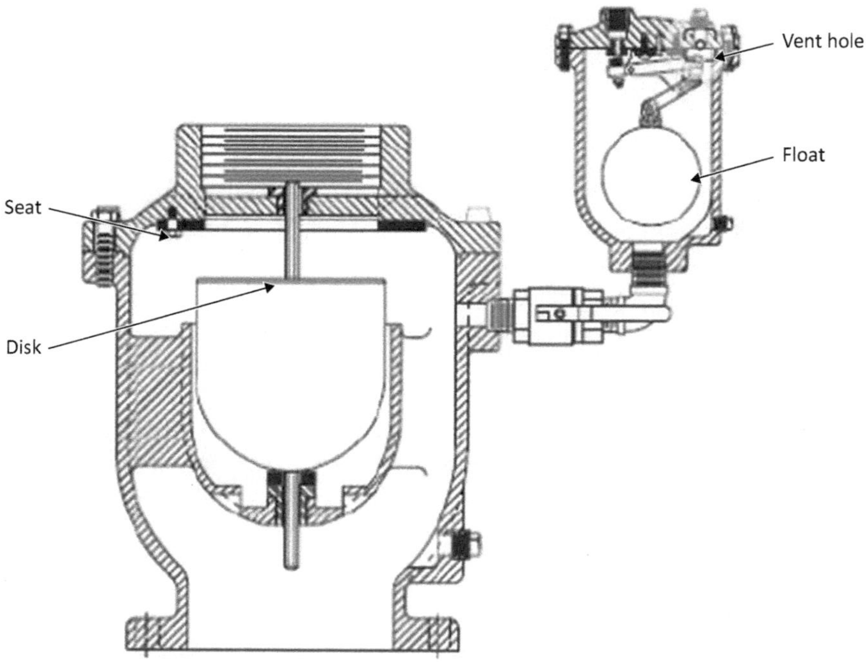

FIGURE 2.27 Illustration of a vacuum breaker (*left*) and an air release valve (*right*).

PART 3: VALVE COMPARATIVE ANALYSIS

This comparative analysis summarized in Table 2.1 highlights the key differences and advantages of various valve types based on their *flow control capabilities, pressure drop, operational speed, material compatibility*, and *application suitability*.

Key Notes

- *Flow Control Capability:* Describes the precision with which the valve regulates fluid flow, ranging from on/off control (ball, gate valves) to fine throttling (needle, globe valves).
- *Pressure Drop:* Indicates the loss of pressure as the fluid flows through the valve. A higher pressure drop can reduce system efficiency.
- *Operational Speed:* Refers to how quickly the valve can be opened or closed. Some valves like ball and butterfly valves are quick to operate, while others like globe and needle valves take more time.
- *Material Compatibility:* Identifies which materials are suitable for use with the valve based on the types of fluid or gas in the system (e.g., corrosive, high-temperature).

TABLE 2.1
Valve Types Comparison: Key Parameters and Application Suitability

Valve Type	Flow Control Compatibility	Pressure Drop	Operational Speed	Material Compatibility	Application Suitability
Ball valve	On/off control, limited flow regulation	Medium to high (depending on design)	Fast opening/closing	Compatible with a wide range of materials (metals, plastics, etc.)	Suitable for general shutoff applications (both low and high pressure) and both clean and particle-containing (dirty) services
Butterfly valve	On/off and also good for throttling and flow regulation	Medium to high	Fast opening/closing	Compatible with a wide range of metallic materials	Suitable for general shutoff and flow regulation applications, mainly in low-pressure applications. However, butterfly can be used in high pressure with both clean and dirty services
Gate valve	On/off	Low due to linear flow path	Slow to moderate speed (requires several turns)	Compatible with a wide range of metallic materials	Best for *on/off* applications, suitable for both particle-containing and clean services in both low- and high-pressure applications depending on the type of gate valve
Plug valve	On/off control, limited throttling capability	Medium to high	Fast opening/closing	Compatible with a wide range of metallic materials	Ideal for on/off applications in particle-containing fluid (dirty) services in both low- and high-pressure applications—limited throttling applications

(*Continued*)

TABLE 2.1 (*Continued*)
Valve Types Comparison: Key Parameters and Application Suitability

Valve Type	Flow Control Compatibility	Pressure Drop	Operational Speed	Material Compatibility	Application Suitability
Globe valve	Precise flow control, throttling	High (due to flow path design)	Slow (requires multiple turns to open/close)	Compatible with a wide range of metallic materials	Ideal for precise flow regulation in both low and high pressure and clean and dirty services
Needle valve	Very fine flow control (precise adjustments)	Low (due to smaller orifice)	Slow (requires manual adjustment)	Compatible with a wide range of metallic materials	Used for fine adjustments in both low-and high-pressure applications and suitable for only clean services
Diaphragm valve	On/off and precise control for erosive liquids	Moderate to high (depending on diaphragm design)	Slow to moderate	Compatible with a wide range of metallic materials	Used for particle containing services in low-pressure applications
Check valve	Prevents backflow, no flow control	Low (no major obstructions to flow)	Instant (automatically closes with reverse flow)	Compatible with a wide range of metallic materials	Suitable for both clean and dirty services and both low- and high-pressure applications
Choke valve	Regulates flow to control pressure or flow rate in high-pressure systems	Medium to high (depending on application)	Slow to moderate	Usually made of materials capable of handling high-pressure and abrasive (dirty) fluids	Used in oil and gas production, especially in wells
Safety valve	Provides automatic shutoff in case of excessive pressure	Low to medium	Fast (opens instantly under high pressure)	Compatible with a wide range of metallic materials: Must be compatible with pressure equipment materials	Suitable for both clean and dirty services and both low- and high-pressure applications

PART 4: VALVE SELECTION CRITERIA

Choosing the right valve for a particular application is critical to ensuring system reliability, performance, and efficiency. This section provides a practical guide to selecting the most suitable valve, taking into account key factors such as *fluid properties*, *operating pressure and temperature*, and *system design requirements*. Additionally, it highlights common mistakes made during valve selection and provides guidance on how to avoid them.

Key Considerations

1. *Fluid Properties:* The first step in valve selection is understanding the *fluid properties* such as whether it is a liquid, gas, or slurry, as well as its *viscosity*, *density*, and potential for being *abrasive* or *corrosive*. For corrosive fluids, materials like stainless steel or special alloys may be required. For abrasive fluids, a valve with a hardened material and appropriate sealing is essential.
2. *Operating Pressure and Temperature:* Make sure that the valve is rated for the *maximum pressure* and *temperature* expected in the system. Valves not suited for high-pressure systems may fail, causing catastrophic damage. Similarly, extreme temperatures can affect the valve's material properties, causing leaks or failures if not properly accounted for.
3. *System Design Requirements:* The size of the valve and its ability to fit into the existing system should be carefully evaluated. It is also crucial to consider the *actuation method*—manual, pneumatic, or electric—based on the application's needs. Additionally, some systems may require *precise flow control*, which would necessitate valves like *globe* or *needle valves*.
4. *Maintenance and Durability:* The selected valve should align with the expected *maintenance frequency* and be *durable* enough for long-term operation. Some valves require more frequent maintenance (e.g., *globe valves*), while others, like *check valves* or *ball valves*, require minimal maintenance.
5. *Cost and Budget:* Finally, the cost of the valve, in terms of both *initial installation* and *long-term operation*, should be considered. While a cheaper valve might seem appealing upfront, its higher operational or maintenance costs could make it less economical in the long run. Chapter 13 provides more information about cost of the valves.

Common Pitfalls to Avoid

- *Underestimating Pressure and Temperature:* Always ensure that the valve is rated for both the maximum pressure and temperature. Neglecting these factors can lead to valve failure.

- *Overlooking Fluid Compatibility:* Make sure the valve material is compatible with the fluid, especially in corrosive or abrasive applications. For example, carbon steel should not be selected for corrosive sea water services.
- *Incorrect Valve Sizing:* Choosing a valve that is too large or too small for the application can result in inefficient operation, unnecessary wear, or leakage.
- *Inadequate Actuation:* If an automated valve is required, ensure the actuator is properly sized for the application's demands.

Example 1: Overlooking Fluid Compatibility: An example of selecting the wrong valve type based on fluid properties could be using a *ball valve* for controlling the flow of *slurry* (a mixture of solid particles and liquid). Since ball valves have a tight seal that can easily get clogged by solid particles, they are not ideal for slurry applications. Over time, the valve may experience damage, reduced performance, or even failure because it wasn't designed to handle the abrasive and particulate nature of the slurry. A *plug valve* or *pinch valve* would be more suitable for such applications, as they can better handle the flow of solids.

Example 2: Incorrect Valve Sizing: An example of selecting the wrong ball valve for a piping system would be choosing a *small-sized ball valve* for a *high-flow* application. In this case, the valve would be unable to handle the required flow rate, leading to *excessive pressure drop, poor flow control,* and *possible valve damage* due to *cavitation* or *erosion* from the high-speed fluid.

QUESTIONS AND ANSWERS

1. What is the purpose of choke valves in the oil and gas industry?
   A. Increasing the flow rate of fluids
   B. Completely stop the flow of fluid
   C. Control the flow rate and reduce pressure
   D. Remove impurities from fluids

   Answer: Option C is the correct answer.

2. Which of the following factors should be considered when selecting the right valve for an industrial piping system?
   A. Fluid properties (corrosiveness, viscosity, etc.)
   B. System pressure and temperature conditions
   C. Valve size and flow requirements
   D. All of the above
   E. Only valve material compatibility

   Answer: Option D is the correct answer.

3. Which valve is most suitable for precise flow regulation in low-flow applications?
   A. Ball valve
   B. Globe valve
   C. Gate valve
   D. Butterfly valve

   Answer: Option B is the correct answer.

4. Which of the following valves would be most suitable for an application involving slurry or abrasive fluids?
   A. Ball valve
   B. Needle valve
   C. Pinch valve
   D. Globe valve

   Answer: Option C is the correct answer.

5. What is a common limitation of a gate valve?
   A. Inability to handle high pressures
   B. Slower operation and no precise flow control
   C. Limited design and type
   D. Not suitable for on/off applications

   Answer: Option B is the correct answer.

6. A check valve is primarily used to prevent:
   A. Backflow of fluid
   B. Excessive pressure buildup
   C. Flow regulation
   D. Cavitation

   Answer: Option A is the correct answer.

7. In which of the following applications would a butterfly valve be the most appropriate choice?
   A. High-pressure systems with frequent cycling
   B. Large diameter piping with high flow rates
   C. Precise flow control in steam systems
   D. Control of abrasive or viscous fluids

   Answer: Option B is the correct answer.

8. What is the main advantage of a diaphragm valve over other valve types?
   A. Cost-effectiveness for general applications
   B. Precise flow control for corrosive and erosive fluids
   C. Fast on/off control in low-pressure systems
   D. Ability to handle high-temperature fluids

   Answer: Option B is the correct answer.

9. Which valve is the most suitable for handling high-viscosity liquids with minimal turbulence?
   A. Globe valve
   B. Needle valve
   C. Ball valve
   D. Pinch valve

   Answer: Option D is the correct answer.

10. Which of the following is the most suitable valve type for handling slurries or abrasive fluids?
    A. Ball valve
    B. Knife gate valve
    C. Globe valve
    D. Needle valve

    Answer: Option B is the correct answer.

BIBLIOGRAPHY

Crosby Valve Inc. (1997). Crosby Pressure Relief Valve Engineering Handbook. Technical document number TP-V300.

Hellemans, M. (2009). The Safety Relief Valve Handbook: Design and Use of Process Safety Valves to ASME and International Codes and Standards. 1st edition. Butterworth-Heinemann: Oxford, UK.

Nesbitt, B. (2007). Handbook of Valves and Actuators: Valves Manual International. 1st edition. Elsevier: Oxford, UK.

Skousen, W. L. (2012). Valve Handbook. 3rd edition. McGraw-Hill Education: New York, NY.

Sotoodeh, K. (2018). Why are butterfly valves a good alternative to ball valves for utility services in the offshore industry? American Journal of Industrial Engineering, 5(1), 36–40. doi:10.12691/ajie-5-1-6.

Sotoodeh, K. (2021). A Practical Guide to Piping and Valves for the Oil and Gas Industry. 1st edition. Gulf Professional Publishing: Houston, TX.

Sotoodeh, K. (2021). Subsea Valves and Actuators for the Oil and Gas Industry. 1st edition. Gulf Professional Publishing: Houston, TX.

Zappe R.W., & Smith P. (2003). Valve Selection Handbook: Engineering Fundamentals for Selecting the Right Valve Design for Every Industrial Flow Application. 4th edition. Gulf Professional Publishing: Houston, TX.

3 Valve Components

INTRODUCTION

In industrial applications, valves are essential components since they control the flow of gases or fluids in piping systems. Individual components contribute to the overall functionality and performance of a valve. It is crucial to understand these components when choosing a valve for a specific application. We will explore in this chapter the key components that make up a valve, their roles, and how they contribute to its efficiency and reliability.

A

Auxiliary Valve Connections

The auxiliary valve connection refers to additional ports, fittings, or openings on a valve that serve secondary or supporting functions in addition to controlling the flow of a medium. Auxiliary connections on valves facilitate monitoring, control, or other operations. These connections serve a variety of purposes, such as vent and drain, injection and sampling, pressure relief, etc. An example of a drain or vent port is located at the bottom or top of a valve for the purpose of draining fluids or venting air. A valve stem or seat injection point is intended for the purpose of lubricating or repairing seals. It is possible to inject sealant to the damaged seats through these ports to repair the seat temporarily before applying the next scheduled maintenance. In the case of low-pressure class valves, a plug could be used to close the auxiliary connection, and in the case of high-pressure class valves, a flange. Another type of valve accessory is a bypass line which connects both sides of a valve together to equalize the pressure and allow the valve to be opened and closed more easily. Drain and vent plugs on a small-size ball valve less than 4″ size is shown in Figure 3.1a. Figure 3.1b shows drain and vent flanges on a large-size ball valve.

B

Backseat

Backseats in valves are additional sealing mechanisms located near the top of the valve body, which create a secondary seal around the valve stem when the valve is fully open. This feature is applicable to gate and globe valves. When a gate valve or globe valve is closed, there is no or little pressure within the body cavity. As a result, there is little pressure on the stem packing. Nevertheless, if the valve starts open (from 1% opening to full opening), the packing is subjected to the working

DOI: 10.1201/9781003640967-3

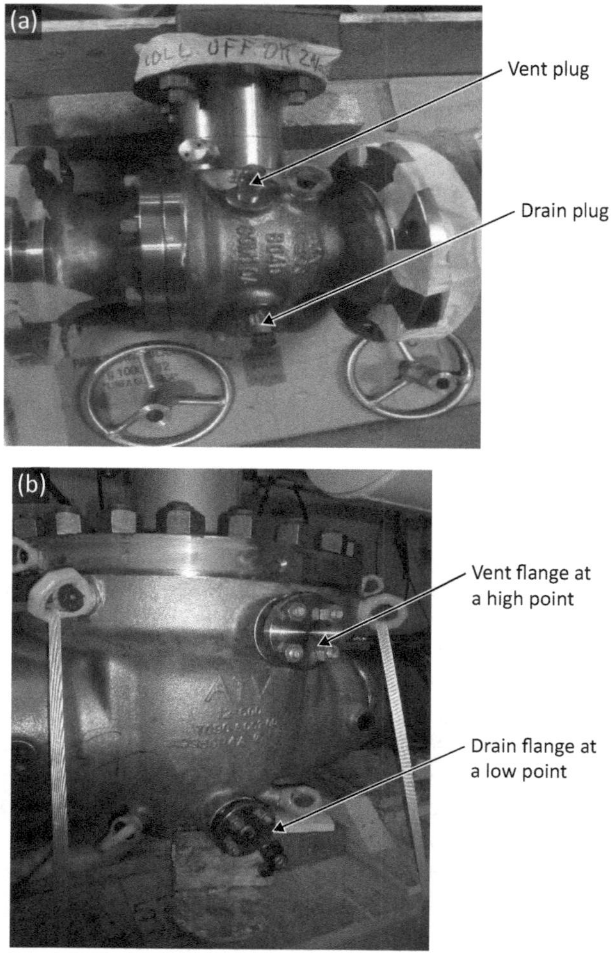

FIGURE 3.1 (a) Drain and vent plugs on the body cavity of a ball valve. (b) Drain and vent flanges on the body cavity of a ball valve.

pressure. In the course of time, the packing can wear and loosen due to contact with the working fluid, fluctuations in temperature, and pressure, which leads to leaks. It is possible to tightly seal the stem and bonnet if a backseat engages the valve body. Therefore, the stem packing is isolated from the working pressure and fluid leakage is prevented. Essentially, this arrangement provides an additional/secondary barrier against external leakage and ensures a tight seal. As a result of the backseat seal, it is possible to perform maintenance on the valve packing (the sealing material surrounding the stem) without the need to depressurize the system, which allows packing replacement or adjustment to be performed. Many end users, however, do not consider packing maintenance without depressurizing and relying on back seating to be a safe practice. Backseats provide an additional

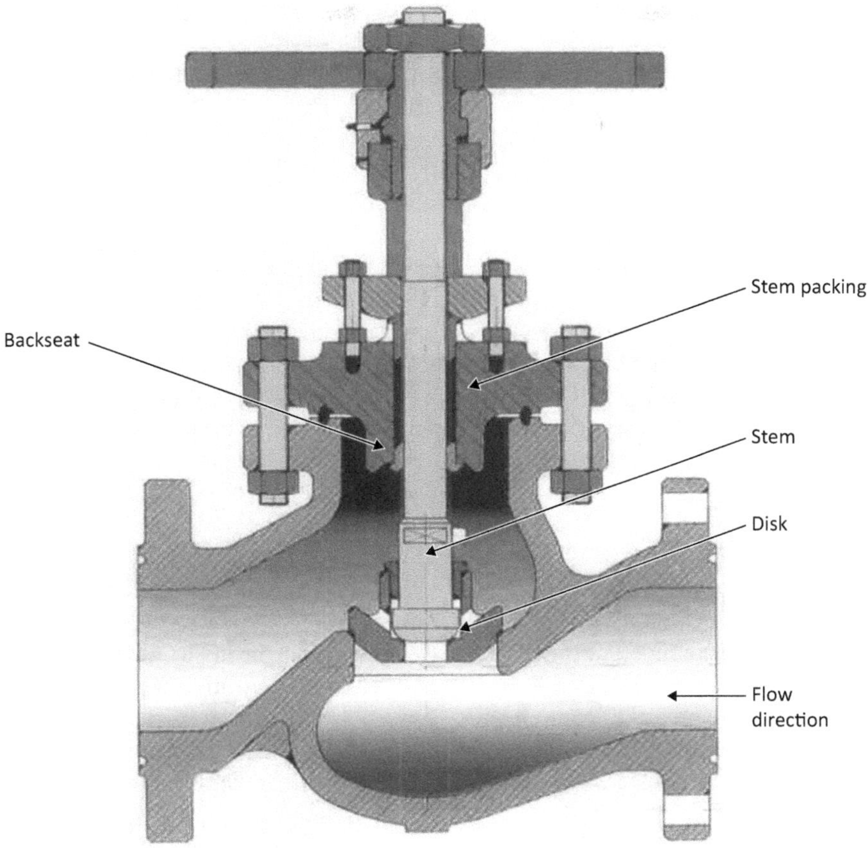

FIGURE 3.2 Illustration of a globe valve in the closed position.

layer of safety, particularly in applications involving hazardous or pressurized fluids, by creating a double barrier to prevent leaks. Figure 3.2 illustrates a globe valve in the closed position, where the fluid is located beneath the valve disk and has no contact with the backseat or stem packing. As the valve begins to open, the fluid flows upward and can potentially reach the backseat area. However, as the stem continues to move upward, it engages with the backseat, creating a seal that prevents the fluid from entering the stem packing area. This backseat mechanism isolates the packing from fluid pressure, providing additional protection and reducing the risk of leakage along the stem.

Ball

In a ball valve, the ball serves as the closure member in valves. The ball is the primary flow control element by having a hole through the center of its spherical shape. Fluid flow is controlled by the rotation of the ball within the valve body.

Upon alignment of the ball hole with the piping or pipeline, the valve opens, allowing fluid to pass through. Rotating the ball 90°, the hole becomes perpendicular to the piping and pipeline, which closes the valve and prevents flow.

Bearing

There are valve bearings, which are circular components usually made of metallic materials with a nonmetallic internal layer. Bearings are installed around the stem of the valve in order to guide the stem, control its linear movement, and prevent its side movement. In summary, the valve bearing provides support and ensures smooth motion of the valve stem or rotating parts. It is typical for the inside of the bearing which is in contact with the stem to be made of a nonmetallic material such as rubber in order to reduce friction and wear (galling) between the metal stem and the bearing. As can be seen in Figure 3.3, the stem and bearing of a ball valve have been highlighted.

Belleville Washer

Belleville washers are spring washers with a conical (dome) shape, commonly used in valves to apply a controlled, consistent force. As a result of pressure and temperature fluctuations, the valve seals could be dislocated or worn. Belleville washers are often used in valve assemblies to maintain tension on seals, gaskets, and other components. They assist in maintaining a reliable seal as well as reducing wear. Belleville washers are typically installed around bolts, stems, and other fastening components that require a controlled, consistent

FIGURE 3.3 An illustration of a ball valve in closed position with the stem and bearing highlighted. (With permission: Shutterstock.)

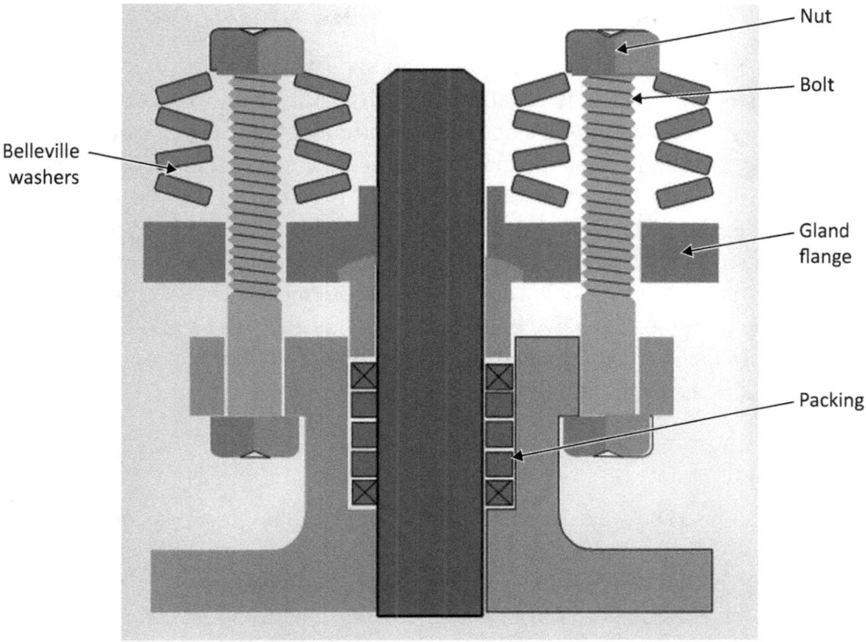

FIGURE 3.4 Belleville washers on the gland flange are used to improve the sealing performance.

preload in valve assemblies. In addition to installing Belleville washers under bolts connecting the valve body and bonnet, they can also be installed behind the seat of a ball valve in order to maintain a tighter seal between the ball and the seat. In many valves, Belleville washers are used around the stem packing gland or on the gland flange in order to maintain a consistent compression on the packing (see Figure 3.4).

Bellows Seal

A *bellows* is a flexible, accordion-like component with multiple convolutions that allow it to expand, contract, or flex under pressure while maintaining a seal. In valves, a bellows seal or bellows seal bonnet provides a highly reliable, leak-tight seal around the valve stem, effectively isolating the valve's internal process fluid from the external environment. Typically, this design is used in systems handling hazardous, toxic, or highly volatile substances, in which fluid leakage needs to be minimized. By expanding and contracting as the valve stem moves, the bellows, which are typically made of stainless steel or corrosion-resistant alloys, form a seal around the valve stem. In this manner, no process fluid can escape from the stem, resulting in a zero-leakage seal. It is especially important when it comes to systems that require high levels of purity, safety, or environmental protection.

Body

The valve body is the primary structural component of a valve. It serves as the housing that holds all the internal parts, such as the stem, disk, and seat. The valve body is designed to withstand the pressure and temperature of the fluid or gas passing through it. It can be made from metallic materials in most cases like carbon steel, stainless steel, or nickel alloys. The body directs the flow of the media and ensures a tight seal to prevent leaks. Depending on the type of valve, the body's shape and internal structure can vary. It is designed to optimize flow characteristics and meet specific operational requirements. The selection of material and the design of the valve body are crucial for the valve's performance and durability. In fact, the valve body is the main pressure-containing component, and if it fails, fluid may leak into the surrounding environment.

Body Seals

The seals are used between the body pieces of valves. Some valves consist of two or three body pieces and require sealing between these parts. For example, the valve shown in Figure 3.5 is made of three pieces body bolted together. Seals may consist of graphite rings or spiral wound gaskets containing metallic rings filled with graphite.

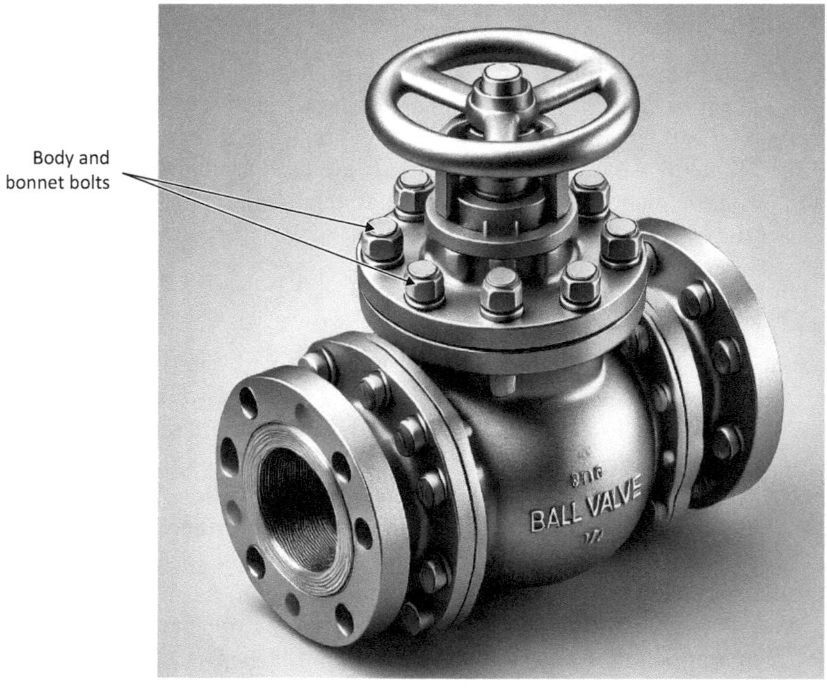

FIGURE 3.5 Highlighting the bolts that connect the body to the bonnet of a ball valve.

Bolting

Bolting is the process of attaching two valve components to each other, such as the body and bonnet, using bolts and nuts. Figure 3.5 illustrates the bolts connecting the body and bonnet of a ball valve.

Bonnet

The *bonnet* is a crucial component of a valve that covers and seals the internal parts, such as the stem, disk, and seat. In addition, it provides a means of access for maintenance and repair. Typically bolted or welded to the valve body, the bonnet protects the internal components from external elements and pressure. It serves as the containment boundary, preventing leakage of fluid or gas from the valve body. The bonnet can also house additional mechanisms, such as the actuator or spring, depending on the valve design. Material selection for the bonnet is important, as it must withstand the same environmental conditions and pressures as the valve body. Bonnet design plays a significant role in the valve's ease of maintenance, as it often allows for disassembly and replacement of components without needing to remove the entire valve from the piping system. As shown in Figure 3.6, the bonnet of a ball valve is highlighted.

FIGURE 3.6 Illustration of a ball valve with its bonnet highlighted.

Bottom Cover

Valve bottom covers, also known as valve bottom flanges or bonnet bottoms, are detachable components located at the lower end of certain valve types like ball valves. As part of the design, operation, and maintenance of the valve, it serves a variety of purposes. The bottom cover encloses the lower part of the valve body, ensuring the internal components, such as the disk or plug, are securely contained. With this feature, it is possible to inspect, repair, or replace the valve's internal components (e.g., the seat, disk, and stem) without having to remove the entire valve from the pipeline. Debris, sediments, and fluids from the valve body can be removed from some bottom covers through a drain port. In systems that are prone to clogging, it is particularly useful. The connection point is sealed with gaskets or O-rings to ensure a tight seal and prevent leakage.

Bushing

A valve bushing is a simple cylindrical sleeve that provides support, reduces friction, and prevents wear between moving parts. It is usually installed around the valve stem or in areas where parts slide or rotate. It is important to note that bushings are types of bearings, specifically plain bearings. In addition to reducing friction and supporting movement, it also performs many of the same functions as bearings. A bushing may be made of both metallic and nonmetallic materials, or a combination of both. In some globe valves, bearings or bushings guide the movement of the disk as it aligns with the seat, as illustrated in Figure 3.7. The bushing reduces wear by preventing the disk from causing galling or excessive friction against the body and bonnet.

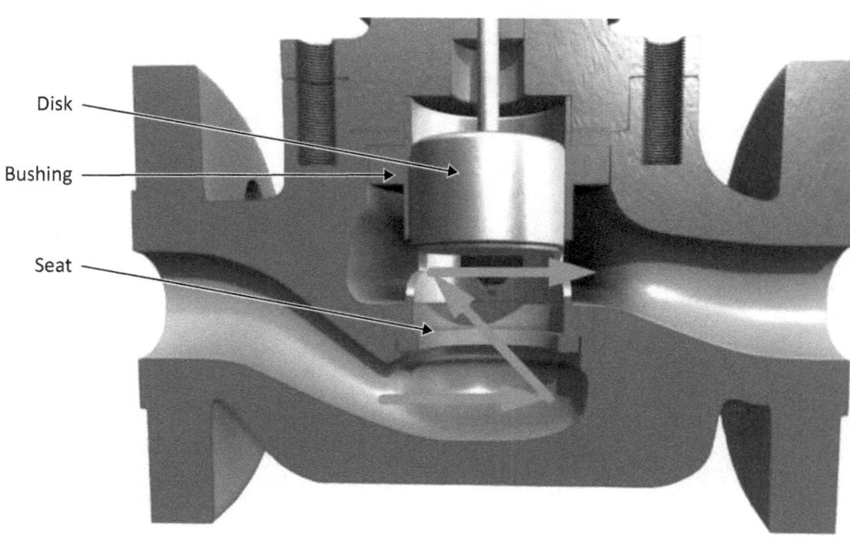

FIGURE 3.7 A bushing used in the globe valve to guide the movement of the disk.

C

Cage

A cage in a valve is an internal component typically found in control valves. It surrounds the valve's plug and guides its movement. The primary function of the cage is to control the flow characteristics. Additionally, by managing the pressure drop across the valve, it helps mitigate cavitation and its consequences, such as noise and vibration. Cages are commonly used in globe valves and high-performance control valves to ensure precise flow control. They are often constructed from durable materials, such as stainless steel or other high-strength alloys, due to the significant pressure and exposure to harsh fluids they must endure. Figure 3.8 shows a control valve cage that affects the characteristic of the flow passing through the valve.

Cavity

A valve cavity is the enclosed space inside the valve body and bonnet. The body cavity in a valve is the enclosed space located between the body, bonnet, seats, and the internal components (e.g., ball, disk, or plug). During normal operation, this cavity can trap fluids, gases, or solids. Trapped substances can lead to adverse effects on the valve's performance and safety. Some potential issues include increased pressure buildup, contamination, or mechanical damage. In this area,

FIGURE 3.8 A control valve cage.

FIGURE 3.9 Body cavity in a ball valve, the area between the ball, body, and bonnet. (With permission: Shutterstock.)

fluid can become pressurized and potentially damage both the internals and the body of the valve. Additionally, if a viscous fluid accumulates, it may negatively impact the valve's functionality, particularly its ability to open and close effectively. The body cavity in a ball valve is illustrated in Figure 3.9.

Clamp Connection (with Hubs)

In demanding industries such as oil and gas, chemical processing, and offshore applications, clamp connections with hubs rather than ASME flanges are used when strong, high-pressure, and leak-proof connections are needed. In these connections, hubs (flanged or ring-shaped components) are attached to the ends of two pipes or a valve and connected piping, which are then aligned and secured using a clamp or bolted clamp assembly, compressing the gasket between the hubs. Using hubs instead of flanges has the advantage of saving weight, space, and money. There are three brands of clamp connections with hubs known for their high performance: Grayloc®, G-lok®, and Techlok®. A hub and clamp connection is shown in Figure 3.10.

Closure Member

The *valve closure member* is the key component responsible for controlling the flow of fluid through the valve. It can take various forms depending on the type of valve, such as a disk in a globe valve, a ball in a ball valve, a gate in a gate valve, or a plug in a plug valve. When the valve is actuated, the closure member moves

FIGURE 3.10 A hub and clamp connection.

to either allow fluid to pass, block the flow entirely, or regulate the flow depending on the type of valve and its intended application. The design and material of the closure member are critical for maintaining a tight seal and preventing leaks. In addition, closure member requires a reliable performance under different pressure, temperature, and flow conditions. The closure member's movement is typically driven by an actuator or a manual operation. The role of closure member is essential for achieving precise flow regulation or complete shutoff in various industrial applications.

D

Dash Pot

In the valve industry, a dashpot is a mechanical device that cushions or slows down the movement of valve components, especially during opening and closing. These devices are commonly used in safety valves, check valves, and large industrial valves to prevent sudden movements that could cause damage or lead to system instability. For example, dash pot in a swing check valve can be used during operation to slow down the valve closure and prevents slamming and water hammering effects. Water hammering is explained in more detail in Chapter 8. Slamming and non-slam check valves are explained in more detail in Chapter 5.

Disk

The *disk* (or *disc*) is a key component of certain valve types, such as globe valves, and plays a crucial role in regulating or stopping the flow of fluid.

Typically made of durable materials like stainless steel, the disk is a flat, circular piece that moves within the valve body to either block or allow the passage of fluid. When the valve is open, the disk is lifted away from the seat, permitting flow through the valve; when the valve is closed, the disk presses against the seat to create a seal and stop the flow. The design and material of the disk are critical to its ability to provide a reliable seal under high-pressure and high-temperature conditions. Additionally, the disk's geometry and surface finish influence the valve's flow characteristics, such as flow rate. In some valves, such as globe valves, the disk can be adjusted to modulate (regulate) the flow, providing precise control over fluid movement.

Drain

A drain in piping and valves refers to a specific opening, connection, or valve designed to remove unwanted fluids (liquid or condensate) from a system. Drains are essential for maintenance, safety, and operational efficiency in piping systems and valves.

F

Flange

Flanges are prominent rings or connectors used to connect valves to pipes or other equipment in piping systems (see Figure 3.11). The flange provides a leak-proof connection and simplifies installation, maintenance, and disassembly of the valve.

A method of connecting a valve to a pipe using flanges, which are bolted together to form a tight seal is called flange connection. Figure 3.12 illustrates a flanged end valve that is connected to piping system from both sides through flanges.

G

Gasket

A gasket in a valve is a sealing component located between two mating surfaces, such as the valve body and bonnet. Its primary function is to prevent fluid or gas leakage. Gaskets are typically made from materials like rubber, PTFE (polytetrafluoroethylene), or metal, which are selected based on the temperature, pressure, and chemical properties of the fluid being transported in the piping system.

There are several types of gaskets, including *flat gaskets*, *spiral wound gaskets*, *ring joint gaskets*, and *metal gaskets*. Each type is designed for specific applications related to pressure, temperature, and fluid type. They are manufactured from various materials to meet different requirements for pressure, temperature, and fluid resistance.

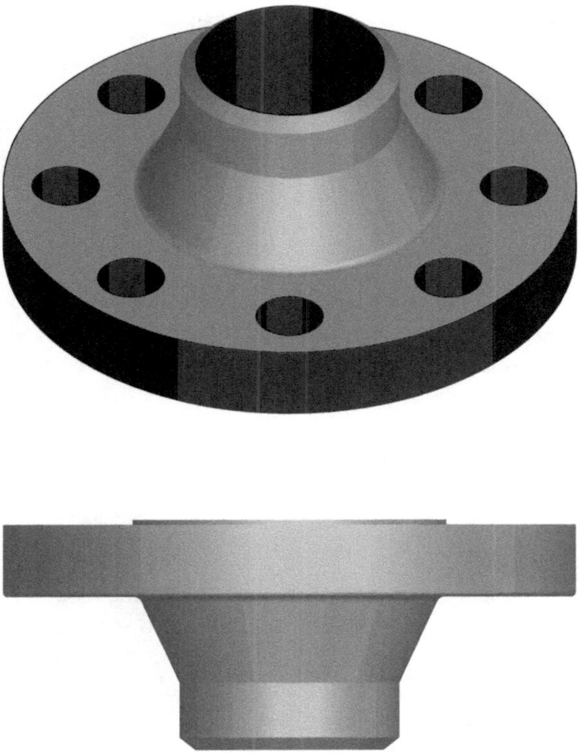

FIGURE 3.11 Weld neck flange.

FIGURE 3.12 Flange connections in piping and valves. (With permission: Shutterstock.)

GATE

A *gate* in a valve is a flat component that slides up and down in a gate valve to control the flow of fluid by either blocking or allowing passage through the valve. It provides a tight seal when fully closed, making it ideal for on/off applications with minimal flow resistance when open.

GEARBOX

A *gearbox* for valves is a mechanical device used to transmit torque from the handwheel to the valve stem, enabling smooth and controlled movement of the valve's closure member. For large valves or high-pressure classes, high force is required from the operator, and the gearbox increases the force produced by the operator by multiplying the rotation of the handwheel, making it easier to operate the valve under demanding conditions.

GLAND

In a valve, the gland is a component of the gland packing assembly that provides additional pressure on the steam sealing or packing to seal the area between the valve stem and bonnet and prevent leakage. Normally, glands work in conjunction with packing and gland followers (or gland flanges) to form a tight seal around the stem, preventing process fluid from escaping. The assembly plays a crucial role in high-pressure applications as well as valves that control hazardous or corrosive fluids.

GLAND FLANGE

The gland flange in a valve is responsible for securing the gland packing, which seals the valve stem to prevent fluid leaks. Along with other components of the gland assembly, the gland flange compresses the packing material around the valve stem, ensuring a tight seal. This component is crucial for valves that handle fluids where leakage could pose safety risks or environmental concerns. These fluids may include flammable substances, high-pressure liquids, and toxic materials, such as high-pressure oil and gas, as well as fluids containing hydrogen sulfide.

Regular adjustment of the gland flange is important to maintain optimal sealing and to extend the life of the valve. In Figure 3.13, a gland flange is shown on the left, as well as a gland flange installed on a valve on the right.

H

HANDWHEEL

A valve handwheel is a mechanical device used to manually operate certain types of valves, allowing users to turn the wheel to open or close the valve. The handwheels are usually circular and can vary in size, and they typically have a hub

FIGURE 3.13 Gland flange.

at the center that enables them to be connected to the valve stem (see Figure 3.14). Typically, handwheel are made from cast iron, or stainless steel, or alternative materials depending on the application and environment. The handwheel rotates the valve stem, which, in turn, moves the valve's internal components (such as a gate, ball, or plug) to control or start/stop fluid flow. By using handwheels, operators are able to control valve positions without relying on automated systems or actuators.

A valve handwheel consists of spokes and rims that allow the user to operate the valve by rotating the wheel. A spoke is a bar or rod that connects the central hub of the handwheel (where it is attached to the valve stem) to the outer rim. Multiple spokes are often arranged in a radial pattern on handwheels to provide

FIGURE 3.14 A valve handwheel with highlighted spoke and rim.

stability and strength. By distributing force evenly, the spokes make it easier to turn the wheel and operate the valve. To turn the handwheel, the user grips the rim, which is the outer circular part of the wheel. A textured or ridged surface is typically used to improve grip. By means of the spokes, the rim is connected to the central hub and is the part that is directly manipulated in order to open and close the valve. In addition to giving the handwheel its shape, the spokes and rim enable efficient transmission of the user's turning motion to the valve mechanism. Figure 3.14 illustrates a handwheel with highlighted spoke, rim, and hub.

HINGE

A *hinge* in a valve is a mechanical component that allows the valve's *closure element* (such as a *disk* or *flap*) to rotate around a fixed axis, enabling the valve to open and close. It is commonly found in *butterfly valves* and swing *check valves*, where the hinge facilitates smooth movement and helps maintain a tight seal when the valve is closed.

HINGE PIN

A *hinge pin* in a valve is a component typically found in *butterfly valves* and other types of valves like check valves that have a rotating disk or plate. It acts as the central pivot point around which the valve's disk or flap rotates to control the flow of fluid. Figure 3.15 shows a swing check valve in an open position highlighting disk, hinge, and hinge pin.

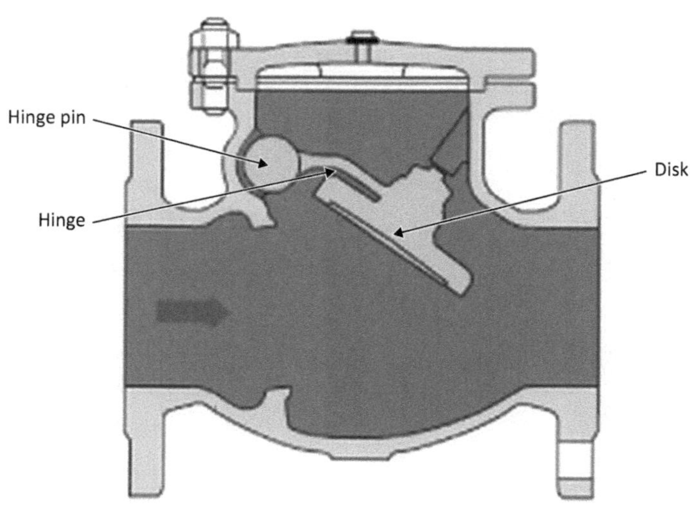

FIGURE 3.15 A swing check valve in the open position, with the flow moving from left to right. The highlighted components include the disk, which swings to allow fluid flow, the hinge that supports the disk's movement, and the hinge pin, which serves as the pivot point for the disk's rotation.

Hub

When a hub and clamp connection is used, a hub is an end fitting that is specially machined and flanged. Hub assemblies are aligned and pressed together with a gasket in between, and a clamp assembly (often a bolted clamp) is used to secure the connection. Hub-ended valves are typically used in large-size and/or high-pressure class piping to save weight and space compared to the standard flange.

L

Lever

A *valve lever* is a mechanical device used to manually operate a valve by providing leverage to open, close, or regulate the flow of fluid. It is commonly attached to the valve stem and allows the user to exert greater force with minimal effort to control the valve's position. A *valve lever* is generally compact and suited for smaller valves or quarter-turn operations, such as ball or butterfly valves, where quick and efficient control is needed. A *handwheel*, on the other hand, is often used for larger valves.

Lifting Lug

A lifting lug is a valve component designed to facilitate the safe lifting and handling of industrial valves and their components during manufacturing, testing, shipping, construction, and installation. The lifting lugs can be integrated with the valve body, or they can be welded or screwed to the valve body. In Figure 3.16,

FIGURE 3.16 A non-slam check valve is lifted by chains attached to lifting lugs integrated into the valve body during the final stage of manufacture.

a non-slam check valve is lifted by chains attached to lifting lugs integrated into the valve body during the final stage of manufacturing.

Liner

A *liner* in a valve refers to a *protective lining* or *coating* placed inside the valve body to enhance its resistance to corrosion, abrasion, and chemical attack. Liners are typically made from materials like *rubber*, *plastic*, or *ceramics*. Liners are typically used in valves handling aggressive fluids, high-pressure environments, or extreme temperatures. The liner helps to extend the *valve's lifespan* and improve its sealing capabilities. A good example of a valve that uses a liner is a *concentric butterfly valve*, where the liner ensures smooth operation and prevents wear and tear of the valve body from harsh or corrosive fluids.

Lip Seal

In a valve, a lip seal is a type of dynamic or static seal that consists of a flexible, outward- or inward-facing lip that creates a tight seal against the mating surface. This type of sealing is commonly used in valves to prevent the leakage of fluids or gases, particularly in high-pressure or high-temperature applications. In addition, lip seal has fire safe characteristic since it has metallic part which is recommended for fire-safe design valves. Fire-safe design valves are explained in more detail in Chapter 5. Lip seal can be used as a stem seal (dynamic seal due to movement of stem) or seat and body seal (static seal due to stationary positions of both seat and body) on industrial valves. A seal is composed of two parts: one part is a soft material such as PTFE (Teflon) and the other part is a metallic spring such as Inconel 625 or Elgiloy (Cobalt alloy) that energizes the soft material part. Figure 3.17 illustrates lip seals made of PTFE in white and with Inconel 625 springs in gray.

Locking Devices

Locking devices in valves are safety mechanisms designed to prevent unauthorized operation or accidental opening or closing of a valve. These devices help to ensure that the valve remains in a specific position, particularly in critical systems where maintaining flow control is essential for safety and operational integrity. Common types of locking devices include *padlocks*, which secure the valve handle to prevent movement; *chain locks*, where a chain is attached to the valve and locked to restrict movement; and *visual locks*, which provide a clear visual indication that the valve is in a locked or secured position. Other types of locks, like *keyed locks*, can be used to restrict access to authorized personnel only who has a key, ensuring that valves are operated under the right conditions. Figure 3.18 shows a valve secured with a chain lock to prevent unauthorized operation.

FIGURE 3.17 Lip seal made of Teflon and Elgiloy spring.

FIGURE 3.18 An industrial valve with a handwheel, secured by a chain and key lock, ensuring it remains in the closed position to prevent unauthorized use. (With permission: Shutterstock.)

O

Operator

A valve operator refers to the mechanism or device that opens, closes, or adjusts the position of a valve in order to control the flow of a process fluid. Manual or automatic input is translated into movement of the valve's internal components, such as the disk, ball, plug, or gate by the operator. An actuator, a lever, a handwheel, and a gear all are considered operators. Alternatively, the term operator can be used to refer to a person who operates (opens and closes) the valve.

O-Ring

An *O-ring* in a valve is a *circular sealing component* typically made of elastomeric materials, used to prevent leakage between the valve's moving parts, such as the stem and bonnet. It is placed in a groove and provides a *tight seal* to withstand pressure and temperature, ensuring leak-free operation.

P

Packing

Packing in a valve refers to a sealing material placed around the valve stem to prevent leakage of fluids or gases. It creates a tight seal between the stem and the valve body, helping to stop any leakage. Selecting the appropriate packing material is crucial, as it can reduce friction with the stem, making it easier to open and close the valve and ensuring smooth operation.

Packing Box

A packing box, also known as a stuffing box, is a chamber within a valve designed to house the packing material. The packing box forms a seal around the valve stem to prevent fluid or gas leakage while allowing the stem to move (rotate or slide) for valve operation.

Plug

A *plug* is a type of closure member used in certain valves, specifically in *plug valves*. It is designed to regulate or stop/start the flow of fluid in a piping system. The plug is a tapered or cylindrical element that fits inside the valve body and rotates to control the flow path. The plug has a passage or hole (called a flow port) through which fluid can flow when the valve is open. As the plug rotates, the flow port aligns with the valve body's inlet and outlet, allowing or stopping the flow of fluid.

Plunger

A *plunger* in the context of valves is a cylindrical or piston-like component that moves linearly within the valve body to regulate the flow of fluid. It is commonly used in *needle valves* where the plunger's movement allows precise control over the opening and closing of the flow path.

Positioner

The purpose of a positioner is to control the position of the actuator of a valve in order to ensure that the valve opens precisely to the position required for accurate flow control. During this process, it compares the actual position of the valve (feedback) with the desired position (setpoint) and adjusts the actuator accordingly. The positioner is an essential component of automated systems, particularly those requiring precise control of fluid flow, pressure, or temperature. A positioner may operate on air, 4-20 MA electricity, or digitally. Figure 3.19 shows a control valve automated with an actuator and has a positioner. The positioner is an important component of control valves, but their application is not limited to these devices.

Position Indicator

It is a device or feature that shows the current position of a valve, whether it is open, closed, or partially open, visually or electronically. A position indicator helps operators and automated systems ensure that the valve is properly positioned for safe and efficient operation, by monitoring and controlling the status of the valve. Typically, mechanical indicators consist of a visible pointer, dial, or indicator plate mounted directly on the valve or actuator. Electronic indicators utilize sensors and digital displays to display the position of the valve.

FIGURE 3.19 A control valve with an actuator and a positioner. (With permission: Shutterstock.)

Pressure-Containing Parts

A valve's pressure-containing components are those that are designed to withstand and contain the internal pressure of the process fluid. During normal operation, these parts serve as the primary pressure boundary of the valve, ensuring that fluid remains contained within it without leakage or failure. The valve body and bonnet are the main parts that contain pressure. In the event that these parts fail to function, leakage will result.

Pressure-Controlling Parts

Parts that regulate, adjust, or control the pressure of fluid passing through a valve are known as pressure-controlling parts. In a piping system, these parts play a crucial role in managing flow dynamics and maintaining pressure levels. A valve disk/ball/plug (closure member), seats, and stem are considered pressure-controlling parts or components.

Process-Wetted Parts

Same as "Wetted parts."

Pup Piece

A valve pup piece is a short section of pipe installed on either side of a valve to ensure proper connection and installation within a piping system. These pup pieces are typically welded to the connected piping. The pup piece must be long enough to keep the nonmetallic components of the valve safe from the heat generated during welding. This is important because high welding temperatures can melt nonmetallic materials, such as seals in the valve.

S

Sealing

Sealing in a valve refers to the mechanism or material used to prevent leakage of fluid through the valve's body, stem, or seating area. It ensures that the valve maintains a tight shutoff and reliable operation, using components such as gaskets, O-rings, packing, or seat rings, depending on the valve type and application.

Seal Pocket

In the valve body or seat area, the seal pocket is a recessed area or groove that is designed to hold a seal. The type of seal could be an O-ring or a gasket. Seal pockets enhance the sealing ability of seals by keeping them in place within the groove.

Seat

The seat is the surface that the valve disk or plug rests against when the valve is in the closed position. It ensures that there is no leakage when the valve is fully closed. In addition to providing a barrier against leaks when the valve is closed, the seat works in conjunction with the closure element to prevent or regulate flow through the valve. It is possible to make seats from a variety of materials, depending on the service conditions. Seat materials could be metallic, PTFE (Teflon), elastomers, and other wear-resistant and resilient materials.

Seating Surface

The seating surface in a valve refers to the area where the moving internal component (e.g., gate, disk, ball, or plug) comes into contact with the fixed seat to form a tight seal and prevent fluid leakage when the valve is in the closed position.

Seat Pocket

The seat pocket of a valve is a recess or machined groove in the valve body that holds and secures the valve seat in its place. In order to ensure proper seat positioning and alignment, it plays an important role. As a result, a reliable seal is provided between the valve's closure element and the seat of the valve.

Shaft

A shaft is a cylindrical component typically used to transfer torque or rotation in rotary motion valves such as butterfly valves and ball valves. Unlike stems, shafts do not move linearly; their primary function is rotational.

Spring

A *spring in a valve* is a mechanical component used to provide force or tension, ensuring proper operation of the valve's moving parts. Springs are commonly found in check valves, relief valves, where they help return the valve to its default position (open or closed) or maintain pressure thresholds. In ball and gate valves, *springs are often used to push the seat against the ball or gate*, ensuring a tight seal and preventing fluid leakage when the valve is closed.

It is important to know that there are other types of springs that are commonly seen in ball valves. The springs are known as *antistatic springs*. These are typically springs or grounding balls that are placed between the stem and the ball or body. Consequently, any charge or static electricity that builds up is dispersed away from the inner parts of the valve as long as continuous electrical contact is maintained. More information about anti-static springs is provided in Chapter 5, under heading "Anti-Static Design."

Stem Key

Stem key is like a bar connected to the corner of the valve stem. The stem key is used to connect the stem to the valve operator (e.g., gearbox, actuator, or handwheel) by transmitting torque from the operator to the valve stem. This ensures that the stem can effectively open or close the valve. Furthermore, if the gear box or actuator is removed during maintenance, the direction of the key is parallel to the hole inside the ball. Thus, it is possible to identify the position of the valve (e.g., open or close).

Stem Seal

A *stem seal* in a valve prevents fluid or gas from leaking around the valve stem. It typically consists of materials like packing, O-rings, lip seals, or gaskets, providing a tight seal while allowing the stem to rotate or move linearly during operation.

Stem Thread

The *stem thread* in a valve is the threaded section on the valve stem. It converts rotational motion (when the handwheel or actuator turns) into linear motion, allowing the valve's disk, plug, or gate to move up or down, thereby opening or closing the valve.

Stuffing Box

Similar to the packing box. For further details, refer to the "Packing Box" section.

T

Throat

Valve throats are the narrowest passages in the valve, and can be found in both control valves and axial (check) valves. Throttling valves (like globes or control valves) have their highest velocity and lowest pressure at the throat. In high-pressure drop applications, choked flow (sonic velocity) happens at the throat. The heading "Choked Flow" in Chapter 8 explains choked flow.

Thrust Washer

A *thrust washer* in a valve is a flat ring-shaped component designed to *absorb axial loads* and *reduce friction* between moving parts. They are typically placed between a moving and stationary part of the valve. It helps ensure smooth operation by preventing direct metal-to-metal contact, minimizing wear, and distributing forces evenly across the valve assembly. Thrust washers are commonly used in various types of valves, including ball, butterfly, and gate valves. Thrust washers are typically made from materials such as PTFE, PEEK bronze, stainless steel, or composite materials.

Example: In a ball valve, thrust washers are positioned between the stem and the top of the ball or between the stem and the valve body housing, often near the packing gland area. Although the stem rotates, fluid pressure inside the valve can create axial forces that push the ball and stem assembly upward or downward. The thrust washer absorbs these forces and prevents excessive upward (stem blow out) or downward movement of the stem. Moreover, during rotation, the stem may experience friction against the packing or actuator assembly. The thrust washer provides a low-friction interface between the stem and packing, reducing operational torque and wear.

Trim

The *valve trim* refers to the internal components of a valve that come into direct contact with the fluid and control its flow, such as the disk, seat, stem, and sometimes the cage or guide. These parts are critical for the valve's performance and are often made from durable materials to resist wear, corrosion, and other operational stresses.

Trunnion

A trunnion is a support that allows rotational movement of the ball, typically found ball valves. In the context of valves, particularly trunnion-mounted ball valves, the trunnion is a fixed support mechanism that helps anchor the ball and absorb forces from fluid pressure. Trunnion in the form of either flange or plate is typically found in larger and higher pressure class valves. Opposition to trunnion-mounted design is floating design. Chapter 5 provides more information about floating and trunnion-mounted balls in a ball valve.

W

Washer

A *washer in a valve* is a small, flat, and typically round component used to distribute load, prevent leakage, or provide a sealing surface in various parts of the valve assembly. Washers are commonly placed under bolts, nuts, or packing glands to ensure a secure connection and even pressure distribution, which helps maintain the valve's structural integrity and sealing performance.

Wetted Parts

Wetted parts in valves refer to all the components of a valve that come into direct contact with the process fluid (liquid, gas, or slurry) during operation. These parts are crucial to consider when selecting a valve for specific applications, as their material must be compatible with the chemical properties, temperature, and pressure of the process fluid to prevent corrosion, erosion, contamination, or failure. The body, bonnet, seats, stem, disk, gate, or ball are typically all wetted parts. Process wetted parts are same as wetted parts.

Y

YOKE

Yoke in a valve is positioned *above the bonnet or body*. It serves as a structural bridge that supports the valve stem and connects it to the actuator or handwheel. The yoke ensures proper alignment of the stem during operation, minimizing wear, and maintaining the valve's efficiency. This design is typical in valves like gate and globe valves.

QUESTIONS AND ANSWERS

1. What is the primary function of a valve stem?
   A. To control the flow of fluid through the valve
   B. To provide a sealing mechanism
   C. To connect the actuator or handwheel to the valve's internal components
   D. To protect the valve from corrosion

   Answer: Option C is the correct answer.

2. Which component of a valve is responsible for creating a tight seal when the valve is closed?
   A. Stem
   B. Seat
   C. Body
   D. Bonnet

   Answer: Option B is the correct answer.

3. What is the primary role of a valve bonnet?
   A. To regulate fluid pressure
   B. To house and protect the internal components of the valve
   C. To adjust the flow rate
   D. To act as a valve actuator

   Answer: Option B is the correct answer.

4. In a ball valve, what is the function of the ball?
   A. To provide a rotating mechanism to stop or start the flow
   B. To filter particles from the fluid
   C. To create a spring-loaded seal
   D. To absorb pressure shocks

   Answer: Option A is the correct answer.

5. What is the purpose of packing in a valve?
   A. To prevent leakage along the valve stem
   B. To adjust the flow rate of fluid
   C. To connect the valve to the piping
   D. To reduce vibration in the valve

   Answer: Option A is the correct answer.

6. Which component of a globe valve directly regulates the flow of fluid?
   A. Stem
   B. Seat
   C. Disk
   D. Bonnet

   Answer: Option C is the correct answer.

7. In a check valve, what prevents backflow?
   A. Stem
   B. Spring
   C. Hinge and hinge pin
   D. Disk

   Answer: Option D is the correct answer.

8. In a valve equipped with a gearbox, what is the primary function of the gearbox?
   A. To convert electrical energy into mechanical motion
   B. To reduce the torque required to operate the valve manually
   C. To increase the speed of valve operation
   D. To act as a position indicator for the valve

   Answer: Option B is the correct answer.

9. What is the primary purpose of a valve positioner in a control valve assembly?
   A. To directly control fluid flow through the valve
   B. To prevent backflow in the valve
   C. To improve the response and accuracy of the valve actuator
   D. To measure the pressure drop across the valve

   Answer: Option C is the correct answer.

10. What is the primary purpose of a valve liner?
    A. To provide structural support to the valve body
    B. To act as a sealing surface and protect the valve body from corrosion or abrasion
    C. To reduce the weight of the valve for easier installation
    D. To increase the flow rate through the valve

    Answer: Option B is the correct answer.

BIBLIOGRAPHY

American Petroleum Institute. (2021). Specification for Pipeline and Piping Valves. API SPEC 6D. API: Washington, DC.

Nesbitt, B. (2007). Handbook of Valves and Actuators: Valves Manual International. 1st edition. Elsevier: Oxford, UK.

Skousen, P. L. (2011). Valve Handbook. 3rd edition. McGraw-Hill Professional: New York, NY.

Smith, P., & Zappe, R. W. (2004). Valve Selection Handbook: Engineering Fundamentals for Selecting the Right Valve Design for Every Industrial Flow Application. 5th edition. Elsevier: Amsterdam, The Netherlands.

Sotoodeh, K. (2020). Using valve locking devices: A review and lessons learned. Valve World Magazine, 1, 64–67.

Sotoodeh, K. (2021). A Practical Guide to Piping and Valves for the Oil and Gas Industry. 1st edition. Gulf Professional Publishing: Houston, TX.

Sotoodeh, K. (2021). Prevention of Valve Fugitive Emissions in the Oil and Gas Industry. 1st edition. Gulf Professional Publishing: Houston, TX.

Sotoodeh, K. (2022). Coating Application for Piping, Valves and Actuators in Offshore Oil and Gas Industry. 1st edition. Taylor & Francis Ltd.: Philadelphia, PA.

4 Materials of Construction

A

Acid Corrosion

Acid corrosion occurs when materials, particularly metals, are exposed to acidic environments, resulting in chemical reactions that degrade the material. This type of corrosion is common in valves handling acids like hydrochloric acid (HCl), sulfuric acid (H_2SO_4), or nitric acid (HNO_3) in industrial processes. The acids react with the metal surface, dissolving the material and forming soluble metallic salts. The salts often lead to uniform or localized corrosion, such as pitting. Acid corrosion is particularly aggressive on carbon steel and austenitic stainless steel without protective lining. Alloys with lower amounts of alloying elements such as molybdenum (Mo) and nickel (Ni) are generally more prone to acid corrosion. To mitigate acid corrosion, materials like Inconel, Hastelloy, or PTFE-lined valves are often used, along with proper inhibitors and pH control in process streams.

Acid Pickling

Acid pickling is a surface treatment process that uses acid solutions to remove impurities, such as scale, rust, or oxide layers, from the surface of a material, typically metals. This process cleans and prepares the material for subsequent treatments, such as coating or welding, by creating a clean and reactive surface. For example, stainless steel components like valve bodies may undergo acid pickling. Acid pickling removes contaminants such as mill scale, welding oxides, rust, and other impurities from the stainless steel surface. After pickling, stainless steel is exposed to air or an oxidizing solution, allowing chromium in the alloy to react with oxygen and form a new, uniform, and corrosion protective chromium oxide layer. Common acids used include hydrochloric acid, sulfuric acid, or nitric acid, often combined with inhibitors to minimize material loss.

American Society for Testing and Materials (ASTM)

ASTM International, originally known as the American Society for Testing and Materials, is a globally recognized organization that develops and publishes technical standards for materials, products, systems, and testing methods. Founded in 1898, ASTM provides guidelines that ensure the quality, safety, and performance

DOI: 10.1201/9781003640967-4

of materials across a wide range of industries, including manufacturing, and oil and gas. These standards are widely used for material specification, such as ASTM A240 for stainless steel or ASTM A105 for carbon steel, as well as for testing procedures, like tensile testing (ASTM E8) or corrosion evaluation (ASTM G48). By standardizing material properties and testing methods, ASTM promotes consistency and reliability, making it an essential reference for engineers and manufacturers worldwide.

Antiexplosive Decompression (AED)

It refers to the ability of materials, typically elastomers like seals or O-rings, to resist damage caused by rapid decompression in high-pressure systems. It is an important consideration in industries such as oil and gas, which experience rapid pressure changes. A high-pressure gas permeates an elastomer (e.g., a seal) during operation, causing explosive decompression (ED). In the event that system pressure drops rapidly (for example, during a shutdown or venting), the trapped gas within the elastomer expands rapidly. This sudden expansion can damage the elastomer and cause cracks and blisters, leading to seal failure and potential system leaks. Materials used in AEDs are designed to minimize gas absorption at high pressures. In addition, AED materials are able to withstand the stress caused by rapid gas expansion. Ensure that they maintain their structural integrity during and after decompression. Materials certified for AED are tested according to standards like NORSOK M-710 standard "Qualification of nonmetallic sealing materials and manufacturers." Lip seals that contain PTFE with a spring inside a PTFE ring, for example, are considered AED sealing materials.

Austenite

Specifically, in the valve industry, austenite refers to a specific crystal structure of steel and iron alloys. Austenitic and super austenitic stainless steels and some nickel alloys like Alloy 625 and 718 and Hastelloy C-276 have austenite phase. Austenitic phase occurs when atoms are arranged face-centered cubically (FCC), giving the material unique mechanical and corrosion-resistant properties. The strength, toughness, and corrosion resistance of austenitic stainless steels with austenite structure make them a popular choice in valve manufacturing. The FCC structure is a type of crystalline arrangement in which atoms are positioned at the corners and the centers of all six faces of a cube-shaped unit.

Austenitic Stainless Steel

1. Structure

The austenitic stainless steel is one of five types of stainless steel based on their crystalline structure. The other four types being ferritic, martensitic, duplex, and precipitation hardened. Due to its primary crystalline structure (face-centered cubic), steels are not hardenable by heat treatment and are not magnetic.

2. Chemical Composition

The chemical composition of austenitic stainless steel is as follows: Chrome is typically present at a concentration of 16–26%, which provides corrosion resistance. Nickel is usually present at a concentration of 6–22%, which provides ductility and toughness. It is possible to add molybdenum (up to 7%) to steel to improve its resistance to pitting and crevice corrosion. Carbon is generally present in small amounts (up to 0.08%), contributing to its strength. Due to their excellent corrosion resistance, austenitic stainless steels are suitable for a wide range of environments, including mild or low concentration and acidic and alkaline conditions. Since they are extremely ductile, they are capable of being formed, shaped, and welded in a variety of shapes without compromising their structural integrity. Austenitic stainless steels retain good toughness at low temperatures, making them suitable for cryogenic applications (extremely cold temperatures). They are generally nonmagnetic in annealed condition. Austenitic stainless steel valves are not common in offshore environment due to external chloride stress corrosion cracking (SCC).

3. Grades

Austenitic stainless steels like *304*, *304L*, *316*, *316L*, *321*, and *347* are highly versatile and widely used due to their excellent corrosion resistance and good mechanical properties. The grade *304* is the most common grade, offering general corrosion resistance with a chromium (18%) and nickel (8%) composition, while *304L* has a lower carbon content, reducing the risk of carbide precipitation during welding. The grade *316* enhances corrosion resistance with added molybdenum (2–3%), making it suitable for chloride-rich environments, and *316L* offers the same properties with lower carbon for improved weldability. The grade *321* incorporates titanium to prevent carbide precipitation and enhance high-temperature strength and weldability, while *347* achieves similar stabilization with niobium, making it resistant to intergranular corrosion (IGC) in harsh thermal conditions. These grades are commonly used in industries like food processing, chemical plants, and marine environments.

B

Base Metal (Base Material)

The base metal refers to the primary material from which a valve component is manufactured, forming the structural foundation for its mechanical and chemical properties. In valves, common base metals include *carbon steel*, *stainless steel*, *duplex stainless steel*, and *nickel alloys*. These materials are selected based on the application's pressure, temperature, fluid service, and corrosive environment. The base metal determines the valve's ability to withstand mechanical stress, resist corrosion, and endure long-term exposure to aggressive media. For example, carbon steel is widely used for general-purpose valves due to its strength and cost-effectiveness. Stainless steel or nickel-based alloys like Inconel are preferred for handling corrosive fluids. The base metal is often enhanced with coatings, linings, or heat treatments to optimize performance in demanding industrial environments. As an example, carbon steel could be used for a corrosive fluid, but it is welded overlaid by 3 mm of Inconel 625 (lining) to enhance corrosion resistance. Additionally, lining is much more cost-effective than making the entire valve from Inconel 625.

Brittleness

Brittleness is a material property characterized by the tendency to fracture or fail suddenly under stress without significant deformation or ductility. Brittle materials, such as cast iron or certain hardened steels, absorb minimal energy before breaking. Thus, they are unsuitable for dynamic or high-impact applications. In valve materials, brittleness can lead to catastrophic failure under high-pressure, low-temperature, or shock-loading conditions. Factors contributing to brittleness include improper heat treatment, material composition, or exposure to low temperatures (brittle fracture). To minimize brittleness, materials like *ductile iron*, *stainless steel*, or alloys with high toughness are selected. Moreover, heat treatments like tempering are employed to enhance material performance in demanding environments.

Buna-N

It is same as NBR. *NBR* (nitrile butadiene rubber) is the technical name, while *Buna-N* is a common trade name.

C

Carbon Dioxide

Carbon dioxide is a naturally occurring gas in oil reservoirs. When dissolved in water, it forms carbonic acid (H_2CO_3), leading to *CO_2 corrosion*. This type of corrosion called sweet corrosion attacks carbon steel. Carbon dioxide corrosion causes material degradation through uniform or localized metal loss, particularly in piping and pipelines, valves, and other equipment exposed to CO_2-rich environments.

Carbon Dioxide Corrosion

CO_2 corrosion, also known as sweet corrosion, occurs when carbon dioxide dissolves in water to form carbonic acid (H_2CO_3), lowering the pH and promoting metal dissolution. The acid reacts with steel, forming iron carbonate ($FeCO_3$), which may act as a protective layer. However, under high flow rates or turbulence, this layer can erode, leading to localized pitting or uniform metal loss. Carbon steel is very commonly material that can be corroded by carbon dioxide in oil and gas environments.

Carbon Equivalent

Pipeline and valve carbon equivalents (CE) are calculated values that indicate the hardenability and weldability of steel. This number combines the effects of various alloying elements into a single value, allowing engineers to assess a steel's potential for cracking or other problems during welding. As a result, reliable welds are essential for maintaining structural integrity under high pressures and temperatures, particularly in the oil and gas industry. The higher the carbon equivalent, the harder the material as a consequence of welding, which makes it more susceptible to cracking. The carbon equivalent of a weld must be reduced below certain limits, such as 0.4, in order to prevent cracking, especially in sour services. To ensure that materials are safe and reliable for use in critical applications, many standards specify maximum CE limits, such as NACE MR0175 for sour service environments. The most commonly used formula for calculating carbon equivalent is that of the IIW (International Institute of Welding):

Equation 4.1: Carbon equivalent calculation

$$CE = C + \frac{Mn}{6} + \frac{Cr + Mo + V}{5} + \frac{Cu + Ni}{15}$$

where C denotes carbon steel, Mn denotes manganese, Cr denotes chromium, Mo denotes molybdenum, V denotes vanadium, Cu denotes copper, and Ni denotes nickel.

Carbon Steel

The body of the valve can be made out of carbon steel. In industrial applications, carbon steel is one of the most commonly used materials for valve bodies, particularly in the oil, gas, petrochemical, and power industries. The high tensile strength of carbon steel makes it suitable for use in valve applications due to its ability to withstand mechanical stresses. It is this strength that ensures the structural integrity of the valve under varying pressures and flows. Standard carbon steel can handle temperatures from −29°C up to around 400°C (750°F), depending on the specific carbon steel grade. As a result, it may not be suitable for extreme temperatures (e.g., cryogenic), as it becomes brittle at very low temperatures and loses strength at very high temperatures. Because of its durability, mechanical strength, and cost-effectiveness, it is ideal for handling a wide range of fluids at moderate temperatures and pressures and is less corrosive. The natural properties of carbon steel limit its corrosion resistance, especially in environments containing moisture, acids, or salts. A protective layer (such as epoxy or paint) is often applied to carbon steel valve bodies in order to prevent corrosion. In corrosive fluid services, valve bodies may also be lined with other materials (e.g., Inconel 625) to prevent corrosion. Because it is cheap and low corrosion-resistant, it is not very common for offshore valves. Offshore environments are corrosive with moisture and chloride that can cause different kinds of corrosion

FIGURE 4.1 A corroded carbon steel body valve in an onshore plant. (With permission: Shutterstock.)

such as pitting and chloride SCC especially for carbon steel. Besides, offshore oil and gas services are typically more corrosive than refineries or petrochemical plants where crude oil and gas are processed and contain fewer corrosive compounds. A corroded carbon steel body valve in an onshore plant is shown in Figure 4.1. A carbon steel valve in an onshore environment can rust externally, especially if it is exposed to moisture, humidity, and corrosive elements.

Cast Iron

The cast iron group is composed of iron–carbon alloys with a carbon content exceeding 2%, Cast iron has high-carbon content which gives it distinctive properties such as hardness and wear resistance. Known for its excellent castability, it is widely used in the manufacture of complex shapes and components, including valves. Furthermore, cast iron has a good machineability. As a result of the graphite content in cast iron, which acts as a lubricant during cutting, cast iron is relatively easy to machine once it has been cast.

Casting

Casting is a manufacturing process in which molten metal is poured into a mold to produce the desired shape of a valve or its components. Once the metal has cooled and solidified, the mold is removed, leaving a metal valve body or part. With casting, complex shapes and intricate designs can be produced that would

FIGURE 4.2 Casting a valve body in a foundry.

otherwise be difficult or costly to produce. To summarize, casting is an essential process in the valve industry that allows the creation of durable, complex valve components. Particularly for larger parts, it is a cost-effective method for producing valves that meet specific demands for flow, pressure, and durability. Casting a valve body in a foundry is shown in Figure 4.2.

Cathodic Protection (CP)

Cathodic protection is a corrosion prevention method applied to valves and other metallic components in environments where they are exposed to corrosive elements, such as seawater or soil. In CP systems, the valve acts as the *cathode* of an electrochemical cell, and an external sacrificial anode (galvanic CP) or an impressed current system provides a continuous supply of electrons to prevent the valve from corroding. CP is widely used in subsea oil and gas applications, where valves are exposed to chloride-rich seawater and aggressive conditions. It is particularly critical for protecting valve bodies and fittings made from materials like carbon steel. However, improper CP design can lead to hydrogen generation at the valve surface, increasing the risk of hydrogen embrittlement (HE) or hydrogen-induced stress cracking (HISC) in susceptible materials like duplex stainless steel. CP systems are essential for extending the lifespan and ensuring the reliability of valves in pipelines, subsea systems, and buried infrastructure.

Cavitation

1. Definition

In valves used for flow regulation or throttling, such as globe or butterfly valves, cavitation is a significant operational problem. Cavitation is a phenomenon in which rapid changes in pressure cause tiny vapor bubbles to form within liquids. For example, this can occur when a liquid's pressure drops below its vapor pressure, often as a consequence of sharp pressure drops across control valves or pump impellers. Whenever these bubbles flow through areas of higher pressure, they collapse or implode, creating powerful shock waves. Cavitation in control valves and pumps can cause physical damage, noise and vibration, and reduced efficiency. Typically, cavitation is minimized through careful design adjustments such as reducing the fluid velocity, increasing the downstream pressure to avoid high pressure drop, or using cavitation-resistant materials such as Stellite (cobalt alloy) in control valves and pumps. An illustration of cavitation in a globe valve plug is shown in Figure 4.3.

2. Connection to Material

Cavitation is related to material because it has significant implications for the *selection*, *performance*, and *degradation of materials* used in valves. While cavitation is primarily a mechanical phenomenon related to fluid dynamics, its effects directly impact the *durability and integrity of valve materials*, making it a crucial consideration in material engineering.

FIGURE 4.3 A valve cavitation.

CHARPY IMPACT TEST

Charpy impact tests are mechanical tests used to determine the toughness of valve materials, especially their ability to absorb energy and resist brittle fracture at different temperatures. In valves with extreme operating conditions, such as low temperatures or high impact forces, this test is essential. This test ensures the valve material will be able to withstand sudden impacts or shocks without breaking. It is common for industries such as oil and gas to require Charpy impact testing as part of valve material certification. Charpy impact test is required especially in order to comply with API 6A, NACE MR0175, and ASTM A370, which specify the toughness requirements for valves. Preparation of specimens is the first step in this test. It is necessary to prepare a small, notched sample of the valve material. To simulate weaknesses that could lead to cracking under stress, the notch creates a point of stress concentration. In the Charpy testing machine, a pendulum hammer is used to strike the notched side of the sample horizontally (see Figure 4.4). Upon striking the sample, the hammer swings down, breaking it as shown in Figure 4.4 (left side). The amount of energy absorbed by the sample during the break is measured in joules (J). The energy absorption value of a material indicates its impact toughness—a higher value indicates greater toughness. A comparison is made between the amount of energy used in the test and the value provided in the material standard (e.g., ASTM A216 standard for carbon steel cast body material) for the material to determine whether it should be accepted or rejected.

CHLORIDE

Chlorides, commonly found in seawater, oil, and gas, are highly aggressive ions that can cause severe corrosion in valve materials. In particular, chlorides promote *pitting corrosion*, *crevice corrosion*, and SCC in susceptible materials like austenitic stainless steels (e.g., 304, 316). For valves exposed to chloride-rich environments, materials with high *Pitting Resistance Equivalent Number* (PREN), such as *super duplex stainless steels*, *Inconel 625*, or *Hastelloy*, are preferred for

FIGURE 4.4 Charpy impact test.

their superior resistance. Protective measures like coatings, linings, or cathodic protection are also used to mitigate chloride-induced corrosion. The presence of chlorides is a critical factor in material selection for valves in offshore, subsea, and chemical processing industries, where durability and corrosion resistance are paramount.

Chloride Stress Cracking Corrosion (CLSSC)

CLSSC is a form of localized corrosion that occurs in materials, particularly austenitic stainless steels, when exposed to tensile stress and chloride-rich environments. It involves the combined action of mechanical stress, often from residual (e.g., due to welding) or operational loads, and the corrosive effect of chloride ions, leading to the formation and propagation of cracks. This type of corrosion is accelerated in high temperatures and humid conditions. It is very common in applications like desalination plants, seawater systems, and chemical processing. CLSCC can cause sudden and brittle failure, even in corrosion-resistant materials such as austenitic stainless steels. Mitigating this type of corrosion requires careful material selection, stress relief, and environmental control (e.g., reducing the operating temperature or using dichlorination process to remove chloride) to prevent it.

CLSCC is a common type of corrosion for austenitic stainless steels. CLSCC is a specific type of SCC caused by chloride. SCC can be caused by other types of media, such as amine and caustic media. For SCC to occur, three conditions must be met: corrosive environment, vulnerable material, and tensile stress (see Figure 4.5). Figure 4.6 illustrates chloride stress cracking of a stud bolt used for flange connection in the offshore environment.

Cladding

Cladding is the process of bonding a corrosion-resistant material, such as *Inconel 625*, to a less expensive base metal, like *carbon steel*. This is done to enhance the valve's resistance to corrosion and wear while reducing costs. The cladded material forms a protective layer on the internal surfaces exposed to aggressive fluids. This ensures the valve's durability in harsh environments. For example, in valves used in the *oil and gas industry*, *Inconel 625 cladding* is often applied to the wetted surfaces of carbon steel valve bodies to resist chloride-induced pitting, crevice corrosion, and sour gas corrosion. This combination leverages the mechanical strength and cost-effectiveness of carbon steel with the superior corrosion resistance of Inconel 625, making it ideal for high-pressure and high-temperature applications in subsea and offshore systems.

Coating

Coating in valves refers to the application of a protective layer on the valve's surface to prevent corrosion, enhance wear resistance, and extend the valve's

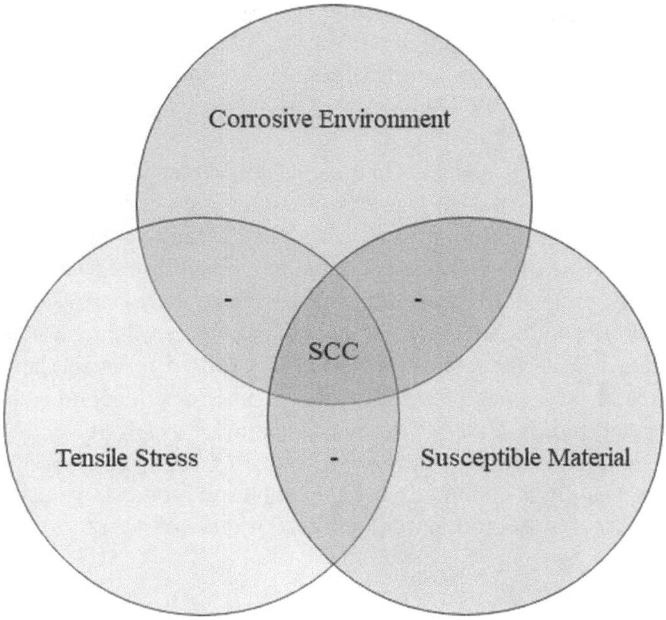

FIGURE 4.5 Contribution of parameters for SCC to occur.

FIGURE 4.6 Stud bolt CLSCC in the offshore environment. (With permission: Shutterstock.)

lifespan in challenging environments. Coatings are especially critical for valves installed underground and in the offshore. Coating is applicable in both offshore and onshore applications. There are various types of coating such as *thermal spray aluminum* (TSA), *phenolic epoxy*, and *zinc-rich epoxy*.

TSA, *phenolic epoxy*, and *zinc-rich epoxy* are advanced coating technologies used in the oil and gas industry to protect valves from corrosion. *TSA* involves spraying molten aluminum onto valve surfaces, creating a durable, corrosion-resistant layer ideal for high-temperature and offshore environments. It provides excellent protection against atmospheric and subsea corrosion, especially in high-chloride conditions. *Phenolic epoxy* coatings are chemical-resistant and heat-tolerant, making them suitable for valves exposed to harsh chemicals and temperatures in processing plants. They offer superior protection against acidic and caustic substances. *Zinc-rich epoxy*, containing a high percentage of zinc, provides cathodic protection to underlying steel by acting as a sacrificial layer. This type of coating is commonly used on valves in moderately corrosive environments and serves as a primer for additional protective layers.

Copper–Nickel Alloys

Copper–nickel alloys, typically containing *70–90% copper and 10–30% nickel*, are widely used for valve components in marine and seawater applications due to their excellent resistance to corrosion, erosion, and biofouling. These alloys, such as *CuNi 90/10 (UNS C70600)* and *CuNi 70/30 (UNS C71500)*, offer superior durability in chloride-rich environments. As a result, these alloys are ideal for valves in desalination plants, shipbuilding, and offshore oil and gas systems. Copper–nickel alloys' inherent resistance to biofouling reduces maintenance needs in seawater systems. In addition, its good mechanical properties, including ductility and toughness, support reliable performance in high-flow and abrasive conditions. One limitation of Cu–Ni alloys is it is less strong than alternatives like nickel aluminum bronze (NAB) and duplex. Copper–nickel alloy is often chosen for its balance of corrosion resistance, manufacturability, and cost-effectiveness in moderately aggressive environments.

Corrosion

Corrosion occurs when metals and other materials deteriorate due to chemical reactions with their environment. Metals are often subjected to this process when exposed to oxygen and moisture. This results in oxidation, causing them to weaken, which can ultimately lead to structural failure. It is possible for corrosion to occur internally as a result of contact with fluids under operating pressure and temperatures, or externally as a result of contact with corrosive environments such as underground soil or humid environments offshore containing chlorides. A valve external corrosion in the offshore environment is shown in Figure 4.7.

FIGURE 4.7 Valve external corrosion in an offshore environment. (With permission: Shutterstock.)

Corrosion Allowance

Corrosion allowance is the additional material thickness added to valve components to account for material loss due to corrosion over the valve's service life. It is commonly specified as *1 mm*, *1.5 mm*, or *3 mm*, depending on the operating environment and design standards. For example, *NORSOK standards* recommend a *3-mm corrosion allowance* for carbon steel in offshore oil and gas applications. While some texts and standards define corrosion allowances up to *6 mm*, such high values should be avoided as they significantly increase the valve's weight, making it less practical. Corrosion allowance is not applicable for *corrosion-resistant alloys* (CRAs) like stainless steels or nickel-based alloys, as these materials are inherently resistant to corrosion and do not require extra thickness. This approach balances reliability with practicality in valve design. Furthermore, corrosion allowances are added as a strategy to combat carbon dioxide (sweet) corrosion, because metal loss is the mechanism responsible for this type of corrosion.

Corrosion Control

Corrosion control in valves refers to methods and practices used to prevent or minimize valve materials deterioration due to chemical reactions with their environment. Since valves often operate in harsh conditions with exposure to moisture, chemicals, and temperature fluctuations, controlling corrosion is essential to maintain their integrity, performance, and lifespan. Material selection, coating

FIGURE 4.8 Inconel 625 weld overlay (lining) inside the carbon steel body valve to prevent internal corrosion. (With permission: Shutterstock.)

and lining, and cathodic protection are some approaches to controlling or mitigating corrosion. Figure 4.8 shows a valve in carbon steel material and it has lining of Inconel 625 overlaid inside the valve in 3 mm to prevent internal corrosion.

Corrosion Inhibitor

A corrosion inhibitor is a chemical compound added to a fluid or gas to reduce or prevent the corrosion of valve materials, particularly in aggressive environments like those found in the oil and gas industry. These inhibitors work by forming a protective film on the valve's internal surfaces, blocking corrosive agents like oxygen, H_2S, CO_2, or chlorides from reacting with the material. For example, amine-based inhibitors are used to mitigate CO_2-induced corrosion, while sulfide scavengers help in H_2S environments. Corrosion inhibitors are critical for extending the lifespan of valve components made of carbon steel, stainless steel, or other alloys. In addition, they reduce maintenance costs and ensure reliable performance in corrosive environments such as sour service and high-chloride fluid conditions. Their application is especially vital when material upgrades to more corrosion-resistant alloys, such as Inconel or duplex steels are not economically viable.

Corrosion Inspection

An inspection for corrosion in valves is the process of examining valves for signs of corrosion or degradation in order to ensure that they remain in a safe, reliable operating condition. The corrosion inspection process has several advantages, such as ensuring structural integrity, preventing leaks and failures, maintaining valve performance, and ensuring safety. Important valve components require inspection of the body, bonnet, seat, closure member, bolts, and nuts. The various types of corrosion inspection include visual inspection, ultrasonic and radiography tests, magnetic particle tests, and dye penetration tests. These inspection techniques are explained in Chapter 6.

Corrosion-Resistant Alloy

As the name implies, corrosion-resistant alloys are metals or alloys that are designed to withstand corrosion due to environmental factors such as moisture, chemicals, or salt. An alloy containing nonferrous metals is typically added to corrosion-resistant alloys to increase their corrosion resistance. These alloys are titanium, nickel, cobalt, chromium, and molybdenum in which any one or a combination of the elements exceeds 50% mass fraction. There are a number of common CRAs, including stainless steel, nickel alloys, titanium alloys, and copper–nickel alloys. It would be possible to make the valve's body and internal components from CRA materials in order to prevent corrosion.

Corrosion under Insulation (CUI)

Corrosion under Insulation is a type of localized corrosion that occurs on the metal surface beneath thermal insulation, typically due to the accumulation of moisture. This phenomenon is prevalent in insulated piping, valves, and equipment operating at intermediate temperatures, where condensation or water ingress can occur. CUI is accelerated by factors such as temperature cycling, presence of chlorides, and acidic or alkaline contaminants in the trapped moisture. It is particularly a hidden attack because it remains hidden under the insulation, making detection and mitigation challenging. Common mitigation strategies include using moisture-resistant insulation, protective coatings on metal surfaces under the insulations, and regular inspection techniques such as ultrasonic testing. Figure 4.9 illustrates damage of piping insulation on a piping branch connection and CUI.

Creep

In the valve industry, creep is the slow, permanent deformation of materials in valve components, such as valve bodies, stems, or seats, as a result of prolonged exposure to high temperatures and stress. If not managed properly, creep can affect the durability and reliability of valve materials and can ultimately lead to failure in high-temperature applications. Especially in high-temperature environments such as steam and power generation, creep can cause the valve body and bonnet to warp. As a result of this warping, misalignment and seal failure may occur, resulting in leaks. In summary, creep in the valve industry refers to a gradual deformation of materials as a result of prolonged exposure to high temperatures and high stress. An example of creep is where the body is deformed

FIGURE 4.9 Damaged cold conservation insulation and CUI on a pipe branch area. (With permission: Shutterstock.)

and the stem is elongated. An elongated stem refers to a valve stem that has been stretched or deformed as a result of prolonged exposure to high levels of stress and elevated temperatures. This condition may lead to creeping of the stem material over time, resulting in thinned stems.

Crevice Corrosion

The process of crevice corrosion occurs as a result of a fluid becoming stagnant in narrow, confined spaces (crevices). These spaces may exist between two metal surfaces, or between a metal and a nonmetallic surface, such as a gasket or seal. Metals that are normally resistant to corrosion, such as stainless steel, are frequently affected by this phenomenon. Crevice corrosion can initiate other types of corrosion such as pitting. Figure 4.10 shows a crevice corrosion on the grooves of a flange face.

Cryogenic Materials

Cryogenic materials are specifically designed to maintain their mechanical properties, such as toughness and ductility, at extremely low temperatures, typically below *−150°C (−238°F)*. These materials, such as *austenitic stainless steels* (e.g., *304, 316*), *aluminum alloys*, and *nickel-based alloys*, are essential for valves and components used in cryogenic applications like liquefied natural gas (LNG), liquid hydrogen, or liquid oxygen handling. At cryogenic temperatures, many materials become brittle and prone to fracture, making the selection of cryogenic-compatible materials critical for ensuring the safety and reliability of valves.

FIGURE 4.10 Crevice corrosion on a flange face made of SS316. (With permission: Shutterstock.)

The ability of these materials to withstand thermal cycling and low-temperature impact stresses is a key factor in their performance.

D

Digital Tools (for Material Selection and Failure Prediction)

Digital tools are transforming the way materials are selected and valve failures are predicted. Engineers now leverage artificial intelligence (AI), machine learning (ML), and advanced simulation software to optimize material performance and reliability. Material selection tools integrate extensive databases and performance parameters, enabling engineers to evaluate materials based on properties like corrosion resistance, mechanical strength, and cost. Tools such as CES Selector and ANSYS allow the simulation of real-world operating conditions to predict how materials will perform under extreme temperatures, pressures, or corrosive environments.

Predictive maintenance, powered by AI and ML, has become a cornerstone of modern valve management. Embedded sensors in valves collect operational data, which is analyzed to identify patterns in wear or stress that indicate impending failures. This approach minimizes unplanned downtime and maintenance costs while improving safety in critical operations like oil and gas. For example, predictive algorithms can monitor valves under cyclic stresses, providing early warnings about potential component fatigue or degradation.

Simulation software has also advanced failure prediction. Using historical failure data and real-time operational parameters, these tools model the lifespan of valve components, identifying vulnerabilities under high-pressure or high-temperature conditions. Digital twin technology further enhances this capability by creating a virtual replica of physical valve systems. These digital twins monitor performance, simulate operational scenarios, and predict failures, offering a proactive approach to maintenance. They are especially valuable in offshore or subsea environments where physical inspections are challenging. More information about various types of maintenance is provided in Chapter 9.

Ductility

It is the ability of a material to undergo significant plastic (permanent) deformation before breaking that determines its ductility. To put it simply, a ductile material can be stretched or drawn into thin wires or elongated shapes without breaking. In materials engineering, ductility is an important characteristic, particularly in applications where materials are subject to forces such as bending, stretching, and pulling as well as low-temperature applications.

Duplex Stainless Steel

The duplex stainless steel is a type of stainless steel that comprises austenitic and ferritic microstructures, typically containing approximately 50% of each phase. Because of this unique composition, duplex stainless steel is suitable for a wide range of applications, including harsh environments. As a result of its excellent resistance to pitting and crevice corrosion, duplex stainless steel is ideal for marine and chemical environments. Due to its combination of austenitic and ferritic structures, duplex stainless steel produces higher yield strength than standard austenitic stainless steels. This allows for thinner-walled structures and weight reduction. As compared to austenitic stainless steels, duplex stainless steels are more resistant to SCC in chloride-rich environments. Standard welding techniques can be used to weld these materials, but the heat input must be controlled to prevent unfavorable microstructures known as the sigma phase when welding. Duplex stainless steel is suitable for use at temperatures as low as −46°C, making it ideal for low-temperature applications. However, in offshore environments, it is not recommended for use above 100°C due to the risk of external corrosion caused by environmental factors (humid and chloride-containing offshore environment). For higher temperature applications, protective coatings are necessary to ensure its durability and resistance to corrosion. The maximum operating temperature for duplex stainless steel in onshore environments typically depends on the specific grade and the environmental conditions. The maximum operating temperature for duplex is around *250°C (482°F)* and *350°C (572°F)* for onshore applications where the risk of external corrosion is lower compared to offshore environments.

E

Economic Considerations (in Material Selection)

Economic analysis in valve material selection involves evaluating the trade-offs between material performance and cost to achieve optimal reliability and affordability. For instance, carbon steel is cost-effective for less corrosive environments. However, carbon steel lacks the corrosion resistance required in aggressive conditions. This makes materials like stainless steel or duplex stainless steel more suitable for very corrosive conditions despite higher initial costs. Over the life cycle of a valve, materials with superior durability and resistance to wear and corrosion, such as nickel alloys or super duplex stainless steels, often prove more economical. That is due to reduced maintenance, downtime, and replacement costs. Economic analysis also considers factors like manufacturability, availability of materials, and operational efficiency to ensure cost-effective solutions for specific applications.

Elasticity

1. Introduction

Valve elasticity refers to the ability of the components of a valve, such as seals, gaskets, or even the metal body, to deform under stress and then return to their original shape once the stress is removed. The presence of this property ensures proper sealing, reliable operation, and long-term durability of the valve under varying pressure, temperature, and mechanical load conditions. It is important to note that elastic deformation contributes to the maintenance of the seal despite minor wear or irregularities. As temperature, pressure, or vibrations change, certain parts of a valve, such as seals or gaskets, can stretch or squeeze slightly. As the temperature rises, parts may expand slightly, and when they cool back down, they shrink. Additionally, pressure changes or vibrations can push and pull these parts. Due to their elasticity, these parts can adjust to these changes without breaking or losing their shape. This ensures the valve remains sealed tightly and functions properly, even under difficult conditions.

2. Case Study

Lack of Seal Elasticity Leading to Valve Failure

Application In a chemical processing plant, a *butterfly valve* is equipped with an *elastomeric EPDM seal.* The butterfly valve was used to control the flow of chemical vapors. The operating conditions included high temperatures, reaching approximately 180°C. This high temperature exceeds the EPDM material's recommended temperature tolerance of 150°C. Initially, the valve performed well, providing tight shutoff and reliable flow control. However, over time, the seal was exposed to continuous high temperatures and frequent pressure cycling. These conditions gradually caused the elastomeric seal to lose its *elasticity* due to thermal degradation.

Failure Mechanism As the EPDM seal hardened and became brittle, it could no longer conform properly to the valve seat, creating gaps between the sealing surfaces. These gaps allowed chemical vapors to leak past the valve, compromising its shutoff capability. The escaping vapors further eroded the seal, exacerbating the leakage issue.

Consequences The chemical vapor leak led to operational inefficiencies in the plant. Eventually, the valve's performance deteriorated to the point where it could no longer control flow, forcing the plant to shut down the affected section. This unplanned downtime caused significant production losses and additional repair costs.

Solution To address the issue, the operator replaced the failed valve with one designed for high-temperature applications. The new valve was equipped with a *fluorocarbon elastomer* (FKM) seal, which has a much higher temperature tolerance. Additionally, regular maintenance and seal inspections were implemented to prevent similar failures in the future.

Elastomer

Elastomers are flexible, rubber-like materials used in valve sealing applications to achieve effective seals between components, thereby preventing leaks in various industrial systems. As a result of their elasticity, and ability to regain their original shape after being compressed, these materials are ideal for creating tight seals in valves under a variety of operating conditions. In the oil and gas industry, elastomers such as nitrile rubber (NBR), EPDM (ethylene propylene diene monomer), Viton (FKM—fluoroelastomer), silicone rubber, and neoprene are used in valves.

Elgiloy

1. Definition

An alloy of cobalt with (Unified Numbering System) UNS R30003 (the meaning of UNS is explained in more detail in this chapter). Materials such as this are used as spring materials in valves or as metallic springs inside lip seals. Elgiloy (Co, Cr, Ni) is a cobalt alloy containing 40% cobalt, 20% chromium, 15% nickel, and 7% molybdenum. The corrosion resistance of this material in general, as well as its resistance to seawater, is very good.

2. Case Study

Material Failure of Seat Springs in a Ball Valve for Seawater

Application A *ball valve* was installed in a seawater service system, responsible for isolating the flow of seawater in a marine facility. The valve design included springs located behind the seats to ensure proper sealing under varying pressure conditions.

Initial Design The springs behind the valve seats were manufactured from stainless steel 316 (SS316). This material was selected for its corrosion resistance and mechanical properties under typical industrial conditions. At first, the valve operated as expected, providing a reliable seal in the challenging marine environment.

Failure Over time, the SS316 springs began to corrode due to the highly aggressive nature of seawater, which contains chloride ions that are particularly damaging to stainless steel. The corrosion led to *spring weakening* and loss of elasticity, reducing the pressure applied to the valve seats. As a result, the valve began to exhibit *leakage* and inconsistent sealing. The leakage compromised the performance of the seawater system, leading to inefficiencies and requiring frequent maintenance. The degraded springs also caused uneven wear on the valve seats, necessitating additional repairs and increased downtime.

Solution To address the issue, the SS316 springs were replaced with *Elgiloy*, a cobalt–chromium–nickel-based alloy known for its exceptional corrosion resistance in chloride-rich environments. Elgiloy's superior mechanical strength and elasticity ensured consistent performance in the harsh seawater conditions. Once the springs were replaced, the valve returned to normal operation, with no further issues reported.

EPDM

EPDM (ethylene propylene diene monomer) is a synthetic rubber known for its excellent resistance to heat, weathering, ozone, and a wide range of chemicals, including acids, alkalis, and water. It has outstanding flexibility and durability, even at high temperatures (up to 150°C) and low temperatures (−40°C). In valves, EPDM is commonly used for seals, gaskets, and seat materials in applications requiring chemical and water resistance, such as in potable water systems, HVAC systems, and wastewater treatment plants. However, EPDM is not suitable for use with hydrocarbons, oils, or fuels, making it unsuitable for certain oil and gas applications.

Erosion

Erosion in valves is the process of material loss or wear on valve components due to the mechanical action of fast-moving fluids with or without abrasive particles passing through the valve. This erosion gradually wears down the valve's internal surfaces, particularly in high-velocity flow areas. Erosion potentially leads to leaks, reduced efficiency, and eventual failure of the valve. It has been shown that erosion can accelerate corrosion, especially in harsh environments, when erosion and corrosion occur simultaneously. It has been concluded that erosion is caused by the movement of fluids or by the impact of particles or liquids at high speeds. Both corrosive and mechanical forces may be involved to accelerate erosion. Particles can cause erosion when they are carried by a high-speed fluid (like water, oil, or gas) and impact a surface. Over time, these particles remove material from the surface, causing erosion. In this case, erosion is driven by abrasive forces caused by suspended particles within the fluid. To mitigate erosion, in addition to using erosion-resistant materials like Stellite, it is also possible to incorporate an erosion allowance by increasing the material thickness.

Explosive Decompression

In valves, ED is the rapid release of high-pressure gas that can damage elastomeric seals and other components. Due to a sudden pressure drop, gas trapped within the elastomer material expands rapidly, potentially causing the material to crack, blister, or even disintegrate. During the operation of a valve, elastomeric seals (such as O-rings or gaskets) are exposed to high-pressure gas. Gas molecules penetrate the surface of the elastomer and dissolve into its structure under high pressure. When the system pressure decreases suddenly (e.g., during depressurization or a process upset), the dissolved gas in the elastomer expands rapidly. Rapid expansion of gas creates internal stresses, which may result in cracking, blistering, or rupture of the elastomer.

F

Fatigue

When materials are repeatedly or cyclically loaded below their ultimate tensile strength, fatigue will cause progressive and localized structural damage. In time, this repetitive stress can result in microscopic cracks developing, growing, and eventually leading to failure, even if the applied load is well within the nominal strength of the material.

Fatigue Strength

Fatigue strength is the maximum stress a material can endure for a specified number of cycles without failing due to repeated or fluctuating loads. It is a critical property for valve materials exposed to cyclic stresses in high-pressure or high-vibration environments, such as pipelines and rotating machinery. Materials with high fatigue strength, like stainless steels and nickel-based alloys, are preferred in applications where durability and resistance to crack initiation and propagation are essential.

Fatigue Stress

Fatigue stress refers to the fluctuating or cyclic stress that a material experiences during repeated loading and unloading. Unlike static stress, which is constant, fatigue stress varies over time and can lead to fatigue failure, even when the applied stress levels are significantly below the material's ultimate tensile strength.

Ferrite

The term ferrite refers to a specific microstructural phase of iron present in certain metallic alloys, particularly steels. Among the key phases of steel metallurgy, it plays a crucial role in determining the mechanical and corrosion-resistant properties of valve materials. A body-centered cubic (BCC) crystal structure is characteristic of ferrite. In BCC structure, atoms are at each of the eight corners of the cube and one atom is at the center of the cube. It is relatively soft and ductile compared to other microstructural phases like austenite or martensite. Magnetic properties can be a distinguishing feature of some valve materials, such as ferrite. Ferritic stainless steels contain 10.5–30% chromium and little or no nickel. Besides, ferrites are less hard than phases such as martensite due to their low carbon solubility (up to 0.02% at room temperature). Ferrite is the dominant phase in ferritic stainless steels. Ferrite and austenite phases appear in roughly equal proportions (50%) in duplex stainless steel. Ferrite appears in carbon and some alloy steels in the microstructure.

Ferritic

Ferritic refers to a category of materials, particularly steels, that are primarily composed of the ferrite phase in their microstructure.

Ferritic Stainless Steel

The ferritic stainless steel type is primarily composed of iron and chromium, with little or no nickel. This material has a ferritic microstructure, which is characterized by a body-centered cubic (BCC) crystal structure. In comparison to other types of stainless steel, this type of stainless steel is known for its excellent corrosion resistance, particularly against oxidation and SCC.

FKM

FKM is same as *Viton*. See Viton for more information.

FKM is the generic designation for fluoroelastomers, while *Viton* is a trademarked brand name for FKM manufactured by *DuPont* (now *Chemours*).

Flashing

More information about flashing is provided in Chapters 8 and 9, under the heading of "Flashing."

Forging

In the valve industry, forging is the process of shaping metal into valve components by applying compressive forces, typically using a hammer or press. Unlike casting, which involves pouring molten metal into a mold, forging involves shaping solid metal at high temperatures, resulting in a stronger, denser, and more durable product. In high-pressure or high-temperature applications, forging is often used for critical parts such as valve bodies, bonnets, and stems. The process of forging minimizes porosity and internal voids that can be present in cast components, which makes forged parts more reliable in environments with high pressures.

G

Galling

More information about galling is provided in Chapter 9, under the heading of "Galling."

Galvanic Corrosion

1. Definition

When two dissimilar metals come into contact with each other in the presence of an electrolyte (such as water with dissolved salts), galvanic corrosion occurs. A galvanic cell is formed when these metals are connected, and an electric current flows between them. Consequently, the more reactive metal (anode) corrodes more quickly, whereas the less reactive metal (cathode) corrodes more slowly or not at all. Various examples of galvanic corrosion can be found in piping and valves. If stainless steel bolts are used with carbon steel flanges in a wet environment, galvanic corrosion may occur at the carbon steel. Copper pipes joined directly to steel pipes may experience accelerated corrosion. There are several methods of mitigating galvanic corrosion, including selecting two metals with similar electrochemical potentials, applying coatings or insulation between the metals, using cathodic protection (sacrificial anodes), and controlling the environment in which metals are in contact.

2. Case Study

Galvanic Corrosion in a Valve Assembly

Application A *butterfly valve* was installed in a seawater intake system at a desalination plant. The valve body was made from *NAB*, chosen for its corrosion resistance in marine environments, while the disk was made from *stainless steel* (SS316). The valve was used to regulate the flow of seawater under continuous exposure to high chloride content.

Failure Mechanism Within months, the stainless steel disk began to exhibit *pitting corrosion* due to its limited resistance to the high chloride content in seawater. This damage was exacerbated by *galvanic corrosion* between the stainless steel disk (acting as the *anode*) and the NAB valve body (acting as the *cathode*) in the galvanic couple. The stainless steel disk corroded faster than expected, leading to surface roughness and erosion, which impaired the valve's sealing performance.

Consequence The corrosion on the disk caused *leakage* through the valve, reducing its efficiency and reliability. Over time, the surface damage became severe enough to prevent the valve from fully closing, requiring the desalination plant to perform emergency repairs. The unplanned downtime resulted in significant operational disruptions and financial losses.

Solution To resolve the issue, the stainless steel disk was replaced with a disk made from *NAB* which offers much greater resistance to pitting and crevice corrosion in chloride-rich environments. The *NAB* body was retained as it exhibited minimal corrosion during the initial period.

Galvanizing

1. Definition

Galvanizing is a corrosion protection process that involves coating steel or iron with a layer of *zinc* to prevent rust and extend the material's lifespan. The zinc coating acts as a physical barrier and provides *sacrificial protection*, meaning it corrodes preferentially to the base metal when exposed to the environment. The most common method is *hot-dip galvanizing* (HDG), where the material is immersed in molten zinc, forming a robust metallurgical bond. Galvanized coatings are widely used in valves for onshore oil and gas applications. However, galvanizing is not suitable for high-temperature or highly corrosive environments, as the zinc layer *can* degrade under extreme conditions.

2. Example

Galvanizing on Valve Bolts

HDG is often applied to *valve bolts* made of *low alloy steel*, such as ASTM A193 Grade B7, to enhance their corrosion resistance in outdoor or wet environments. For instance, in water treatment facilities or cooling systems, bolts used to secure valve flanges are hot-dip galvanized to protect them from rust and atmospheric corrosion.

Grain Boundary

Grain boundaries are the interfaces where individual crystalline grains in a metal meet. These boundaries play a crucial role in a material's mechanical and corrosion properties, as they are often sites of weakness or higher chemical activity. In valves, grain boundaries can affect performance, particularly in processes like IGC, where chemical attacks occur along the boundaries due to impurities or sensitization. For more information, see the "Intergranular Corrosion" section in this chapter.

Graphite

A graphite is a form of carbon that is stable at ordinary conditions of temperature and pressure. The appearance of this material is black, shiny, and nonmetallic. Its hardness is low. As graphite can tolerate extreme temperatures without breaking down, it is an ideal material for applications in high-temperature environments, such as those in the oil, gas, and petrochemical industries. In contrast to other materials that may degrade or melt under similar conditions, it maintains its integrity and sealing ability even when exposed to temperatures exceeding 450°C (850°F). Graphite also has the advantage of having low friction. Due to its natural slippery nature, graphite has low friction and lubricating properties. This self-lubricating quality is especially useful in dynamic applications where parts move or rotate frequently. Due to this property of graphite, valve stems and other moving components are less likely to wear out. Since graphite is installed around the valve stem, its low friction properties are extremely important since friction between the packing and the valve stem increases the amount of force that must be applied to open and close the valve. An increase in force is required for valve operation, which requires the installation of a larger gearbox or actuator on the valve, increasing the cost of the valve assembly. The third advantage of graphite is its corrosion resistance. In harsh environments where chemical resistance is essential, graphite is suitable for applications because it does not react with most chemicals, acids, or corrosive agents. Graphite also has the advantage of being electrically conductive. In the oil and gas industry, it is important for the valves to be designed in a fire-safe manner. Static electricity can be retained inside nonmetallic components such as elastomers and thermoplastics, which increases the risk of valve malfunctions in the event of a fire. The advantage of graphite is that it conducts the static electricity, making it a good fire-resistant material. Furthermore, graphite can be compressed to form a tight, resilient seal, which makes it a suitable packing material for valves. Due to this, it is able to maintain a seal under high pressure conditions and adjust to fluctuations in pressure. Four graphite packing rings are shown in Figure 4.11. In addition, graphite is considered a fire-resistant material.

H

Hard-Facing

Hard-facing is a surface engineering process in which a wear-resistant material is deposited onto the surface of a component to enhance its durability and resistance to abrasion, erosion, or corrosion. For example, *Stellite*, a cobalt-based alloy, is commonly used for hard-facing valve seats and closure member sealing surfaces in high-wear environments, such as oil and gas applications. This extends the service life of the component and reduces maintenance needs.

FIGURE 4.11 Several graphite packing rings.

Hardness

1. Definition

Hardness is a material property that measures its resistance to localized plastic deformation, such as indentation, scratching, or abrasion. It is a critical parameter in valve materials, influencing wear resistance and durability under high-pressure or abrasive conditions. Common hardness measurement scales include *Brinell* (HB), *Rockwell* (HR), and *Vickers* (HV), each suited for different applications and material types.

2. Example

Hardness Limitation for Valve Material in the Sour Service

In *sour service environments* containing hydrogen sulfide (H_2S), materials used for valves must adhere to strict hardness limits to prevent *sulfide stress cracking* (SSC). For example, *NACE MR0175/ISO 15156* specifies that the hardness of carbon steel used in sour service, such as *ASTM A105* for valve bodies, should not exceed *22 HRC (248 HV)*. Exceeding this hardness can make the material more brittle and susceptible to cracking under the combined effects of H_2S and tensile stress. To meet these requirements, materials are typically normalized and tempered to control hardness while maintaining sufficient strength for high-pressure applications. This limitation ensures valve reliability and safety in sour gas environments.

Hastelloy

Hastelloy is a trademarked name for a group of corrosion-resistant metal alloys primarily composed of nickel, along with various other metals like molybdenum, chromium, and iron. These alloys are designed to offer excellent resistance to corrosion, especially in harsh environments such as those found in chemical processing and power generation. Hastelloy C-276 and Hastelloy C-22 are the most common materials in the oil and gas industry. Hastelloy is primarily composed of nickel, molybdenum, and chromium, with smaller amounts of iron, tungsten, and cobalt.

Heat-Affected Zone (HAZ)

HAZ is a region in the base material surrounding a weld that undergoes thermal changes during the welding process. The high temperatures in this area alter the material's microstructure and mechanical properties without melting it, potentially leading to issues like reduced toughness, increased hardness, or corrosion susceptibility. For example, in austenitic stainless steels, the HAZ is prone to sensitization, where chromium carbides precipitate along grain boundaries. Sensitization depletes chromium, making the material vulnerable to IGC. HAZ areas are shown in Figure 4.12 around the welding and are different in color from the base metal.

Heat Treatment

1. Definition

Heat treatment is a controlled process used to alter the microstructure and mechanical properties of valve materials, ensuring optimal performance in

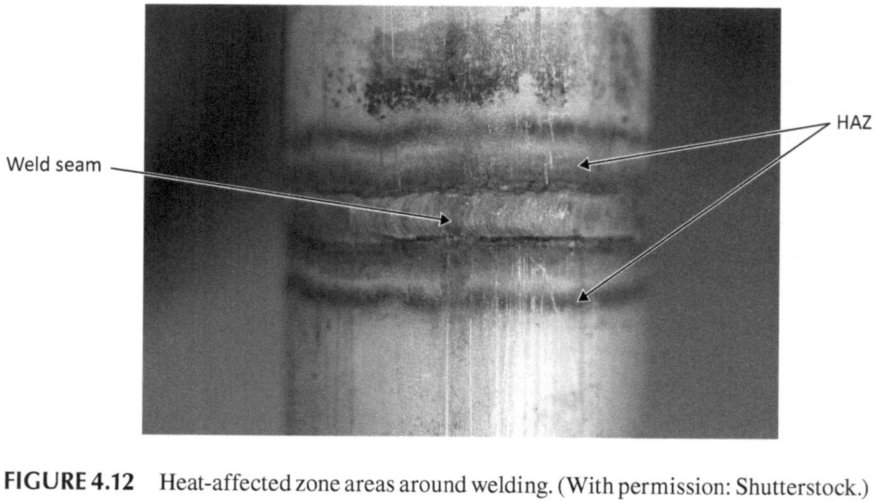

FIGURE 4.12 Heat-affected zone areas around welding. (With permission: Shutterstock.)

the required applications. In valves, heat treatment is crucial for achieving desired properties such as improved corrosion resistance, enhanced strength, and resistance to high pressures and temperatures. Proper heat treatment ensures reliability in critical environments like offshore oil and gas or chemical processing.

2. Types

Common heat treatment techniques for valves include *annealing*, to soften the material and improve machinability; *quenching and tempering*, to enhance strength and toughness; and *solution annealing*, often applied to stainless steels to dissolve precipitates and restore corrosion resistance. Precipitation hardening is also used for materials like 17-4 PH to increase hardness and wear resistance. *Normalizing* is a heat treatment process where the material is heated to a temperature above its critical range (usually 850–950°C for carbon steel) and then cooled in air. Normalizing refines the grain structure, improves toughness, and relieves internal stresses introduced during manufacturing. *Tempering* is done after normalizing or quenching, tempering is performed to reduce brittleness and enhance ductility and toughness.

3. Example

In a valve with a *carbon steel body*, *austenitic stainless steel trim*, and a *17-4 PH stem*, each material undergoes specific heat treatment to optimize its performance. The *carbon steel body*, such as ASTM A216 WCB, is typically normalized and tempered to improve its toughness and resistance to cracking under high-pressure applications. The *austenitic stainless steel trim*, such as 316L, undergoes solution annealing to dissolve chromium carbides, restoring corrosion resistance and ensuring longevity in harsh environments. The *17-4 PH stem* is subjected to precipitation hardening (aging) after solution treatment to achieve high strength and wear resistance while maintaining good corrosion resistance. This combination of heat treatments ensures the valve components perform reliably under corrosive and high-pressure conditions commonly encountered in industrial and offshore applications.

HNBR

HNBR (hydrogenated nitrile butadiene rubber) is a synthetic elastomer derived from *NBR* through a hydrogenation process that enhances its performance. The hydrogenation significantly improves HNBR's resistance to heat, ozone, and weathering, making it suitable for harsh environments. It offers excellent resistance to oils, fuels, and hydrocarbons, as well as superior mechanical properties like tensile strength and abrasion resistance. HNBR operates effectively in a wide temperature range, typically from −40°C to 150°C, and is widely used in oil and gas, automotive, and industrial applications. In valves, HNBR is commonly used for seals, O-rings, and gaskets in systems requiring durability under high pressures and temperatures, or exposure to aggressive media.

Hot-Dip Galvanize

For more information, refer to "Galvanizing."

Hydrogen Embrittlement

This type of corrosion is related to hydrogen sulfide corrosion. HE refers to the reduction in ductility and strength of a material caused by hydrogen diffusion, making it more prone to cracking. H_2S environments accelerate hydrogen absorption, increasing the risk of embrittlement, especially in high-strength alloys.

Hydrogen-Induced Cracking (HIC) Corrosion

It is a type of sour corrosion. HIC occurs when atomic hydrogen, generated during H_2S corrosion, diffuses into steel and accumulates in internal voids or inclusions. The trapped hydrogen forms molecular hydrogen, creating internal pressure that leads to blistering and cracks along the steel's structure. HIC is prevalent in carbon steels exposed to sour service environments.

Hydrogen-Induced Stress Cracking Corrosion

HISC is a form of hydrogen embrittlement that occurs in duplex stainless steels when exposed to cathodic protection in the presence of H_2S. It combines the effects of tensile stress and hydrogen diffusion, leading to localized cracking.

HISC can occur in valves installed in the *subsea oil and gas industry*, particularly those made from *duplex stainless steels* or other susceptible materials. Subsea environments require cathodic protection to prevent corrosion. However, this protection can lead to hydrogen generation at the valve's surface, which may diffuse into the material and cause HISC. Figure 4.13 shows HISC in low-alloy valve bolting material.

Hydrogen Sulfide

Hydrogen sulfide is a chemical compound that contains two atoms of hydrogen and one atom of sulfur (see Figure 4.14). An undesirable by-product of oil, it is highly toxic, corrosive, and flammable. Human life can be endangered by a concentration of this compound as low as 50 parts per million. The smell of hydrogen sulfide is similar to that of rotten eggs.

Hydrogen Sulfide Corrosion

Same as "Sour corrosion." Refer to "Sour Corrosion" for details.

FIGURE 4.13 HISC in low-alloy steel fasteners for subsea valves.

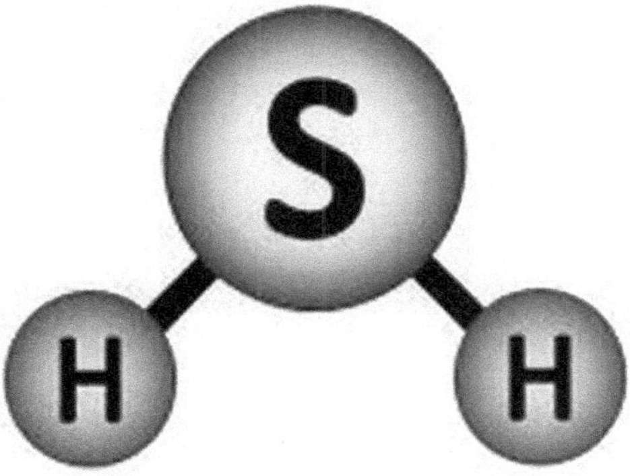

FIGURE 4.14 An illustration showing two hydrogen (H_2) atoms and an "S" representing hydrogen sulfide.

I

Inconel

Inconel is a trademarked name for a family of nickel–chromium-based superalloys that are highly resistant to corrosion, oxidation, and high temperatures. These alloys are widely used in extreme environments, such as oil and gas, aerospace, chemical processing, and power generation industries, where conventional materials would fail due to heat, pressure, or corrosive conditions. Inconel 625 is widely used in oil and gas industry. Inconel 625 is a nickel-based superalloy known for its exceptional resistance to corrosion, oxidation, and high temperatures, along with excellent mechanical strength and weldability.

Intergranular Corrosion

IGC is a localized form of corrosion that occurs along the grain boundaries of a material, leaving the grains themselves largely unaffected. It is typically caused by the precipitation of chromium carbides (Cr_3C_6) at grain boundaries in stainless steels during improper heat treatment or welding, a process known as sensitization. This depletes chromium from the surrounding areas, reducing the material's corrosion resistance and making it susceptible to attack in corrosive environments. IGC is particularly problematic in high-temperature applications or acidic environments. However, it can be mitigated by using low-carbon grades (e.g., 304L, 316L) or stabilized stainless steels (e.g., 321, 347) that resist carbide formation.

L

Leachable

It is important to note that leachables in graphite packing refer to substances or impurities that may be extracted or released from the material when it comes into contact with certain fluids (such as water, steam, or chemicals). In a graphite packing, chloride and fluoride are known leachable compounds. These substances may be by-products of the manufacturing process, fillers, or additives in the packing material. In order to prevent corrosion and leakage from the valve stem packing, the amount of leachable material should be limited. Graphite packing is often used in valves, pumps, and other industrial equipment to provide reliable sealing under high temperatures and pressures. However, the presence of leachables can affect its performance and compatibility with the process medium.

Lifecycle and Sustainability

Lifecycle analysis (LCA) evaluates the environmental and economic impact of valve materials over their entire lifespan. The valve life starts from raw material extraction to manufacturing, operation, and finally disposal. Materials like stainless steel and nickel alloys are more expensive initially. However, these materials often have lower environmental footprints due to their recyclability and long service life. As a result, choosing stainless steel and nickel alloys reduces waste and replacement needs compared to carbon steel. LCA considers factors such as energy consumption during production, emissions from corrosion or degradation, and end-of-life recyclability. For instance, stainless steel is highly recyclable, with over 80% of its content reused globally, making it a sustainable choice for industries like oil and gas. Additionally, selecting materials with extended life cycles minimizes environmental impacts associated with frequent replacements, aligning with modern sustainability goals and cost-efficiency.

Lining

Lining in valves refers to the application of a protective layer to the internal surfaces of the valve body and components. Lining materials can be made from nonmetallic materials like *PTFE*, *rubber*, or *ceramics*, or metallic materials such as Inconel 625 or other corrosion-resistant alloys. This lining acts as a barrier between the valve's base metal and the fluid, protecting against corrosion or erosion in aggressive environments. Lined valves are commonly used in industries such as chemical processing, and water treatment where handling acids, alkalis, or abrasive slurries can degrade metal surfaces. PTFE-lined valves, for instance, are ideal for high-corrosion applications, while rubber linings are effective for abrasive (erosive) media. Lining extends the valve's lifespan, reduces maintenance, and allows cost-effective materials like carbon steel to be used as the base metal in harsh environments.

Example: Usage of Lining for a Valve: Rubber-Lined Concentric Butterfly Valves

A *concentric butterfly valve* with a *natural rubber lining* is a common example of how linings are used to enhance valve performance. In this design, the valve body is made from *carbon steel*, providing the necessary structural strength. In addition, the natural rubber lining covers the internal surfaces of the valve body and disk. The rubber lining acts as a protective barrier, preventing the carbon steel from erosion or reacting with the fluid. In fact, lining offers excellent flexibility and sealing performance. This type of valve is widely used in water distribution systems and wastewater treatment. Thus, the natural rubber lining provides resistance to water and mild chemicals while ensuring a cost-effective and somehow durable solution.

Liquid Metal Embrittlement (Zinc Embrittlement)

Zinc negatively affects (corrodes) stainless steel at high temperatures. Liquid metal embrittlement (LME) occurs when molten zinc penetrates the grain boundaries of stainless steel, leading to brittle fracture under stress. It is common in welding or high-temperature applications where zinc-coated materials are involved. Zinc embrittlement is a general term referring to the weakening and cracking of stainless steel due to the diffusion of zinc at elevated temperatures. For example, using hot-dip galvanized bolts with stainless steel valves in high-temperature applications, especially when insulated, could be a significant problem.

Localized Corrosion

Localized corrosion refers to a concentrated form of corrosion that occurs in specific areas of a valve's surface. This type of corrosion often leads to severe damage despite the majority of the material remaining intact. In valves, this type of corrosion is common in environments with aggressive fluids containing chlorides, sulfides, or acids.

Forms of localized corrosion include *pitting corrosion*, *crevice corrosion*, and *SCC*. Localized corrosion is particularly problematic for materials like carbon steel or austenitic stainless steels (e.g., 304, 316) in chloride-rich environments. To prevent localized corrosion caused by chloride, materials with high PREN, such as duplex stainless steels or nickel-based alloys, are often selected. The other approach to mitigate localized corrosion is to use coatings or cathodic protection are applied.

Low-Alloy Steel

1. Definition

Low-alloy steel is a type of steel that contains a small percentage (typically less than 5% by weight) of alloying elements such as chromium, molybdenum, nickel, or vanadium, in addition to carbon and iron. These alloying elements are added to enhance specific properties like strength, toughness, hardness, wear resistance, and corrosion resistance.

2. Application

Low-alloy steels, such as chromium–molybdenum (Cr-Mo) steels, are specialized materials containing small amounts of chromium and molybdenum (typically 1–5%) along with carbon and other elements. These steels are commonly used in the oil and gas industry for *high-temperature services*, such as steam applications, due to their exceptional strength, toughness, and resistance to creep and corrosion at elevated temperatures. The chromium provides oxidation resistance, while molybdenum enhances high-temperature strength and thermal stability. For example, *ASTM A217 WC6 and WC9* grades are widely used for valve bodies in high-pressure steam systems, ensuring reliability and durability in extreme operating conditions.

Low-Temperature Carbon Steel (LTCS)

LTCS is a type of carbon steel specifically designed for applications requiring toughness and durability at low temperatures, typically down to *−46°C (−50°F).* Standard carbon steel can generally withstand minimum temperatures of *−29°C (−20°F)*, but at lower temperatures, it becomes brittle and susceptible to cracking. LTCS, often specified as *ASTM A333 Grade 6* or similar, has a modified chemical composition with controlled levels of carbon and manganese to improve its impact resistance and ductility at colder conditions. Its ability to retain toughness and resist brittle fracture at temperature range between *−29°C* to *−46°C*, makes LTCS a reliable choice for ensuring safety and performance in low-temperature conditions.

M

Martensitic Stainless Steel

Martensitic stainless steels are widely used in valve applications for their high strength, hardness, and moderate corrosion resistance. Grades like *17-4 PH (UNS*

S17400) are commonly used for valve stems in refineries and petrochemical plants due to their excellent mechanical properties and resistance to wear, achieved through precipitation hardening heat treatment. In offshore environments, *13Cr (UNS S41000)* and *13Cr-4Ni (UNS S41500)* are favored for their good corrosion resistance in mildly aggressive environments and high strength, making them suitable for components like valve trims. These materials are particularly effective in applications requiring resistance to erosion, wear, and moderate levels of CO_2 or chloride-induced corrosion. Martensitic steels offer a cost-effective solution for valves in demanding environments, combining strength and durability with adequate corrosion resistance. However, in general, martensitic stainless steels are not ideal for parts exposed to high-chloride environments because they lack sufficient pitting and crevice corrosion resistance, which can lead to premature failure. For critical offshore applications with high chloride levels, duplex or super duplex stainless steels or nickel alloys are often preferred over martensitic grades.

Example: Valve Stem Material Upgrade in Offshore Environment

In an offshore oil and gas platform, a *control valve stem* made of *17-4 PH (UNS S17400)* stainless steel experienced significant corrosion and pitting after prolonged exposure to a high-chloride environment. Despite the material's excellent strength and wear resistance, it lacked adequate resistance to chloride-induced pitting and crevice corrosion. This led to operational issues such as reduced sealing integrity and frequent maintenance requirements. To resolve the problem, the valve stem was replaced with *Inconel 718 (UNS N07718)*, a nickel-based alloy with superior corrosion resistance, especially in chloride-rich and sour service environments. After the material upgrade, the valve operated without any corrosion-related issues, demonstrating the effectiveness of Inconel 718 in mitigating chloride-induced corrosion in offshore applications. This example highlights the importance of selecting materials with suitable corrosion resistance for critical offshore components.

Material Selection

Material selection for valves is a critical aspect of valve design and performance, as it directly influences the valve's durability, functionality, and ability to withstand operating conditions. The right material ensures compatibility with the fluid being handled and its operating conditions, resists stresses, and meets safety and regulatory requirements. The selection process requires careful consideration of several factors, as outlined below.

1. Operating Conditions

The valve material must withstand the operating *temperature* and *pressure* of the system. Materials like carbon steel are commonly used for moderate temperatures and pressures, while stainless steel or superalloys such as Inconel or Hastelloy are chosen for high-temperature, high-pressure applications. Cryogenic applications require materials like austenitic stainless steels due to their low-temperature toughness. Valves for steam systems often use low-alloy Cr–Mo to handle elevated pressures and temperatures.

2. Corrosion Resistance

Corrosion resistance is a key factor, especially in applications involving aggressive fluids such as acids, alkalis, or seawater. Duplex and super duplex stainless steels and nickel-based alloys offer excellent corrosion resistance, making them ideal for harsh chemical or marine environments. For less aggressive fluids, coated materials or standard carbon steel with protective linings may suffice.

3. Fluid Type and Characteristics

The type of fluid passing through the valve heavily influences material selection. For abrasive fluids (e.g., slurries or sand-laden water), hard materials like tungsten carbide or Stellite hard-facing alloys are preferred to resist wear. Noncorrosive fluids, like clean water or air, may allow the use of more economical materials like carbon steel. For flammable or hazardous fluids, materials must comply with safety standards to minimize risks.

4. Mechanical Properties

Mechanical properties, such as tensile and yield strength, impact resistance, and hardness, are critical for ensuring the valve can handle the stresses it will encounter during operation. For example, hardness of metallic materials in sour services must be limited to the values given in the National Association of Corrosion Engineers (NACE) standard.

5. Temperature Compatibility

Valves operating in high-temperature environments, such as those in power plants or industrial furnaces, require materials that maintain their strength and functionality at elevated temperatures. Materials like stainless steel, alloy steel, or carbon steel are ideal for such applications. Conversely, low-temperature and cryogenic conditions require materials like austenitic stainless steel, or nickel alloys which retain ductility and toughness in extreme cold.

6. Erosion and Wear Resistance

Applications involving high-velocity fluids, abrasive particles, or frequent valve cycling require materials with excellent wear resistance. Hard-facing materials like Stellite or tungsten carbide are often applied to valve seats and closure members to enhance durability and reduce erosion caused by fluid flow.

7. Compliance with Industry Standards

Valves must comply with industry standards and certifications, such as API, ASME, or ISO, which dictate material properties and performance. In many cases, valve materials are in accordance with ASTM standards. Chapter 5 provides more information about different standards applicable to industrial valves.

8. Environmental Factors

Environmental conditions, such as humidity, exposure to saltwater, soil, or the presence of contaminants, can affect material performance. For outdoor or marine environments, materials like duplex stainless steel, titanium, or specially coated steels are chosen for their ability to resist corrosion and environmental degradation. External coatings are applied to underground and several offshore valves.

9. Maintenance and Longevity

Selecting materials that minimize wear and require less frequent maintenance enhances the valve's life cycle. For critical systems where downtime is costly such as offshore applications, premium materials like titanium, or duplex stainless steel may be chosen despite their higher upfront cost due to their extended service life and reduced maintenance needs.

10. Availability

Material availability is a crucial factor in material selection, as the chosen material must be readily obtainable in the required quantities and forms to meet project deadlines and budgets. Limited availability or long lead times for specialized materials can delay production, increase costs, and necessitate reconsideration of alternative materials that provide comparable performance.

11. Cost

While performance is vital, material costs also play a crucial role in selection. Balancing performance requirements with budget constraints is necessary. For example, carbon steel is cost-effective for general noncorrosive applications. Superalloys or exotic materials are reserved for extreme conditions even they have higher cost.

Mechanical Properties

Mechanical properties of valve materials refer to the physical characteristics that determine the material's ability to withstand forces, stresses, and environmental conditions during operation. These properties include *strength* (such as tensile and yield strength, indicating resistance to deformation and failure under load), *ductility* (the ability to deform without breaking, crucial for forming processes and absorbing shock), *hardness* (resistance to surface indentation and wear), *toughness* (the ability to absorb energy and resist fracture under impact), and *elasticity* (ability to return to original shape after deformation). For valves, these properties ensure reliable performance under varying pressures, temperatures, and corrosive conditions. Mechanical properties are critical in material selection for specific applications in industries like oil and gas, chemical processing, and power generation.

Microbial Corrosion

MIC (*microbiologically influenced corrosion*) refers to the accelerated degradation of materials, including valve components, caused by the activities of microorganisms such as bacteria, algae, or fungi. These organisms can create localized corrosive environments by producing acids and sulfides. For example, *sulfate-reducing bacteria* (SRB) are common in oil and gas pipelines and valves; they generate hydrogen sulfide as a by-product, leading to severe pitting corrosion in carbon steel valves. MIC is particularly problematic in environments like water injection systems or seawater-handling applications, where stagnant or low-flow conditions allow biofilm formation. Proper material selection, such as using corrosion-resistant alloys (e.g., duplex stainless steel), and regular cleaning or biocide treatments are essential to mitigate MIC in valves.

Note: Biofilm in MIC refers to a complex layer of microorganisms (such as bacteria, fungi, or algae) that adhere to a metal surface.

Microstructure

Microstructure refers to the arrangement of grains, phases, and defects within a material, visible under a microscope. It determines the material's mechanical, physical, and chemical properties, such as strength, toughness, and corrosion resistance. Microstructure is influenced by factors like composition, manufacturing processes, and heat treatment, making it crucial in tailoring materials for specific applications.

Monel

Monel is a family of nickel–copper alloys known for its excellent corrosion resistance, particularly in marine and chemical environments. Typically composed of around 63% nickel and 28–34% copper, with small amounts of iron and manganese, Monel alloys exhibit exceptional resistance to seawater, hydrofluoric acid, and other harsh chemicals. Monel is highly durable, with good mechanical strength and toughness, even at subzero temperatures, making it suitable for cryogenic applications. The most commonly used grade, *Monel 400 (UNS N04400)*, is widely employed in the oil and gas, marine, and chemical processing industries for valve components, pumps, and piping systems where both corrosion and erosion resistance are critical.

N

National Association of Corrosion Engineers

NACE, now part of *AMPP* (*Association for Materials Protection and Performance*), is an organization that develops standards and guidelines for materials used in corrosive environments, particularly in the oil and gas industry. In relation to valves, NACE standards like *NACE MR0175/ISO 15156* specify requirements for material selection to ensure resistance to *sour service conditions*, where H_2S is present. These standards help prevent corrosion-related failures such as hydrogen embrittlement, SSC, and HIC in valve components. Compliance with NACE ensures that valve materials, such as carbon steels, duplex stainless steels, and nickel alloys, meet safety and reliability requirements for H_2S-rich and other aggressive environments. These guidelines are critical for designing valves that operate reliably in sour oil and gas applications.

NBR

It is an elastomeric material used for valve seals. NBR is known for its excellent resistance to oils, fuels, and hydrocarbons, NBR is commonly used in oil and gas valves for sealing applications. It performs well in moderate temperatures but has limited resistance to ozone, weathering, and strong acids.

Nickel Alloys

1. Definition

Nickel alloys are a group of highly corrosion-resistant materials designed for use in extreme environments, such as high temperatures, aggressive chemicals, and chloride-rich or sour conditions. Their excellent mechanical properties, including high strength, toughness, and resistance to oxidation and SCC, make them ideal for demanding industrial applications. Nickel alloys achieve their superior performance through their high nickel content, often combined with elements like chromium, molybdenum, iron, and cobalt to enhance specific properties. They are widely used in the oil and gas, chemical processing, and marine industries, where reliability and durability are critical under harsh operating conditions.

2. Common Nickel Alloys in the Valve Industry

In the valve industry, several nickel alloys are frequently used for components exposed to aggressive environments. *Inconel 625 (UNS N06625)* offers excellent resistance to pitting, crevice corrosion, and oxidation, making it suitable for seawater applications. *Inconel 725 (UNS N07725)* provides similar corrosion resistance with enhanced strength due to its age-hardenable properties. *Monel 400 (UNS N04400)*, a nickel–copper alloy, is well-known for its resistance to seawater and hydrofluoric acid. *Incoloy* alloys, such as 825, are used in sour service due to their ability to resist SCC. *Hastelloy C-series* alloys, like *C276 (UNS N10276)*, provide exceptional corrosion resistance in highly acidic and oxidizing environments. These alloys are essential for critical valve components such as stems, bodies, and trims, ensuring reliable performance in the most challenging industrial applications.

Nickel Aluminum Bronze

1. Definition and Application

NAB is a high-strength copper-based alloy containing aluminum (5–11%), nickel (up to 5%), and iron (up to 5%), designed for exceptional corrosion and wear resistance. This alloy is particularly well-suited for valves used in marine, offshore, and chemical processing environments due to its resistance to seawater, erosion, and biofouling. NAB's excellent mechanical properties, such as high toughness and good fatigue resistance, make it ideal for critical valve components like bodies, disks, and seats in applications involving high flow rates or abrasive media. Its ability to withstand pitting and crevice corrosion further enhances its reliability in harsh operating conditions.

2. Example: NAB Valves Selected for Offshore Seawater Application

In an offshore oil and gas platform, valves for seawater injection systems were initially designed with *titanium* components due to their exceptional corrosion resistance in chloride-rich environments. However, titanium's high material and manufacturing costs prompted a review of alternative options. *NAB* was selected as a cost-effective replacement, offering excellent resistance to seawater corrosion, erosion, and biofouling while meeting the required performance standards. Although NAB does not match titanium's corrosion resistance in every aspect, its superior strength, fatigue resistance, and significantly lower cost made it an ideal choice for these valves. The switch to NAB achieved substantial cost-savings without compromising reliability. This material change demonstrates the importance of balancing material performance and cost in valve material selection for offshore environments.

Nondestructive Testing (NDT)

Nondestructive testing (NDT) refers to a set of techniques used to evaluate the integrity, properties, and defects of materials or components without causing damage. For valve materials, NDT ensures reliability and compliance with standards in critical applications like oil and gas, chemical processing, and power generation. Common nondestructive examination (NDE) methods include *positive material identification* (PMI) for verifying material composition, *Ultrasonic testing* (UT) to detect internal flaws or thickness variations, *radiographic testing* (RT) for identifying weld defects or voids, *magnetic particle testing* (MPT) for surface and subsurface crack detection in ferromagnetic materials, and *dye penetrant testing* (DPT) for highlighting surface defects like cracks or porosity. These techniques play a crucial role in detecting potential weaknesses, ensuring safety, and maintaining performance without causing damage. These tests verify compliance with standards like ASTM, NACE, and API, ensuring valves meet safety and performance criteria in critical applications. Chapter 6 provides more information about NDT and its different techniques.

O

Oxidation

Oxidation in materials is a chemical reaction where a material, typically a metal, reacts with oxygen in the environment to form an oxide layer on its surface. This process can either protect the material, as in the case of stainless steel, or lead to degradation, like rusting in carbon steel. Oxidation is influenced by factors such as temperature, exposure to moisture, and the material's composition.

P

Passivation/Passivation Layer

Passivation is the process of forming a protective oxide layer on the surface of a material, typically a metal, to prevent further corrosion. This layer is often created naturally through exposure to air or chemically induced by treatments like acid baths. Stainless steel, for example, becomes corrosion-resistant due to the formation of a thin chromium oxide film during passivation.

A *passivation layer* is a thin, protective oxide film that forms on the surface of a material, typically metals, to shield it from corrosion. This layer is usually stable, adherent, and self-healing, as seen in stainless steel, where a chromium oxide layer prevents further oxidation or chemical attack. It enhances the material's durability, especially in harsh environments.

PEEK

PEEK (polyether ether ketone) is a high-performance thermoplastic renowned for its exceptional mechanical strength, chemical resistance, and ability to withstand high temperatures (up to 260°C). In valve applications, PEEK is commonly used for seats, seals, and thrust washers. PEEK is used especially in environments requiring durability under high pressures, temperatures, and exposure to aggressive chemicals. Its resistance to wear, abrasion, and hydrolysis makes it ideal for demanding industries such as oil and gas, chemical processing, and power generation. PEEK also offers excellent dimensional stability and low friction, ensuring reliable operation in dynamic applications like ball and butterfly valves. Its combination of properties provides a lightweight and cost-effective alternative to metal components in specific valve designs.

Piezoelectric Materials

Piezoelectric materials are advanced functional materials that generate an electric charge when subjected to mechanical stress or, conversely, deform when an electric field is applied. Common types include *quartz, ceramics like lead zirconate titanate* (PZT), and *polymeric materials like PVDF (polyvinylidene fluoride).* In valve applications, piezoelectric materials are primarily used in *precision control valves,* where they serve as actuators for fine movement or as sensors to monitor pressure, flow, or vibration. For example, piezoelectric actuators enable rapid and precise adjustments in microvalves.

Pitting Corrosion

Pitting corrosion is a localized form of corrosion that occurs on valve surfaces exposed to aggressive environments, such as those containing chlorides in oil and gas operations. It manifests as small, deep pits or cavities, often penetrating the material and compromising its integrity. Pitting typically affects materials like carbon steel and austenitic stainless steel, especially in the presence of high

FIGURE 4.15 Pitting corrosion on a metal surface. (With permission: Shutterstock.)

chloride levels or under stagnant fluid conditions. Once initiated, pits can grow rapidly, leading to leakage, structural failure, and reduced performance of the valve. To mitigate pitting corrosion, materials like duplex stainless steels, super duplex, or nickel-based alloys (e.g., Inconel 625) are often used. These materials provide higher resistance to such localized attacks. Proper material selection, coating, and maintenance are critical to minimizing pitting corrosion in oil and gas valves. Figure 4.15 illustrates pitting corrosion on a metal surface.

Pitting Resistance Equivalent Number

PREN is a measure used to predict the resistance of stainless steels and other alloys to pitting corrosion, particularly in environments with high chloride concentrations. PREN is calculated using the formula:

$$\text{PREN} = \%\text{Cr} + 3.3(\%\text{Mo}) + 16(\%\text{N}),$$

where chromium (Cr), molybdenum (Mo), and nitrogen (N) are the key alloying elements that enhance pitting resistance.

Higher PREN values indicate greater resistance to pitting corrosion. For example, duplex stainless steels like 2205 have a PREN of approximately 35, while super duplex grades like 2507 can exceed 40, making them suitable for aggressive offshore and subsea environments. The PREN formula is a critical tool for material selection in the oil and gas industry, especially for chloride-containing services.

Positive Material Identification (PMI)

PMI is a nondestructive testing method used to verify the chemical composition of materials in order to ensure that they meet specified standards (e.g., ASTM). In industries such as oil and gas, power generation, and chemical processing, where valves are subject to high pressures, temperatures, and corrosive environments, PMI is very important. In order to reduce the risk of failure and maintain safety standards, the purpose of this testing is to confirm that the valve is constructed from materials suitable for the intended application. In order to verify that the valve's material composition meets the required specifications, PMI is utilized by means of X-ray fluorescence or optical emission spectroscopy. Figure 4.16 illustrates an operator performing a PMI test on a valve component using a handheld X-ray fluorescence (XRF) analyzer.

Post-Weld Heat Treatment (PWHT)

PWHT is a controlled heating and cooling process applied to a welded component after welding to improve its mechanical properties and reduce residual stresses. PWHT relieves stresses induced by the welding process, reduces the risk of cracking, and can enhance toughness and corrosion resistance by refining the

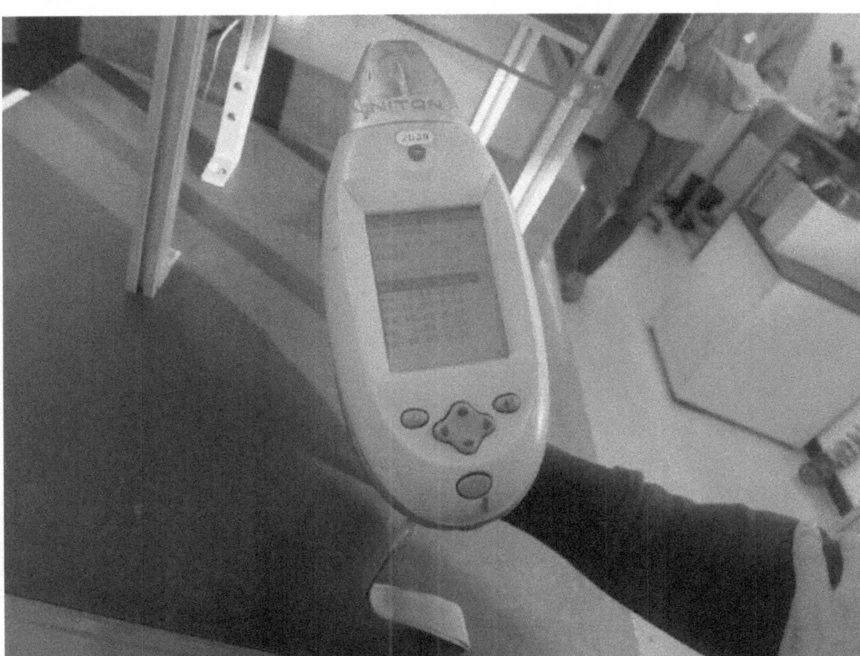

FIGURE 4.16 PMI test on a valve by an operator.

microstructure. It is particularly critical for materials like carbon steels and alloy steels used in high-pressure or high-temperature applications, such as valves and pressure vessels. For example, in a welded valve body, PWHT ensures durability and reliability by minimizing the risk of failure during operation.

Preheat

Preheat is the process of heating the base material to a specific temperature before welding to reduce the risk of defects such as cracking or distortion. Preheating helps in minimizing thermal gradients between the weld and the surrounding material and in reducing residual stresses. In addition, it prevents rapid cooling that could lead to brittle microstructures. This process is particularly critical for thick sections, high-carbon steels, or materials prone to HIC. For example, in welding a valve body made of carbon steel, preheating ensures a more controlled and defect-free weld, improving overall structural integrity and performance in demanding applications.

PTFE

PTFE (polytetrafluoroethylene) is same as "Teflon." For more information, refer to "Teflon."

R

Residual Stress

1. Definition

Residual stress in valve materials refers to internal stresses that remain locked within the material after it has been manufactured, processed, or subjected to external forces, even when no external load is applied. These stresses can arise from uneven cooling during casting or welding, machining, heat treatment, or plastic deformation processes.

2. Case Study

Scenario

Austenitic stainless steel (e.g., 316L) is commonly used for valves due to its corrosion resistance. However, during manufacturing, processes like welding or machining can introduce high residual stresses in the material.

Failure

When the valve is exposed to a chloride-rich environment under operational stresses, these residual stresses combine with environmental factors. This condition leads to chloride SCC, a brittle fracture mechanism. Cracks initiated by residual stress propagate over time, resulting in valve leakage and catastrophic failure.

Mitigation

Proper post-weld heat treatment or stress-relief annealing can reduce residual stresses and prevent such failures. Material alternatives, like duplex stainless steels, are also more resistant to SCC under similar conditions.

Rusting

Rusting is a specific type of corrosion that occurs when iron or steel reacts with oxygen and moisture in the environment to form iron oxides. Rusting weakens the structural integrity of iron and steel by consuming the base material, often leading to flaking and brittleness (see Figure 4.7).

S

Salt Spray Test

The *salt spray test* is a standardized accelerated corrosion test used to evaluate a material or coating's resistance to corrosion when exposed to a saline environment. During the test, specimens are placed in a chamber where a fine mist of saltwater solution is continuously sprayed, simulating harsh conditions like marine or industrial environments. It is commonly used for quality control and comparing the performance of protective coatings, such as those on valve materials.

Self-Healing Polymers

Self-healing polymers are an emerging technology with potential applications in oil and gas valves, particularly in coatings and sealing systems. These polymers can autonomously repair cracks or damage caused by wear, abrasion, or corrosion. Self-healing ability maintains continuous protection and functionality. In valves, self-healing polymer coatings can be applied to internal surfaces or sealing areas to prevent leakage and extend operational life.

An example of a *self-healing polymer application in valves* for the oil and gas industry is the use of *epoxy-based coatings with embedded microcapsules* containing healing agents. These coatings are applied to the internal surfaces of valves used in pipelines or processing systems exposed to harsh conditions, such as high pressures and corrosive fluids. If a crack or damage occurs, the microcapsules break open, releasing a healing agent (e.g., resin or hardener) that reacts to seal the damaged area, restoring the protective barrier.

Sensitization

Sensitization is a metallurgical phenomenon in which a material, typically austenitic stainless steel, becomes susceptible to IGC due to the formation of chromium carbides along grain boundaries. This occurs when the steel is exposed to temperatures between 450°C and 850°C (the sensitization range), causing chromium to combine with carbon and deplete the grain boundaries of chromium. As a result, the affected areas lose their corrosion resistance. Sensitization is a common concern in welding or heat treatment processes and can be mitigated by using low-carbon grades (e.g., 304L, 316L) or stabilized grades (e.g., 321, 347) of stainless steel.

Shape Memory Alloys (SMA)

SMAs are a class of metallic *smart materials* that can return to their original shape after being deformed, upon exposure to specific conditions like heat or pressure. This behavior, known as the *shape memory effect*, occurs due to a reversible phase transformation between the material's martensite (low-temperature) and austenite (high-temperature) crystal structures. SMAs, such as *Nitinol (nickel–titanium alloy)*, are widely used in applications like actuators and valves, where adaptability and precise movement are essential. For example, SMA can be used as a type of spring in both valves and actuators to operate them in response to change of temperature.

These materials dynamically expand or contract in response to temperature fluctuations, ensuring a tight seal even in extreme conditions such as subsea or sour gas applications. SMAs also provide self-healing capabilities for minor damage, enhancing reliability and reducing leakage risks. Additionally, SMA components in valve seats can adjust to wear and dimensional changes over time, maintaining consistent sealing performance without requiring frequent maintenance.

Smart Materials

1. Introduction

Smart materials are advanced materials that can respond to external stimuli—such as heat, pressure, electricity, or pH changes—and adapt their properties, such as shape, stiffness, or conductivity, in real time. Examples include *shape memory alloys*, *piezoelectric materials*, and *self-healing polymers*.

2. Application for Valves

Smart materials and coatings are redefining the capabilities of valves by providing self-adaptive, protective, and enhanced performance features. Self-healing polymers are among the most promising advancements. These materials can repair damage autonomously through embedded microcapsules or vascular networks that release healing agents when cracks or abrasions occur. For example, a valve coating with self-healing properties can significantly reduce leakage risks and extend service life in environments exposed to corrosive fluids or abrasive media. However, challenges such as cost and limited operational ranges (temperature and pressure) still need to be addressed for widespread adoption.

Nanocoatings represent another cutting-edge development in valve technology. By using materials at the nanoscale, these coatings deliver exceptional properties like wear resistance and corrosion protection. Superhydrophobic nanocoatings, for instance, are ideal for offshore valves as they combat biofouling and chloride-induced corrosion in saltwater environments. Similarly, anti-wear nanocoatings reduce friction and extend the lifespan of moving parts such as stems and seals. Despite their benefits, challenges such as uniform application and environmental concerns regarding nanoparticle runoff remain areas of ongoing research.

Smart alloys, such as *shape memory alloys* (SMAs), are gaining attractions in valve applications. SMAs like Nitinol can revert to their original shape upon exposure to specific conditions, such as temperature changes. This makes them suitable for actuation in remote or inaccessible locations. These materials reduce dependency on external power sources, offering a simpler and more efficient alternative for valve control systems.

Sour Corrosion

In the presence of hydrogen sulfide, sour corrosion can result in corrosion of pipes, valves, and equipment. Hydrogen sulfide can originate from the degradation of organic materials during the formation of fossil fuels. Oil and gas can be considered sour if they contain significant amounts of hydrogen sulfide. The corrosion mechanism associated with sour corrosion is usually a crack or blister. The presence of water is necessary for sour corrosion to occur. As H_2S comes into contact with water, it causes acid and becomes corrosive. After contact with metal surfaces, it may cause electrochemical reactions that accelerate corrosion. In the presence of H_2S, iron sulfide (FeS) can form on metal surfaces, which can either protect or damage the underlying metal, depending on the environment. As hydrogen sulfide is corrosive and toxic, it is considered an impurity in oil and gas that should be removed by a sweetening process. The presence of hydrogen sulfide can result in the corrosion of carbon steel, stainless steel, and nickel alloys. Pipes and valves made of carbon steel are extremely susceptible to hydrogen sulfide corrosion. It is true that stainless steel materials are more resistant to sour corrosion than carbon steel materials, but they are still susceptible to this type of corrosion. It is possible to mitigate or control the corrosion caused by hydrogen sulfide in a variety of ways. It is necessary to select more resistant materials to hydrogen sulfide during the design phase of engineering if the service is sour. As a result, it is common to select materials based on recommendations from the NACE and/or upgrade the materials to more corrosion-resistant alloys. A second approach to preventing or minimizing sour corrosion is to inspect and monitor the system. As a third option, corrosion inhibitors can be added to the fluid service to prevent corrosion. In addition, pH should be maintained at a nonacidic level and water should not be allowed to enter the fluid service. As a conclusion, sour corrosion poses a serious threat to safety, equipment integrity, and environmental protection in the oil and gas industry. Figure 4.17 illustrates a corrosion on a valve stem made of 17-4PH material that has been corroded in the form of cracks by hydrogen sulfide.

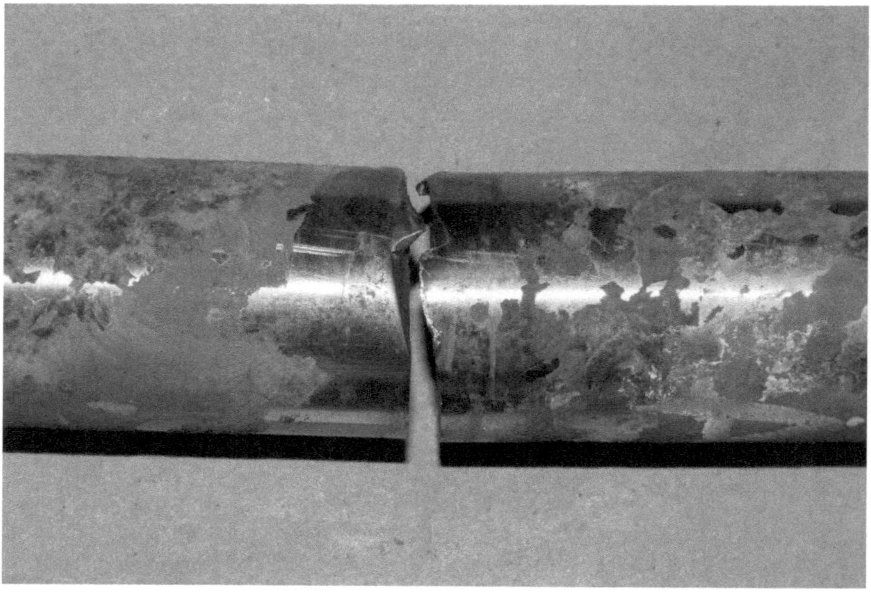

FIGURE 4.17 A valve stem in 17-4PH material is corroded in the form of crack by hydrogen sulfide. (With permission: Shutterstock.)

Sour Service

Sour service refers to the fluid service containing hydrogen sulfide. Equipment, valves, pipelines, and pressure vessels are exposed to process conditions (environments) containing H_2S, typically found in oil and gas production.

Stainless Steel

1. Definition

Stainless steel is an alloy primarily composed of iron, with a minimum of 10.5% chromium content by mass. A thin layer of chromium oxide forms on the surface as a result of the chromium content, preventing further corrosion and oxidation.

2. Types and Grades

Stainless steels are widely used in the oil and gas industry due to their corrosion resistance and mechanical properties. *Austenitic stainless steels*, such as 304, 316, and 316L, are common for general applications requiring good corrosion resistance, while *super austenitic grades* like 6Mo and 254 SMO handle more aggressive environments, including seawater. *Martensitic stainless steels*, such as 13Cr and 17-4 PH, are used in components like valve stems and trims for their high strength and wear resistance but are less suited for high-chloride conditions. *Duplex stainless steels*, such as 2205, and *super duplex grades*, like 2507, offer superior resistance to pitting, crevice corrosion, and SCC in harsh environments, including sour service. Finally, specialized alloys like *ferritic stainless steels* are occasionally used for their resistance to SCC in moderate conditions. Each type is chosen based on its specific balance of cost, strength, and corrosion resistance.

Stellite

As a cobalt–chromium alloy, Stellite is known for its outstanding wear resistance, hardness, and resistance to high temperatures and corrosion. Typically, it is used for valve internals, particularly for components such as seats, and closure members which need to withstand extreme erosive and abrasive conditions. Stellite is widely used in industries such as oil and gas, chemical processing, and power generation as a result of its properties. Stellite 6 is the most commonly used Stellite alloy for valves, offering a balance of wear resistance, hardness, and machinability. Stellite 21 has a slightly higher impact resistance than Stellite 6 and is often used in applications that require slight deformation under pressure. Several methods can be used to apply Stellite, including welding (hard-facing), and overlay coatings on valve surfaces to increase their lifespan. Due to its use, frequent maintenance and replacement are significantly reduced in conditions of harsh service.

Sulfide Stress Cracking Corrosion

This type of corrosion is associated with hydrogen sulfide or sour corrosion. SSC is a form of hydrogen embrittlement that occurs when hydrogen diffuses into steel under tensile stress, causing brittle fractures. It is common in high-strength steels and materials exposed to H_2S environments and can lead to sudden and catastrophic failure.

Sulfuric Acid Corrosion

Sulfuric acid corrosion occurs when materials are exposed to H_2SO_4. Sulfuric acid is a highly aggressive chemical widely used in industries like chemical processing. This type of corrosion can lead to rapid material degradation, especially in metals like carbon steel, which reacts with H_2SO_4 to form iron sulfate, causing thinning and potential failure. Corrosion severity depends on factors such as acid concentration, temperature, and the presence of water and impurities. For example, acid solution that has been mixed with a large amount of water, reducing its concentration and making it less aggressive. Materials like stainless steel, high-nickel alloys, or nonmetallic linings (e.g., rubber or Teflon) are often used to resist sulfuric acid corrosion in industrial equipment like storage tanks, pipelines, and valves.

Super Austenitic Stainless Steel

Super austenitic stainless steels are a class of high-performance alloys with a fully austenitic microstructure, offering exceptional corrosion resistance and mechanical strength. These steels are characterized by high levels of chromium (Cr), nickel (Ni), and molybdenum (Mo), often with additional elements like nitrogen (N) or copper (Cu) to enhance pitting and crevice corrosion resistance.

Super austenitic grades, such as *AL-6XN*, and *6Mo (6% molybdenum) stainless steels such as 254 SMO (UNS S31254)*, are specifically designed for demanding environments, including highly acidic or chloride-rich conditions, where standard stainless steels like 316L or 317L would fail. Their superior resistance to SCC, pitting, and crevice corrosion makes them ideal for applications in chemical processing, desalination, and offshore oil and gas industries.

Super duplex Stainless Steel

Super duplex stainless steels are a high-performance class of duplex alloys with a *50:50 austenitic–ferritic microstructure*, offering superior strength and corrosion resistance compared to standard duplex grades. They are characterized by higher levels of *chromium (25–27%)*, *molybdenum (3–4%)*, and *nitrogen*. This chemical composition significantly enhances their resistance to *pitting*, *crevice corrosion*, and *SCC* in aggressive environments, including seawater and high-chloride conditions.

Super duplex grades, such as *UNS S32750 (2507)* and *UNS S32760*, are widely used in the oil and gas, chemical processing, and marine industries, where both mechanical performance and corrosion resistance are critical. They provide a cost-effective alternative to nickel-based alloys while maintaining excellent durability in demanding applications.

Surface Treatment

Surface treatment refers to processes applied to the surface of a material to enhance its properties, such as corrosion resistance, wear resistance, appearance, or adhesion to coatings. These treatments can include mechanical methods, like polishing or shot peening, and chemical processes like passivation. For example, stainless steel valves may undergo passivation to form a protective oxide layer that resists corrosion in aggressive environments. Surface treatments are crucial in extending the life and performance of materials, especially in demanding applications like oil and gas, and chemical processing.

Sweet Corrosion

Refer to "Carbon Dioxide Corrosion" for more information. Sweet corrosion occurs due to the presence of carbon dioxide (CO_2), which reacts with water to form carbonic acid, leading to metal degradation in oil and gas environments.

T

Tack Weld

1. Definition

A tack weld is a temporary, small weld used to hold two pieces of metal together before performing the final, full weld. It serves as a placeholder to keep the parts in position and prevent them from shifting during the welding process.

2. Tack Weld and Material Properties

Tack welds play a crucial role in valve fabrication by temporarily holding components in place for precise alignment before full welding is performed. The effectiveness of a tack weld depends significantly on the material's weldability and its behavior under localized heat. For example, materials like carbon steel and stainless steel generally allow for strong and reliable tack welds, while alloys such as duplex stainless steel require careful control of welding parameters to prevent HAZ issues, such as cracking or loss of corrosion resistance. Proper selection of compatible filler materials and pre-welding treatments, like cleaning and preheating, is essential to ensure that tack welds do not compromise the structural integrity or material properties of the valve components.

Teflon

Teflon (the brand name for polytetrafluoroethylene, or PTFE) is a synthetic fluoropolymer widely used in valve components as a sealing and soft seat material due to its excellent chemical resistance, low friction, and high-temperature tolerance. In applications where a tight seal and resistance to harsh chemicals are essential, Teflon is preferred.

Tensile Strength

Tensile strength is the maximum amount of stress a material can withstand while being stretched or pulled before breaking. It measures the material's ability to resist deformation and failure under tension. Tensile strength is a critical property for materials used in load-bearing applications, such as valves and pipelines, to ensure they perform reliably under operating stresses.

Thermoplastic

Thermoplastic materials are polymers that become melted when heated and solidify upon cooling. This property makes them ideal for molding into precise shapes. In valve applications, thermoplastics are commonly used for *soft seats* due to their excellent sealing properties, chemical resistance, and low friction. Examples include *PTFE (polytetrafluoroethylene)*, *nylon*, *PEEK (polyetheretherketone)*, and *UHMWPE (ultra-high-molecular-weight polyethylene)*.

These materials are widely applied in industries requiring corrosion resistance, such as chemical processing, food and beverage, and water treatment.

Titanium

As a corrosion-resistant, titanium is frequently used in valves and internals in highly corrosive seawater services that are subject to corrosion types caused by chloride. Titanium is very resistant to various types of corrosion, including pitting and chloride SCC. This material is used in various industries such as offshore oil and gas, chemical processing, and desalination that experience extreme corrosive environments due to presence of seawater. The two most common grades of titanium material are grades 2 and 5. Grade 2 titanium (commercially pure titanium) is known for its excellent corrosion resistance, lightweight characteristics, and moderate strength. Grade 2 titanium is often used in applications that require resistance to seawater and other corrosive substances. Due to its weldability, it is a popular choice for valves in marine and chemical environments. A titanium alloy composed of titanium, aluminum, and vanadium, Titanium Grade 5 (Ti-6Aluminium-4Vanadium) offers a higher strength than commercially pure titanium. Often, it is used in applications requiring a high degree of strength, such as high-pressure valves.

Toughness

Toughness in materials refers to the ability to absorb energy and deform plastically without fracturing, even under sudden impacts or extreme stress. It reflects a material's capacity to resist crack initiation and propagation, combining both strength and ductility. Tough materials, such as certain steels used in valve bodies, are essential for applications exposed to dynamic loads or harsh conditions, where brittleness could lead to catastrophic failure.

Tribaloy

Tribaloy is a family of cobalt-based alloys designed for exceptional wear and corrosion resistance, especially in high-temperature and abrasive environments. *Tribaloy* can serve as a replacement for *Stellite* in seawater applications, particularly for valve components exposed to high wear and corrosive conditions. That is because Stellite is not recommended for corrosive seawater services by metallurgists.

Tungsten Carbide

It is widely used in valves, especially for seats, and closure members. Tungsten carbide is an extremely hard, wear-resistant material. As a composite material made of tungsten and carbon, it offers a unique combination of durability and strength. When the seat of a ball valve is metallic, metal-to-metal contact occurs between the ball and seat during opening and closing, leading to corrosive and eroding conditions. Consequently, tungsten carbide is commonly applied to the contact surfaces of the ball and metal seat ball valves. One of the benefits of tungsten carbide is its reduced maintenance requirements. Its durability reduces the need for frequent replacements, which in turn reduces downtime and maintenance costs. It enhances the life expectancy of valve components in a way that components coated with tungsten carbide are less likely to be damaged, thus extending their lifespan. Since tungsten carbide is extremely strong, it is an ideal material for valves used in challenging environments, such as those found in oil and gas, and petrochemical industries.

U

Ultimet

1. Definition

Ultimet is a cobalt-based alloy known for its exceptional combination of wear, corrosion, and galling resistance. It contains a balanced composition of cobalt, chromium, molybdenum, and nickel, making it highly resistant to harsh environments, including high temperatures and chemical exposure. Ultimet is often used in applications such as valve components, pump parts, and chemical processing equipment. Its unique properties make it an excellent alternative to traditional materials like stainless steels or Stellite in both corrosive and erosive environments. *Ultimet* offers several advantages over Stellite, particularly in applications where enhanced corrosion resistance and weldability are critical such as chloride-rich conditions, where Stellite may be more susceptible to pitting or crevice corrosion.

2. Application

In butterfly or ball valves handling seawater, Ultimet can be used as a hardfacing material for valve seats and disks or balls, where its excellent corrosion and wear resistance ensure reliable performance over extended periods. Stellite, while highly wear-resistant, may suffer from localized corrosion in seawater, leading to reduced service life. Ultimet's better weldability also simplifies maintenance and repairs, making it the material of choice for critical seawater valve components.

Unified Numbering System

A UNS is a standardized system for identifying materials used in manufacturing and engineering, particularly metals and alloys. Based on the composition and properties of materials, it assigns a unique alphanumeric code which simplifies communication and ensures consistency across industries globally. A UNS number consists of a letter followed by five digits. The letter indicates the material category (e.g., stainless steel, copper alloys). The digits indicate the chemical composition or other unique characteristics of the material. Stainless steel grade

316, for example, is designated UNS S31600. It is used in engineering drawings, material procurement, and quality assurance documents like NDT. This ensures that all parties (suppliers, engineers, fabricators) are referring to the same material.

Uniform Corrosion

Uniform corrosion is a type of corrosion that occurs evenly across the entire exposed surface of a material, resulting in a consistent reduction in thickness over time. In reality, uniform corrosion is rare. In fact, most corrosion processes tend to be localized rather than perfectly uniform. While uniform corrosion is a commonly discussed concept in corrosion science, real-world environments introduce complexities that lead to nonuniform corrosion.

However, sweet or carbon dioxide corrosion is a kind of corrosion that is modeled as a uniform corrosion. This assumption makes this type of corrosion the most predictable form of corrosion, as it proceeds at a relatively uniform rate. As a result, this type of corrosion can be mitigated by calculating material loss and incorporating corrosion allowances in design.

V

Viton

The Viton brand of fluoroelastomer (a synthetic rubber) is widely used for valve seals and gaskets. Viton offers exceptional resistance to high temperatures, aggressive chemicals, and hydrocarbons, making it ideal for valves in chemical processing, oil and gas, and industrial systems. It can withstand up to 200°C but has limited performance in low temperatures.

W

Wear Resistance

Wear resistance is the ability of a material to withstand gradual material loss due to mechanical interaction, such as abrasion, erosion, or adhesion, during operation. High wear resistance is crucial for components exposed to repetitive motion, friction, or contact with abrasive fluids. Selection of wear resistance materials for such applications extends the service life, reduces maintenance, and ensures reliable performance.

1. **Materials with High Wear Resistance**
   - *Hard Metals:* Tungsten carbides, Stellite 6 and 21, are commonly used in applications requiring high wear resistance. Applying a layer of wear-resistant material (e.g., Stellite or tungsten carbide) on a component's surface to withstand high abrasion and erosion. This is common in valves and pumps.

- *Ceramics:* Ceramics such as silicon carbide exhibit exceptional hardness and wear resistance, making them ideal for high-abrasion environments. Ceramic lining that could involve using thin films of silicon carbide enhances surface wearing resistance.
- *Composites:* Fiber-reinforced polymers offer a balance of wear resistance and lightweight properties.
- *Surface-Treated Metals:* Nitride steels gain wear resistance through surface-hardening processes.

Weldability

Weldability refers to the ability of a material to be joined or fused with another material using a welding process while achieving a strong, defect-free, and durable weld. It indicates how easily a material can undergo welding without issues like cracking, porosity, or loss of mechanical properties. Weldability is a critical consideration in material selection and design for structures, pipelines, pressure vessels, and other welded components. The chemical makeup of a material, especially the content of elements like carbon, sulfur, and phosphorus, plays a critical role in weldability. Low-carbon content reduces the hardenability of the steel, making it less likely to form brittle microstructures like martensite during cooling and causes high weldability. High-carbon content increases the hardenability of the steel, making it prone to forming martensite or other brittle phases when cooled rapidly after welding. These steels are more susceptible to HIC (cold cracking) and other weld defects which reduces weldability. Carbon equivalent less than 0.4 indicates a good weldability. More information about carbon equivalent is provided earlier in this chapter.

Weld Decay

It is a form of IGC specifically localized around the *HAZ* of a weld in unstabilized stainless steels like 304 or 316. It occurs because the high temperatures in the HAZ cause sensitization (precipitation of chromium carbides), leaving grain boundary areas chromium-depleted and prone to attack in corrosive environments.

Welding

1. Definition

Welding is a fabrication process that joins two or more materials, typically metals or thermoplastics, by melting their surfaces and fusing them together, often with the addition of a filler material. This process can create strong, permanent bonds and is commonly used in industries like construction, manufacturing, and oil and gas.

2. Common Types

The most common welding processes include *metal inert gas* (MIG), *tungsten inert gas* (TIG), and *shielded metal arc welding* (SMAW), each suitable for different applications and materials. *MIG welding*, also known as gas metal arc welding (GMAW), uses a continuous wire electrode fed through a welding gun while a shielding gas protects the weld pool, making it efficient for large-scale, high-speed welding. *TIG welding*, or gas tungsten arc welding (GTAW), uses a nonconsumable tungsten electrode and provides precise, high-quality welds, ideal for thin materials and critical applications like aerospace or stainless steel. *SMAW*, often called stick welding, employs a consumable electrode coated with flux, which creates a protective slag, making it versatile and suitable for outdoor or less controlled environments. These methods vary in complexity, cost, and the quality of welds they produce, catering to a wide range of industrial needs.

Work Hardening

Work hardening, also known as strain hardening, is the process by which a material becomes stronger and harder due to plastic deformation caused by mechanical stress. For example, in valves, work hardening can occur in stainless steel components like valve stems.

Y

Yield Strength

Yield strength is the stress at which a material begins to deform plastically, meaning it will not return to its original shape once the stress is removed. It represents the point where elastic behavior ends and permanent deformation starts. Yield strength is a critical property in material and mechanical engineering, as it determines the maximum load a material can withstand without undergoing irreversible damage (plastic deformation). For example, in valve materials, the yield strength ensures the components can handle operating pressures without distortion, making it a key factor in material selection and design for safety and durability.

QUESTIONS AND ANSWERS

1. Which of the following materials is commonly used for valves in refineries and petrochemical plants?
   A. Cast iron
   B. Carbon steel
   C. Brass
   D. Polyethylene

   Answer: Option B is the correct answer.

2. What is the primary alloying element in stainless steels that provides corrosion resistance?
   A. Nickel
   B. Molybdenum
   C. Carbon
   D. Chromium

   Answer: Option D is the correct answer.

3. In an offshore project, the valve stem was fabricated from Inconel 718, while the stem key was made from carbon steel. The stem key serves as a critical mechanical component that connects the valve stem to its actuator or handwheel, ensuring effective torque transmission. It also functions as an indicator of the valve's internal ball or plug position when the actuator is removed, facilitating proper alignment during operation or maintenance. When the valve actuator was removed, both the Inconel 718 stem and the carbon steel stem key were exposed to the corrosive offshore environment. Under these conditions, what kind of corrosion could happen?
   A. Galvanic corrosion
   B. Pitting corrosion
   C. Sour corrosion
   D. Both options A and B are correct.

   Answer: Option D is the correct answer.

4. Which valve material is most suitable for handling hydrogen sulfide (H_2S) in sour service?
   A. Monel
   B. Aluminum
   C. Low alloy steel
   D. Carbon steel

   Answer: Option A is the correct answer.

5. What is the main advantage of using Stellite as a valve material?
   A. High strength
   B. Low cost
   C. Exceptional wear resistance
   D. Lightweight properties

   Answer: Option C is the correct answer.

6. Which material is used for valves in cryogenic applications?
   A. Cast iron
   B. Carbon steel
   C. Inconel 625
   D. Brass

   Answer: Option C is the correct answer.

7. Which type of corrosion occurs when two dissimilar metals are in electrical contact in a corrosive environment?
   A. Pitting corrosion
   B. Crevice corrosion
   C. Intergranular corrosion
   D. Galvanic corrosion

   Answer: Option D is the correct answer.

8. Which of the following is the most effective method for preventing pitting corrosion in stainless steel valves?
   A. Applying a zinc coating
   B. Increasing molybdenum content in the alloy
   C. Painting the surface
   D. Using carbon steel instead of stainless steel

   Answer: Zinc coatings primarily protect steel through cathodic protection and are not designed to prevent pitting corrosion on stainless steel. Stainless steel relies on its passive oxide layer for corrosion resistance, and a zinc coating can interfere with this natural mechanism. Moreover, in chloride-rich environments, zinc can degrade and develop pits itself, providing only limited protection. Thus, option A is not correct. Option B is the correct answer. Increasing the molybdenum content in stainless steel is the most effective method for preventing pitting corrosion because molybdenum significantly improves the alloy's resistance to localized attacks, especially in chloride-rich environments. Pitting corrosion occurs when protective oxide layers on stainless steel are compromised, allowing aggressive ions like chlorides to penetrate and attack the underlying metal. Molybdenum enhances the stability and repairability of the passive oxide layer, making it more resistant to such attacks. For example, stainless steel grades like 316L, which contain around 2–3% molybdenum, exhibit much higher pitting resistance compared to 304 stainless steel, which lacks molybdenum. This makes molybdenum-enriched alloys ideal for applications in marine, chemical processing, and other harsh environments. It should be noted that adding Mo increases PREN. Option C is not correct as paint acts as a barrier to protect against general corrosion but is not highly effective at preventing pitting, especially in aggressive environments like seawater. Any minor defect or scratch in the paint exposes the stainless steel, making it susceptible to pitting corrosion. Option D is not correct either because carbon steel is much more prone to corrosion, including pitting, in environments with chlorides or other aggressive agents.

9. Which of the following standards specifically addresses material selection for H_2S-containing environments?
   A. ASME Section VIII
   B. ISO 15156
   C. API 598
   D. ASTM A216

   Answer: Option B is the correct answer.

10. What is the purpose of passivation in valve materials?
    A. Increase strength
    B. Improve weldability
    C. Enhance corrosion resistance by forming a protective oxide layer
    D. Reduce the weight of the material

    Answer: Option C is the correct answer.

BIBLIOGRAPHY

American Petroleum Institute. (2015). Metallic and Non-Metallic Materials for Valves. API 20A. API: Washington, DC.

American Petroleum Institute. (2021). Specification for Wellhead and Christmas Tree Equipment. API 6A, 21st edition. API: Washington, DC.

American Society for Testing and Materials. (2019). Standard Specification for Nickel and Nickel Alloy Forgings. ASTM B564-19. ASTM: West Conshohocken, PA.

American Society for Testing and Materials. (2020). Standard Specification for Castings, Austenitic, for Pressure-Containing Parts. ASTM A351/A351M-20. ASTM: West Conshohocken, PA.

American Society for Testing and Materials. (2021). Standard Specification for Carbon Steel Forgings for Piping Applications. ASTM A105/A105M-21. ASTM: West Conshohocken, PA.

American Society for Testing and Materials. (2021). Standard Specification for Alloy-Steel and Stainless-Steel Bolting Materials for High-Temperature or High-Pressure Service and Other Special Purpose Applications. ASTM A193/A193M-21. ASTM: West Conshohocken, PA.

American Society for Testing and Materials. (2021). Standard Specification for Chromium and Chromium–Nickel Stainless Steel Plate, Sheet, and Strip for Pressure Vessels and for General Applications. ASTM A240/A240M-21. ASTM: West Conshohocken, PA.

American Society for Testing and Materials. (2021). Standard Test Methods and Definitions for Mechanical Testing of Steel Products. ASTM A370-21. ASTM: West Conshohocken, PA.

American Society for Testing and Materials. (2021). Standard Test Methods for Pitting and Crevice Corrosion Resistance of Stainless Steels and Related Alloys by the Use of Ferric Chloride Solution. ASTM G48-21. ASTM: West Conshohocken, PA.

American Society for Testing and Materials. (2022). Standard Specification for Steel Castings, Carbon, Suitable for Fusion Welding, for High-Temperature Service. ASTM A216/A216M-22. ASTM: West Conshohocken, PA.

American Society for Testing and Materials. (2022). Standard Specification for Steel Castings, Martensitic Stainless and Alloy, for Pressure-Containing Parts, Suitable for High-Temperature Service. ASTM A217/A217M-22. ASTM: West Conshohocken, PA.

American Society for Testing and Materials. (2022). Standard Test Methods for Tension Testing of Metallic Materials. ASTM E8/E8M-22. ASTM: West Conshohocken, PA.

Bodaghi, M., & Zolfagharian, A. (2022). Smart Materials in Additive Manufacturing, Volume 1: 4D Printing Principles and Fabrication. Elsevier: London, UK.

De Kanter, A. J., Jongsma, K. R., Bouten, C. V. C., & Bredenoord, A. L. (2023). How smart are smart materials? A conceptual and ethical analysis of smart lifelike materials for the design of regenerative valve implants. Science and Engineering Ethics, 29(5), 1–18. doi: 10.1007/s11948-023-00453-1.

International Organization for Standardization. (2015). Petroleum and Natural Gas Industries: Materials for Use in H_2S-Containing Environments in Oil and Gas Production. ISO 15156-1, -2, -3. ISO: Geneva, Switzerland.

Jawaid, M., Khan, A., & Khan, M. M. (2020). Smart Coatings: Fundamentals, Preparation, and Applications. Wiley-VCH: Germany.

Makhlouf, A. S. H. et al. (2016). Handbook of Materials Failure Analysis with Case Studies from the Oil and Gas Industry. 1st edition. Elsevier: Oxford, UK.

NACE International. (2015). Petroleum and Natural Gas Industries: Materials for Use in H_2S-Containing Environments in Oil and Gas Production. NACE MR0175/ISO 15156. NACE: Houston, TX.

NACE International. (2016). Materials Selection for Oil and Gas Corrosion Environments. NACE MR0103. NACE: Houston, TX.

Nesbitt, B. (2007). Handbook of Valves and Actuators: Valves Manual International. 1st edition. Elsevier: Oxford, UK.

NORSOK M-001. (2002). Materials Selection. 3rd edition. Lysaker: Norway

NORSOK M-501. (2012). Surface Preparation and Protective Coating. 6th edition. Lysaker: Norway

NORSOK M-710. (2014). Qualification of Non-Metallic Sealing Materials and Manufacturers. 3rd edition. Lysaker: Norway

Okeremi et al. (2009). Preventing pitting and crevice corrosion of offshore stainless steel tubing. *World Oil Magazine*, pp. 4.73–4.80.

Raja, V.S., & Shoji, T. (2011). Stress Corrosion Cracking: Theory and Practice. Woodhead Publishing Ltd.: Sawston, UK.

Rorvik, G. et al. (2014). Fasteners in subsea applications: End users experiences and requirements. International Conference of Offshore Mechanics and Artic Engineering. American Society of Mechanical Engineers. Paper # OMAE2014-24520.

Skousen, P. L. (2011). Valve Handbook. 3rd edition. McGraw-Hill Professional: New York, NY.

Smith, P., & Zappe, R. W. (2004). Valve Selection Handbook: Engineering Fundamentals for Selecting the Right Valve Design for Every Industrial Flow Application. 5th edition. Elsevier: Amsterdam, The Netherlands.

Sotoodeh, K. (2018). Valve failures, analysis and solutions. Valve World Magazine, 23(11), 48–52.

Sotoodeh, K. (2020). Optimized material selection for subsea valves to prevent failure and improve reliability. Life Cycle Reliability and Safety Engineering, 10, 173–182.

Sotoodeh, K. (2021). A Practical Guide to Piping and Valves for the Oil and Gas Industry. Gulf Professional Publishing: Amsterdam, The Netherlands.

Sotoodeh, K. (2022). Coating Application for Piping, Valves and Actuators in Offshore Oil and Gas Industry. CRC Press: Boca Raton, FL.

Speight, J.G. (2014). Oil and Gas Corrosion Prevention: From Surface Facilities to Refineries. Elsevier Science: London, UK.

Winnik, S. (2016). Corrosion under Insulation (CUI) Guidelines. 2nd edition. European Federation of Corrosion. Elsevier: London, UK.

Yadav, L., Sihmar, A., Kumar, S., Dhaiya, H., & Vishwakarma, R. (2024). Review of nano-based smart coatings for corrosion mitigation: Mechanisms, performance, and future prospects. Environmental Science and Pollution Research. doi: 10.1007/s11356-024-33234-9.

5 Valve Design and Standards

A

Actuated Valves

Actuated valve designs incorporate an actuator, a mechanical or electrical device that automates valve operation. The valve can be opened, closed, and moved completely by an actuator rather than by the hand force of the operator. Different types of actuators such as pneumatic, hydraulic, electrical, gas, electromechanical, and more exist. In Chapter 11, you will find more information about actuators. Figure 5.1 shows a ball valve equipped with a hydraulic actuator. The actuator control panel, located on the side, is responsible for controlling the flow of hydraulic oil into the actuator. Figure 5.2 illustrates a control valve to control the flow and it is equipped with a pneumatic diaphragm actuator.

Anti-blowout Stem

Under high-pressure conditions, an anti-blowout stem design prevents the valve stem from being ejected or blown out. As a result of this design feature, valve stems remain attached to the valve body and bonnet even when there is high internal pressure. This maintains the integrity of the valve and safeguarding personnel

FIGURE 5.1 A ball valve equipped with a hydraulic actuator. The actuator control panel, located on the side, regulates the flow of hydraulic oil into the actuator, enabling precise valve operation. (With permission: Shutterstock.)

DOI: 10.1201/9781003640967-5

FIGURE 5.2 A control valve used for flow control and equipped with a pneumatic diaphragm actuator. (With permission: Shutterstock.)

while performing maintenance on the valve. In order to prevent the stem from being ejected, a physical barrier or locking mechanism is incorporated into the design of the stem. Typically, a flange or step on the stem rests against the inside of the valve body or bonnet, which locks the stem in place and prevents upward movement. These types of applications may be appropriate for valves under high pressures or that require a high level of safety. The ball valve shown in Figure 5.3 illustrates the concept of an anti-blowout stem. Let us assume that the stem seal is

FIGURE 5.3 Illustration of an anti-blowout stem in a ball valve. The stem step located beneath the bonnet ensures operator safety by preventing stem ejection during stem seal change, even when the valve cover is removed. (With permission: Shutterstock.)

worn and needs to be replaced. In order to gain access to the stem seal, the operator must remove the cover from the top of the valve. As a result, it is important to ensure that the operator will not be subject to any risks of injury due to the ejection of the stem. The stem step is located under the bonnet, and the bonnet prevents the stem from being ejected out and injuring the operator.

Antistatic Design

Valve designers use the antistatic design concept to ensure the fire safety of their products. This concept refers to the measures taken during the design process to prevent static electricity from building up within a valve. As a result of friction between internal components, static electricity may accumulate inside the valve, which can result in a fire or explosion, especially if the fluid service in the valve is flammable. There is a possibility that this static electricity charge could ignite combustible fluids and pose a significant risk to the valve's safety and reliability. The most common *antistatic devices* or components are *springs* or grounding balls placed between a stem and a ball and/or between a stem and the body. By creating an electrical continuity between ball, stem, and body in a ball valve, any charge that builds up is directed away from the inner parts. Antistatic requirements are outlined in a number of industry standards, including API 6D (for pipeline valves) and ISO 10497 (fire testing standard). There are two antistatic springs shown in Figure 5.4—one is located between the stem and ball at a lower

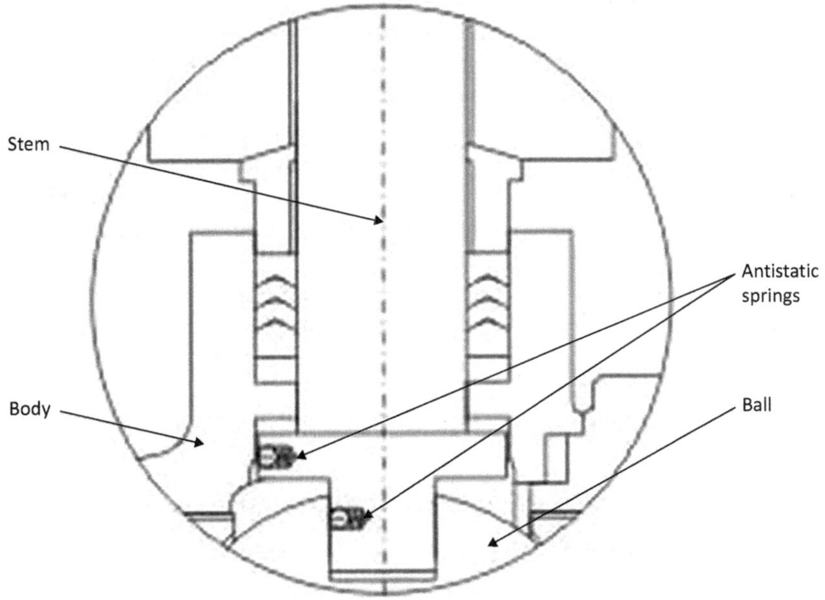

FIGURE 5.4 Application of antistatic springs in a ball valve to achieve antistatic and fire safe design.

level and the other is located between the stem and the body at a higher level. It should be noted that antistatic design is closely related to fire safety design. The concept of fire safety design encompasses a wide range of aspects of valve design, including antistatic design.

API (American Petroleum Institute)

API is a group of standards that are used in the oil and gas and renewable energy industries for a variety of equipment and components, including valves for industrial use. The API is one of the largest trade associations in the United States, with its headquarters located in Washington, DC. In its Energy Framework, API outlines its approach to supporting renewable and low-carbon energy initiatives. API plays an essential role in setting safety and operational guidelines for the petroleum and renewable industries.

Some of the most important API valve standards addressing valve design are listed as follows:

API 600: Steel Gate Valves—Flanged and Butt-Welding Ends, Bolted Bonnets
API 602: Gate, Globe, and Check Valves for Sizes DN 100 (NPS 4) and Smaller for the Petroleum and Natural Gas Industries
API 603: Corrosion-Resistant, Bolted Bonnet Gate Valves—Flanged and Butt-Welding Ends
API 608: Metal Ball Valves – Flanged, Threaded and Welding Ends
API 609: Butterfly Valves: Double-Flanged, Lug- and Wafer-Type
API 6D: Pipeline and Piping Valves
API 6DSS: Subsea Pipeline Valves
API 6A: Specification for Wellhead and Christmas Tree Equipment
API 594: Check Valves: Flanged, Lug, Wafer, and Butt-Welding Ends
API 623: Steel Globe Valves—Flanged and Butt-Welding Ends, Bolted Bonnets
API 624: Type Testing of Rising Stem Valves Equipped with Flexible Graphite Packing for Fugitive Emissions
API 641: Type Testing of Quarter-Turn Valves for Fugitive Emissions
API 607: Fire Test for Soft-Seated Quarter-Turn Valves (not applicable to valve design)
API 6FA: Specification for Fire Test for Valves
API 599: Metal Plug Valves—Flanged, Threaded, and Welding Ends

ASME (American Society of Mechanical Engineers)

There is a long history of ASME promoting art and science, as well as multidisciplinary engineering practices. It was founded in 1880 and is one of the oldest and most prestigious engineering societies in the world. The application of ASME is not only limited to mechanical engineering. In fact, ASME covers a wide range of engineering disciplines. ASME is widely known for developing internationally recognized codes and standards that help engineers to design, manufacture, construct, and maintain mechanical systems, including piping and valves. ASME's most famous code is the ASME Boiler and Pressure Vessel Code (BPVC), which specifies the design and construction requirements for boilers, pressure vessels, valves, and nuclear power plants. Moreover, ASME organizes seminars, conferences, and technical workshops in order to exchange knowledge by presenting research and innovations in the engineering field. Furthermore, it provides engineers with opportunities for professional development and training. ASME standards play a significant role in the design and improvement of safety for engineers, including those working on valves and pressure vessels. Some of the essential ASME codes and standards addressing industrial valves are highlighted as follows:

ASME B16 Series (Piping Components Standards)

1. **ASME B16.34**: Valves—Flanged, Threaded, and Welding End
2. **ASME B16.10**: Face-to-Face and End-to-End Dimensions of Valves
3. **ASME B16.5**: Pipe Flanges and Flanged Fittings
4. **ASME B16.11**: Forged Fittings, Socket-Welding, and Threaded
5. **ASME B16.47**: Large Diameter Steel Flanges

ASME BPVC (Boiler and Pressure Vessel Code)

6. **ASME BPVC Section I**: Rules for Construction of Power Boilers (Safety Valves and Pressure-Relief Valves)
7. **ASME BPVC Section VIII**: Rules for Construction of Pressure Vessels (Pressure-Relief Valves)
8. **ASME BPVC Section III**: Nuclear Facility Components (Valves for Nuclear Applications)

ASME PTC (Performance Test Codes)

9. **ASME PTC 25:** Pressure Relief Devices

ASME B31 Series (Piping Standards)

10. **ASME B31.1**: Power Piping
11. **ASME B31.3**: Process Piping
12. **ASME B31.4**: Pipeline Transportation Systems for Liquids and Slurries
13. **ASME B31.8**: Gas Transmission and Distribution Piping Systems

ASME Pressure Class or Pressure Rating

A valve's ASME Class refers to its pressure–temperature ratings defined by ASME B16.34, which specifies the maximum allowable working pressure (MAWP) for valves in different temperature ranges. The ASME pressure class designations of 150, 300, 600, 900, 1500, and 2500 indicate the valve's ability to withstand increasing pressure levels, with higher classes designed for higher pressure levels. Valve pressure classification is determined by the type of valve material and the operating temperature, as well as the system pressure under which the valve operates. The Class 150 valve can handle lower pressures, while the Class 2500 valve can handle high pressures. Due to material strength reduction at high temperatures, each class's pressure capacity decreases as the temperature increases.

Class 150: Handles up to *285 psi (19.6 bar)*
Class 300: Handles up to *740 psi (51 bar)*
Class 600: Handles up to *1,480 psi (102 bar)*
Class 900: Handles up to *2,220 psi (153 bar)*
Class 1500: Handles up to *3,705 psi (255 bar)*
Class 2500: Handles up to *6,170 psi (425 bar)*

As a point of clarification, the pressure rating is different from the ASME pressure–temperature rating or classes since the pressure rating is a single value, while the ASME pressure classes represent a range of pressures and temperatures.

ASTM (American Society for Testing and Materials)

In the valve industry, ASTM standards are a set of technical guidelines and specifications published by the American Society for Testing and Materials. Throughout the world, these standards are widely recognized and widely used to define the quality of materials in valve design and testing procedures. ASTM standards are used in various industries such as oil and gas, petrochemical, and power generation. ASTM standards ensure that valves meet specific material quality and requirements. Various ASTM standards specify the requirements for material composition, mechanical properties, and heat treatment. Examples of valve ASTM material grades are ASTM A216 (for cast carbon steel valve bodies), ASTM A351 (for cast stainless steel valve bodies), and ASTM A105 (for forged carbon steel valve bodies). According to standards such as ASTM A370, valve materials are tested to determine their mechanical properties such as tensile strength, impact resistance, and hardness. ASTM standards play an important role in material quality, safety requirements, and standardization of materials globally.

B

Balanced Safety Valve

Balanced safety valves are designed to maintain consistent performance regardless of fluctuations in backpressure (the pressure at the valve's outlet). A feature such as this is essential in systems where there may be variations in backpressure as a result of piping configurations, downstream equipment, or other process conditions. In industries such as oil and gas, chemical processing, and power generation, balanced safety valves are commonly used to prevent overpressure without getting affected by the backpressure effect. High backpressure can affect the ability of standard safety valves to open at the set pressure and compromise their relief capacity when high backpressure pushes against the valve seat. Balanced safety valves are equipped with secondary pistons or bellows that counteract the effects of backpressure on the valve disk, allowing them to open and close consistently at the set pressure regardless of fluctuations in backpressure. In order to balance the forces acting on the valve disk, a bellows or piston is installed within the valve. As a result of the bellows, the process fluid is isolated from the spring chamber, allowing the valve to operate independent of backpressure.

Bellows Design (Safety Valve)

In the context of pressure relief valves, bellows safety valves are pressure relief valves equipped with flexible metal bellows to isolate valve internal components, especially spring mechanisms from backpressure and potentially corrosive process fluids. It is important to note that backpressure can change the set point and reliability of safety valve operation. The internal area of a safety valve covered with bellows is highlighted in Figure 5.5.

Bellows Seal Design

In terms of valve stem sealing, a bellows seal provides a highly reliable, leak-tight seal around the valve stem (between the stem and bonnet). This design effectively isolates the process of fluid flowing inside the valve from the external environment. This design is typically used for piping systems handling hazardous (e.g., toxic) fluids. Bellows are typically made of stainless steel or corrosion-resistant alloys, and they form a tight seal between the valve stem and bonnet. The bellows

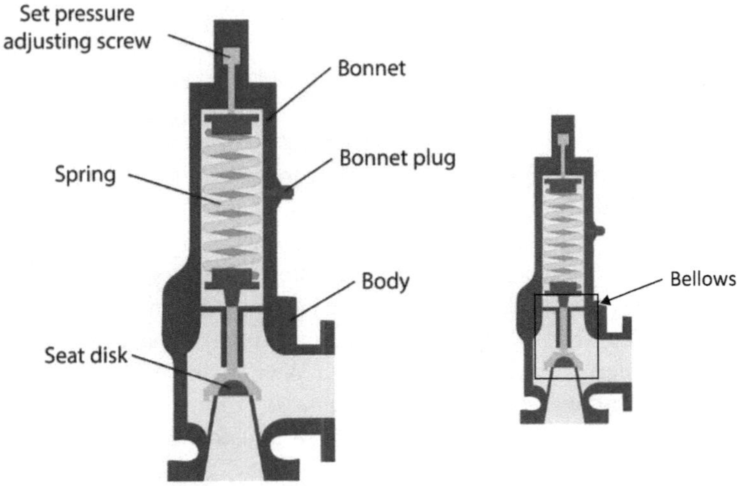

FIGURE 5.5 A safety valve with key components and highlighted area where bellows are installed. (With permission: Shutterstock).

in stem sealing have an accordion-like appearance, characterized by a series of thin-walled corrugations that allow for flexibility and expansion/contraction.

Bidirectional Seat

Bidirectional seat is same as "Double-Piston Effect Seat". More information is available in this chapter under section "Double-Piston Effect Seat."

Bidirectional Valve

Flow can be handled in both directions by bidirectional valves (such as standard ball valves, or gate valves). This means that they have no preferred flow direction during operation. Additionally, these valves can be connected from both sides to the piping by rotating the valve 180°. The opposite of bidirectional valves are *unidirectional valves.*

Blowout Proof Stem

Same as anti-blowout stem.

Body and Bonnet Joint Design

A variety of body and bonnet joint designs are used, including bolted, screwed, and welded joints.

Bonnet Design (safety Valve)

Safety valve bonnets, also called spring housings or spring casings, are the top parts of the valve that attach to the valve body through flanges. Figure 5.6 illustrates a design for a couple of safety valves with a *closed bonnet* at the top and an *open bonnet* at the bottom.

From the outside, the spring located inside the bonnet can be viewed due to the open bonnet design. An open bonnet design also exposes the spring to the atmosphere. When the valve is opened, fluid service or media is discharged through the hole in the bonnet into the environment. As a result of releasing fluid into the environment when the bonnet opens, and exposing the spring to the weather, open bonnet designs appear unsuitable for toxic and harmful services. Based on the application and the properties of the fluid being handled, open-bonnet safety valves or closed-bonnet safety valves should be selected. A safety valve with an open bonnet is typically used for high-temperature applications, such as steam systems, where heat dissipation is crucial to preventing spring overheating. In fact, an open bonnet design is suitable for nontoxic, noncorrosive fluids and environments in which the discharged fluid does not pose a threat to health or the environment. In contrast, closed bonnet safety valves are designed for handling hazardous, toxic, or corrosive fluids, since they prevent fluid or vapor from escaping into the atmosphere.

FIGURE 5.6 (a) A couple of closed bonnet safety valves. (b) An open bonnet. (With permission: Shutterstock.)

C

Cavity-Free Design

It is generally considered that cavities in ball and through conduit gate valves are weak points, as they can serve as a place where fluid can be trapped and overpressure can cause damage to the valve body and internals. Cavity-free valves are those with minimal or no internal cavities or void spaces where fluid, particles, or other contaminants can accumulate. It is possible to design the ball with precise tolerances in order to ensure that it fits tightly within the valve body, thereby reducing the cavity size. The fact that some valves, such as globe valves, wedge gate valves, and butterfly valves, are cavity-free is important to be aware of. More information about cavity in valve is given in Chapter 3.

CE Marking

The CE marking indicates that a product complies with the health, safety, and environmental protection standards set by the European Union for products sold within the European Economic Area (EEA). In order to obtain a CE certification, the manufacturer must declare that the product complies with all applicable EU directives and regulations, including the Pressure Equipment Directive (PED), through a conformity assessment process. Regarding product category and risk level, this assessment can vary from self-certification (low-risk equipment) to third-party verification by a Notified Body (high-risk equipment). In order to demonstrate compliance, the manufacturer must compile and maintain technical documentation (often referred to as a technical file) such as calculations, drawings, and material certificates. If the product is considered high-risk, such as high-pressure valves, Notified Bodies (independent organizations authorized by the EU member states) must inspect, and certify it. This ensures the product meets the required standards and validates the manufacturer's statement. The manufacturer then issues a declaration of conformity as proof and a document that the product is in compliance with the PED or EU directive.

Code

An industry code is a set of standards and regulations that are established by recognized organizations to ensure safety, reliability, and quality when it comes to designing, manufacturing, testing, and installing valves. Codes are essential for compliance in industries such as oil and gas, chemical processing, and power generation, where valves must meet specific requirements in order to function safely under various conditions. The valves industry is governed by a number of codes, such as ASME. As a result, valve codes serve as comprehensive guidelines for valve specifications, testing, and performance, helping manufacturers and end users to ensure that valves are safe, reliable, and comply with industry standards.

Compliance

In the valve industry, compliance with codes and standards refers to a valve meeting the requirements set forth by recognized industry standards and regulations. Valve compliance refers to ensuring that the design, materials, manufacturing, testing, and performance of valves are compliant with established safety, quality, and operational standards.

Concentric Butterfly Valve

Concentric butterfly valves do not have any offset between essential valve components, so they are referred to as zero-offset butterfly valves. To clarify, the valve disk is centered within the valve body, while the stem passes through the center of the disk. Concentric butterfly valve is commonly known as a rubber-lined valve due to its soft seat, which is typically made of rubber, PTFE, or other resilient materials. It is the eccentric butterfly valve that is the opposite design, which is described further in this chapter. The butterfly valve shown in Figure 5.20 is a concentric butterfly valve with no offset.

Cryogenic Valve Design

The use of cryogenic valves is restricted to extremely cold temperatures such as −150°C or even lower. In order to design the valve in accordance with this condition, multiple factors must be taken into consideration. In the first instance, there is the material used in the construction of the valve, which is typically austenitic stainless steel grades 304, 316, or nickel alloy, for example. Moreover, stem extensions are very common to keep the valve operator away from cryogenic fluid passing through the valve. Sealing materials must be suitable for cryogenic applications, which is why lip seals containing a layer of PTFE with a metallic spring inside are commonly used. In order to prevent the cryogenic fluid from being heated, the valve may be insulated to prevent the loss of cold temperature to the environment. There may be a reduction in size of valve internal components due to cryogenic service, which can cause leaks either internally or externally through the valve. By using finite element analysis (FEA), it is possible to ensure that shrinkage of valve components is compensated and does not lead to leakage. Additionally, it is important to include a pressure relief mechanism, such as a plug or safety valve, in order to release the cryogenic fluid if it becomes trapped in the cavity and is under pressure.

D

Data Sheet

A valve data sheet is a detailed document prepared by valve engineer, instrument or process engineer, that provides essential information and specifications for a valve. This document ensures that the valve meets the design, operational, safety, and compatibility requirements for its intended application. During engineering, procurement, and construction (EPC) projects, valve data sheets are commonly used to standardize valve selection and installation. The valve data sheet contains information such as valve size, pressure class, operating and design pressure and temperature, as well as valve components requirements such as materials for body, bonnet, stem, gaskets, and sealings.

Design Life

A valve's design life refers to the period of time it is expected to operate reliably and effectively under specified conditions before requiring major repairs or replacement. This is the estimated lifespan that a valve is engineered to achieve under specified operating conditions. A valve's lifespan is determined by factors such as its materials, construction, operating conditions, and intended application. A valve's design life is typically expressed in terms of years or cycles. High-quality industrial valves are typically designed to last for a number of decades if they are properly maintained. It is important to note that the design life of a valve and its service life are not exactly the same, although they are closely related. See valve service life further in Chapter 8 for more information.

Design Pressure

An engineered valve, or any component of a process system, is designed to withstand a maximum pressure under specified circumstances, which is known as design pressure. As a safety margin, the design pressure is the maximum pressure a valve can safely withstand, set above the operating pressure. Typically, valves are constructed using specific metallic materials and designs so that they can withstand design pressure safely. This allows the structure to withstand occasional high-stress events without compromising its structural integrity. However, it is the operating pressure that determines the valve's durability on a day-to-day basis. In valves and other components of systems, the relationship between design pressure and operating pressure is based on the margin of safety that design pressure provides over operating pressure. Operating pressure is the normal pressure at which a system or valve operates under typical conditions, whereas design pressure is the maximum pressure the valve is designed to handle safely, including occasional pressure surges or extreme conditions. In the event of unexpected pressure increases, the design pressure is set above the operating pressure to ensure reliable operation. As an example, the operating pressure in a valve could be 8 bar while the design pressure is 12 bar.

Design Temperature

The design temperature of a valve refers to the maximum temperature that the valve is designed to withstand safely during operation. It is the maximum temperature that the valve's materials and components can withstand without compromising structural integrity, performance, or safety. A valve's design temperature plays an important role in determining its suitability for various applications, particularly in systems that may experience occasional high temperatures. Design temperatures include a safety buffer above the operating temperature in order to accommodate occasional surges or abnormal operating conditions. By providing this safety margin, the valve is capable of handling temporary temperature increases without deterioration or failure. The valve in a steam system has an operating temperature of 180°C and a design temperature of 250°C. A safety buffer is provided by setting the system design temperature (250°C) higher than the operating temperature to provide protection against unexpected temperature spikes or system upsets.

Differential Pressure

The differential pressure in valve design refers to the difference in pressure between the upstream and downstream sides of the valve. Flow control, pressure drop, energy efficiency, and the size of the valve operator (e.g., actuator) are affected by this parameter. The value of this parameter affects valve sizing, actuator design, and selection, as well as operational issues such as cavitation, noise, and vibration. A higher pressure drop affects the force that is required for actuators to operate valves because it becomes more difficult to operate the valves under higher pressure drops. There is a possibility that a high-pressure drop could affect the opening and closing of a check valve, causing noise and chattering. There are several issues that can arise due to high differential pressure, including cavitation (formation of vapor bubbles in liquids), noise, vibration, and erosion of globe valve components. In the case of pressure relief valves, differential pressure is considered a good sign, since it triggers the release of excess pressure. However, in the case of isolation valves, differential pressure is not considered a good sign, since it limits the isolation valve's ability to achieve tight shutoff. For valve design to be successful, differential pressure must be properly evaluated and managed.

Dimension Nominal (DN)

A DN is a standardized size designation used in piping and valves to indicate the size of the pipe or valve in millimeters. DN provides a simple, consistent method of specifying sizes among different piping components, regardless of their actual dimensions or wall thickness. A pipe size in millimeters is represented by DN. DN50, for example, refers to a pipe or valve with a nominal size of 50 mm. For regions using the Imperial System, NPS (nominal pipe size) is the equivalent size designation. As an example, DN50 is roughly equivalent to NPS 2. Examples of DN sizes are given in Table 5.1.

DIN

DIN (Deutsches Institut für Normung) refers to a set of German and European engineering standards containing specifications for valve dimensions, materials,

TABLE 5.1
Examples of DN Sizes

DN Size	Approximate NPS Equivalent	Typical Size (mm)
DN 15	NPS 1/2″	~15
DN 25	NPS 1″	~25
DN 50	NPS 2″	~50
DN 100	NPS 4″	~100

performance, and testing. As a result, DIN standards are widely used in Europe and other regions. They are widely recognized for their rigor in defining quality, reliability, and safety requirements for valves in a wide range of industries, including oil and gas, water treatment, and chemical processing.

Double Block and Bleed (DBB)

A DBB valve configuration (see Figure 5.7) is used to achieve secure, leak-proof isolation in piping systems, often for safety, testing, or maintenance purposes. A DBB setup consists of two isolation valves in sequence with a small bleed valve positioned between them. By using this design, a more reliable seal can be achieved. The DBB configuration includes two main valves (either ball, gate, or plug valves) that work together to isolate the upstream and downstream sections of the piping. Both isolation valves must be closed to create a double barrier against the flow. A small bleed valve is placed between the two isolation valves. If both main valves are closed, the bleed valve can be opened to release trapped pressure or fluid between the two barriers. An example of a bleed valve would be a needle valve. Opening the bleed valve can detect any leakage past the primary isolation valve. This provides a safety check and allows prompt action if either valve fails to isolate completely. DBB valves provide more reliable isolation than a single valve by adding an extra redundant valve.

Double Isolation and Bleed (DIB)

Chapter 10 provides more information about DIB under the heading "Double Isolation and Bleed (DIB)."

FIGURE 5.7 A DBB design including two block valves and a bleeder between.

Double-Piston Effect Seat

In contrast to self-relieving seats (also called single-piston effect seats) that release pressure into the valve cavity, double-piston effect seats do not release the overpressure in the valve cavity. In the case of a valve with one DPE seat and one SR seat, the cavity pressure will be released through the SR seat. In the event that there are two DPE seats on a valve, a safety plug or a pressure safety valve should be installed on the valve cavity in order to release the overpressure. Typically, DPE seats are used in ball and conduit gate valves.

Double-Offset Butterfly Valve

More information about double offset butterfly valve is provided in this chapter under section "Eccentric Butterfly Valve."

E

Eccentric Butterfly Valve

Eccentric butterfly valves are butterfly valves that have an offset stem to the disk (first offset) and an offset disk to the body (second offset), which results in a valve disk that is not located in the center of the valve body or the flow path. A third offset is achieved by conical seating surfaces. In the case of a butterfly valve that only has the first two offsets, it is referred to as a double-offset butterfly valve. Butterfly valves with all three offsets are referred to as triple-offset butterfly valves. Unlike double- and triple-offset designs, single offsets (stem offset to valve disk) are not commonly used in the oil and gas industry.

End Connections

The valve end connection refers to the method by which a valve is connected to the adjoining pipes or equipment in a piping system. A valve end connection must be selected in such a manner as to ensure a secure, leak-free connection. A *flanged end* connection is the most common type of valve end connection, in which the valve flange is bolted to the corresponding pipe flange in order to make installation and maintenance easier. A *threaded end* is used on smaller valves with a size of 2″ or less, as well as utility services such as air and water in the low-pressure class for securing directly to pipe threads (NPT or BSP). For large dimensions above 2″, *butt welds* (smooth, beveled edges) are available. For smaller sizes, *socket welds* (recessed areas for pipe insertion before welding) are available. For small valves where fittings compress around the pipe, *compression fitting ends* are used; *wafer ends* and *lug ends* are designed for lightweight valves sandwiched between two flanges, such as swing check and butterfly valves. As an alternative to standard ASME flanges, *hubs and clamps* (*mechanical joints*) are commonly used in offshore piping for high-pressure class piping to save weight

FIGURE 5.8 A ball valve with a buttweld end that is attached to a meeting pipe by welding.

and space. Figure 5.8 illustrates a ball valve with a buttweld end that is attached to a meeting pipe by welding.

End-to-End Dimension

As a measure of distance between the ends of the valve where it will be welded to the pipeline, the end-to-end dimension is typically used for butt-welded or socket-welded valves. In order to ensure consistency and interchangeability between different manufacturers, this dimension has been standardized to allow for easy installation and replacement in piping systems. *Face-to-Face Dimension* is used when referring to flanged valves, indicating the distance between the gasket contact surfaces (flange faces). Flanged valves face-to-face are standardized by norms such as ASME B16.10 to ensure consistency.

Extended Bonnet

An extended bonnet is a feature of a valve in which the valve's bonnet (the component that houses the stem packing and seals) is lengthened to provide more distance between the valve's body and fluid flowing in the valve body and its actuator or stem. Specifically, this extension is designed to protect the valve stem and packing from extreme operating conditions, such as high temperatures, cryogenic temperatures, or corrosive environments.

Extended Stem

Extended stem is explained in this chapter under the heading "Stem Extension."

F

Face-to-Face Dimension

For more information, refer to "End to End dimension."

Fire-Safe Design

The fire-safe valve is specially designed to maintain valve functionality and prevent or minimize leakage in case of a fire. For valves that handle flammable or hazardous fluids, fire test design is generally a top priority. In the event of a fire, leakage of hazardous or flammable fluids can quickly escalate the fire. For the design of a fire-safe valve, there are several important elements to consider, including using firebox around the valve (see Figure 5.9), the primary and secondary seals, the use of graphite seals, which are more resistant to fire and high temperatures, and the use of antistatic springs. A double sealing system may consist of a primary seal made of soft materials such as PTFE, which has greater seal-ability than graphite or metal seals, but is not fire-resistant, and a secondary seal made from metallic or graphite materials which remains functional during a fire. As a result of burnout or degradation of the primary seals, secondary seals

FIGURE 5.9 Designed around valves, fireboxes serve the primary purpose of protecting valves and ensuring that they continue to function during a fire. (With permission: Shutterstock.)

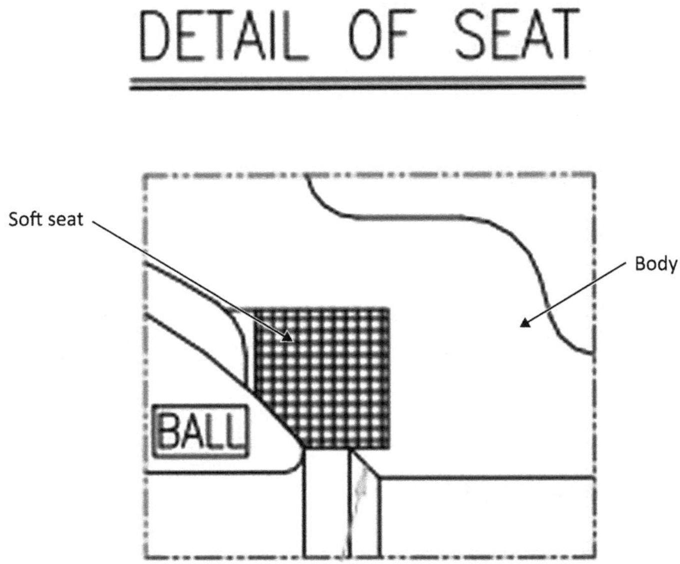

FIGURE 5.10 As part of the fire-safe design of the ball valve seat, if the soft seat is removed in the event of a fire, a secondary metal-to-metal seal is achieved between the valve ball and the body of the valve.

prevent leakage. There is an antistatic device installed between the stem and the closure member and/or the stem and body in order to provide a path for the discharge of static electricity. As can be seen in Figure 5.10, the seat in a ball valve has been designed in such a way that if, in the event of a fire, the soft seat melts and is removed, the secondary metal-to-metal seal between the ball and the body is achieved.

Flange Face

In valves, a flange face refers to the surface where the flange connects to another flange, usually on a pipe, fitting, or another valve. An important attribute of the flange face is that it is related to the sealing surface of the flange, which is vital for creating a seal to prevent leaks in the connection. When combined with a gasket and an appropriate bolting process, the flange face ensures a tight and reliable seal. There are different types of flange faces, such as flat faces, raised faces, ring-type joints (RTJs), tongue and groove faces, and male and female faces.

Flexible Wedge

A flexible wedge is a type of gate valve disk designed with a small cut or groove in the middle, allowing for a degree of flexibility. This flexibility enables the wedge to adapt to thermal expansion and contraction that may occur due to high temperatures. Without this flexibility, thermal expansion could lead to excessive stress on the valve body and internals, potentially causing jamming, deformation, or damage. In fact, flexible wedge prevents binding or sticking of the wedge or disk within the valve body due to thermal expansion of components. There is a weakness in this design in terms of corrosion due to the possibility of fluid being trapped within the groove of the wedge.

Floating Ball

As the name implies, a floating ball is a design in which the ball is not fixed or supported by a trunnion (a structure such as a flange or plate that provides support). Instead, the ball is held in place by two valve seats located on either side. As a result of this design, the ball is free to move slightly within the valve body. A floating ball valve (see Figure 5.11), typically used for smaller valve sizes and lower pressure classes, is a cheaper design than a trunnion ball design due to the lack of trunnion support. In general, floating ball valves are not used for large valves because large valves require a greater amount of force to operate due to their larger and heavier stem and ball. As a result, a trunnion valve would make sense for such valves since the trunnion can be used to control the additional movement of the ball, which may cause more friction between the ball and the seat. For large-size ball valves typically greater than 6″, using a floating ball requires the operator to exert considerable force, resulting in a very expensive operator, such as a gearbox or actuator. More information about trunnion ball is provided further in this chapter.

FIGURE 5.11 An illustration of a floating ball valve with a floating ball without trunnion support. (With permission: Shutterstock.)

Flow Coefficient

1. Flow Coefficient

Valve coefficients (often represented as C_v for imperial units or K_v for metric units) measure the capacity of a valve to allow fluid to flow. It indicates how much fluid (usually water) can pass through a valve at a specific pressure drop. Using this coefficient, engineers are able to determine the ideal valve size and type for a particular application, ensuring optimal performance and control. In the case of a valve, C_v is the flow rate measured in gallons per minute (GPM) at a pressure drop of 1 psi across the valve at a temperature of 60°F. For instance, a valve with a C_v of 10 will allow 10 GPM of water to flow through with a 1 psi drop. K_v is a similar quantity, but uses metric units. In cubic meters per hour, it indicates the flow rate of water at a pressure drop of 1 bar across the valve.

The flow coefficient (C_v) for liquids is calculated from Equation 5.1.

Equation 5.1: Flow Coefficient (C_v) Calculation for Liquids

$$C_v = Q\sqrt{\frac{\text{SG}}{\Delta P}}$$

where

C_v: Flow coefficient $\left(\frac{\text{US gal per minute (gpm)}}{\text{psi}}\right)$
Q: Flow rate $\left(\frac{\text{US gal}}{\text{min}}\right)$
SG: Fluid specific gravity (dimensionless) calculated by Equation 5.2
$\varnothing P$: Differential pressure across the valve (pound per square inch (psi)).

Note: Water specific gravity is equal to 1.

Equation 5.2: Specific Gravity (SG) Calculation

$$\text{SG} = \frac{\rho_{\text{fluid}}}{\rho_{\text{water}}}$$

where

ρ_{fluid}: Density of the fluid passing through the valve
ρ_{water}: Water density equal to $1\frac{\text{g}}{\text{cm}^3}$ or $1{,}000\frac{\text{kg}}{\text{m}^3}$.

Fluid Service

Fluid service refers to the classification or description of a fluid (liquid or gas) that flows through a piping system, valve, or other equipment, as well as the associated conditions under which it operates. It is a key concept in engineering and piping design, ensuring that the materials, components, and design of the system are suitable for the fluid's properties and operating conditions.

Full Bore

This type of valve design has an internal diameter that is approximately equal to the internal diameter of the pipe to which it is connected. By using this design, flow restrictions are minimized and pressure drops across the valve are reduced. A reduced bore design is the opposite of a large bore design. Some valves are designed with full bores, such as ball valves and valves with through conduit gates, while butterfly valves are always designed with reduced bores. As a general rule, full-bore valves are more expensive than reduced-bore valves in the same size and pressure class because the internal components of the valve, such as the ball and gate, are larger. In general, full-bore valves are used in four situations: First, if the pressure drop is of concern to the process engineer. Second, ball valves should be installed before and after pressure safety or relief valves in order to facilitate the rapid release of overpressure fluid through the piping. Third, in a pipeline where a piping inspection gadget (PIG) is being used for inspection or cleaning, the full bore must be utilized in order to facilitate the passage of the PIG. Finally, small valves of 2″ or less must have full bores in order to prevent pressure loss. In Figure 5.12, a full-bore ball valve is shown where the bore of the valve is the same at both ends of the valve as it connects to the mating flange.

FIGURE 5.12 A full-bore ball valve.

Fugitive Emission Prevention Design

The design of valves for the prevention of fugitive emissions must minimize the unintentional release of gases or fluids into the atmosphere, especially hazardous services (e.g., toxic and flammable). The use of low-emission packing (stem sealing) arrangements and materials, such as graphite or PTFE, is essential to minimize leakage. In order to minimize stem leakage, it is important to certify the packing materials according to fugitive emission standards, such as API 622. The sealing or packing of the valve stem is an important factor because leakage from the stem directly enters the surrounding environment. In addition, bellows seals are essential for applications that require near-zero emissions, since they create a tight seal around the valve stem. The live-loaded packing systems described in this chapter should be used to apply constant force to the stem sealing to improve seal performance under various operating conditions. Extra sealing around the stem, such as double or triple O-rings, provides an additional layer of protection. A corrosion-resistant alloy (CRA) should be selected for seals and valve parts, especially the body and bonnet, in order to ensure their durability under harsh conditions. As a result of compliance with testing standards like API 624 and ISO 15848, the valve is certified to meet stringent emission control requirements.

G

Gear Ratio

There is a term known as *gear ratio* that refers to the gearbox installed on the valve to facilitate the operation of the valve. Essentially, gear ratio refers to the number of teeth on two meshing gears. It is responsible for determining how torque and speed are transferred between gears in a mechanical system. For example, if a driving gear has ten teeth and it drives a gear with 20 teeth, then the gear ratio is 1:2. As a result, the driving gear will rotate twice for every rotation of the driven gear. This will increase torque twice while reducing speed. The higher the gear ratio, the greater the torque, but the lower the speed; whereas the lower the gear ratio, the greater the speed, but the less torque.

Gearbox-operated Design

Manually operated valves may be operated by levers or handwheels in combination with gearboxes. For large valves and/or high-pressure classes, gearboxes are typically used in addition to handwheels for valve operation. To operate these valves, the operator must apply a great deal of force, which cannot be accomplished without using a gearbox and simply rotating the handwheel. To put it simply, a gear-operated design refers to a valve operation mechanism that utilizes a gearbox to facilitate valve opening and closing. A gearbox-operated ball valve with handwheels is shown in Figure 5.13.

FIGURE 5.13 Gearbox operated ball valve installed on the piping system in an offshore platform. (With permission: Shutterstock.)

Globe Valve Types Design

During the selection of globe valves, three types of valve designs are considered. The T-type design is the standard design in which the flow is rotated around two 90° angles, causing a high-pressure drop. As a result, a Y-type globe design may be used as an alternative in which the flow rotates two 45° and there is a much smaller pressure drop. A Y-TYPE globe valve may be the preferred choice in applications where pressure drop is an issue. Last but not least, there is the angle globe design, whereby the globe valve provides a 90° rotation of the fluid between the inlet and outlet ports. As a result of the angle globe design, it is no longer necessary to have a 90° elbow to change the flow direction after the globe valve. Chapter 2 provides additional information regarding each of these globe valve types or designs.

Guided Disk Globe Valve

A guided disk globe valve is a type of globe valve in which the disk (the flow control element) is guided during its movement in order to ensure precise alignment and sealing with the valve seat. Through the use of this design, the valve performs better, particularly in demanding (e.g., corrosive and erosive) applications, as wear and misalignment are minimized. Guided disks are fitted with guides (metal or nonmetallic) that slide along a path to ensure the disk remains perfectly aligned throughout operation. The guides are incorporated into the valve body or bonnet and prevent the disk from moving laterally or wobbling. The guided disk globe valve is an advanced version of a standard globe valve, offering enhanced performance and reliability in demanding applications.

H

High-Strength (Yield) Bolting

In contrast to standard bolts, high strength or high yield bolts are designed to withstand greater amounts of stress, pressure, and load. These bolts are made from high-strength materials and are designed to resist deformation, fracture, and wear, making them ideal for use in demanding applications requiring high mechanical strength. In accordance with ASME B31.3, "Process piping code" bolts with a yield strength greater than 30 Ksi are referred to as high-yield or high-strength bolts. For high-pressure piping, high-yield bolts should be used. High-pressure piping systems have a pressure value of at least CL600 (CL stands for class) equal to a pressure nominal (PN) of 100 bar. High-yield bolting is another name for high-strength bolting.

I

Inside Screw and Yoke (IS&Y)

In an IS&Y design, the threads of the valve stem are located inside the valve body. The *yoke* is an external structure that connects the valve body to the actuator or handwheel. Because of disadvantages such as exposing the valve stem thread to the fluid service and high pressure inside the valve, this design is less common than the *outside screw and yoke* (OS&Y) design. Typically, IS&Y valves have a *non-rising stem*, meaning the stem does not move up or down externally. The gate or disk inside the valve moves instead. A globe valve with IS&Y threads or screws located in the lower part of the stem inside the pressure boundary is shown in Figure 5.14. Due to its weakness, this design is not used in the oil and gas industry.

Integrated (Integral) Seat

Integrated seat can be seen in gate and globe valves. An integrated seat is part of the valve body and cannot be removed or replaced separately. If it gets damaged, the entire valve body or a large part of the valve must be replaced, making repairs more costly and difficult.

International Organization for Standardization (ISO)

ISO has developed several standards specifically relating to valves. In industrial applications, ISO standards are important for ensuring consistency in quality, safety, and performance. A number of aspects are covered by these standards, including valve design, materials, testing, and installation. Various industries, such as oil and gas, chemical processing, power generation, and more, require valves to meet ISO international standards. There are a number of ISO standards

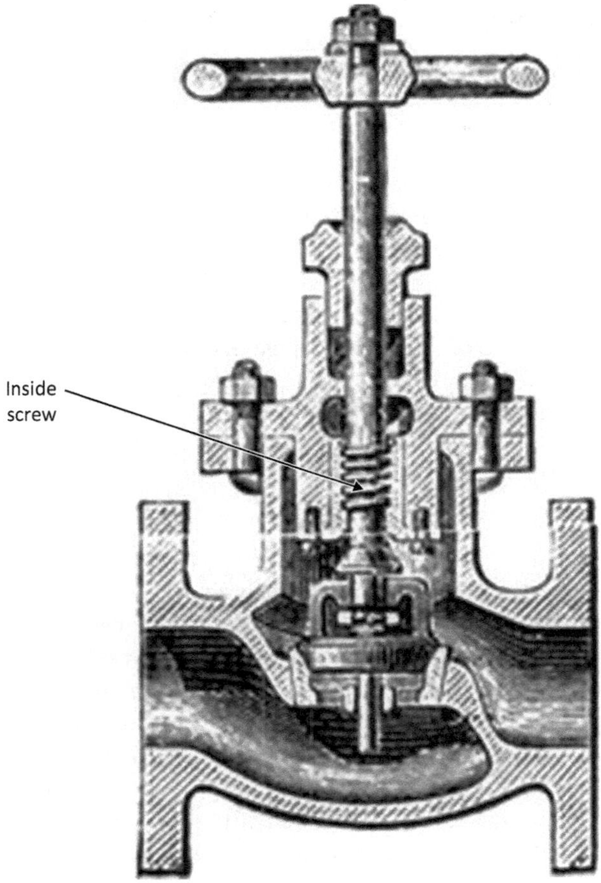

FIGURE 5.14 Inside screw and yoke globe valve with stem threads located at the lower part, in the pressure boundary. (With permission: Shutterstock.)

related to the valve industry, such as ISO 5208, ISO 17292, and ISO 15848. The ISO 5208 standard specifies the procedures for pressure testing industrial valves to determine their performance under specific pressure conditions. According to this document, testing requirements include the types of tests (e.g., seat tests and shell tests) and the pressure levels that valves must be able to withstand in order to be certified. A standard called ISO 15848 specifies the methods for measuring fugitive emissions from industrial valves. The standard provides a method for testing valves that are used in systems in which gas or liquid may leak from valve seals or components. This standard controls and minimizes valve emissions. Fugitive emission prevention is important for environmental protection and regulatory compliance, particularly in the oil and gas industry. ISO 17292 is an international standard that addresses ball valves.

J

Jacketed Design

Jacketed valves or piping systems are composed of an outer layer called a jacket that surrounds the valve body and connected piping. Using the jacket, a heating or cooling medium (such as steam, hot oil, or chilled water) can circulate around the internal valve or pipe, maintaining a consistent temperature. The use of jacketed designs is common in applications where temperature control is essential. A jacket is created by welding or bolting an external casing around the valve body or pipe. By providing a space for the heated or cooled medium to circulate, the jacket ensures that the fluid inside the valve or pipe does not solidify (applies to sulfur, wax, or heavy oil fluids) or overheat. Depending on the situation, the jacket may fully or partially cover the valve.

L

Leakage Class

In a valve, leak/leakage occurs when fluid (liquid or gas) escapes through areas that should be sealed off, either internally or externally. There may be issues with valve components, wear, improper sealing, or damage resulting in this undesirable flow. The leakage class is primarily used for control valves. Although leakage class and leakage rating are sometimes used interchangeably in valve testing, they have distinct meanings when it comes to standards for allowable leakage. Leakage classes are generally predefined categories defined by industry standards that specify leakage limits for different types of valves. As it pertains to leakage rate, it describes the actual or permitted amount of leakage that can be measured for a particular valve during a test.

For example, in ANSI/FCI 70-2, the leakage classes for control valves are as follows:

- **Class I**: No test requirements; typically, the highest allowable leakage
- **Class II**: Small leakage allowed; generally used for valves with metal seats
- **Class III**: Industrial class; allows slightly lower leakage
- **Class IV**: Metal-to-metal tightness with low leakage, often used in moderate applications
- **Class V**: Low-leakage class for high-pressure valves, suitable for critical applications
- **Class VI**: Bubble-tight standard, requiring nearly zero leakage, often achieved with soft-seated valves

Example: If a valve has a design requirement for *Class IV leakage*, the end user might request a lower leakage rate (stricter than Class IV), essentially aiming for a tighter shutoff to improve system integrity. This means that although the valve is designed for a specific leakage class, during factory acceptance testing (FAT) or site acceptance testing (SAT), lower leakage rates may be required.

Lever-Operated Valves

It is possible to operate manual valves with a lever or gearbox and handwheel. An explanation of the operation of gearbox valves was provided earlier. In general, lever-operated valves are used on small, low-pressure valves since they do not require a great deal of force by the operator.

Lifting Lever (safety Valve)

The lifting lever design in a safety valve is a manual mechanism located at the top of the valve. This design allows the operator to partially lift the valve disk off its seat without the need for full-system overpressure to open the valve. The purpose of this feature is to verify the functionality of the valve. Furthermore, it ensures that it does not become stuck or corroded during routine maintenance or inspections. It is usually located externally for ease of access and is connected to the valve stem so that its force is transmitted to the valve disk. It is common to find lifting levers in applications requiring regular operational checks, such as boilers and pressure vessels. It is important to use caution when operating this device, as improper use of the lever could result in unintended discharge of the pressurized fluid. Figure 5.18 shows a safety valve equipped with a lifting lever.

Live Load Gland Flange

A live load gland flange is a type of design that is applicable to wedge gates and globe valves. According to this design, a spring-loaded spring or washers, also known as Belleville washers, are installed on the gland flange and under the gland flange bolts (see Figure 5.15). By using these springs, the bolts can provide the required pressures on the gland and gland flange for pressing the stem seal over a long period of time. The stem seal receives linear force from the gland and gland flange, which is converted by seals into lateral sealing on the stem. In spite of changes in operational parameters such as pressure and temperature, the gland flange ensures that the live load force is evenly distributed to the packing material.

Live Load Packing

A live load packing system is used to maintain consistent sealing in valves, pumps, and other equipment. In spite of fluctuating conditions, such as thermal cycling, pressure variations, or mechanical movement, the sealing provided does not change over time. In this case, spring-loaded washers are used (commonly known as Belleville washers or springs). A constant force is exerted on the packing material by these washers, ensuring a tight seal between the valve stem and

FIGURE 5.15 Live load gland packing by installing a spring under the gland bolt.

the packing. A live load gland flange and a live load packing are closely related concepts, but they are not the same. As a mechanical system, the live load gland flange ensures that the live load packing performs optimally over time by maintaining consistent pressure.

Lubricated/Nonlubricated Plug Valve

Lubricated plug valves are valve designs that reduce friction and enhance sealing by injecting a specialized lubricant between the plug and valve body. Lubrication injection by the operator ensures smoother operation and minimizes wear. The injected lubricant may be a grease of high viscosity compatible with the process fluid. Lubricant acts as a sealant and reduces the torque required to operate the valve. A lubricated plug valve is particularly suitable for handling fluids that are corrosive and abrasive.

A nonlubricated plug valve is a type of plug valve that achieves a tight seal and smooth operation without the need for grease or external lubrication. Instead, it uses a self-lubricating material, such as a polymer sleeve or elastomeric liner, which acts as a barrier between the plug and the valve body. This lining not only facilitates the plug's rotation but also provides a tight, leak-proof seal, reducing maintenance requirements compared to lubricated plug valves. The absence of grease injection makes these valves cleaner to operate, without any need for a human operator to inject sealant, and particularly suited for applications where contamination from lubricants is undesirable. Although some valve suppliers consider this type of plug valve to be suitable for particle-containing (dirty) services, some engineers believe that the liner can be damaged by the particles in dirty services, so these valves are not recommended.

Lug-Type Design

A lug-type valve is one with a specific configuration of flanges or valve bodies, such as those used in butterfly valves. A lug-type valve or flange is characterized by protruding lugs or tapped holes around its circumference. There are two different bolts and nuts connecting the lugs to both sides of the flange. The design of the lugs makes it possible for the valve to be bolted to one end of the flange even when the other end flange is removed. As a result of this design, it is possible to bolt the valve independently to the flanges on either side of the valve, without the valve being dependent on the piping on the other side. It is a common practice for operators to disconnect the piping on one side of butterfly valves when performing maintenance or replacing piping. It is in this situation that the lugged butterfly valve can serve as a block valve, isolating the flow and pressure from the rest of the pipeline. A lug-type butterfly valve is shown in Figure 5.16 connected to vertical piping system by couple of flanges on both sides and cap screws.

M

Material Selection

Material selection plays a critical role in valve design since it impacts the valve's performance, safety, and durability. Material selection for valves is influenced by

FIGURE 5.16 A wafer lug butterfly valve installed between two flanges with the possibility to keep the valve on the line after removing the connected flange on each side. (With permission: Shutterstock.)

several factors, including compatibility with the process fluid, temperature, pressure, environmental conditions. The materials must be selected in a way to prevent corrosion, erosion from erosive fluids, and mechanical failure. For example, a corrosion resistance material such as stainless steel is a good choice for corrosive fluids rather than carbon steel. In addition, super duplex and titanium are suitable for seawater services that contain chloride and cause pitting and chloride stress cracking corrosion. Exotic alloys such as Hastelloy and Inconel are suitable for highly aggressive chemicals and high-temperature conditions. A detailed description of valve materials, including both metallic and nonmetallic materials, can be found in Chapter 4. It is possible to experience leaks, system failures, and costly maintenance costs as a result of improper material selection. According to Figure 5.17, the valve body and bonnet have been heavily corroded as a result of the corrosive offshore environment.

Maximum Allowable Stem Torque (MAST)

MAST is defined as the maximum torque that a valve stem can safely withstand without mechanical failure. Under all operating conditions (e.g., opening, closing, or regulating flow), the actuator must provide sufficient torque to operate the valve. However, it should not exceed the MAST to prevent overloading and damaging the valve stem. A valve stem or related components (e.g., packing) may fail if actuator torque exceeds MAST. This can result in operational downtime, safety

FIGURE 5.17 Heavy external corrosion on the wedge gate valve body and bonnet due to corrosive marine environment. (With permission: Shutterstock.)

risks, or catastrophic failures. If a valve's MAST is 1,000 Nm and the actuator provides 800 Nm, the design is safe because 800 Nm is below the MAST.

Maximum Allowable Working Pressure (MAWP)

More information about MAWP is provided in Chapter 10 under the heading of "Maximum Allowable Working Pressure (MAWP)."

Metal Seat (Metal-to-Metal Seat)

A metal seat in a valve indicates that the valve seat is made of metal. Although metal seats are more costly and less leak-tight than soft seats, they are often used in applications where high temperatures, high pressures, or abrasive media make soft seats unsuitable. While ball and butterfly valves can have metal or soft seats, globe valves and wedge gate valves typically have metal seats. The metal-to-metal seat refers to a valve design in which both the valve seat and the closure member (e.g., ball, gate, or disk) are made of metal rather than soft materials such as PTFE or PEEK. In most cases, a *metal seat* is synonymous with *metal-to-metal seating.* That is because both terms refer to valve designs in which the sealing surfaces of the seat and the closure element (such as a ball, gate, or disk) both are made of metal.

Modularity

Modularity in valve design means using interchangeable parts so that individual components can be replaced, or upgraded without replacing the entire valve. This makes maintenance easier and more cost-effective since worn-out parts like seats can be changed instead of throwing away the whole valve just because of the damaged seat. Modular designs extend valve life, reduce waste, and support sustainability by minimizing material consumption and the need for new valve production.

Multistage Trim

A multistage trim is a specialized design feature of control valves. In this design, high-pressure drops in fluid flow are managed while noise, vibration, and cavitation are minimized. There are multiple flow paths or stages within the valve. As the fluid flows through the valve, these paths gradually reduce the pressure. As a result of this staged pressure reduction, abrupt pressure drops are prevented. Especially in high-pressure or high-velocity applications, excessive pressure drop can cause cavitation or excessive noise.

A good example is a renewable valve seat, which can be pressed, screwed, or fitted into the valve body and easily replaced when worn out. This design contrasts with an integrated seat, which is permanently part of the body and requires changing the whole valve body and seat in case of damage to the seat. Renewable or integral seats are common in gate and globe valves.

Multi-turn Valve

In a multi-turn valve, the closure element (e.g., gate or disk) is rotated multiple times in order to move it from the fully open to the fully closed position, or vice versa. It is common to find this design in gate valves and globe valves, which are considered the opposite of quarter-turn valves such as ball and butterfly valves.

N

National Pipe Thread (NPT)

NPT is a standard for the threading of pipes and fittings used in the United States covered by ASME B1.20.1. It specifies the shape, pitch (the distance between threads), and diameter of the threads used for connecting pipes and fittings in a secure and leak-tight manner. The NPT standard is commonly used in plumbing, industrial piping, and gas systems. Common NPT sizes include 0.5″, 1″, 2″, and so on. It is common for NPT threads to be tapered, which means that the thread diameter decreases slightly along the length of the threaded portion. In addition to providing a wedging effect, this taper ensures a tight, leak-proof seal when the threads are tightened.

Non-rising Stem

The stem of a non-rising stem valve does not rise or fall during operation. Instead, the plug (closure member) is used to raise or lower the stem. It is necessary for the stem to remain stationary externally while the gate or disk rotates inside the valve body in order for the valve to function. The threads of the valve stem are located inside the valve body and are exposed to the process fluid, which is a disadvantage. As this design does not permit upward movement of the stem above the valve operator, a certain amount of space can be saved. See the "Rising Stem" section in this chapter which gives alternative (opposite) stem design.

Non-Slam Check Valve

The slamming of a check valve occurs when the flow reverses abruptly, causing noise and vibration, which can lead to excessive wear on the disk and seat of the valve. Additionally, it can result in water hammering, which is a hydraulic shock to the system as a result of the sudden stoppage of the fluid. The term non-slam check valve refers to a type of check valve that is specifically designed to minimize slamming, which can occur when a valve closes too rapidly as a result of a sudden reversal of flow and/or fast movement of the valve disk due to gravity forces. As opposed to swing check valves, dual-plate check valves, also known as low slam check valves, operate by a spring, and they use two disks instead of one, which reduces the weight of each disk as it closes, and also does not rely on gravity. In additional information, the disk of a swing check valve is closed by gravity, which intensifies the slamming effect. Using an axial flow check valve, however, prevents slamming, due to the use of springs and short distances between the disk and the seat. Thus, the design of an axial check valve is known as non-slamming. As a second important point, dashpots or other external mechanisms connected to the swing check valve disk can control (reduce) the speed of the disk during closure, resulting in a low-slamming swing check valve.

Nozzle Design (Safety Valve)

There are two types of safety valves, *semi-nozzles* and *full-nozzles*, which refer to the inlet portion of the valve through which the fluid is introduced. A full-nozzle design, as illustrated in Figure 5.18, includes a separate, removable nozzle that

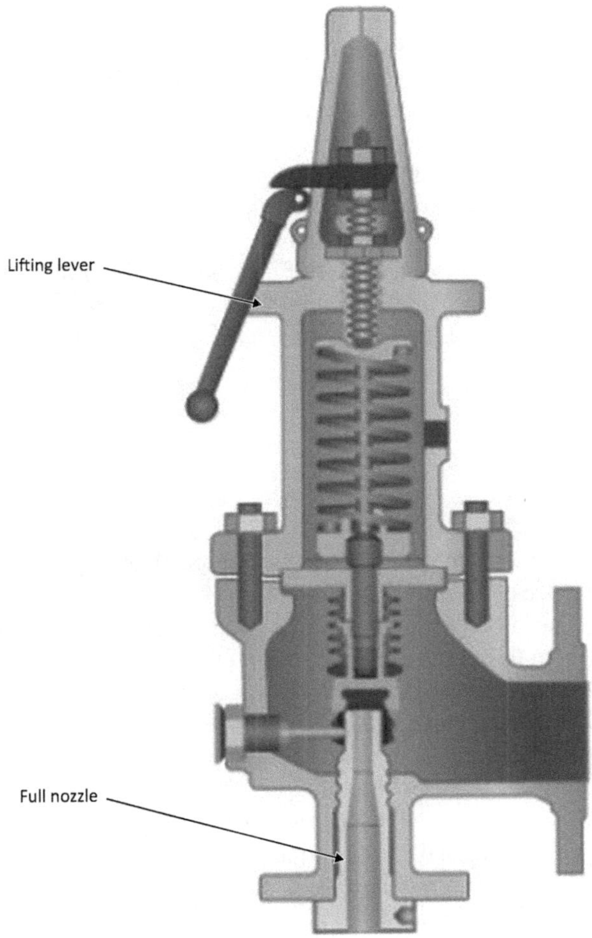

FIGURE 5.18 Full-nozzle safety valve. Enhanced durability and corrosion resistance for high-performance applications. (With permission: Shutterstock.)

extends from the inlet flange to the seating area, with the seat integrated into the nozzle. Because this design isolates the valve body from the process fluid, it is ideal for corrosive applications, as it protects the body from corrosive materials. In contrast, a semi-nozzle design has the nozzle seat machined directly into the valve body, exposing the body to the process fluid. While semi-nozzle designs are more cost-effective and simpler, they are less suitable for corrosive fluid applications due to increased wear on the valve body. As nozzle design determines how much of the valve body is exposed to fluid during static conditions, it is particularly important when the valve is closed.

O

Outside Screw and Yoke

The OS&Y design is typically used for wedge-type gate valves and globe valves. The OS&Y is a valve design feature in which the stem threads are located outside the valve body. This design is a rising stem which allows the stem to move through and inside the yoke, which is a structural component that supports and guides the stem. There are threads on the valve stem that are located outside the valve's body, making it easier to inspect and lubricate the stem (a great benefit for inspection and maintenance). Furthermore, the stem rises above the valve body when the valve is opened (see Figure 5.19) and lowers when the valve is closed. This visually indicates the valve's position for operating personnel. Furthermore, the threads of the stem are located outside the fluid passage and the temperature and pressure boundaries.

P

Pipeline Valve

A pipeline valve is a mechanical device used in pipelines to control the flow of fluids (liquids, gases, or slurries) through the system. It achieves this by opening, closing, or partially obstructing various passageways in the pipeline. Pipeline valves are essential for maintaining the safety, efficiency, and operability of fluid transport systems, especially in industries such as oil and gas, water distribution, and chemical processing. Pipeline valves are covered by API 6D standard. A ball valve and a through conduit gate valve are two of the most common types of pipeline valves. Pipeline valves should allow pipeline PIGs (devices used for cleaning, inspection, or maintenance) to pass without being obstructed. Therefore, the valves should have an internal diameter matching the pipe diameter. As a pipeline valve, sharp edges or features must be avoided to prevent damage to PIGs or obstruction to their movement. The valve must be able to withstand the maximum operating pressure and temperature of the pipeline according to API 6D and ASME B16.34. All materials used must resist corrosion, erosion, and chemical reactions with the fluid being transported. It is necessary for valves to ensure zero

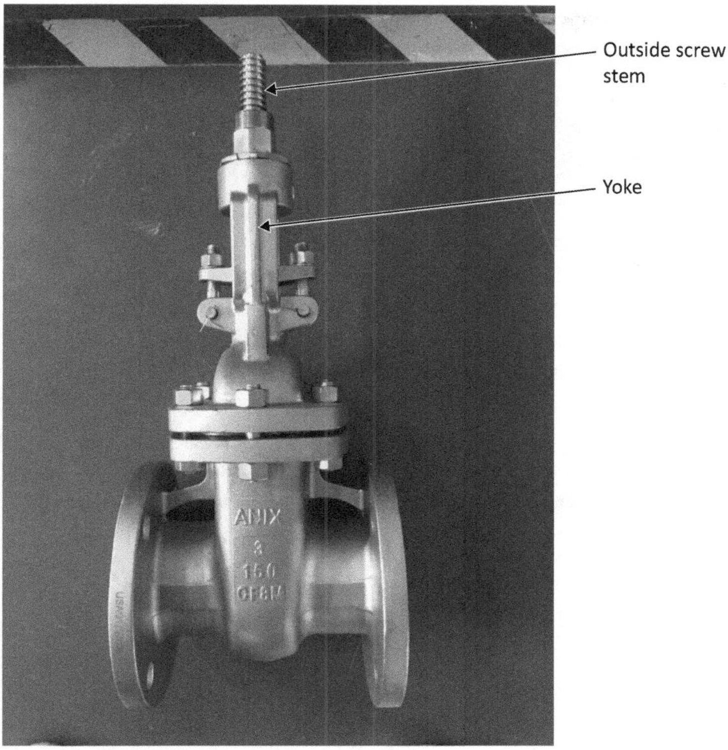

FIGURE 5.19 An outside screw and yoke wedge gate valve with threads located on top of the stem outside the valve body. A visible stem thread indicates that the valve is in the open position and the stem is rising.

or minimal leakage in order to prevent product loss or environmental damage. It is necessary for valves to remain operable during and after a fire and to prevent leakage.

Pressure Class

In valve terminology, a valve's pressure class describes the maximum pressure at which the valve can safely be operated at a given temperature. It is a standard classification system used to determine the suitability of valves for the pressures and temperatures in the system in which they are used. A valve's pressure class is defined by industry standards such as those established by the ASME, the API, or the ISO.

Pressure classes are denoted by numbers, such as Class 150, Class 300, Class 600, 900, 1500, and 2500. However, the numerical value of the class does not directly indicate the maximum pressure in psi. For example, a Class 150 valve is capable of withstanding *285 psi at 100°F* for carbon steel materials, while a Class 300 valve can handle *740 psi at 100°F.* The actual pressure rating decreases as the temperature increases. It should be noted that pressure class, pressure rating, and pressure–temperature rating are highly connected but they are not the same.

Pressure Drop

A valve's pressure drop is the amount of pressure reduction that occurs when a fluid passes through it. For example, if a fluid has 3 bar pressure and its pressure is reduced to 2 bar after passing through the valve, then the valve's pressure drop will be 1 bar. Valve pressure drop is important because valves with a high-pressure drop result in lower energy optimization and efficiency, resulting in energy waste. Compared to a full-bore ball valve, a reduced bore ball valve is less expensive, but it causes a greater pressure drop. Nevertheless, if the process engineer requires a low-pressure drop in the piping system for energy efficiency, it is necessary to utilize a full-bore ball valve with a lower pressure drop. Calculations of pressure drop are crucial to the selection of valves. A valve's flow coefficient (C_v) or flow velocity (K_v) is also a significant factor in determining pressure drop.

Pressure Equipment Directive (PED)

Directive 2014/68/EU, also known as the PED, is a regulation of the European Union. In accordance with this regulation, pressure equipment and assemblies, including valves, intended for the European market must be designed, manufactured, and conformity assessed. Equipment with a maximum allowable pressure greater than 0.5 bar is subject to PED. By requiring compliance with essential safety requirements (ESRs), PEDs ensure safety.

According to the PED, equipment is classified into categories (I – IV) based on factors such as pressure, volume, and fluid type (gas, liquid, steam, or hazardous). During the conformity assessment process, the category determines the level of assessment, ranging from self-certification for lower risk equipment to mandatory involvement of Notified Bodies for higher risk equipment. The CE marking indicates compliance with the PED. In addition, CE marking shows that the equipment is safe and environment-friendly. It is also necessary for manufacturers to provide detailed technical documentation, perform pressure testing, and ensure traceability in order to maintain compliance. Valve PED compliance is essential for legal placement on the European market, making it an important consideration during the design and production process.

Pressure Rating

The pressure rating refers to the MAWP that a valve, flange, or fitting is capable of handling under specified conditions, for example, at normal temperatures (e.g., 20°C or 68°F). It is expressed typically in psi or bar. As a point of clarification, the pressure rating is different from the ASME pressure classes since the pressure rating is a single value, while the ASME pressure classes represent a range of pressures and temperatures. For example, the *pressure class* of a Class 150 valve specifies the actual MAWP at different temperatures, depending on the valve's material and design. It can be 285 psi pressure rating at 100°F.

Pressure Relief Mechanism

In a valve, the pressure relief feature refers to a design element or mechanism that is integrated into the valve to relieve excess pressure. The purpose of pressure relief in valves is normally to reduce the buildup of pressure within the valve cavity. There are a number of pressure relief mechanisms available for valve cavities, such as self-relieving seats to release the cavity pressure inside the valve and pipe, or plugs or pressure relief valves to release the cavity pressure externally.

Pressure–temperature Rating

The pressure–temperature rating indicates how the MAWP changes with increasing operating temperature. It represents a range of pressure values corresponding to different operating temperatures for a given pressure class and material. For example, a Class 300 valve made of carbon steel (WCB) has 740 psi at 20°C (68°F), 570 psi at 200°C (392°F), and 355 psi at 400°C (752°F). Increasing the temperature reduces the strength of the valve material, which decreases the pressure that a valve can safely handle as the temperature increases. There is a difference between a pressure rating and a pressure–temperature rating. The reason for this is that a pressure rating represents a single value representing the maximum pressure that can be applied at a certain temperature (e.g., 20°C). The pressure–temperature rating, however, is a table or range that illustrates how the allowable pressure decreases with increasing operating temperature. It is important to know that *ASME pressure classes* (e.g., Class 150, 300, 600) represent the *pressure–temperature ratings* for valves, flanges, and fittings as defined in standards like *ASME B16.34* and *ASME B16.5*.

Q

Quarter-Turn

This type of valve design requires only a 90° (one quarter of a full rotation) turn of the valve operator, stem, and closure member to open or close it. In this design, which is mainly applicable to ball, butterfly, and plug valves, fast operation is provided. *Quarter-turn valve* can be considered the opposite of a *multi-turn valve.*

R

Reduced Bore (Reduced Port)

A *reduced-bore valve* (also known as a *reduced port valve*) is a valve design in which the internal diameter of the valve (bore) is smaller than the internal diameter of the connected piping. In comparison to a full-bore valve, this design

FIGURE 5.20 This illustration shows a reduced-bore ball valve with a diameter of 3″ connected to the pipe and an internal diameter of 2″. (With permission: Shutterstock.)

creates a restriction in the flow and a high-pressure drop. The use of reduced-bore valves (e.g., ball valves) is generally not recommended in three applications. In the first instance, if pressure drop in the piping system is a concern that is typically raised by the process engineer. The second scenario occurs in pipelines which are cleaned, inspected, or maintained using PIG. A reduced-bore valve cannot be piggable and cannot permit safe passage of the PIG. Piggable valves ensure minimal obstructions and smooth pig movement without causing damage to the pig or the valve. In order to be piggable, valves should not have a reduced bore, and their internal diameters must be very close to those of the pipeline. The third application is for the ball valves installed before and after the pressure safety valves. Ideally, these valves must have a full bore so that the overpressure fluid or gas can be quickly released and the piping system can be depressurized. As shown in Figure 5.20, a reduced bore ball valve is illustrated with an internal port that is considerably smaller than the pipe that connects to the valve. It is important to know that a full-bore valve in 3″ size has only one size, while a reduced-bore valve has two sizes $\left(3'' \times 2''\right)$.

Renewable Seat

A renewable valve seat is common in gate and globe valves. It can be pressed, screwed, or fitted into the valve body and easily replaced when worn out. In contrast, an integral seat is part of the valve body and cannot be removed. If the seat gets damaged, the entire body or a large part of the valve must be replaced, which is a disadvantage. Renewable seats make maintenance easier, cheaper, and more sustainable.

Rising Stem

It is a commonly used stem design in valves, most commonly used in gate valves and globe valves. In a rising stem valve, the stem moves up and down as the

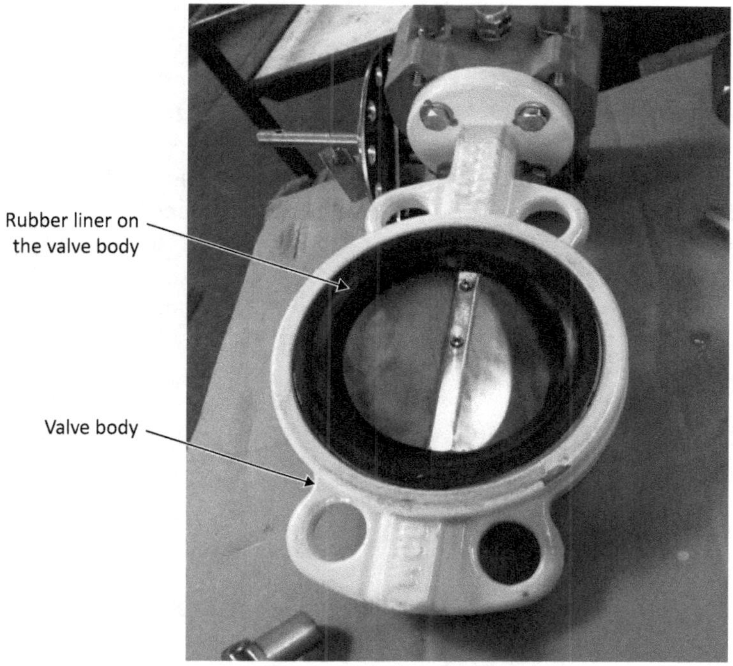

FIGURE 5.21 A rubber-lined butterfly valve.

valve is opened or closed. It is common for stem threads to be located outside the valve body. The stem rises when the valve is opened and lowers when it is closed. Because the threads are external (outside the valve body), they are more easily maintained, cleaned, and lubricated. See the "**Non-Rising Stem**" section in this chapter which gives alternative (opposite) stem design.

Rubber-Lined Butterfly Valve

Concentric butterfly valves typically use a soft seat, as illustrated in Figure 5.21, that is made from an elastomeric material such as rubber, PTFE, or another resilient material. The soft seat forms a tight seal with the disk, preventing leaks.

S

Safety Factor

In valve design, the safety factor is an important margin to consider in order to ensure that the valve is able to handle loads, stresses, as well as pressure scenarios beyond its normal operating and design conditions. When safety factors are incorporated into valve design, they improve the reliability and safety of the valve during operation. A safety factor is a measure that takes into account abnormalities such as material imperfections, manufacturing errors, unexpected pressure surges, and environmental factors such as temperature fluctuations and external corrosion. As an example, if a valve is designed for a working pressure of 100 bar with a safety factor of 2.0, its components are designed to withstand 200 bar before failure. A margin of safety prevents catastrophic failure, leakage, or deformation under extreme or unexpected conditions. Safety factors vary according to the application (e.g., pipeline or nuclear plant) that demands high levels of safety, industry standards (e.g., ASME B16.34, API 6D), and the valve designer's and operator's risk tolerance. The proper design of safety factor in valves ensures the durability of the valves, as well as enhancing their safety and reliability.

Self-Relieving Seat

Refer to the "Single-Piston Effect" section in this chapter for more information.

Shutoff Pressure

In the fully closed position, the shutoff pressure is the maximum pressure that a valve can withstand or hold without leaking through the valve seats. This parameter is essential to valve design for several reasons: First, it ensures that the valve's seating mechanism (like seals or seats) will be able to withstand the pressure without failing or leaking. Furthermore, it determines whether the valve is suitable for high-pressure applications and how effectively it is able to isolate flow.

Side-Entry Valve

Contrary to a top-entry valve, a side-entry valve provides access to the valve internals through the side flange, making it possible for the operator to repair and maintain the valve. Figure 5.22 illustrates a side entry ball valve in a valve workshop, showing that it is possible to gain access to the valve's interior by removing the side flange bolts (bolts and nuts).

Slab Gate Design

Slab gate valves have a single, flat (slab-like) rectangular or circular plate that moves up and down within the valve body to stop or start fluid flow. As the stem does not protrude out of the valve body during operation, the valve is more compact. The slab gate is also often fitted with a through-conduit bore. This makes the flow path free of obstructions and matches the pipeline bore without pressure drop or turbulence. There are two seats on the valve (upstream and downstream) that seal against the gate. The gate could be fixed (sealed by seats) or floating (moving to one seat at a time for sealing).

FIGURE 5.22 2 A side entry ball valve in a valve workshop, showing the possibility to gain access to the valve's interior by removing the side flange bolts (bolts and nuts). (With permission: Shutterstock.)

Smart Valve Design

Smart valve design refers to integrating advanced technologies into traditional valves, such as sensors, actuators, and digitalization (digital control systems). With these advanced technologies and modern designs, valves are capable of performing intelligent functions such as real-time monitoring of process parameters like pressure, temperature, and flow, dynamic and constant movement based on the process need, predictive maintenance, and remote access.

Soft Seat

This feature is commonly used in ball and butterfly valves where the seat is made of nonmetallic materials, such as Teflon (PTFE), or polyether ether ketone (PEEK). The opposite design is known as a metal seat design. Seats made of soft materials provide a tighter seal than seats made of metal and are less expensive. However, soft seats have a limited pressure and temperature range compared to metal seats and are not suitable for dirty (particulate-containing) services or valves with high-pressure drops.

Solid Wedge

It represents a wedge gate valve of which the solid metal component is one piece and does not have a flexible component. It is a standard wedge used in gate valves.

SPE Seat

A SPE seat, also known as a self-relieving seat, is a valve seat design commonly used in ball valves and through conduit gate valves. As a result of this design, accumulated fluid pressure in the body cavity can be released to the valve bore and piping system. As a result, this release of body cavity pressure prevents damage to the valve caused by excessive pressure within the body cavity. SPE design is the opposite of DPE design, which has been explained previously in this chapter.

Split-Body Design

The split-body design in valves refers to a valve body that consists of at least two separate pieces that are assembled together by bolting (bolts and nuts). The design is very common for ball valves, but it can also be used for other types of valves. Assembling, maintaining, and replacing internal components like the ball, seats, and stem is made easier with the split-body design.

Spring-Loaded Design

Typically, a spring-loaded valve refers to a valve that uses a spring to control the movement of its internal components, such as its disk and gate. A spring is typically used to control the opening and closing of a valve, primarily in check valves (dual plate and axial flow) and safety valves.

Stem Extension (Extended Stem)

Stem extensions, or extended stems, are important valve design features that extend or elongate the stem so that the valve operator (handwheel or actuator) is kept away from the fluid flowing inside the valve. Typically, this type of design is used for three types of valves: valves that are used for cryogenic (extremely cold) and high-temperature applications, and valves that are installed underground, and valves which require insulation. Cryogenic and hot temperature service extensions are necessary for operator safety so that his hand does not come into contact with fluid flowing through the valve. Underground valves, however, are extended to have the valve operator above ground level. As insulation adds additional thickness to the valve, it may be necessary to extend the stem to keep it out of the insulation box.

Stem Torque

The torque represents the force required to rotate and move the valve stem. As a result of this parameter, the force or torque required by an operator or actuator to operate the valve (e.g., opening and closing) is affected. Stem torque is an important aspect of valve design since it determines the valve's functionality and ease of operation. Valve stem torque determines the design and selection of the actuator in order to ensure the actuator can generate sufficient force to move the valve stem and closure member in order to change the position of the valve. Despite this, the actuator torque should not exceed the valve structural limitations regarding the MAST. High MAST and stem torque can also determine the requirement for manual or actuated operability, as excessive MAST requirements may make the valve difficult or impossible to operate by hand, and an actuator could be used for such a valve.

T

Top-Entry Valve

In a top-entry valve, the operator has access to the valve's internals from the top, such as seats and closure members. It is a side-entry valve as opposed to a top-entry valve, which allows access from the side. Top-entry valves have several advantages over side-entry valves, but the most significant advantage is the ability to access the internals of the valve without removing the valve from the piping system for maintenance or repair. It is important to note that removing a valve from a piping system could lead to more downtime and production loss than online maintenance in which the valve is connected to the piping while being maintained. As shown in Figure 5.23, a top-entry ball valve which will be installed on a pipeline on top of an oil and gas platform.

Triple Offset Butterfly Valve

More information about double and triple butterfly valve is provided in this chapter under section "Eccentric Butterfly Valve."

FIGURE 5.23 A large size top-entry ball valve to be installed on an offshore platform in the oil and gas industry.

Trunnion Ball

In a trunnion ball valve, the ball is supported by a trunnion in the form of a flange or plate beneath the ball. This design prevents the ball from moving sideways, causing less friction between the ball and the seats. In this manner, the valve operator exerts a minimal amount of force and torque. The trunnion ball valve is ideal for larger valve sizes and higher pressures. A trunnion ball is designed in the opposite way as a floating ball, as explained in this chapter under the floating ball heading. In Figure 5.24, a trunnion mounted ball valve is illustrated, highlighting the trunnion flange that is attached to the valve beneath the ball.

U

Unidirectional Seat Design

Unidirectional seat is same as *self-relieving seat* and *single-piston effect seat.* More information is available in this chapter under sections "Self-Relieving Seat" and "Single-Piston Effect Seat."

FIGURE 5.24 A trunnion mounted ball valve highlighting the trunnion flange that is attached to the valve beneath the ball.

Unidirectional Valve

A unidirectional valve (such as a check valve or globe valve) is designed to allow flow in only one direction. A valve's body, flange, or name plate normally indicates the flow direction.

V

Valve Configuration

A valve's configuration refers to its physical structure (type, size, body design, side- or type-entry), its connection type, its flow configuration, such as two-way or three-way, flow control, stop and start flow, actuation method, and internal mechanisms (e.g., double-piston effect or single-piston effect seat). Various aspects of a valve's design and functionality are considered in order to ensure that it is designed to fulfill specific application requirements in a piping system or industrial application.

Valve End

Valve ends refer to the design or configuration of the points where a valve connects to a pipeline or other equipment. In order to ensure a secure and leak-proof connection between the valve and the piping system, these ends are crucial. A valve may have a flanged end, a buttweld or beveled end, a socket weld end, or a threaded end.

Valve Piggability

The piggability of a valve refers to its ability to allow the safe and efficient passage of a pipeline PIG during operations such as cleaning, inspecting, or maintaining the pipeline. Pipeline PIGs are tools that are inserted into pipelines to perform tasks such as cleaning, measuring, and inspecting the pipeline for defects. The piggable valve ensures minimal obstruction and smooth PIG movement without causing damage to the PIG or the valve itself. In order to be piggable, valves should not have a reduced bore and must have an internal diameter that is very close to the internal diameter of the pipeline.

Valve Selection

Valve selection refers to the process of choosing the appropriate type of valve for a specific application in piping systems, processes, or equipment. In valve selection, various parameters are considered in order to select the most suitable valve for the application, including the type of fluid that passes through the valve, the pressure and temperature of the fluid, the application of the valve regarding start or stop the fluid, control the fluid, safety and preventing the fluid from returning, as well as the size and connection of the valve. The valve selection process could be carried out by a variety of department engineers, including valve engineers, piping and layout engineers, and process engineers.

Valve Sizing

The valve sizing process is an important step in the valve design process in order to select the correct valve size and associated dimensions, such as the internal bore, in order to achieve the desired amount of flow through the valve. A valve's correct sizing affects its performance and reliability as well as ensuring that it can handle the required flow rate, pressure drop, and process conditions without causing cavitation, noise, vibration, or excessive energy losses. In order to size valves, valve designers and engineers utilize a number of parameters, including C_v (flow coefficient), which represents the valve's flow capacity, fluid type (liquid, gas, or steam), temperature, inlet and outlet pressures, and flow velocity. An undersized valve may restrict flow and cause pressure loss, while an oversized valve may result in unnecessary expenditures.

Valve Testing

There is a close relationship between valve testing and valve design. In contrast to valve design, which focuses on the theoretical and engineering aspects (e.g., material selection, pressure classes, flow characteristics, sealing mechanisms, actuator sizing), valve testing ensures that the valve design will perform as intended in real-life conditions.

W

Wafer Design

A wafer valve is a compact type of valve designed to fit between two pipe flanges. Unlike flanged valves, wafer valves do not have their own bolt holes for mounting; instead, they rely on the bolts from the flanges of the connected pipes to hold them in place. Wafer valves are commonly used in applications requiring lightweight, space-saving solutions, such as in butterfly or check valves. Wafer valves have a compact profile, making them lighter and more space-efficient than the fully flanged valves. The valve is sandwiched between two pipe flanges, and bolts pass through the flanges to secure the assembly. The simpler design reduces material and manufacturing costs compared to fully flanged valves. For wafer design, butterfly valves are very common (see Figure 5.25).

Wedging Effect

The wedging effect refers to the mechanical force generated by stem when a wedge-shaped component, such as a gate or plug in a gate valve especially, is pressed into a mating surface to create a tight seal. This principle is commonly

FIGURE 5.25 Wafer-type butterfly valves between flanges. (With permission: Shutterstock.)

used in valve designs, especially wedge gate valves, where the gate has an angled or tapered profile that interacts with the valve seats to ensure a reliable sealing.

QUESTIONS AND ANSWERS

1. What is the primary purpose of valve material selection in design?
   A. Cost reduction
   B. Enhancing appearance of the valve
   C. Reducing valve flow capacity
   D. Enhancing the safety and reliability of the valve

   Answer: Option D is the correct answer.

2. What does C_v standards for?
   A. Pressure rating
   B. Flow coefficient
   C. Control volume
   D. Stem torque

   Answer: Option B is the correct answer.

3. What is the function of pressure–temperature ratings in valve design?
   A. Calculating actuator size
   B. Determining valve material cost
   C. Ensuring valve strength under various conditions
   D. Predicting the valve flow characteristics

   Answer: Option C is the correct answer.

4. Which valve design is most appropriate for quick shutoff applications?
   A. Globe valve
   B. Check valve
   C. Ball valve
   D. Needle valve

   Answer: Option C is the correct answer.

5. What design feature makes a valve "fire-safe"?
   A. Explosion-proof actuator
   B. Secondary sealing mechanism in case of fire
   C. Reducing the pressure drop across the valve
   D. Additional heat shields

   Answer: Option B is the correct answer.

6. What is the purpose of an anti-blowout stem in a valve design?
   A. To prevent fluid leakage through the stem
   B. To increase flow capacity
   C. To prevent the stem from being ejected under high pressure
   D. To reduce stem friction during operation

   Answer: Option C is the correct answer.

7. What pressure class (rating) designation is used for valves in high-pressure pipelines?
   A. Class 150
   B. Class 300
   C. Class 600 and above
   D. Pressure Nominal (PN) 10

   Answer: Option C is the correct answer.

8. What does a double block and bleed valve design reduce the risk of?
   A. Contamination between two different fluids
   B. Stem blowout
   C. Pressure drop across the valve
   D. Cavitation

   Answer: Option A is the correct answer.

9. In pressure-containing valve components, why is stress analysis critical?
   A. To determine flow capacity
   B. To ensure no failure occurs under design pressure
   C. To prevent overpressure in the pipeline
   D. To improve valve operation speed

   Answer: Option B is the correct answer.

10. Which of the following is a key characteristic of an Outside Screw and Yoke (OS&Y) valve design?
    A. The stem threads are exposed to the flow medium.
    B. It is primarily used for low-pressure systems.
    C. The stem threads are located outside the valve body.
    D. The yoke is integrated inside the valve body.

    Answer: Option C is the correct answer.

BIBLIOGRAPHY

American Petroleum Institute (API). (2021). Check Valves: Flanged, Lug, Wafer, and Butt-Welding Ends (API Standard 594). 7th edition. API Publishing Services: Washington, DC.

American Petroleum Institute (API). (2021). Metal Plug Valves: Flanged, Threaded, and Welding Ends (API Standard 599). 8th edition. API Publishing Services: Washington, DC.

American Petroleum Institute (API). (2021). Bolted Bonnet Steel Gate Valves for Petroleum and Natural Gas Industries (API Standard 600). 13th edition. API Publishing Services: Washington, DC.

American Petroleum Institute (API). (2021). Gate, Globe, and Check Valves for Sizes DN 100 (NPS 4) and Smaller for the Petroleum and Natural Gas Industries (API Standard 602). 9th edition. API Publishing Services: Washington, DC.

American Petroleum Institute (API). (2021). Corrosion-Resistant, Bolted Bonnet Gate Valves: Flanged and Butt-Welding Ends (API Standard 603). 10th edition. API Publishing Services: Washington, DC.

American Petroleum Institute (API). (2021). Fire Test for Soft-Seated Quarter-Turn Valves (API Standard 607). 8th edition. API Publishing Services: Washington, DC.

American Petroleum Institute (API). (2021). Metal Ball Valves: Flanged, Threaded, and Welding Ends (API Standard 608). 7th edition. API Publishing Services: Washington, DC.

American Petroleum Institute (API). (2021). Butterfly Valves: Double-Flanged, Lug- and Wafer-Type (API Standard 609). 9th edition. API Publishing Services: Washington, DC.

American Petroleum Institute (API). (2021). Type Testing of Process Valve Packing for Fugitive Emissions (API Standard 622). 3rd edition. API Publishing Services: Washington, DC.

American Petroleum Institute (API). (2021). Type Testing of Rising Stem Valves Equipped with Graphite Packing for Fugitive Emissions (API Standard 624). 2nd edition. API Publishing Services: Washington, DC.

American Petroleum Institute (API). (2021). Type Testing of Quarter-Turn Valves for Fugitive Emissions (API Standard 641). 1st edition. API Publishing Services: Washington, DC.

American Petroleum Institute (API). (2021). Pipeline and Piping Valves (API Standard 6D). 25th edition. API Publishing Services: Washington, DC.

American Petroleum Institute (API). (2021). Subsea Pipeline Valves (API Standard 6DSS). 4th edition. API Publishing Services: Washington, DC.

American Petroleum Institute (API). (2021). Specification for Wellhead and Christmas Tree Equipment (API Standard 6A). 21st edition. API Publishing Services: Washington, DC.

American Petroleum Institute (API). (2021). Specification for Fire Test for Valves (API Standard 6FA). 4th edition. API Publishing Services: Washington, DC.

American Society of Mechanical Engineers (ASME). (2020). Pipe Flanges and Flanged Fittings (ASME Standard B16.5-2020). ASME: New York, NY.

American Society of Mechanical Engineers (ASME). (2020). Face-to-Face and End-to-End Dimensions of Valves (ASME B16.10-2020). 2020 edition. ASME Publishing: New York, NY.

American Society of Mechanical Engineers (ASME). (2020). Forged Fittings, Socket-Welding, and Threaded (ASME B16.11-2020). 2020 edition. ASME Publishing: New York, NY.

American Society of Mechanical Engineers (ASME). (2021). Valves: Flanged, Threaded, and Welding End (ASME Standard B16.34-2021). ASME: New York, NY.

American Society of Mechanical Engineers (ASME). (2020). Large Diameter Steel Flanges (ASME B16.47-2020). 2020 edition. ASME Publishing: New York, NY.

American Society of Mechanical Engineers (ASME). (2020). Power Piping (ASME B31.1-2020). 2020 edition. ASME Publishing: New York, NY.

American Society of Mechanical Engineers (ASME). (2020). Process Piping (ASME B31.3-2020). 2020 edition. ASME Publishing: New York, NY.

American Society of Mechanical Engineers (ASME). (2020). Pipeline Transportation Systems for Liquids and Slurries (ASME B31.4-2020). 2020 edition. ASME Publishing: New York, NY.

American Society of Mechanical Engineers (ASME). (2020). Gas Transmission and Distribution Piping Systems (ASME B31.8-2020). 2020 edition. New York, NY: ASME Publishing.

American Society of Mechanical Engineers (ASME). (2021). Rules for Construction of Power Boilers: Safety Valves and Pressure-Relief Valves (ASME BPVC Section I). 2021 edition. New York, NY: ASME Publishing.

American Society of Mechanical Engineers (ASME). (2021). Rules for Construction of Pressure Vessels: Pressure-Relief Valves (ASME BPVC Section VIII, Division 1). 2021 edition. ASME Publishing: New York, NY.

American Society of Mechanical Engineers (ASME). (2021). Nuclear Facility Components: Valves for Nuclear Applications (ASME BPVC Section III, Division 1). 2021 edition. ASME Publishing: New York, NY.

American Society for Test and Materials. (2022). Standard Test Methods and Definitions for Mechanical Testing of Steel Products (ASTM A370-22). ASTM International: West Conshohocken, PA.

Fluid Control Institute (FCI). (2006). Control Valve Seat Leakage (ANSI/FCI 70-2-2006). Fluid Control Institute: Cleveland, OH.

International Organization for Standardization (ISO). (2015). Industrial Valves: Measurement, Test and Qualification Procedures for Fugitive Emissions (ISO 15848-1). 2nd edition. ISO Publishing: Geneva, Switzerland.

International Organization for Standardization (ISO). (2010). Testing of Valves: Fire Type-Testing Requirements (ISO 10497:2010). 3rd edition. ISO Publishing: Geneva, Switzerland.

International Organization for Standardization (ISO). (2015). Metal Ball Valves for the Petroleum, Petrochemical, and Allied Industries (ISO 17292). 2nd edition. ISO Publishing: Geneva, Switzerland.

International Organization for Standardization (ISO). (2015). Industrial Valves: Pressure Testing of Metallic Valves (ISO 5208). 3rd edition. ISO Publishing: Geneva, Switzerland.

Nesbitt, B. (2007). Handbook of Valves and Actuators: Valves Manual International. 1st edition. Elsevier: Oxford, UK.

Skousen, W. L.. (2012). Valve Handbook. 3rd edition. McGraw-Hill Education: New York, USA.

Sotoodeh, K. (2019). Valve operability during a fire. American Society of Mechanical Engineers (ASME). Journal of Offshore Mechanics and Arctic Engineering, 141(4), 044001. https://doi.org/10.1115/1.4042073.

Sotoodeh, K. (2019). The importance of maximum allowable stem torque in valves. Springer Nature Applied Sciences, 1, 433. https://doi.org/10.1007/s42452-019-0445-0

Sotoodeh, K. (2020). Optimized valve stem design in oil and gas industry to minimize major failures. Journal of Failure Analysis and Prevention, 20, 634–640. doi: https://doi.org/10.1007/s11668-020-00891-0

Sotoodeh, K. (2020). A review of valve stem sealing to prevent leakage from the valve and its effect on valve operation. Journal of Failure Analysis and Prevention, 21, 9–16. doi: https://doi.org/10.1007/s11668-020-01050-1

Sotoodeh, K. (2021). A Practical Guide to Piping and Valves for the Oil and Gas Industry. 1st edition. Elsevier (Gulf Professional Publishing): USA

Sotoodeh, K. (2021). Subsea Valves and Actuators for the Oil and Gas Industry. 1st edition. Elsevier (Gulf Professional Publishing): USA.

Sotoodeh, K.. (2022). Cryogenic Valves for Liquified Natural Gas. 1st edition. Elsevier (Gulf Professional Publishing): USA.

Sotoodeh, K. (2024). Safety Valves in Oil and Gas Plants: Essential Design Considerations. 1st edition. Wiley: New York, NY. ISBN: 9781394294916

Zappe, R. W., & Smith, P.. (2003). Valve Selection Handbook: Engineering Fundamentals for Selecting the Right Valve Design for Every Industrial Flow Application. 4th edition. Gulf Publishing: Oxford UK.

6 Valve Manufacturing, Testing, and Quality Assurance

INTRODUCTION

This chapter provides a comprehensive overview of key and fundamentals of *valve manufacturing, testing*, and *quality assurance* (QA). Proper valve manufacturing, testing, and QA are the key to maintain industrial valve safety and reliability. The chapter is divided into three main sections for more clarity. The *valve manufacturing section* includes the key activities during the valve production, such as casting, forging, and machining. This part emphasizes the importance of manufacturing and assembling valve components. A section on *valve testing* discusses essential testing methods, such as hydrostatic, fire-safe, and fugitive emissions (FE) testing. Further, this section provides an overview of some of the most important valve testing standards. Finally, the *quality assurance* section emphasizes the importance of inspections, nondestructive testing (NDT) techniques, material traceability, and compliance with international standards such as ISO 9001. ISO 9001 is a globally recognized standard for quality management systems (QMS) and plays a vital role in the manufacturing and QA of valves. QA implementation indicates that the highest standards are maintained throughout the manufacturing and testing processes. As this chapter addresses these three topics, it provides a clear and comprehensive overview of the manufacturing, testing, and QA processes for industrial valves.

PART 1: VALVE MANUFACTURING

A

ASSEMBLY

When it comes to valve manufacturing, assembly is the process of integrating all the parts of the valve into a complete, functional assembly, such as the body, bonnet, stem, seat, disk, seals, and actuators. The process involves precise alignment and secure fitting of components to ensure proper sealing and movement, thus achieving the valve's intended function. Depending on the type of valve, assembly may involve threading, bolting, welding, or adhesive bonding (e.g., coating with epoxy).

DOI: 10.1201/9781003640967-6

B

Bolt Tightening

Bolt tightening is a critical process in valve manufacturing which is done through proper valve assembly. It involves connecting or assembling components such as valve bodies, bonnets, gaskets, and flanges using bolts which are tightened to the amount required in the valve specifications. Proper tightening creates a uniform sufficient force between components, preventing leaks and maintaining sealing surfaces integrity. This process is essential for valves to withstand operating pressures, temperatures, and other mechanical loads during operation. It is common for the tightening process to follow a specific sequence and procedure. To achieve precise bolt tightening, torque wrenches, hydraulic tensioners, or bolt stretch gauges are commonly used tools. Additionally, proper documentation and QA checks are essential. More information about bolt tightening methods is provided in Chapter 7.

Example: Gland Flange Bolt Tightening

Generally, gland bolt torque refers to the amount of rotational force (torque) applied to the bolts that secure glands (or stuffing boxes) in mechanical equipment and components such as pumps, valves, or compressors. Bolts are an integral part of gland flanges in valves. Torque or load applied to the gland flange is important since it exerts axial force on the packing rings in order to seal the stem area. As the axial load on the packing is converted to radial expansion, the packing covers and seals the area between the valve stem and bonnet. For an effective seal to be achieved and leaks to be prevented, gland bolt torque must be properly applied. A bolt that is overtightened may damage the packing material or cause unnecessary stress on the equipment, while a bolt that is undertightened may result in an inadequate seal, causing leaks.

Boring (Line Boring)

Boring is a machining process used in valve manufacturing to enlarge the internal diameter of predrilled or cast holes within valve components, such as the body or bonnet. It is common practice to bore inside a valve body during the manufacture of a ball valve. When the valve body is cast, the internal cavity in which the ball is placed may be slightly undersized. Boring is used to machine this cavity to the precise diameter required for the ball to fit securely. The process ensures that the ball cavity inside the body is concentric and sized properly. Additionally, this procedure allows the ball to rotate freely while maintaining a tight seal with the seats when the valve is closed. Through the use of specialized boring tools, such as a milling machine, material is removed to achieve tight tolerances and concentricity. *Line boring* is a specific type of boring that is used to machine multiple holes or a long bore. Line boring is applicable to multiport such as three-way or four-way ball valves.

Brazing

As part of valve manufacturing, brazing involves heating metal components to temperatures typically below 450°C (below the base metals melting points) and using a filler metal that melts and flows into the joint by capillary action in order to bond them together. As a result of the filler metal, which typically has a lower melting point than the base metals, a strong, leak-proof bond is formed when it is cooled. On the other hand, welding fuses base metals together by melting them, creating a stronger joint but with increased heat typically above 450°C. A brazing process is commonly used in the manufacturing of valves in order to assemble components such as valve stems, seats, and bodies, or valve bodies and their connections to tubing. For example, valve seats could be connected to the valve body by brazing to have leak tight joint. In control valves, stem and disk can be connected by brazing to have robust connections. In valve manufacturing, brazing is often used instead of welding to avoid melting the base materials and prevent deformation.

C

Casting

Casting is a manufacturing process in which molten metal is poured into a mold to produce the desired shape of a valve or its components. Once the metal has cooled and solidified, the mold is removed, leaving a metal valve body or part. With casting, complex shapes and designs can be produced that would otherwise be difficult or costly to produce. To summarize, casting is an essential process in the valve industry that allows the creation of durable, complex valve components. Particularly for larger parts such as larger valve body and bonnet, it is a cost-effective method for producing valves. Casting valve parts are shown in Figure 6.1.

Chamfering

The process of chamfering is used in the manufacture and assembly of valves to create an angled or beveled edge at the intersection of two surfaces. In valve manufacturing, chamfering is often applied to prepare surfaces for assembly. This includes ensuring smooth transitions for threaded connections, flanged joints, or seating areas. During assembly, chamfering facilitates the easier installation of bolts, seals, and gaskets as well as promoting better alignment. In addition, chamfering can eliminate sharp edges that can cause friction or other problems during operation. Chamfering is usually carried out using tools such as chamfer mills or specialized cutting machines. For precise and consistent chamfering, CNC machines are widely used in modern manufacturing.

Two common examples of chamfering applications in the valve industry are chamfering on valve seats to provide a correct alignment between seats

FIGURE 6.1 Casting valve parts.

and ball without damaging their sealing surfaces and chamfering the valve pup piece (piece of pipe at the end of the valve to be connected to the pipe) with an edge of 37° in accordance with ASME B16.25 standard for butt-weld end preparation.

Cleaning

Cleaning in valve manufacturing is the process of removing contaminants such as grease, oil, dirt, and other residues from valve components. Valve cleaning is performed to ensure that contamination, corrosion, or malfunction will not occur during valve operation.

CNC (Computer Numerical Control) Machining

In CNC machining, computer-controlled machines are used to automate the cutting, shaping, and drilling processes on raw materials or semifinished components. It is used to replace manual machining in order to achieve higher levels of precision, repeatability, and efficiency. Valve manufacturers use CNC machines to precisely machine valve components, such as bodies, bonnets, and stems. By using CNC machining, valves can be manufactured to meet tight tolerances and complex shape specifications for critical applications.

Coating

In valve manufacturing, coating refers to the application of a protective layer to the exterior surfaces of the valve for enhanced resistance to corrosion, wear, and environmental damage. As an alternative, lining involves applying protective materials, such as rubber, PTFE, or metallic materials, such as Inconel 625, to the internal surfaces of the valve to prevent corrosion or erosion caused by fluid service. The coating is usually the last step in the valve manufacturing process before packing and preservation. Coating is applied on the valve external surfaces after machining, assembly, and testing. Figure 6.2 shows a coated ball valve with thermal spray aluminum (TSA) coating in a coating workshop. It is common knowledge among engineers that painting and coating are the same. Coating and painting overlap, but they are two separate processes. The purpose of coatings is primarily functional, focusing on protection and performance, while the purpose of painting is primarily decorative.

Corrosion Resistance Application

In the valve industry, corrosion resistance applications involve the design, selection of materials, and application of surface treatments to protect valves from corrosive environments and extend their service lives.

FIGURE 6.2 A small size ball valve coated with thermal spray aluminum for external corrosion protection.

D

Degreasing

In valve manufacturing, degreasing involves removing oils, greases, and other contaminants from valve components. These contaminants may be a result of machining operations, lubricants, or handling during production. In order to ensure the cleanliness of valves, degreasing is a crucial step, particularly when they are intended for cryogenic applications. Degreasing improves valve reliability by preventing contamination (grease) from interfering with sealing, or valve operation. If grease accumulates on the valve body and the valve requires coating, it is necessary to remove the grease before coating the valve in order to ensure the coating adheres to the valve. Various methods are available for degreasing, including *solvent-based cleaning*, *aqueous cleaning*, and *ultrasonic cleaning*. In solvent-based degreasing, chemicals dissolve grease and remove it, while in aqueous degreasing, water-soluble detergents are used. A process of ultrasonic degreasing involves immersing components in a cleaning solution and using high-frequency sound waves to agitate and remove contaminants from surfaces and hard-to-reach areas.

Desiccant (Drying Agent) Placement

The placement of desiccants (drying agents) is intended to mitigate or prevent humidity in the valve components and sealing during transportation and storage. Desiccants are substances that absorb moisture. The placement of desiccant is not a direct part of valve manufacturing, but it is an integral part of the packaging, storage, and transportation of valves. Common drying agents include *silica gel* and *molecular sieves*. Figure 6.3 illustrates silica gel packs placed inside a wooden box used for valve transportation. These packs absorb moisture, preventing corrosion during transit and storage.

Dimensioning

In valve manufacturing, dimensioning refers to the process of identifying and specifying the geometric and dimensional properties of a valve and its components. Key dimensions include face-to-face and end-to-end dimensions, height, flange dimensions, bore size, and wall thickness.

Drilling

Drilling is the process of creating precise holes in valve components, such as the body, bonnet, flanges, or stems. A hole can be drilled for several purposes, including accommodating fasteners (in flange connections), allowing fluid flow (in valve bodies), or facilitating assembly.

FIGURE 6.3 Silica gel packs placed in a wooden box to absorb moisture and protect the valve from corrosion during transportation and storage.

E

Electropolishing

Electropolishing is a surface preparation process used in valve manufacturing. It involves the use of an electrochemical reaction to remove a thin layer of metal from the surface of the valve. Consequently, the surface becomes smooth, shiny, and corrosion-resistant. In this process, the valve or its components are immersed in an electrolytic solution. Electric current dissolves the metal surface high points. It is commonly used on stainless steel valves for a variety of reasons, including improving cleanliness, reducing surface roughness, and increasing corrosion and biofouling resistance. Food and beverage processing industries require high standards of hygiene and cleanliness, which makes electropolishing especially important.

F

Fitting

Fitting is the process of aligning individual components of a valve, such as the body, bonnet, stem, seat, disk, and seals, before assembly. To ensure the performance and functionality of valves, this process is carried out in order to maintain proper assembly of the valve components.

Flange Drilling

In the manufacturing of flanged end valves, flange drilling refers to the process of creating bolt holes in the flanges. As a result of flange drilling, flanged end valves can be securely attached to piping systems or other equipment. It is essential that the drilling pattern comply with specific standards such as ASME B16.5, ASME B16.47, or API 6A. Standard flange drilling is important because if the flanged valve needs to be changed later, the new valve manufacturer can make the same flange connection based on the standard, making it easier to replace the valve. The drilling of flanges requires precise alignment and spacing of bolt holes in order to meet the requirements for bolt circle diameters (BCDs) and hole sizes. In this way, a leak-tight connection can be achieved, and installation and maintenance can be eased. In the manufacturing of valves, flange drilling is a critical step. In piping and valve systems, improper drilling can result in misalignment, leakage, or mechanical failure.

Flange Installation

Flange installation during valve manufacturing involves attaching and securing flanges such as flanged bodies or bonnets to the valve body or central piece. This task is essential since it affects the valve's sealing and structural integrity. As long as the body flange is forged or cast as a part of the valve body (like one-piece body valve), there is no separate installation process for the body flange. However, many valves, especially ball valves, are made of two or three body pieces bolted together. In addition, body flanges can be welded to the bonnet and body. Flanges can be welded together using methods such as butt-welding or socket-welding, which could be considered flange installation. It is common for welded valves to be used in underground applications or in special situations requiring the elimination of leak risk, such as the use of hazardous fluids. Generally speaking, however, flanged end valves of large size are the most common type of flange installation. A flange is typically installed around a valve for pressure testing purposes rather than a test bench in this instance. Some valves are too large to fit on the test bench, so they are installed between two flanges during testing.

Forging

In the valve industry, forging is the process of shaping raw material (metal) into valve components by applying compressive forces, typically using a hammer or press. Forging is a completely different process from casting. The reason for this is that in casting, molten metal is poured into a mold. The term forging, however, refers to the process of shaping solid metal at high temperatures in order to create a product that is stronger, denser, and more durable. Forging is often used for critical parts such as valve bodies, bonnets, and stems. The process of forging minimizes porosity and internal voids that can be present in cast components, which makes forged parts more reliable in environments with high pressures.

Forge Masters

The forge masters are specialized suppliers or companies that produce forgings and supply them to the valve manufacturers during the project. Different processes are used by them, such as open-die forging, and they work with a variety of materials, including carbon steel and stainless steel.

Foundry

During valve manufacturing, a foundry is a specialized location and facility where molten metal is poured into molds to create valve components. Valve components such as bodies, bonnets, and other parts can be made by casting in the foundry. This step is an integral part of the production process for cast valves.

G

Gasket Fitting

The gasket fitting process in valve manufacturing involves installing a gasket between the valve's mating surfaces. In order to ensure a tight, leak-proof seal, gaskets are typically installed between the body and the bonnet or flange. The gasket serves as a compressible barrier that prevents leakage of fluids or gases under pressure.

Grinding

In valve manufacturing, grinding is a machining process used to achieve precise dimensions, smooth surfaces, and tight tolerances. Abrasive wheels are used to remove material from the surface of a workpiece. In Figure 6.4, a metal worker grinds away welded metal spatter and cleans the surface of a welded surface, creating a spray of sparks in the process. Grinding can be performed on body, bonnet, closure member, seats, and stem of a valve to remove imperfections and make smooth surfaces.

H

Hard-Facing

A hard-facing process involves welding a layer of hard, wear-resistant material onto the surface of a valve component. The purpose of this process is to increase the material resistance of valve components to abrasion, wear, erosion,

FIGURE 6.4 A metal worker grinds away welded metal spatter and cleans the surface of a welded surface. (With permission: Shutterstock.)

and galling. It is commonly used as a component in critical valve components, such as seats and closure members. In particular, Stellite 6, a cobalt-based alloy, is widely used to harden wedge gate valve seats and gates due to its excellent wear resistance. It is also common to use tungsten carbide for the ball and seat of metal-seated ball valves in applications requiring superior abrasion resistance. As a result of hard-facing, valve components have a longer service life and maintenance costs are reduced.

Heat Treatment

Heat treatment is typically performed *after forging and casting* in valve manufacturing. Heat treatment changes or enhances the mechanical properties of materials in order to meet required standards, such as hardness, strength, and toughness. In this process, a metal, typically steel or an alloy, is heated and cooled under controlled conditions. The purpose of this is to achieve desired characteristics such as increased hardness, strength, ductility, and corrosion resistance. Different types of heat treatment are used, including annealing, solution annealing, normalizing, hardening, quenching, and tempering. The following are examples of heat treatments:

Annealing: Metal is heated to a specific temperature and then cooled slowly, usually in a furnace, in order to soften it and improve its ductility. It is common for austenitic stainless steels to undergo annealing.

Normalizing: It is necessary to heat the metal to a temperature above its critical range, and then to cool it in air. As a result, a uniform grain structure is created. The critical range of a material indicates a temperature range that causes major structural changes in the material. Therefore, these changes affect the mechanical and physical properties of the material, including its hardness, strength, and ductility. For example, in carbon steel, the critical range typically occurs between 723°C and 910°C.

Quenching: The metal is heated to a high temperature and then rapidly cooled in a medium such as water. As a result, a martensitic structure is formed. This structure is used in valve stems to increase hardness and strength. Martensitic stainless steels such as 13%Cr (Chromium) and 17-4PH containing 17Chromium are common martensitic stainless steel stems used for valves in refineries and petrochemical plants achieve their strong mechanical properties by quenching.

Tempering: After quenching, the material is reheated to a lower temperature and then cooled, typically in air, to eliminate the material brittleness.

Stress Relieving (Stress Relief): The material is heated to a temperature below its critical range and then cooled slowly to relieve internal stresses without significantly altering its structural composition. It is usually performed following the welding of valve components in order to release the residual stress created during welding.

Solution Annealing: To dissolve carbide precipitates in austenitic stainless steels such as SS304 and SS316, it is necessary to heat them to a high temperature and then rapidly cool them. Consequently, stainless steel is more resistant to intergranular corrosion. For more information about this type of corrosion, please refer to Chapter 4.

L

Lapping

Lapping is the process of rubbing the valve closure member and seats against each other using diamond paste in ball or through conduit gate valves. During this process, irregularities and porosities between the valve closure member and seats are reduced, and the valve seal is improved during operation. This process is used for critical and special applications that require very tight sealing. During the lapping process, metal seats and the closure member are given extremely smooth and flat sealing surfaces. Lapping is generally not done for soft-seated valves, as the soft seats provide tight sealing with the closure member without the need for the fine surface finish that lapping provides.

Loctite

Loctite in the context of valves refers to a brand of adhesive, sealant, or thread locker commonly used to enhance the sealing, locking, and performance of valve components. Using Loctite on drain plugs in a valve, for example, would improve its sealing and reduce leakage risk. Loctite products, manufactured by Henkel in Germany, are widely used in the industrial and mechanical fields, including valve maintenance and assembly.

Lubrication

It is necessary to apply specialized lubricants to critical valve components during assembly or maintenance in order to reduce friction, prevent wear, and ensure smooth operation of the valve. It should be noted that lubrication is also used during machining for a variety of reasons, including reducing heat and friction, extending the life of the machine tool, and improving the surface finish. Lubricants are important in processes such as turning, milling, drilling, and grinding, which are commonly used to manufacture valve components.

M

Machining

Valve manufacturing involves machining by shaping, cutting, and finishing valve components to the required dimensions using a variety of machine tools in consideration of the required tolerances. During the process, raw materials, such as cast or forged valve bodies, bonnets, disks, and stems, are transformed into functional components. The position of the machining tool and the component being machined can be used to categorize machining processes. Below is a list of some of the most important machining processes:

1. *Turning:* Turning involves rotating the workpiece while a stationary cutting tool removes material. This process is typically used to machine flanged-shape components like valve body.
2. *Milling:* Milling uses a rotating cutting tool (milling cutter) to remove material from a stationary or moving workpiece.
3. *Drilling:* Drilling creates round holes in a workpiece using a rotating drill bit.
4. *Grinding:* Grinding uses an abrasive wheel to remove material and achieve fine surface finishes.
5. *Lapping:* Lapping involves using an abrasive material like diamond paste and a rotating lap between two contact surfaces such as ball and

seats in ball valves to create an extremely fine surface finish to provide perfect fit.

6. *Boring:* Boring is the process of making a hole or cavity larger within a workpiece by means of a single-point cutting tool. It is possible to use this process to make a hole in the valve body.

In Figure 6.5, machining is shown on a valve bonnet. The machining tool is fixed in position while the valve bonnet is rotating, as can be seen in the picture. As explained earlier in this chapter, this type of machining is known as turning.

Material Certificates

The material certificates are used by inspectors to verify the chemical composition, mechanical properties, and compliance with standards and end user requirements of valve materials. Consequently, these certificates can show that the raw materials meet specific industry standards and customer requirements. In this regard, it is necessary to provide material certificates from the valve manufacturer or the raw material supplier. The purpose of this is to ensure that the finished valve materials meet the required specifications. The EN 10204 3.1 certificate is the most common type of certificate issued by the quality department of a manufacturer. In this certificate, the chemical composition, mechanical properties, and test results of the material are confirmed. A certificate issued under EN 10204 3.2 provides similar information, but is verified by a third party to ensure a higher level of reliability than a certificate issued under EN 10204 3.1. Material certificates must be thoroughly checked by the inspector during valve testing.

FIGURE 6.5 Turning machining process on a valve bonnet. The machining tool remains fixed while the valve bonnet rotates.

Milling

Milling is a type of machining process that uses a rotating cutting tool (milling cutter) to remove material from a stationary or moving workpiece.

Milling Slots

When manufacturing valve components, milling slots refer to the process of creating narrow, elongated recesses or grooves (slots) using a milling machine. The topic of milling is discussed in greater detail under the "Machining" section. A slot is often made by milling the stems of valves in order to attach them to a manual operator or actuator by a stem key. It is also possible to mill slots in the valve body or seat to install seals in the form of O-rings or gaskets around the seat to provide a tight seal between the valve body and the seat.

O

Over-torque

A high amount of torque may damage the bolts, deform the flange, or crush the gasket, thereby reducing its ability to seal and increasing the risk of leakage. The opposite of over-torque is "*under-torque*" that is explained further in this chapter.

P

Packing (Packaging)

Valve packing refers to the process of securely enclosing valves in protective materials and wooden boxes to ensure their safe transport. The valve is placed within a sturdy wooden box or crate, which is often customized to fit its size and shape. Supports or fasteners are used to secure the valve during transportation in the box in order to prevent it from moving. As shown in Figure 6.6, a non-slam check valve is packed in a wooden box before transportation to the restriction yard.

Packing Installation

The packing installation in valves involves the placement of packing material around the valve stem or shaft to prevent fluid leakage between the stem and valve body. Packing material is usually cut into rings that fit snugly around the stem. It is imperative that the packing rings be cut to the correct length in order to form a continuous seal when placed around the valve stem in the packing chamber. The packing rings must be placed one by one around the valve stem in the packing chamber. To ensure that there are no gaps between the rings, each ring is carefully positioned and compressed individually.

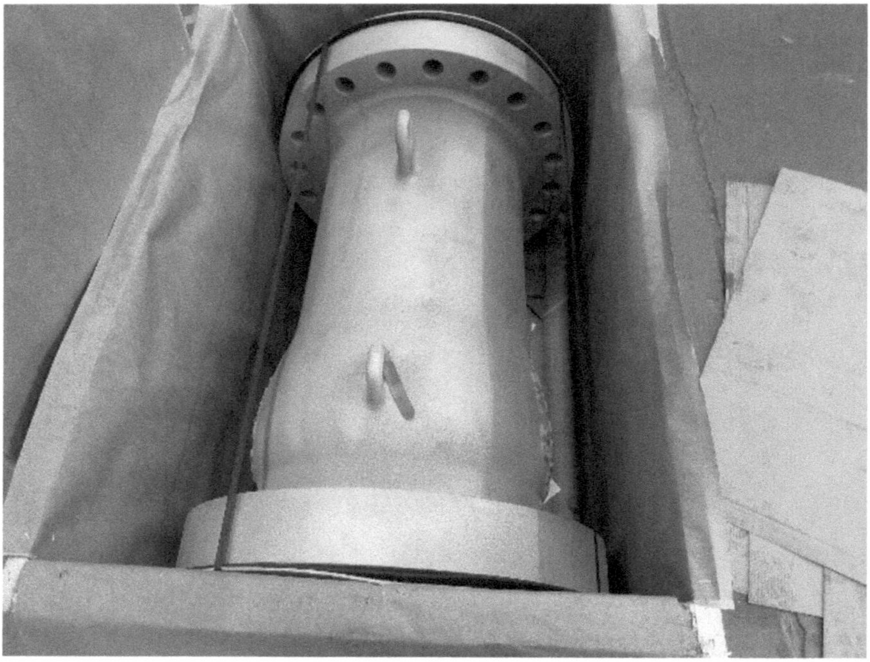

FIGURE 6.6 Packing a non-slam check valve in the wooden box before transportation to the construction yard.

Painting

It is important to note that painting and coating have different purposes in the manufacture of valves, although they may be used in similar ways. By applying a colored layer, such as red paint, painting is primarily used for decoration and identification. In this case, the red color identifies the function or association of a valve, such as its role in a fire safety system. In contrast, a coating is applied to the valve surface to prevent corrosion. Epoxy is a type of coating used in this application. An epoxy coating may be applied to a valve in a corrosive environment to prevent rust, and the valve is then painted red to allow easy identification as part of a fire system. Using a coating process enhances the durability and performance of the valve, whereas painting enhances its appearance and makes use of color coding.

Part Alignment

Part alignment during the valve manufacturing refers to the accurate positioning and assembly of valve components. Part alignment is performed to provide proper functionality, performance, and durability. An alignment process provides a tight seal is maintained, wear is minimized, and smooth operation is achieved.

Components not aligned correctly may leak, increase friction, wear unevenly, or prematurely fail.

Passivation

In the manufacture of valves, passivation is a chemical process that enhances the corrosion resistance of stainless steel and other alloys. In addition to removing surface contaminants, such as iron particles, it also forms a protective oxide layer. A solution of acid is typically used in this process for immersing valve components. During manufacturing processes such as machining, grinding, or polishing, iron particles may remain on the surface of a material, such as stainless steel. In most cases, these particles are a result of the use of tools, equipment, or the handling of materials during production.

Polishing

During valve manufacturing, polishing is used to create a smooth finish for valve components to improve their appearance and functionality. It is commonly applied to internal surfaces, such as valve bodies or flow passages, to reduce friction and enhance fluid flow. A variety of methods are used for polishing valves, including *mechanical* (abrasive tools such as sandpaper), *chemical solutions*, or *electrochemical* methods.

Porting

When it comes to the manufacture of valves, porting refers to the process of designing, machining, and finishing the flow passages (or ports) within the valve body. Valve bodies are equipped with ports that are used to direct fluid or gas flow through the valve. Inlet and outlet ports of a valve are connected by ports.

Positive Material Identification (PMI)

PMI is a NDT method used for determining the chemical composition of materials. Essentially, PMI is used to verify that materials comply with the specifications and requirements of the end user. PMI is extremely important in many industries where valves are exposed to high pressures, temperatures, and corrosive environments. The purpose of this testing is to verify that the valve is constructed from materials suitable for the intended application in order to reduce the risk of failure and maintain safety standards. PMI employs X-ray fluorescence or optical emission spectroscopy to verify that the valve's material composition meets the required specifications. It is typical for the operator to use a tool that is placed on the material and the chemical composition of the material is displayed on the device's screen. Thus, the inspector is able to compare the test result with the material certificates and standards. Figure 6.7 illustrates an operator performing a PMI test on a valve component using a handheld X-ray fluorescence (XRF) analyzer.

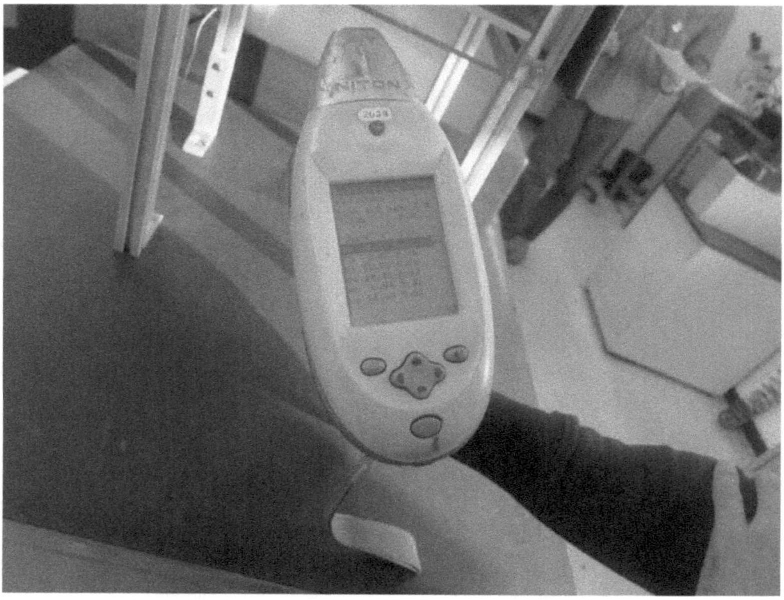

FIGURE 6.7 PMI test on a valve by an operator.

Post-Weld Heat Treatment (PWHT)

PWHT is an essential process for relieving stresses induced by welding, improving mechanical properties, and increasing the lifespan of the welding. Welding generates residual stresses in the heat-affected zones (HAZ) of valve components, which can result in cracking and distortion. Through the use of PWHT, these stresses can be reduced. In addition, PWHT enhances ductility and toughness, reducing the chance of stress cracking corrosion. In general, a single heating cycle is required for PWHT. In some cases, however, multiple heating cycles may be required due to material properties, design criteria, or welding conditions. The material will be cooled after one or more heating cycles.

Preparation for Shipment

Preparation for shipment in valves defines the necessary procedures and actions regarding packing and preservation of the valves before being transported to their final destination. This process has several purposes. In addition to maintaining valve's functionality, it prevents damage, contamination, or deterioration during storage and transportation.

Preservation

The aim of valve preservation is to protect valves from corrosion, contamination, and damage. Valve preservation is typically performed prior to packing. Applying a correct preservation indicates the valves remain in optimal condition during transportation and storage. It should be kept in mind that valves may be stored for a long period of time after transportation and may be exposed to different environmental conditions. Typical preservation tasks include cleaning to remove contaminants, applying rust inhibitors to metallic surfaces, and sealing openings with plugs or caps to prevent dirt, moisture, or debris from entering. In humid environments, moisture absorbers such as silica gel may be required to absorb moisture from valves. Additionally, moving parts such as stems and actuators are lubricated to prevent corrosion and friction. Preserving a valve also involves marking it with the details of the preservation method, such as the name of the coating or lubricant, and the chemical compounds used. It is also important for valve manufacturers to provide storage guidelines to their clients and end users regarding packing and preservation. More detailed information about valve preservation, including methods for protecting valves during storage, transportation, and pre-installation, is provided in Chapter 7.

Procedure Qualification Record (PQR)

Detailed information about PQR can be found in the "Documentation" section under the heading of "Welding Documents."

Q

Quality Control (QC)

QC refers to the systematic inspection, testing, and verification of valve products and components. The purpose of this is to ensure that they meet the defined quality standards and specifications. The QC process in manufacturing monitors different stages of production, such as casting or forging, machining, assembly, and final testing. QC is done in order to detect and correct any defects before the product is delivered to the customer. As part of the QC process, visual inspections, dimensional measurements, functional tests, and pressure or leakage tests are performed in industries such as valve manufacturing.

R

Repairing

During valve manufacturing, valve repair refers to the process of recovering damaged or worn valves or their components to their original functionality and performance. There are several tasks that are involved in this process, such as replacing defective components (e.g., seals, seats, or stems), re-machining worn or corroded surfaces, and performing lapping or polishing to ensure proper sealing and operation. In addition to reassembling, pressure testing may also be included in the repair process.

Rough Machining

Rough machining refers to the first stage of the machining process. A large amount of material is quickly removed from a workpiece during rough machining in order to bring it closer to its final shape or dimensions. In most cases, this process is used to prepare the components for subsequent finishing machining processes. Using large cutting depths and durable tools in order to handle significant forces and heat, it prioritizes efficiency over precision. There is no emphasis on tight tolerances or smooth finishes in this process. During valve production, rough machining is used to shape components such as valve bodies and bonnets made from castings or forgings. As an example of rough machining, excess material from a valve body is removed in order to achieve a general shape that will later be machined to precise tolerances during finishing operations.

S

Sandblasting

A sandblasting process is used in valve manufacturing to prepare surfaces for coatings and paints. With the aid of compressed air, abrasive particles, such as sand, are shot at high speed onto the valve's surface. The purpose of this process is twofold. First, it is used to clean valve components by removing rust, scale, old paints, or other contaminants. Additionally, sandblasting creates a roughened surface that improves the adhesion of subsequent coatings. As a result of sandblasting, the surface of the valve is clean and free of imperfections, making it suitable for further painting or coating. In harsh environments, sandblasting is crucial to improving a valve's durability, corrosion resistance (when coating), and aesthetic appeal (when painting).

Shot Blasting

In valve manufacturing, shot blasting is used to clean, smooth, or roughen valve components. In this method, abrasive materials are thrown onto the surface at high velocity (e.g., steel balls, grit, or sand). The purpose of this method is to remove impurities such as rust, scale, or old coatings. Additionally, it provides a clean, uniform surface suitable for painting, coating, or further machining.

Shot Peening

In valve manufacturing, shot peening is used to enhance the strength and durability of valve components. In this process, the surface of the valve or its parts is shot with small, round metal or ceramic balls (called shots) at high speed. As a result, tiny dents are created on the surface, creating a protective layer of compressive stress to prevent cracks from forming during operation. Compression strengthens the surface since cracks usually originate in stretched or tensioned areas. Shot peening prevents cracks from forming or spreading by keeping the surface compressed.

Silica Gel

A silica gel is an anti-moisture (drying) material used to prevent corrosion, rust, and degradation of valve components during storage and transportation. Generally, it is used inside valves or packaging to maintain a dry environment and prevent valves from being damaged by humidity.

Slot Cutting

Cutting slots is a machining process that creates narrow, elongated cuts in a workpiece. In the manufacture of valves, slot cutting is performed to create a keyway (stem key) in the valve stem. A stem key is used to connect the stem to the valve operator. Furthermore, if the gear box or actuator is removed during maintenance, the direction of the key is parallel to the hole inside the ball. Thus, it is possible to identify the position of the valve (e.g., open or close). More information about stem key is provided in Chapter 3, under section "Stem Key."

Surface Finishing

During valve manufacturing, surface finishing refers to the processes that are implemented in order to enhance the surface quality of valve components. Various surface finishing techniques are used in order to improve performance, durability, and appearance of the surface. These techniques include *coating*, *sandblasting*, *polishing*, and *passivation*. A coating provides a protective layer through the application of different coating materials such as phenolic epoxy or TSA coatings in order to resist corrosion and chemical attack. The purpose of sandblasting is to remove scale, rust, or old coatings. The purpose of this technique is also to create a rough surface for better adhesion of coatings and to clean the surface prior to coating application. In order to enhance corrosion resistance in stainless steels, passivation is commonly applied to form a protective layer of chromium oxide. Lastly, polishing can be performed in order to prevent friction and wear during the operation of the valve with other contacting components.

T

Tapping

In the valve manufacturing industry, tapping is the process of creating internal threads in a valve body or component. The purpose of this process is to accommodate bolts, screws, or pipe connections. For example, it is used to create threaded holes for drain and vent plugs in the valve body and bonnet, and form threads for threaded-end valves that connect to piping systems.

Threading

As the name implies, threading is a machining process used to create helical grooves, which are known as threads, on the inside (internal threading) or outside (external threading) of a component. As a result of this process, male and female threaded parts can attach to each other. Examples of valve threading are bolts, nuts, stems, or threaded end valves.

Trimming

The process of trimming in valve manufacturing refers to the installation of internal components (trims) that come in contact with the process fluid, such as the stem, seat, and disk.

Turning

Turning is a machining process in which a fixed cutting tool removes material from a rotating workpiece. The purpose of this is typically to create cylindrical shapes, precise diameters, and smooth finishes.

U

Ultrasonic Cleaning

The use of ultrasonic cleaning is related to the process of degreasing. This method is highly effective for removing grease, oils, and other contaminants from valve surfaces. Ultrasonic cleaning involves creating microscopic bubbles in a cleaning solution using high-frequency sound waves. As the bubbles collapse violently, they produce an intense localized cleaning action that lifts grease, dirt, and other residues from the surface of the object.

Under-Torque

If bolts are not tightened enough, it can lead to leaks, joint separation, or gasket failure due to insufficient pressure on the seal. The opposite of under-torque is over-torque that is explained earlier in this chapter.

V

Valve Calibration

Calibration refers to the process of adjusting and verifying the operational parameters of a valve in order to ensure its performance within specified tolerances. This process is very important for valves used in applications requiring accurate control, such as control valves, safety valves, and actuated valves. As far as actuators are concerned, they must be adjusted in order to function properly. Calibration involves ensuring that the response of the actuator (pneumatic, hydraulic, or electric) coincides with the input signal. Moreover, actuators enable the valves to be moved correctly (e.g., fully open, partially open, or closed).

Valve Labeling

It is possible to label valves both permanently with permanent labels such as tag plates and temporarily with temporary labels. In valve labeling, essential information is attached to or marked on a valve in order to identify the valve's critical characteristics, such as its size, pressure class, type, and direction of installation (if it is unidirectional). It is also possible to find information about the function and operating conditions of the valve on the valve tag plate. This procedure enables the valve to be traceable, in compliance with industry standards, and to be installed and used in a proper manner by operators. Valve marking and valve labeling are related but distinct processes in valve manufacturing. Valves are labeled with attachable labels, tag plates, or temporary tags. Valve marking, on the other hand, refers to permanent information that is engraved, stamped, or cast directly onto the body of the valve or its components.

Valve Marking

Valve marking is the permanent engraving, stamping, or casting of essential identification information on the valve body or other components of the valve. Valve marking facilitates traceability and compliance with industry standards, as well as proper identification during installation and operation. It is typical to include markings such as valve size, pressure class, material grade, manufacturer's name, flow direction (for directional valves), and applicable standards (e.g., API, ASME).

W

Welding

During valve manufacturing, welding is used to join metal components by melting the base materials and adding filler materials to create a strong, durable bond. Typically, it is used to assemble valve parts such as the body, bonnet, seats, and stems, as well as to attach end connections such as flanges and pipe ends. A variety of welding techniques are used, including arc welding (e.g., TIG, MIG, or stick welding), resistance welding, and laser welding, depending on the material, size, and application of the valve.

Weld Overlay

In valve manufacturing, weld overlays are used to apply corrosion-resistant or wear-resistant materials to valve components. An alloy that is corrosion-resistant is usually applied to a base material that is less expensive, such as carbon steel. However, hard-facing materials such as tungsten carbide or Stellite can be applied to corrosion-resistant alloys such as 22Cr duplex. The use of this technique increases the component's durability and resistance to critical operating conditions, such as corrosive and erosive conditions. Generally, a 3-mm thick Inconel

625 weld overlay is applied to carbon steel components in order to increase their corrosion resistance. As a result of this weld overlay, a protective barrier is created that combines the mechanical strength of carbon steel with the exceptional corrosion resistance of Inconel 625.

Welding Procedure Specification (WPS)

Detailed information about WPS can be found in the "Documentation" section under the heading of "Welding Documents."

Wrapping (Protective Wrapping)

When it comes to valve manufacturing, wrapping refers to the application of protective materials, such as tape, plastic wrapping, and corrosion-resistant coatings like VCI (volatile corrosion inhibitor) around valve surfaces. This is done to prevent damage, corrosion, or contamination from moisture, dirt, dust, scratches, or load impacts during storage, transportation, or installation.

PART 2: VALVE TESTING

A

Air Test

The air test is equivalent to the pneumatic test. For more information, please refer to the section "Pneumatic Test" in this chapter.

Actuator Tests

As an essential component of valve automation, actuators provide the force and motion required to open, close, or modulate a valve. In Chapter 11, you will find more information about actuators. A valve actuator test is an essential procedure that is performed in order to verify the functionality, reliability, and performance of the valve actuator under different operating conditions. A series of actuator tests, which are explained in more detail below, are performed during the manufacturing and assembly process.

Endurance Test

During endurance testing, the actuator is subjected to repeated cycles of operation over an extended period of time in order to determine its durability and predict its operational life cycle. As a result of this test, manufacturers are able to identify potential wear and tear and thus improve their design and maintenance schedules. It is common for subsea valves and actuators to undergo endurance testing.

Function Test

A function test is performed to demonstrate and verify that the actuators are capable of moving the valve through its full range of motion (e.g., fully open to fully closed) without experiencing any issues. In order to verify basic functionality, alignment, and response to control signals, this test could be conducted under no-load, low-load, or high-load conditions (depending on the end user's requirements). Furthermore, the actuator could be subjected to varying loads and pressures in a function test in order to ensure that it is capable of performing under operating conditions. The stroke time, and torque, or trust output of valves and actuators are measured and compared against the design parameters of engineering documents such as valves and actuators data sheets.

Pressure Test (Leakage Test)

A hydraulic or pneumatic actuator that contains pressurized hydraulic oil or air so as to be pressure-tested. In order to ensure that they are capable of withstanding the maximum operating pressure, a pressure test is performed. The purpose of this test is to ensure that the actuator body, seals, and connections are in good condition. Besides, these tests demonstrate that the actuators are not subject to leakage while maintaining the pressure test.

B

Backseat Test

Backseats in valves are additional sealing mechanisms located near the top of the valve body, which create a secondary seal around the valve stem when the valve is fully open. This feature is applicable to gate and globe valves. When a gate valve or globe valve is closed, there is no or little pressure within the body cavity. As a result, there is little pressure on the stem packing. Nevertheless, if the valve is open, the packing is subjected to the working pressure. Thus, the function of backseat is to isolate the stem packing from the fluid flowing through the valve when the valve is open. Thus, it is important to test backseat to ensure that it can isolate the stem packing and withstand the fluid pressure. More information about backseat is provided in Chapter 2, under the heading of "Backseat."

Gate and globe valve backseat tests may be performed as part of the factory acceptance test (FAT), in the same manner as body and seat tests. Some standards and procedures include testing the backseat's effectiveness in preventing leaks when the valve is fully open. Tests on backseats are typically conducted to ensure the following: Leak tightness verification that the backseat can create a sufficient seal around the stem, thereby preventing leaks. A backseat's pressure endurance is determined by its ability to withstand the internal pressure, particularly in applications involving high pressures.

Bending Moment Test

Pipeline valves are exposed to high loads from meeting pipeline. The bending moment test for valves on a pipeline is a crucial structural integrity test to ensure that a valve can withstand external forces exerted by connected pipes without mechanical failure or leakage. The valve is installed in a test fixture or pipeline mock-up, typically with flanges or welded joints. A controlled bending force is applied to the valve. Strain gauges are installed on the valve to show the amount of valve deformation under the stresses. The acceptance criteria for a bending moment test on valves depend on the industry standards such as API 6D. In general, the valve body, bonnet, and connections must not exhibit permanent deformation after removing the bending force. In addition, the valve shall not have any internal or external leakage.

Body Test

Valve body testing, also known as shell testing. The purpose of this test is to determine the structural integrity and leak-tightness of the valve's body or shell. In this test, the valve body is pressurized with water, air, and gas, and the external surface of the valve body and bonnet are monitored for possible leaks. A valve's body must be able to withstand the system's pressure without deforming, cracking, or leaking externally during this test. Thus, any leakage during this test means that the valve is rejected. A valve body test is important due to the fact that the valve body provides primary containment for the fluid or gas flowing through the valve. When a valve body fails, hazardous leaks may occur and the safety of the system may be compromised.

Bubble-Tight

The term bubble-tight refers to a level of sealing in a valve that prevents even the tiniest amount of leakage, such that not a single bubble of gas escapes when it is closed. There is no detectable leakage when closed, even under high pressure, with bubble-tight valves. This is the highest standard of sealing performance. It is good to know that bubble-tight is a type of zero leakage. Zero leakage is explained further in this chapter.

Burst Test

In the case of valves, a burst test is a destructive method of determining the maximum pressure a valve can withstand before failing. In this test, the valve's design, material strength, and compliance with industry standards are confirmed by identifying its pressure limits and failure points. This test is conducted by exposing the valve to pressures that are significantly higher than its design or operating pressure. During the test, the valve is pressurized with a fluid (usually water or oil) until it bursts or fails, with the failure pressure and location recorded.

It should be noted that burst pressure refers to the maximum amount of internal pressure that a valve can withstand before jeopardizing the structural integrity and resulting in catastrophic failure, such as cracking, rupturing, or breaking. As a critical parameter for ensuring the safety and reliability of a valve in a given application, it represents the upper limit of the valve's pressure tolerance. The burst pressure indicates the absolute failure limit, not the recommended operating range, of the valve, which is significantly higher than the maximum allowable working pressure (MAWP) or design pressure. If a valve has a MAWP of 10 bar and a burst pressure of 40 bar, the system should be designed to operate at 10 bar. It is important to note that the burst pressure is not a pressure the valve should ever experience under normal conditions.

C

Cryogenic Test

In order to test and validate the functionality, sealing performance, and material integrity of valves at extremely low temperatures, cryogenic tests are conducted. It is possible to consider various temperatures as cryogenic, including −196°C or typically below −150°C (−238°F). Performing cryogenic tests on valves intended for cryogenic applications is of the utmost importance. Fluid services categorized as cryogenic include those used in liquefied natural gas (LNG) plants, and liquid nitrogen. Tests ensure that the valve operates reliably and maintains leak-tightness in conditions that simulate the environment in which it will be used. Performing cryogenic testing is important because materials and components behave differently at extremely low temperatures compared to normal operating conditions at ambient temperatures or even at very high temperatures. A metal component may shrink or become brittle, an elastomeric seal may lose its flexibility, and a leak path may develop due to thermal contraction. Testing procedures for cryogenic valves are often guided by standards such as BS 6364, ISO 28921-1, and API 6D.

This paragraph describes the general procedure for conducting a cryogenic test. It is necessary to cool the valve down to the required cryogenic temperature. It can be frozen by either immersing it in a cryogenic medium, such as liquid nitrogen or liquid helium, or by surrounding it with a controlled cold environment. Once the valve has reached the specified temperature, a series of operational and sealing tests are conducted. The following are among them:

Operational (Functional) Test: The valve is cycled (opened and closed) to ensure smooth operation and to ensure that moving parts do not stick or cause excessive friction.

Sealing Test: The body and seat are pressurized with helium or nitrogen gas, and leak detection equipment measures any leakage from the body, bonnet, stem seals, or seats.

Seat Leakage Test: The valve seat is tested for tightness against internal leakage, often with high-sensitivity equipment to meet stringent leak rate criteria.

In Figure 6.8, a cryogenic actuated ball valve is shown undergoing a cryogenic test. Because of the extremely low temperatures, the external surface of the valve is partially submerged in a cryogenic liquid. The cold liquid may be liquid nitrogen, for example, with visible ice formation on its external surface. Due to the cold environment, the actuator on top of the valve remains functional, as the test evaluates the valve's performance under simulated cryogenic conditions. During the test, pressure gauges and sensors will be used to monitor the valve's operation and leak-tightness.

FIGURE 6.8 Cryogenic actuated ball valve during cryogenic testing. It is submerged in a cryogenic liquid with ice formation on its surface.

E

Endurance Testing

Endurance testing for valves evaluates valve durability, reliability, and performance over an extended period of working cycles. This test involves repeatedly operating the valve (e.g., opening and closing) to determine its ability to withstand wear and fatigue over time. Extended periods of tests can be combined with simulated operating conditions. Operating pressure, temperature, actuation method, and fluid type are often added into the cycle test to simulate actual operating conditions. Endurance testing is crucial for identifying potential failure modes. Besides, it is used to assess the durability of stem seals and stems. Additionally, it ensures compliance with industry standards, such as API 6DSS, which is used for subsea valves. Typically, endurance testing is performed for critical applications, such as subsea, high-pressure, or high-cycle applications.

F

Factory Acceptance Test (FAT)

FAT is a comprehensive testing process performed at the manufacturer's facility prior to valve shipment to the customer. FAT ensures that the valve meets the project specifications, standards, and required performance. Inspections or representatives of the customer or other relevant parties are typically present during a FAT. The extent of the FAT is an important consideration that is typically negotiated between the valve manufacturer and the client. As an example, FAT could be performed on 10% of manual valves, at least on one item, and could be performed on 100% of actuated valves.

Factory Internal Test (FIT)

It should be noted that valve manufacturers must test all valves internally before FAT. When the manufacturer tests everything internally without an inspector present, it is typically called a Factory Internal Test or Manufacturer's Internal Testing.

Fire-Safe Test

To minimize the risk of leaks, structural discontinuities, or other failures that could intensify a fire, a valve should be fire tested. The purpose of this fire test is to determine whether the valve is able to continue sealing and performing its functions under fire and internal pressure. A valve is heated to a temperature between 750°C and 1,382°C (1,382–1,832°F) for a specified period of time. Test duration and temperature vary according to standards and intended applications, but generally last 15–30 minutes. As part of the test, the valve is also subjected

FIGURE 6.9 A valve fire test is performed in order to determine the structural integrity, sealability, and functionality of the valve in the event of a fire.

to internal pressure and simulating operational conditions. The purpose of this pressure is to ensure that the valve's sealing elements are tested under fire conditions. In addition to the fire exposure phase, a cooling phase is conducted to determine whether the valve can reseal or maintain its integrity after it has been exposed to fire and cooled. Tests are conducted to determine whether any leakage occurs through valve components such as stems, bodies, or seats. The valve must maintain a seal under fire conditions within acceptable leakage limits and be operatable in order to pass the test. Fire tests can be conducted in accordance with various standards, such as API 6FA, ISO 10497, or API 607. When the valve passes the test, it indicates that the seals are in good condition and that it has good structural integrity when exposed to fire. Lastly, passing a fire test indicates that the valve can be operated in the event of a fire. Figure 6.9 illustrates a valve fire test to examine its integrity, sealability and functionality during fire.

Flow Testing

The purpose of flow testing in valves is twofold. First, it refers to the process of evaluating a valve's ability to control and regulate fluid flow under specific conditions, which is applicable to valves that are used for flow control, such as globe valves. Further, this testing determines the valve's flow capacity, which is often quantified using the flow coefficient (C_v).

Fugitive Emission Test

FE tests are comprehensive evaluations designed and implemented to assess the potential leakage of hazardous gases or volatile organic compounds (VOCs) from various components of a valve, such as the packing (stem seal), the body joints, bonnet seals, and sometimes the seat itself. In the test, the valve is pressurized with a test gas, such as helium or methane, at specified conditions (e.g., temperature, pressure, and number of cycles), while sensitive detection equipment is used to monitor leakage.

The primary focus of FE tests is often on the valve packing or stem seal, since this is the most common source of leaks due to the dynamic movement of the valve stem. There is a high risk of leakage when the stem moves (e.g., linear or rotational motion). Tests of this type are conducted by pressurizing the valve with helium or methane, as specified in the test procedure. Additionally, during this test, the valve is cycled with several opening and closing cycles to assess the durability of the stem seals. As frequent cycling of the valve leads to wear and friction between the stem and stem seals, it is a good testing method for assessing the seal's durability. Highly sensitive detectors can detect leakage from the stem through sealings. The ISO 15848-1 standard, the API 624 standard, and the API 641 standard provide strict guidelines for measuring and evaluating packing leakage under dynamic and static conditions. A dynamic condition occurs when the stem is moving like during cycling the valve, while a static condition occurs when the stem is stationary.

FE tests are also carried out to assess the robustness of valve body and bonnet seals. Due to the fact that the body and bonnet do not move, the joint between the two is usually static. The body and bonnet as well as their joint will, however, are subjected to the same interior pressure applied to the valve packing (stem sealing). The ISO 15848-2 standard often outlines procedures for assessing emissions from static sealing points, such as those between the body and the bonnet. In addition to testing for FE from external leaks from body and bonnet joint or stem seals, some standards and customer requirements may also include testing for internal leaks through the seats. This test on the valve seats is to ensure no significant internal leakage. Internal seat leakage is less commonly categorized under FE, as it typically does not result in gas escaping to the atmosphere.

Functional Test

Performing a functional or function test on a valve is an assessment of its performance and compliance with its functional requirements before it is installed in a system. In this test, the valve is tested for its movement and operability, including its ability to open, close, and under simulated or real operating conditions. It is essential to perform functional tests on valves and actuators in order to detect manufacturing defects, misalignments, or problems that may affect the valve's performance in service. During the functional test, the valve is cycled multiple times to ensure that the moving parts are aligned and that the valve operates smoothly. In the case of an automated valve, the actuator and control system are also tested to ensure their response appropriately to commands. This includes fully opening or closing the valve within a specified time frame. In addition to valve stroke time, torque or thrust output are key parameters measured and compared to design specifications. In order to reduce the risk of operational failures and downtime, functional tests are essential for guaranteeing that valves will perform reliably in the field.

H

High-Pressure Test/High-Pressure Gas Test

The high-pressure test is a nondestructive method used to test a valve's ability to withstand its maximum operating pressure and to ensure leak-tight performance under extreme conditions (higher than normal operating pressure). Often, high-pressure tests for valves are performed using water or gas (helium, nitrogen, or combination of both) per API 598 or ASME B16.34 standards) at a pressure 1.1–1.25 times (for the valve body and the valve seats) the maximum operating pressure of the valve. A high-pressure gas test is not typically conducted at 1.5 times the design pressure, as this could pose significant safety risks due to the compressibility and stored energy of gases. Instead, gas tests are usually performed at lower pressure multipliers.

Performing a *high-pressure gas test* on valves used in high-pressure gas services is an essential high-pressure test. As a matter of fact, high-pressure gas testing can simulate real-life conditions for these valves. It is imperative to remember that gas leakage in high pressure from valves can be extremely hazardous. Thus, it is essential to test these valves with high-pressure gas to ensure they will not leak during operation. Gas is also compressible and can pass through subtle imperfections in valves and show defects or leakage points. The test medium is helium or nitrogen or a combination of both. Helium has the smallest molecular size among gases. As a result, helium detects very fine leaks that other gases might not reveal. Nitrogen is much cheaper than helium, making it suitable for general high-pressure testing where extreme sensitivity is not required. Both gases are inert and non-flammable, making them safer than combustible gases like hydrogen or oxygen.

Hydrostatic Test (Hydrotest)

The hydrostatic test is a type of pressure test in which water is used as the test medium. The purpose of this test is to verify the strength and leak-tightness of valve components, such as the body and seats. During the test, the valve is filled with water and pressure is applied to it. As part of the body test, the pressure is typically 1.5 times the MAWP of the valve in order to evaluate its structural integrity. Generally, the seat test pressure is 1.1 times the MAWP in order to determine whether the seating surfaces are leak-tight. The valve standard and test procedure prepared by the valve manufacturer specify the duration during which this pressure is held. As part of the inspection, the inspector is looking for signs of leakage, deformation, or failure of the valve. The hydrostatic testing of valves is a critical activity to ensure the valves are of the proper quality and is governed by standards such as API 598, ASME B16.34, and ISO 5208.

Hyperbaric Testing

The purpose of hyperbaric testing is to test valves in high-pressure environments similar to those encountered in deep sea or extreme environments. A hyperbaric chamber simulates the conditions that equipment would face in deep underwater applications. A valve is submerged in pressurized water simulating a subsea water column and monitored for functionality, as well as ensuring there is no leak inside the valve.

L

Leak/Leakage

In a valve, leak/leakage occurs when fluid (liquid or gas) escapes through areas that should be sealed off, either internally or externally. There may be issues with valve components, wear, improper sealing, or damage resulting in this unwanted flow. The presence of leaks in valves can compromise the performance of the system, pose safety hazards, increase operational costs, and pose an environmental threat, especially when dealing with hazardous fluids. Internal leakage is also known as seat leakage, and occurs when fluid flows through the valve despite it being fully closed. Generally, this occurs as a result of worn or damaged seats, erosion, or buildup of debris, which prevents a tight seal. External leakage occurs when fluid escapes from the valve stem, packing, or body joints to the environment.

Leak Detection

Leak detection in valve testing is the process of identifying and measuring leakage rates. Leak detection is done during valve testing to confirm sealing integrity and industry standards. It demonstrates the valve's ability to prevent unintended fluid flow through the seating area, bonnet, or stem under operating conditions. Common methods include *hydrostatic testing* (using liquid under pressure), *pneumatic testing* (using air or gas), *bubble testing* (visual inspection for gas leaks), *pressure decay testing* (monitoring pressure drops), and *helium leak testing* (high-sensitivity detection for critical applications). These tests detect seat leakage, body leakage, and stem or packing leaks.

Leakage Rate

In a valve, the leakage rate is the amount of fluid (liquid or gas) that leaks through the valve when it is fully closed. As a measure of a valve's ability to prevent fluid from passing through its sealing surfaces, this rate is usually expressed as milliliters per minute (ml/min), cubic centimeters per minute (cm^3/min), or standard cubic feet per hour (SCFH) for gases. The type of valve, the design of the valve, the type of seat and its design, and the type of fluid all affect the valve leakage rate. API 598 and ISO 5208 are among the standards that classify valves based on acceptable leakage rates. The limits are based on the type of valve, the application, and the industry requirements. The leakage rate is typically measured during leak testing by applying a specified pressure to the valve and measuring the amount of fluid passing through the valve seat or other sealing points. ISO 5208 defines various leakage rates from A to G. For example, leakage rate A is less than zero leakage and can be achieved by using soft materials in the seat. With leak rate B, there is a small leakage rate, and with leak rate C, there is a greater leakage rate. Leak rate G, for example, allows for high leakage rates.

Leak Testing

Leak testing is performed to identify and measure the presence of leaks in industrial valves to ensure they are performing as expected and are safe. It is also performed to ensure that the valve maintains a tight seal under specified operating conditions (e.g., high pressure). Common methods include hydrostatic testing (using water or liquid under pressure), pneumatic testing (using air or gas), and bubble testing (submerging the valve in water and observing for bubbles).

Low-Pressure Test

When operating at pressures significantly lower than the valve's normal operating pressure, a low-pressure test is used to evaluate the valve's seat performance and leak-tightness. The test of low pressure would include, for example, pressurizing the valve seat with 4–5 barg of air pressure in order to detect leakage from the valve seat.

M

Metal Spray

As part of valve manufacturing, metal spraying is used to apply protective coatings to external parts of the valve, such as the body and bonnet. In this process, molten or semi-molten metal is sprayed onto a surface in order to form a thin, durable layer. In fact, metal spray is a subset of thermal spray techniques that specifically uses *metallic materials* (in wire or powder form) as the coating material. A thermal spray alloy (TSA) can be applied to valves in corrosive environments, for example.

P

Pneumatic Test (Air Test)

The air test, also known as the pneumatic test, is a nondestructive means of verifying the leak-tightness of a component, such as a valve. The pneumatic test, also known as the air test, involves the use of pressurized air as the medium for testing. The air test is typically categorized as a low-pressure test, where the pressure is normally less than 10 bar. It is very common to perform an air test on seats following a water (hydrostatic) test in order to detect any possible defects that cannot be detected by the water test. A component is filled with test air and the pressure is gradually increased to a specified level.

Pressure Gauge

A pressure gauge as illustrated in Figure 6.10 is used to measure the pressure applied to the valve during valve testing. In order to ensure that the valve has been tested at the appropriate pressure, inspectors are able to view the pressure displayed on the pressure test. The pressure gauge shows the water pressure of approximately 20 bar equal to 300 psi.

Pressure Gauge Calibration

Pressure gauge calibration is an important part of valve testing. This task is done to verify and adjust the gauge's accuracy. Implementation of this task indicates

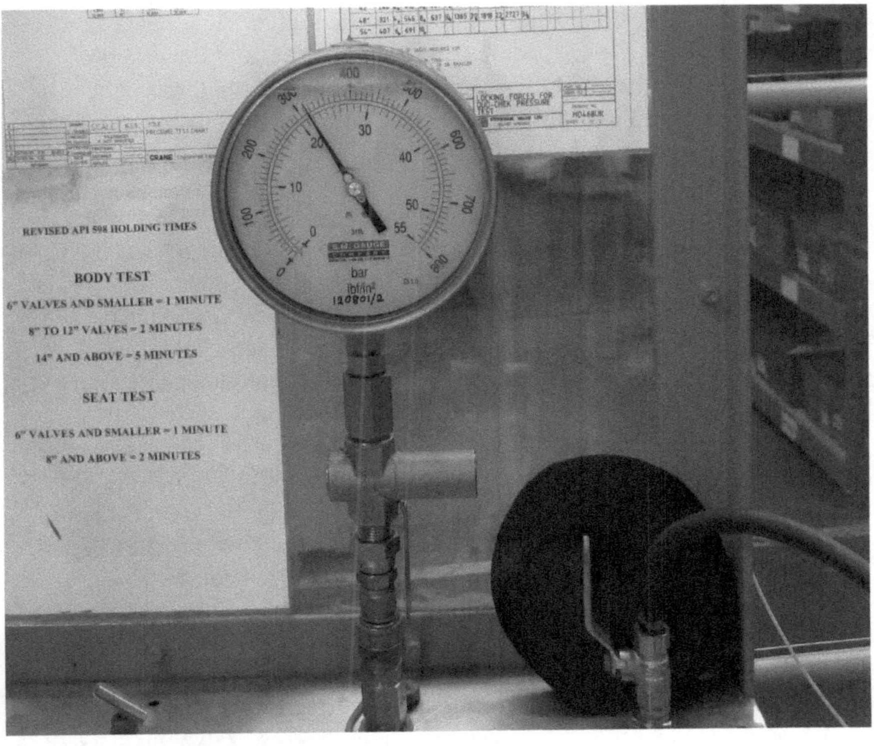

FIGURE 6.10 Pressure gauges during the valve test.

that the pressure gauge is accurate in displaying the test pressure. It is important to have a calibrated and accurate gauge during valve testing. As a result of that, valve manufacturers and inspectors can make sure that results are accurate and comply with industry standards.

Pressure Relief Test

Valve pressure relief tests are essential tests conducted to ensure that safety valves and pressure relief valves operate as intended during operations. The purpose of PRVs and safety valves is to prevent overpressurization of equipment and piping systems. The purpose of this procedure is to prevent damage to equipment, failures, or safety hazards. A number of standards govern this test, including API 527, ASME BPVC Section VIII, and ISO 4126. There are five main tests conducted on pressure relief valves, which are described below in more detail:

Set Pressure Test: The valve is subjected to a gradually increasing pressure in order to determine the exact pressure at which it begins to open (set pressure). By doing so, the valve is ensured that it will relieve pressure at the specified setpoint. It is important to note that if the safety valve fails to open at the set pressure, the pressure in the system can increase beyond its limit, resulting in fire and explosion of piping or equipment.

Leakage Test: Once the valve has been reseated (after being opened), it is tested for leakage at a specified pressure (often 90% of the set pressure) to ensure that it is tightly sealed. The purpose of this procedure is to ensure that minimal fluid loss occurs during normal operation of the system. The loss of fluid through a safety valve results in a loss of production and money.

Blowdown Test: The blowdown of a safety valve is the difference between the set pressure and the pressure at which the valve fully reseats following its opening. In order to verify whether the blowdown falls within acceptable limits defined by the industry, it is measured through the test.

Flow Capacity Test: For the valve to prevent prolonged periods of overpressurization, it may be tested under flow conditions to ensure that it is capable of discharging the required amount of fluid or gas.

Backpressure and Temperature Test: In order to validate the performance of safety valves such as set pressure determination and blowdown value of the valve and other tests mentioned above under different backpressure and temperature conditions, safety valves are tested in different piping systems with varying backpressures or high temperatures.

Pressure Test

In valves, a pressure test is performed to determine whether the valve can withstand a specified internal pressure without leaking. Valve integrity, durability, and the ability to maintain a proper seal under operating conditions must be tested under pressure. A hydrostatic test, an air or a pneumatic test that can be conducted on the valve body or shell, valve seats, or the valve backseat all are considered pressure tests because pressurized fluid or gas is used during such tests.

Production Testing

Production testing is a process used to check the quality and performance of valves during mass production. It ensures that each valve meets the required standards and specifications before being delivered to customers.

Q

Qualification Testing

A qualification test is a test done to prove that a valve meets specific requirements before it can be used. It checks the valve's design and functionality to ensure it works correctly under real conditions. A qualification test is typically performed before the start of main production for a project. On the opposite, production testing is done after manufacturing and before delivery to customers. It is a process used to check the quality and performance of valves after manufacturing. It ensures that each valve meets the required standards and project specifications before shipment.

Quality Control

For more information about QC, please refer to the "Quality Control" section under the heading of "Valve Manufacturing."

R

Reaming

In valve manufacturing, reaming is a precision machining procedure used to slightly enlarge or smooth an existing hole. The purpose of this is to achieve the exact diameter, roundness, and surface finish that is required.

S

Seat Leakage

In valves, seat leakage is the unintended flow of fluid or gas through the valve seat when the valve is closed. It occurs when the valve disk (or plug) does not form a perfect seal with the seat, which permits a small amount of fluid to pass through. It is possible that seat leakage can compromise a valve's ability to isolate or regulate flow, and may also indicate wear, damage, or improper installation.

Seat Test (Seat Leakage Test)

The purpose of a valve seat test is to determine if the seat of a valve is capable of providing a tight seal when the valve is closed. The seat is the surface against which the valve disk, plug, or ball makes contact to stop fluid flow. Valve seat tests confirm that there is no internal leakage (also known as through-seat leakage) or the leakage within the acceptable limits when the valve is closed, which is essential for applications requiring a tight shutoff.

Shell Test

Shell test and body test are the same. For more information, refer to the "Body Test" section in this chapter.

Standard Testing Condition

FAT and QC checks often use atmospheric temperature as a baseline to simulate typical operating conditions. In the valve industry, atmospheric temperature generally refers to the ambient temperature of the surrounding environment, typically between 20°C and 25°C (68°F and 77°F). Performing a test at atmospheric temperature establishes the valve's basic functionality and integrity under normal ambient conditions before subjecting it to higher or lower temperatures as required by the application.

T

Thermal Spray

An extensive category of coating techniques involves heating a material (metal, ceramic, or polymer) to a molten or semi-molten state and spraying it onto a surface as a coating to form a protective or functional layer.

Tightness Test

Tightness testing is a test conducted to verify that a valve seals properly from body and seats and prevent leakage when closed. This test ensures that there is no unacceptable leakage through the valve body, seat, or other sealing surfaces under pressure and temperature conditions. It is often necessary to perform a tightness test in addition to the body and seat tests. A tightness test typically combines both seat leakage and body leakage tests to ensure the valve's overall integrity when it is in a fully closed position. It is necessary to make sure that the entire valve assembly is leak-tight under realistic or operational conditions. The body test evaluates external sealing (e.g., body-to-bonnet joints and stem seals). The seat test evaluates internal sealing (seat-to-closure member). The tightness test verifies that both body and seats seals cooperate without leaking when the valve is assembled. It may use different media (e.g., gas for fine leak detection) or test pressures to simulate real-world scenarios. API 598 or ISO 5208 often require tightness testing.

Torque Testing

Torque testing is a procedure that measures the amount of rotational force (torque) required to operate a valve, such as opening, closing, or running the valve closure member between open and closed positions. This test ensures that the valve's torque value is within acceptable limits for manual and actuation operation. It is now possible to discuss possible problems if the measured torque during the torque test is less or greater than the designed torque.

Example 2: Torque Less than the Designed Torque

In an oil and gas pipeline, a ball valve designed for high-pressure service requires a certain amount of torque in order to open or close fully. If the measured operating torque during torque testing is less than the designed torque, this may indicate a number of issues, including seat wear or damage, a lack of engagement between the seat and ball that results in valve leakage during shutoff, a loose stem, or a lack of proper connection between the stem and the actuator or manual operator.

Example 3: Torque More Than the Designed Torque

The ball valves in a water treatment plant are operated by electric actuators. When the torque required to operate the valve is much higher than expected, it may indicate one of the following reasons. There may be excessive binding or sticking between the valve seat and the ball due to debris inside the seat or insufficient lubrication between the ball and the seat. An overtightened packing could result in excessive wear and friction with the stem results in over-torque scenario. Another possible explanation is that the actuator is oversized and is applying excessive loads.

V

Vacuum Testing

Vacuum testing in valves is a NDT method used to verify the valve's ability to maintain a vacuum without allowing external air or gases to leak into the valve. This test is applicable to valves designed for low-pressure or vacuum service environments. The process involves creating a vacuum inside the valve or its components and monitoring for any air ingress to the valve, which would indicate leakage.

Valve Calibration

Valve calibration is related to all three areas of valve manufacturing, testing, and QA. For more information, refer to "Valve Calibration" in this chapter under the heading of "Valve Manufacturing."

Valve Testing

Testing valves is the process of assessing their performance, functionality, and compliance with industry standards. Valve testing is used to demonstrate the safety and reliability of valves. Various methods are used for this type of testing, including pressure testing (e.g., hydrostatic or pneumatic), leak monitoring and recording, functional testing, and material verification.

Vibration Test

Vibration testing is used to assess a valve's ability to withstand mechanical vibrations and dynamic forces. During the test, the valve is subjected to vibrations in a vibration testing machine or shaker due to a variety of reasons, such as fluid flow, pressure fluctuations, pump operation, or external environmental conditions. Using this machine, frequencies and amplitudes can be simulated during real-world operations and conditions as specified by standards or operational requirements. This test evaluates the performance and durability of valve components under vibration, including the body, seals, seats, and fasteners. A major objective of vibration testing is to ensure that the valve will not experience fatigue, cracks, or failure as a result of vibrations.

The acceptance criteria for vibration testing of valves are explained in this paragraph. To pass the test, the valve must operate reliably and smoothly without misalignment under vibration conditions. These criteria include maintaining structural integrity. Maintaining structural integrity means no visible damage, cracks, or deformation or both internal and external leakage. Additionally, fasteners must remain secure, and the valve should withstand the test's frequency, amplitude, and duration without fatigue or functional degradation.

Z

Zero Leakage

When a valve is closed, zero leakage means that there is no detectable leakage of fluid (liquid or gas) through the valve. This means that the valve prevents any fluid from passing through, thereby creating a perfect seal. A zero leakage rate is especially important in applications where even the smallest amount of leakage can pose a safety risk, a risk to the environment, or hinder the efficiency of the process. Many zero-leakage valves use soft seating materials (e.g., PTFE, elastomers) that can deform slightly to create an airtight seal against the sealing surfaces.

PART 3: QUALITY ASSURANCE

D

Documentation

In QA of valves in a factory, documentation refers to the preparation, review, and maintenance of records and documents pertaining to valves. Documentation is used to demonstrate compliance with quality requirements, standards, customer specifications, and regulatory requirements during the valve manufacturing process. Documentation can be used to trace valves. There are several types of documentation pertaining to the manufacturing of valves:

Material Test Records or Certificates: The certificates demonstrate that the raw materials used to manufacture valve components, such as the body, bonnet, and trim, are in accordance with the required specifications and standards. Inspectors, buyers, and end users can trace valve components to their original batch in the foundry with the assistance of material certificates. This is due to the fact that material heat numbers are engraved on valve parts and are also mentioned on material certificates.

Welding Documentations: This includes records such as *welding procedure specifications* (WPS), *procedure qualification records* (PQR), and welder qualification certificates. It is the WPS that details the specific welding procedures to be followed for a particular welding job. PQR is a document that certifies that a particular welding procedure complies with the applicable codes and standards. There must be at least one PQR that supports the correctness of WPS. When the PQR has been successfully completed and approved, mass welding or production welding can be carried out based on WPS.

Manufacturing Records: Manufacturing records contain *ITP*, *dimensional inspection records*, and *production records (manufacturing and assembly documents)*.

Testing Documentations: Pressure test, functional test, and NDT reports.

Compliance Documentation: Documents showing that valves meet standards such as ASME, API, ISO, as well as compliance with *safety integrity level* (SIL) and with regulatory such as *Pressure Equipment Directive* (PED) and *CE marking*. For more information about PED, please refer to Chapter 5, section "Pressure Equipment Directive (PED)." In addition, more information about CE marking is available in Chapter 5, under section "CE Marking."

Second-Party or Third-Party Inspection Reports

Quality Documents such as QC Check, List of Nonconformities

Final Documentation Package (Final Data Book): This document is a comprehensive collection of all QA documents provided to the client as part of the final documentation package. It is often necessary to prepare a final document as part of the contract and may be associated with the payment milestone. Certificates such as material certificates, test reports, ITPs, and compliance certificates are typically included in this category.

F

Failure Mode and Effect Analysis (FMEA)

FMEA is a systematic method of identifying potential failure modes in valve components, assessing their effects on performance, and categorizing them according to their risk. As a result, it becomes possible to prioritize the mitigation actions for those with the greatest risk. A risk assessment is typically based on the probability of an event occurring as well as its effects and consequences. The risks associated with higher exposure are more likely to be mitigated through proactive measures. It is possible to reduce the risk after taking action and leave it if it is acceptable. The FMEA method systematically analyzes how a valve or its parts may fail (failure modes), determines the consequences of those failures (effects), and evaluates the probability, severity, and detectability of those failures.

Final Data Book

Detailed information about *Final data book* also called *Final documentation package* can be found in the "Documentation" section, the last paragraph.

H

Heat Number

Heat number is explained under the section of "Traceability."

I

Inspection

Inspecting valves is a comprehensive process done during and after valve manufacturing. The purpose of inspection is to determine whether valves meet the required quality standards, design specifications, and performance requirements. This is an important step in the manufacture of valves that confirms that the valve will function safely and reliably under the conditions for which it was designed. There are several types of inspections that are summarized as follows:

Visual Inspection: During this process, some problems in valves are identified visually. In addition to the lack of a tag plate, these problems may also be caused by nonconformity in the materials, such as cracks, porosity, and a lack of coating.

Dimensional Inspection or Check: During the inspection process, the inspector should verify the critical dimensions of the valve, such as the valve bore, height, face-to-face, and end-to-end dimensions, flange end dimensions, and compare them with the dimensions listed in the general arrangement drawings to ensure compliance with ASME standards, such as ASME B16.5 and B16.10, and manufacturer's specifications.

Material Inspection: Typically, an inspector confirms that the valve materials comply with the required ASTM standard. The material checking process with ASTM requires confirming that materials have appropriate compositions and mechanical properties, such as tensile strength.

Function Inspection: The purpose of the functional test is to check and verify the valve's operation, such as opening and closing. Moreover, a functional or function test measures the torque required for valve operation as well as the speed and duration of valve opening and closing. When it comes to valve operation and movement, it is imperative that parts are aligned properly and move smoothly. For example, if the seat does not align properly with the ball during closing, the valve may leak. In the case of an emergency shutdown function of a valve that takes a long time to close in the event of a process malfunction, the valve does not perform its safety function which is to close very rapidly in the event of a process failure.

Pressure Testing: The inspector inspects valves during different types of pressure testing, such as FE, cryogenic testing, and normal body or seat pressure testing.

Nondestructive Testing: This chapter provides a more detailed explanation of NDT. In addition to radiographic (RT), ultrasonic (UT), magnetic particle (MT), and liquid penetration (LT), it involves visual inspection to detect internal or surface defects in the valve body, welds, or other critical areas.

Inspection and Test Plan (ITP)

ITP is a document that is reviewed and approved by the buyer and the end user (client). It is an essential document that outlines the various steps in the manufacture of valves, such as casting, forging, and machining, along with the specific inspections that need to be performed. An ITP is an important QA document that is required during the valve manufacturing process in order to ensure that valves will meet the required standards and client requirements. A QA specification such as ITP outlines each step in the production process, the type of inspection or test to be performed (e.g., visual, dimensional, pressure, or nondestructive tests), acceptance criteria, and the responsible party (manufacturer, client, or third-party inspector).

M

Material Test Records (MTRs) or Material Test Certificates (MTCs)

MTRs or MTCs are official documents issued by the material supplier or manufacturer. The purpose of these documents is to verify the physical and chemical properties of the materials used in the production process. According to these records, the materials conform to specified standards, such as ASTM or EN. An MTR typically includes details such as material grade, heat number, chemical composition, mechanical properties (e.g., yield, tensile strength, hardness), and results of additional tests (e.g., impact or corrosion resistance). In order to ensure the quality of valves, MTRs are critical.

N

Nondestructive Testing

NDT is a method of QA used to examine and verify the integrity and functionality of valve components. By using this method, the material or component is tested without causing damage or altering its structure. Failure of a material or component due to a defect can have serious consequences in many industries. It is therefore essential for valve component materials to undergo NDT tests in order to ensure the absence of defects. NDT contributes to the prevention of potential valve failures by identifying defects that may affect performance. By implementing NDT techniques, maintenance costs during operation can be reduced and safety can be maintained. There are various methods of NDT, including *visual inspection* (VT), *magnetic particle test* (MT), *liquid penetration test* (PT), also called *dye penetration testing* (DPT), *ultrasonic testing* (UT), and *radiography testing* (RT).

Visual Inspection (VT)/Visual Examination

Visual inspection is the easiest and most cost-effective method of detecting welding defects on a material's surface. Detecting welding defects on a material's surface can be done visually or with a magnifier. It should be noted that this method of testing alone is not accurate. Therefore, it is usually combined with other methods, such as RT or UT. This method of inspection can be used to check the overall welding finish and some welding defects on the surface, such as porosity.

Magnetic Particle Test (MT)

When MT is conducted on or near the surface of ferromagnetic materials, one or more magnetic fields are used to detect welding defects. With MT, defects can be detected as deep as 2 mm under the surface. A magnetic field should be applied to the material of inspection or the weld joint in order to detect defects. There is no change in direction of the magnetic flux or magnetic field line in areas where there is no defect. An encounter with a weld defect or discontinuity causes the flux line to produce a magnetic leakage field that is transverse to the direction of travel of the magnetic field.

Liquid Penetration Test (PT)

The liquid penetration test can reveal defects and discontinuities on the surface as well as near the surface, similar to the magnetic particle test. In contrast to a magnetic particle test, a liquid penetration test can be conducted on both magnetic and nonmagnetic materials. Liquid penetration tests suffer from the limitation that they are not very effective on porous materials. There are three types of liquid used in this type of test: The first is a penetrant, a white liquid with very low viscosity and high penetration properties that penetrates into the void spaces. As soon as the penetrant is applied, it is important to wait for a period of time (for example, 15 minutes). Afterward, a red developer is applied to the material, allowing any penetrant from the voids to penetrate the developer and reveal any cracks that may exist. The third liquid, the cleaner, is then used to clean the surface. In Figure 6.11, PT is shown on a weld joint and its results indicate that two defects have been detected—a crack and porosity.

Dye Penetration Test (DT)

Dye penetration test and *liquid penetration test* are the same.

Radiography Test (RT)

During an industrial radiography test, the weldment or other test object is exposed to radiation that passes through the object being examined. X-rays are very

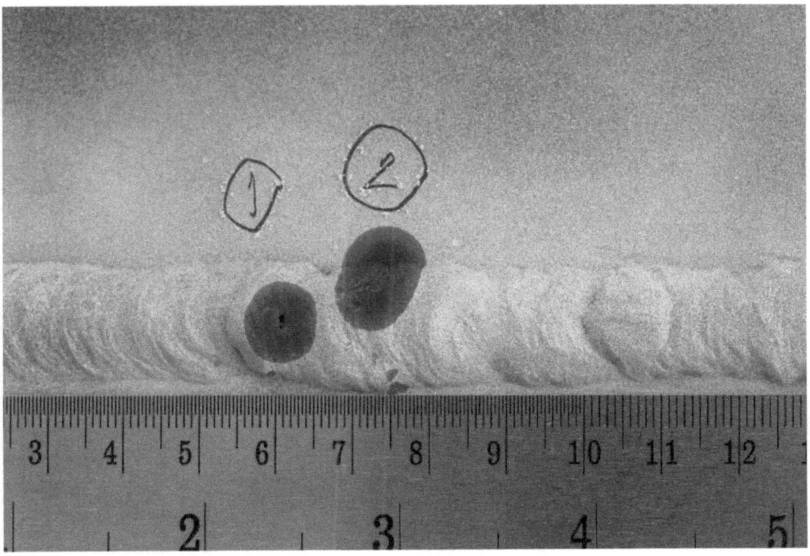

FIGURE 6.11 An LP test is performed on a weld joint by applying penetrant, developer, and cleaner, revealing two defects: Number 1: porosity and Number 2: crack. (With permission: Shutterstock.)

popular for RT examinations; however, thick components cannot be examined with them, so gamma rays may be used instead. X-rays have lower energy compared to gamma rays, which makes it difficult for them to penetrate thick or dense materials effectively. In contrast to VT, MT, and PT, RT is known as volumetric testing, which means that it can detect defects and discontinuities beneath the surface. A further advantage of RT is that it is an accurate test that can be conducted on a variety of materials. During the test, the tested material is placed between the source of radiation and the film. On the film, the RT results are presented, which can reveal defects in the welds, such as porosity. Figure 6.12 illustrates a radiographic image of a weld with porosity defects. In the weld area, the porosity appears as a number of dark, rounded spots that indicate entrapment of gas. Based on the applicable standards and acceptance criteria, these defects may reduce the structural integrity of the weld.

Ultrasonic Test (UT)

An ultrasonic test is another type of volumetric NDT that uses a high-frequency sound wave to propagate within the material. An ultrasonic beam is injected into the material to be tested in this test. Unless it is interrupted or deflected by a defect or discontinuity inside the object, the beam travels within it without loss. It is believed that defects cause a kind of frequency change or fluctuation that is reflected back to the probe as a result of the interruption of the beam.

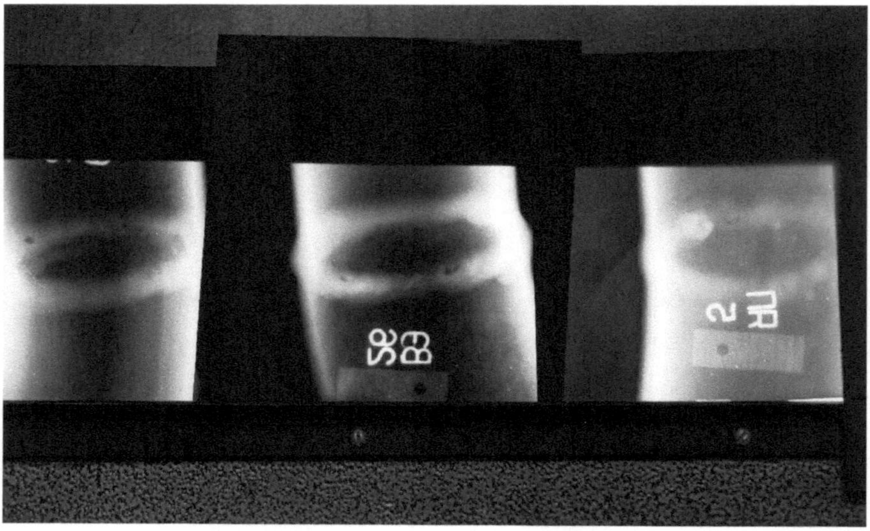

FIGURE 6.12 Radiographic image of a weld showing porosity defects. The dark circular spots are gas voids trapped within the weld metal, indicating areas of reduced density. (With permission: Shutterstock.)

Q

Quality Control

For more information about QC, please refer to the "Quality Control" section under the heading of "Valve Manufacturing."

R

Record-Keeping

Maintaining a record of all maintenance activities, inspections, and tests for future reference.

T

Testing Records

In these documents, the results of calibration or testing are recorded, such as pressure tests, cycle tests, or leak tests, which verify that the valve meets its specifications. The manufacturer of the valve creates testing records after it has performed internal and FATs.

Third-Party Inspection (TPI)

TPI consists of an independent inspection process and product evaluation conducted by a neutral organization that provides inspection services. As with other types of inspection, this is performed on valves and their components to ensure compliance with required quality, safety, reliability, and regulatory standards. Unlike first-party inspections, which are performed by manufacturers, or second-party inspections, which are conducted by buyers or end users, TPI is performed by inspectors of a company that is involved in inspection activities.

Traceability

Traceability in valves refers to the ability of valve manufacturers, inspectors, and end users to track and document the history, application, and material composition of a valve over its lifetime. In industries such as oil and gas, power generation, and petrochemicals, it is a crucial QA and compliance process. There are several documents used for traceability, including *material traceability*, *manufacturing records*, *testing and inspection records*, and the *unique identification number* of the valve.

Material Traceability: To ensure material traceability, materials used in the valve (e.g., body, bonnet, seats, and stem) must meet the required specifications and contain coding or numbers contained in the material certificate. Additionally, the material reports must indicate compliance with relevant standards, such as ASTM. In the context of material traceability, the heat number is an important concept. A *heat number* is a unique identification code related to a batch of metal produced during a casting operation in a foundry. Heat numbers as an important part of material traceability enable manufacturers, suppliers, and end users to trace the origin and composition of a valve component. Typically, heat numbers consist of alphanumeric characters representing the melt sequence, the foundry code, and the date of manufacture.

Manufacturing Records: Manufacturing process documents, such as forging, casting, and machining, are included in manufacturing records. These records assist in identifying possible issues during the production process. In addition, these records show consistent quality throughout the production process.

Testing and Inspection Records: As part of the traceability process, records and reports of inspections, such as radiographic and ultrasonic inspections, are maintained. Moreover, all tests, such as hydrostatic tests and performance tests such as pressure relief and cryogenic tests, must be documented with a traceable document number.

Unique Valve Identification Number: An identifier (such as a serial or tag number) is affixed to each valve in order to link it to its production and quality records. A manufacturer's mark may contain information about the model, size, pressure class, and material of the product.

V

Valve Calibration

Valve calibration is related to all three areas of valve manufacturing, testing, and QA. For more information, refer to "Valve Calibration" in this chapter under the "Valve Manufacturing" section.

Valve Marking

Valve marking is related to areas of both valve manufacturing and QA. For more information, refer to "Valve Marking" in this chapter under the "Valve Manufacturing" section.

W

Welding Inspection

A welding inspection is performed during the manufacturing process of valves in order to ensure the quality and integrity of welded joints. The purpose of this inspection is to identify welding defects such as cracks, porosity, or incomplete fusion that may compromise the performance of the valve. Visual examination and other types of NDT are involved in this process.

Welding Procedure Specification (WPS)

Detailed information about WPS can be found in the "Documentation" section under the "Welding Documents" section.

X

X-Ray Test

It is a kind of radiography test. Radiography test can be performed with either *X-ray* or *gamma ray*. It is applicable to thin components.

QUESTIONS AND ANSWERS

1. Which of the following processes are typically included in the manufacturing of high-performance valves to maintain their reliability in critical applications? High-performance valve is a type of industrial valve that is designed for high-pressure, high-temperature, and corrosive environment. The valve will be installed after a pump and handles large amount of hydrogen sulfide.
   A. Casting or forging of the valve body
   B. Stress relief heat treatment
   C. Ultrasonic or radiographic testing of components
   D. CNC machining of sealing surfaces
   E. Vibration testing to simulate real-world conditions

   Options
   1. A, B, and D
   2. B, C, and E
   3. A, B, C, D, and E
   4. A and C

   Answer: The correct answer is 3. A, B, C, D, and E, as all these processes are essential in the manufacturing of high-performance valves in this condition. Casting or forging (A) is used to create a strong, durable valve body and bonnet that can withstand high pressures. By using a stress relief heat treatment (B), residual stresses from welding or machining can be reduced to prevent cracks, improving the valve's stability and preventing hydrogen sulfide-induced deformation cracks. Ultrasonic or radiographic testing (C) is conducted as more advanced NDT method to detect internal flaws in the valve body or welds. CNC machining (D) ensures the precise sealing surfaces and tight tolerances for leak-free operation that is essential regarding the presence of hydrogen sulfide as a toxic substance in the fluid service. Finally, vibration testing (E) is essential as the valve is installed after pump and subjected to vibration caused by the pump. Together, these processes ensure that the valve meets the high-quality standards necessary for demanding environments like oil and gas, chemical, or power industries.

2. In accordance with API 598, a seat test during valve testing involves which of the following actions?
   A. Pressurizing the body cavity and monitoring the external surface for leaks
   B. Pressurizing one side of the valve while the other side is monitored for internal leakage
   C. Applying a vacuum to the valve body to ensure zero leakage
   D. Cycling the valve through its full range of motion under pressure to evaluate operational integrity

   1. A and B
   2. B only
   3. C and D
   4. A, C, and D

   Answer: The correct answer is B. Seat tests are conducted by pressurizing one side of the valve when it is closed and monitoring the other side (behind the seat) for leaks in the seat.

3. Which of the following are critical for maintaining the sealing performance of a valve during vibration testing?
   A. Precision machining of the seat surfaces
   B. Proper alignment of the valve stem and actuator
   C. Implementation of corrosion-resistant coatings on the body
   D. Sufficient preload on the body-to-bonnet bolts

   1. A and B
   2. B, C, and D
   3. A, B, and D
   4. All are correct

   Answer: The correct answer is option 3—A, B, and D.

   Precision Machining of the Seat Surfaces (A): The seat is the primary sealing surface in a valve, and precision machining of the seat provides a proper fit with the closure member (e.g., disk, ball, or gate). Even minor imperfections can lead to leakage during operation, especially under vibrations that may cause misalignment or gaps. Properly machined seat surfaces reduce leakage risk.

   Proper Alignment of the Valve Stem and Actuator (B): When vibrations occur, the valve stem and actuator can become misaligned, as can the valve stem and the body and bonnet. In the case of misalignment, seals can wear more quickly and leak from the stem seals as a result.

   Sufficient Preload on the Body-to-Bonnet Bolts (D): The bolts securing the body and bonnet must have adequate preload to prevent loosening due to vibrational forces. A loose body-to-bonnet connection can lead to external leaks.

4. Which process is used in the beginning to create the valve body?
   A. Casting
   B. Welding
   C. Machining
   D. Tapping

   Answer: The correct answer is option A.

5. Which process is most commonly used to achieve precise sealing surfaces in valve manufacturing?
   A. Sandblasting
   B. CNC machining
   C. Welding
   D. Stress relief heat treatment

   Answer: The correct answer is option B.

6. During a hydrostatic test, what is the primary purpose of pressurizing the valve body?
   A. To verify the internal leakage through the seat
   B. To test the structural integrity of the valve body and joints
   C. To simulate real operational conditions for the valve
   D. To confirm vibration resistance of the valve

   Answer: The correct answer is option B.

7. Which method is most appropriate for detecting internal defects in the valve body?
   A. Ultrasonic testing
   B. Visual inspection
   C. Torque testing
   D. Bubble testing

   Answer: The correct answer is option A.

8. What is the most critical factor in selecting protective wrapping for valves?
   A. The wrapping must be transparent.
   B. It should enhance the appearance of the valve.
   C. It must be easy to remove during installation.
   D. The wrapping should prevent moisture ingress.

   Answer: The correct answer is option D.

9. How are valves typically preserved to prevent corrosion during long-term storage?
   A. By storing in low-humidity rooms.
   B. By wrapping in transparent plastic sheets.
   C. By applying a volatile corrosion inhibitor (VCI) in wooden boxes used to transport the valves.
   D. All answers are incorrect.

   Answer: The correct answer is option C.

10. What is the primary purpose of stress relief heat treatment in valve manufacturing?
    A. To improve the surface finish of the sealing area
    B. To enhance the corrosion resistance of the valve
    C. To increase the hardness of the valve material
    D. To reduce residual stresses caused by machining or welding

    Answer: The correct answer is option D.

BIBLIOGRAPHY

American Petroleum Institute (API). (2021). Valve Inspection and Testing (API Standard 598). 11th edition. API: Washington, DC.

American Petroleum Institute (API). (2021). Seat Tightness of Pressure Relief Valves (API Standard 527). 8th edition. API: Washington, DC.

American Petroleum Institute (API). (2021). Fire Test for Quarter-Turn Valves and Valves Equipped with Non-Metallic Seats (API Standard 607). 8th edition. API: Washington, DC.

American Petroleum Institute (API). (2021). Type Testing of Rising Stem Valves Equipped with Graphite Packing for Fugitive Emissions (API Standard 624). 2nd edition. API: Washington, DC.

American Petroleum Institute (API). (2022). Fugitive Emissions Type Testing of Quarter-Turn Valves (API Standard 641). 1st edition. API: Washington, DC.

American Petroleum Institute (API). (2022). Specification for Wellhead and Tree Equipment (API 6A-2022). API: Washington, DC.

American Petroleum Institute (API). (2021). Pipeline Valves (API Standard 6D). 25th edition. API: Washington, DC.

American Petroleum Institute (API). (2018). Specification for Fire Test for Valves (API Standard 6FA). 4th edition. API: Washington, DC.

American Society of Mechanical Engineers (ASME). (2021). Pipe Flanges and Flanged Fittings (ASME B16.5-2021). ASME: New York, NY.

American Society of Mechanical Engineers (ASME). (2021). Face-to-Face and End-to-End Dimensions of Valves (ASME B16.10-2021). ASME: New York, NY.

American Society of Mechanical Engineers (ASME). (2020). Large Diameter Steel Flanges: NPS 26 through NPS 60 (ASME B16.47-2020). ASME: New York, NY.

American Society of Mechanical Engineers (ASME). (2021). Valves: Flanged, Threaded, and Welding End (ASME B16.34-2021). ASME: New York, NY.

American Society of Mechanical Engineers (ASME). (2021). Boiler and Pressure Vessel Code: Rules for Construction of Pressure Vessels (ASME BPVC Section VIII, Division 1). ASME: New York, NY.

British Standards Institution (BSI). (1984). Specification for Valves for Cryogenic Service (BS 6364:1984). BSI: London, UK.

International Organization for Standardization (ISO). (2015). Industrial Valves: Measurement, Test, and Qualification Procedures for Fugitive Emissions—Part 1: Classification System and Qualification Procedures for Type Testing of Valves (ISO 15848-1:2015). ISO: Geneva, Switzerland.

International Organization for Standardization (ISO). (2015). Industrial Valves: Measurement, Test, and Qualification Procedures for Fugitive Emissions—Part 2: Production Acceptance Test of Valves (ISO 15848-2:2015). ISO: Geneva, Switzerland.

International Organization for Standardization (ISO). (2022). Safety Devices for Protection against Excessive Pressure: Part 1: General Requirements (ISO 4126-1:2022). ISO: Geneva, Switzerland.
International Organization for Standardization (ISO). (2015). Industrial Valves: Pressure Testing of Metallic Valves (ISO 5208:2015). ISO: Geneva, Switzerland.
International Organization for Standardization (ISO). (2022). Testing of Valves: Fire Type-Testing Requirements (ISO 10497:2022). ISO: Geneva, Switzerland.
International Organization for Standardization (ISO). (2013). Industrial Valves: Isolating Valves for Low-Temperature Applications—Part 1: Design, Manufacturing, and Production Testing (ISO 28921-1:2013). ISO: Geneva, Switzerland.
International Organization for Standardization (ISO). (2015). Quality Management Systems: Requirements (ISO 9001:2015). ISO: Geneva, Switzerland.
Nesbitt, B. (2007). Handbook of Valves and Actuators: Valves Manual International. 1st edition. Elsevier: Oxford, UK.
Skousen, W. L. (2012). Valve Handbook. 3rd edition. McGraw-Hill Education: New York, NY.
Sotoodeh, K. (2016). Valve preservation requirements. *Valve World Magazine*, June 21, pp. 53–56.
Sotoodeh, K. (2021). A Practical Guide to Piping and Valves for the Oil and Gas Industry. 1st edition. Elsevier (Gulf Professional Publishing): USA
Sotoodeh, K. (2021). Prevention of Valve Fugitive Emissions in the Oil and Gas Industry. 1st edition. Elsevier (Gulf Professional Publishing): USA.
Sotoodeh, K. (2022). Cryogenic Valves for Liquified Natural Gas. 1st edition. Elsevier (Gulf Professional Publishing): USA.
Sotoodeh, K. (2020). Subsea valves and actuators: A review of factory acceptance testing (FAT) and recommended improvements to achieve higher reliability. Journal of Life Cycle Reliability and Safety Engineering, 10(2018), 3. doi: https://doi.org/10.1007/s41872-020-00153-w
Sotoodeh, K. (2024). Positive material identification procedure development to prevent valve corrosion failures. Journal of Failure Analysis and Prevention, 24, 1512–1521. doi: hpps://doi.org/10.1007/s11668-024-01951-5
Zappe, R. W., & Smith, P. (2003). Valve Selection Handbook: Engineering Fundamentals for Selecting the Right Valve Design for Every Industrial Flow Application. 4th edition. Gulf Publishing: Oxford UK.

7 Valve Installation and Commissioning

INTRODUCTION

To maintain the reliability, safety, and duration of valves within a piping system during operation, proper installation and commissioning are critical steps. The manufacturing process is done to provide the valve's initial quality and functionality. However, the installation and commissioning phases are between the manufacturing and operation phases. In order to complete installation, it is necessary to carry out a thorough preparation and installation, an extensive testing, and effective documentation. The purpose of tests after installation is to ensure that the valve integrates properly with the piping system and performs as intended under operational conditions. This chapter is divided into six sections to show key tasks and activities in each phase of installation and commissioning.

1. *Preparation and Preinstallation:* Valves are checked during this phase and other required preinstallation tasks are performed. These tasks include cleanliness, storage, and handling. In addition, it is essential to maintain proper alignment between the valves and connected piping systems during installation.
2. *Valve Installation:* Actual tasks that are required to install valves are covered in this section. These tasks contain bolt tightening, flange connections, alignment, and securing actuators.
3. *System Integration:* The concept of system integration focuses on how valves are incorporated into a broader piping system. There are a number of considerations addressed in this section, including pipe stress, system flushing, flow path alignment, and instrumentation consideration.
4. *Testing and Verification:* It contains all the necessary tests such as hydrostatic, seat and functional testing. These tests validate the valve's performance while it is integrated with piping systems. In addition, these tests ensure compliance with operational requirements and industry standards.
5. *Documentation and Records:* In order to provide a clearer and more traceable picture of what has been done during installation and testing, installation and commissioning records, and inspection checklists must be recorded and maintained.
6. *Packing and Preservation:* This section focuses on protecting valves during transport, storage, or delays in commissioning. Detailed packing

DOI: 10.1201/9781003640967-7

and preservation methods are discussed in this section like protective wrapping, moisture prevention, and long-term preservation techniques.

PART 1: PREPARATION AND PREINSTALLATION

Bolt Hole Alignment

A bolt hole alignment refers to the process in which the persons responsible for the installation verify that the bolt holes on the valve flange and the pipe flanges (or other components connected to the valve flange) are aligned correctly. Performing this step is essential to ensure that the connection between the flange-ended valve and the connected flange is leak- and stress-free during operation.

Cleanliness

During the preparation and preinstallation of valves, cleanliness implies the removal of contaminants from all components, including valves especially valve flanges, piping flanges, and gaskets. It is necessary to remove various contaminants such as dirt, oil, grease, rust, scale, welding debris, and other foreign materials from valves and connected piping. It is important to remove contaminants for several reasons, such as causing damage to the seals and internal components of the valve, causing seal malfunction and corrosion, and preventing smooth operation.

Flange Inspection

As part of the preparation and preinstallation of valves, flanges need to be inspected to ensure they are in proper condition. Having a good condition means that they will fit well with the valve and function correctly once it has been installed.

There are several key aspects to be examined when inspecting a flange. By carefully reviewing these key aspects, you can ensure proper installation and performance of the flange during operation. A visual inspection is conducted in order to detect surface damage, corrosion, or contaminants. Furthermore, it is important to verify the flatness of the flange faces where the gasket is installed, in order to ensure an even compression of the gasket. The alignment of bolt holes between two mating flanges is essential to ensuring proper installation and sealing. It is important that the sealing surfaces be smooth and undamaged in order to achieve an effective seal. Bolt holes and threads must be examined for integrity. It is important to maintain control over the material compatibility and pressure rating of the flange. An example of this would be the installation of a flange with class 150 in a class 300 piping system by mistake, which would result in flange damage and leakage. The installation of a carbon steel flange on a stainless steel line may result in flange corrosion. Ensure that the dimensions of the flange comply with ASME B16.5 or any other applicable standard. Inspecting the dimensions of a flange can reveal if there are any misalignments between the flange and the pipe that may result in leaks or excessive stress on the system.

Flow Direction Marking

A flow direction is typically marked on the valve body for indicating the correct direction in which fluid flows through a valve. It is typically marked by arrows on the valve body or flanges. Additionally, it may be displayed on labels, tags, or stickers attached to the valves. It is true that some valves are bidirectional, meaning they can be mounted in both directions by rotating the valve 180° and connecting it to the piping. There are, however, some valves with mandatory flow directions, especially check valves and globe valves. There may be times when the valve manufacturer has preferred requirements regarding flow direction when it comes to butterfly valves. The installation of check valves or globe valves in accordance with the flow direction is essential in order to ensure that the valves will function properly. For instance, if the check valve is incorrectly installed in the line on the opposite side of the flow direction, the flow will keep closing the check valve, which can cause damage to the equipment upstream. Flow direction plays a significant role in the prevention of system malfunctions, equipment damage, and safety hazards. As shown in Figure 7.1, three dual-plate check valves are marked on their bodies with the direction of flow.

Gasket Preparation

When preinstalling and preparing valves, proper gasket preparation is critical to achieving a leak-proof seal between the valve and the flanges that valves are

FIGURE 7.1 Three dual-plate check valves with flow direction markings on their bodies.

connected to. During this process, the appropriate gasket size and material is selected based on factors such as pressure, temperature, and the type of fluid. For example, spiral-wound gaskets should not be used for high-pressure flanged end valves with pressure classes of 600 and higher. Rather, a ring-type joint gasket is suitable for high-pressure applications. The gasket surface and the mating flange surfaces of the valve and pipe flanges must be thoroughly cleaned. Proper cleaning is done to remove contaminants such as dirt, oil, rust, or debris. It is important to inspect the gasket carefully for any damage, including cracks and tears, and to replace it if necessary. Gaskets must be installed correctly and evenly between the flanges, ensuring they cover the entire sealing surface without misalignment (when using flat gaskets or spiral wound gaskets).

Handling

As part of valve preparation and preinstallation, valves and their components must be handled carefully before installation in a system. Proper handling is important to prevent valve damage and contamination. Suppose, for example, that a large valve is lifted and dropped during its transport to the installation site. As a result of the impact, the flange is dented and the sealing surface is scratched. The valve may not be able to form a proper seal when installed as a result of this condition. As shown in the second example, it is important to avoid contamination of valves before they are installed. In a humid environment, valves are stored outside without protective covers. When moisture and dirt enter the internal components of the valve, rust begins to form on the seating surface and the internal components of the valve. The result is corrosion of the valve on the flange face, which ultimately leads to the failure of valve sealing.

Identification Tags

Identification tags in valve preinstallation and preparation are labels or markers attached to valves by valve manufacturers. They provide essential information about the valves such as size, materials, pressure class, and other relevant details. Tags are essential before and during installation because they can show that the valves are installed in the correct location and direction. Additionally, they assist in the traceability of the valves during installation.

In order to identify a dual-plate check valve, a temporary identification tag is attached (see Figure 7.2). It indicates CHAD50R, which is a code defined in the piping material specification (PMS). CH signifies a check valve, A signifies that it is a pressure class 150 valve, D indicates that it is a duplex valve, and R signifies that it has a raised face on the body flange.

Inspection

During valve preparation and preinstallation, valves and their components must be thoroughly inspected. This ensure they meet all specifications, quality standards,

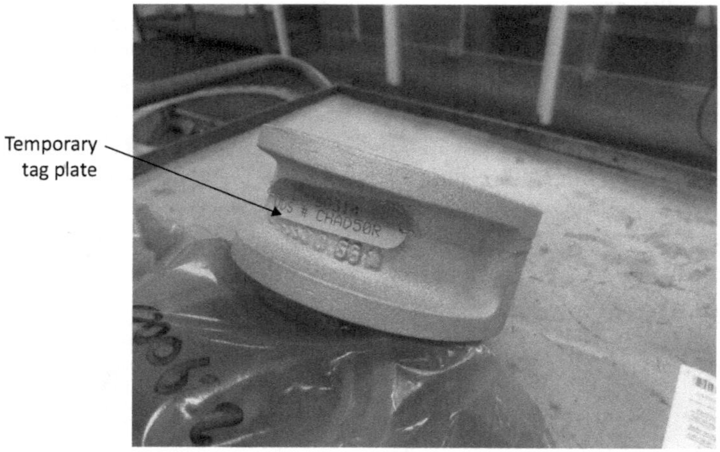

FIGURE 7.2 Temporary identification tag for a dual-plate check valve, indicating the code CHAD50R, as defined in the piping material specification (PMS).

and are ready for installation. In addition to visual inspection for cracks, corrosion, or other damage, dimensional checks, pressure testing to detect leaks, material checks, actuators and accessories, such as control panels, and functional testing all are included in the inspection process.

Lockout/Tagout (LOTO)

A lockout/tagout (LOTO) is performed prior to valve installation in order to ensure the valve is locked and cannot be operated during installation. By placing locks on the valve, it is prevented from being operated. Additionally, tags are attached that warn others not to touch or operate the valve. By doing so, workers are kept safe from accidental releases of pressure, electricity, or other dangerous energies.

Manufacturer Instructions

As the name implies, manufacturing instructions in preinstallation and preparation refer to detailed guidelines provided by valve manufacturers regarding how the valve should be handled, installed, tested, and maintained before it is placed in service. The manufacturer of the valve usually provides these data in a document called the installation, operation, and maintenance manual (IOM). This manual contains instructions on handling, inspection, installation, testing, lubrication, cleaning, and safety precautions.

Material Compatibility

During valve preparation and preinstallation, material compatibility is determined by verifying that the materials used in the valve, including its body, seals, and gaskets, are compatible and suitable for the fluids, gases, temperatures, pressures, and environmental conditions that they will encounter once installed.

Pipe Stress

Several critical and negative effects can be caused by stress during valve preparation and preinstallation. When the valve and mating flange are not aligned properly, excessive stress can be applied to the valve flanges, leading to leaks. Due to temperature changes, thermal expansion of piping and valves can cause the valve to bend or deform, especially if it is not properly supported. Ensure that the valve installation provides enough space for the piping to expand and contract freely without pressure. It is possible that over-torque during flange installation may result in additional stress being applied to the valve body, causing it to become distorted or cracked as a result. Vibrations within the piping system may also cause the components of valves to become fatigued. A pipe support or expansion joint can be installed near the valve to absorb vibration and prevent the valve from moving. Leaks and safety hazards can result from all of these stress-related issues. As part of the valve preparation and preinstallation phase, it is important to manage pipe stress in order to maintain a safe and effective operation of the valve.

Valve Orientation

Valve orientation in valve preparation and preinstallation refers to the correct positioning and alignment of the valve within the pipeline system. This is done to confirm proper operation and prevent leaks, malfunctions, or damage.

PART 2: VALVE INSTALLATION

Valve installation typically takes place during the mechanical completion phase of a project. Mechanical completion is the stage where all physical components, including piping, valves, equipment, and structural elements, are installed, inspected, and verified to be in accordance with design and project specifications.

Accessibility

As part of the valve installation process, valve accessibility refers to the installation of the valve in a location that facilitates easy operation, inspection, maintenance, and repair throughout its lifespan. It is critical to position the valve manual operator (e.g., handwheel) or actuator at a convenient height and location. As a result, operators can easily operate it daily. Sufficient space should be provided around the valves for routine maintenance tasks. Routine maintenance tasks include leakage checking, bolt tightening, replacing seals, or conducting pressure tests. Moreover, there should be sufficient clearance around the valve to facilitate the use of maintenance and installation tools and equipment.

Actuator Alignment

In valve installation, actuator alignment refers to verifying that the actuator is properly positioned and aligned with the stem and body of the valve. In order to ensure smooth, accurate operation of the valve components, the actuator shaft must be aligned with the valve stem. The actuator should be correctly installed on the valve so that it does not apply excessive force or torque on the valve body. An actuator misaligned could either overload or underperform, causing the valve to fail to function properly. When moving, ensure that the actuator does not interfere with any nearby piping, supports, or other components. It is also important to ensure that there is adequate space for the actuator's full range of motion.

Alignment

Alignment in valve installation refers to proper installation of the valve and actuator, as well as the connecting piping, which are properly positioned and aligned within the system. The valve should be aligned with the piping system to prevent stress on the valve flanges or body. Misalignments may result in uneven sealing, leaks, or improper operation. In order for the gasket to properly seal, the flange faces of the valve and mating flange must be parallel to each other. In addition, the valve must be installed in accordance with the flow direction arrow marked on its body or tag plate. To ensure smooth operation, the actuator must be properly aligned with the valve stem.

Bolt Tensioning

A bolt tensioner applies a precise amount of force to the bolts in order to secure them between the flanged end valve and the mating flanges. By using this method, the bolts can be fastened with a greater degree of accuracy than by using a torque tool. Hydraulic tensioning tools are placed over bolts and attached to hydraulic pumps. Hydraulic pressure is applied to the bolt in a precise, controlled manner using the tool. During the process of bolt tensioning, the bolt elongation is measured and directly relates to the tension applied. After bolt elongation, the nut must be turned to ensure that the connection is tight and properly seated. Figure 7.3 illustrates hydraulic tensioning tools being used to tighten bolts on a valve during the manufacturing. The same tools and methods can also be used during valve installation in the construction yard to tighten the bolts between the flanged end valve and mating flanges.

Bolt Tightening

In valve installation, bolt tightening refers to the securely fastening of the valve to the piping system. It is achieved by tightening the bolts on the valve flanges to the required torque specifications. During operation, it is imperative that the bolts are tightened properly in order to achieve proper sealing between the valves and

FIGURE 7.3 Bolt tensioning tools are used to tighten the bolts between body and bonnet of a large ball valve (BV) during valve manufacturing.

their mating flanges. The tightening of bolts can be accomplished using a *torque tool* or by *bolt tensioning* of the bolts.

Connection Type (Valve End Connection)

Valve connection types refer to how a valve is attached to pipes within a system. The most common type of connection is a flanged connection, where the valve is bolted between two flanges. For noncritical fluid applications such as air and water, threaded connections are suitable for small piping less than 2″ with threaded ends that screw directly into the pipe. A welded connection includes buttweld and socket weld (applicable for a maximum diameter of 2″) where the valve is permanently welded to the pipe. It is possible to use buttweld connections for larger valves on the pipeline, such as those that measure 12″ or 24″ or even larger. In addition, hub and clamp connections are used to secure valves using clamps. For high-pressure and large-diameter piping, hub and clamp connections are commonly used in offshore applications since they are lighter and less space consuming than ASME flanges.

Note: Valves which are connected through *buttweld* with piping have *bevel end.* An angled edge on a pipe or fitting that is intended for welding, particularly in buttweld connections, is referred to as a bevel end. To ensure optimal weld penetration and fusion between two joining surfaces, the bevel is typically cut at a specific angle, usually 37.5°. The beveling geometry or buttweld end preparation is explained in ASME B16.25. The use of bevel ends is common in the manufacture of piping and pipelines, especially in high-pressure and industrial applications, so that the weld is strong and durable. *Beveling* is the process of creating an angled edge to prepare a pipe, metal plate, or component for buttwelding.

Flange Connection

A flange connection valve is installed by bolting it between two flanges. Gaskets are placed between the valve and the flanges to create a seal. The flanges are then tightened with bolts in a crisscross pattern to ensure even gasket compression, preventing leaks. In Figure 7.4, several control valves with flange ends are connected to the piping through flanges.

FIGURE 7.4 A control valve with flange connections to the piping system in an onshore plant. (With permission: Shutterstock.)

Leak Prevention

In order to prevent leaks during valve installation, it is important that the valve and mating flanges are properly aligned. The correct bolt torque should also be applied to the valve and mating flange in order to prevent leaks from occurring.

Mounting Bolts

Mounting bolts in valve installation are fasteners used to secure the valve to the piping system, typically through the valve's flanges.

Over-Torque

A high amount of torque may damage the bolts, deform the flange, or crush the gasket, thereby reducing its ability to seal and increasing the risk of leakage.

Pipe Supports

As part of valve installation, pipe supports are structures or devices that prevent excessive stress or movement of the valves. The movement of a valve may be caused by thermal expansion, pressure fluctuations, or external forces.

In a valve, the support refers to structural elements or components that assist in holding the valve in position, stabilizing it within the piping system, and absorbing various stresses during operation. These supports are essential for maintaining valve alignment and preventing vibrations. In addition, using supports ensures that the valve performs reliably, particularly in high-pressure, high-temperature, or heavy-duty applications. Large valves, such as gate or globe valves, and those with heavy actuators, require support to bear their weight, particularly when mounted horizontally. In the absence of support, these valves could place excessive weight on the piping, potentially resulting in pipe deformation or misalignment. There are times when a valve support is an integral part of the valve and is located beneath the valve. In some applications, valves are supported by hangers that suspend the piping and valve from above, which is useful for overhead installations or when space on the ground is limited. Furthermore, bracing supports can be used between the actuator and valve in order to improve the integrity between them and avoid the separation of the actuator from the valve during operation.

Position Indicator

Position indicators are devices used in valve installation to indicate whether the valve is currently open or closed. As a result, operators can easily determine the position of the valve during operation or maintenance. The position indicator can be installed on a valve gearbox or an actuator in order to indicate whether the valve is open or closed. Figure 7.5 shows a position indicator installed on the actuator indicating that the valve is open.

FIGURE 7.5 A valve position indicator that is mounted on the top of the actuator displays the open position of the valve.

Sealing Compound

During valve installation, a sealing compound (PTFE or sealant) is applied to the valve-threaded ends (in the case of threaded connections between the valve and the piping), flanges, or gasket surfaces. A sealing compound improves sealability and minimizes the risk of leakage between a valve and its piping system.

Stem Protection

Stem protection in valve installation refers to measures taken to prevent damage to the valve stem. This protection typically involves covering the stem with a protective cap or sleeve during transportation, handling, and installation. This is to prevent scratches, corrosion, or contamination by debris. Figure 7.6 shows the stem threads of a small wedge gate valve is protected by black tape.

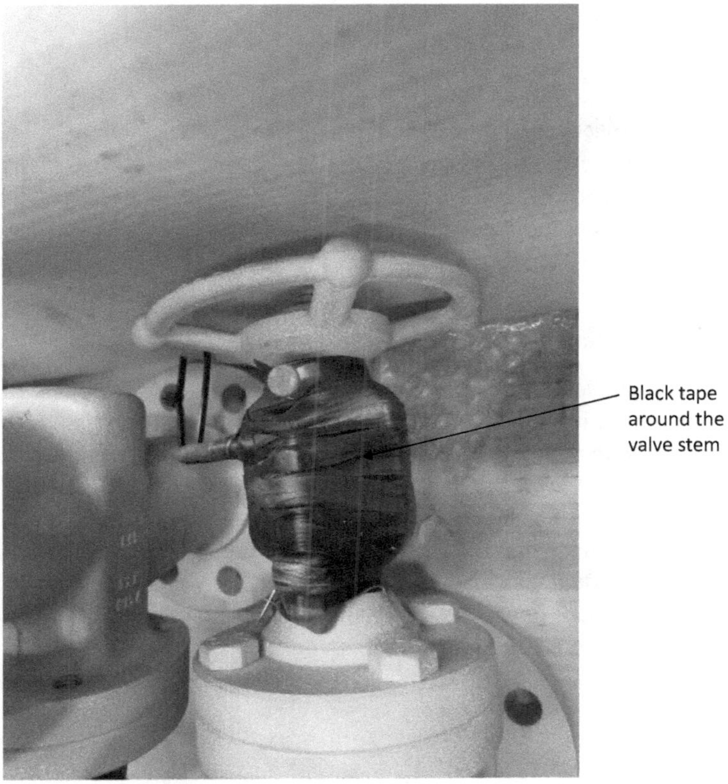

FIGURE 7.6 The stem threads of a small wedge gate valve are protected by black tape to prevent damage during transportation, storage, and installation.

Thermal Expansion

In valve installation, thermal expansion refers to the expansion of the valve and the piping that is connected to it as a result of temperature changes. In response to a rise in temperature, metallic materials expand, causing stress on the valve, its seals, and connections if not properly considered. In order to prevent damage, pipe supports, expansion joints, and flexible connectors are used to accommodate the movement caused by thermal expansion.

Thread Sealant

To prevent leaks in valve installation, thread sealants, such as PTFE, are applied to the threaded connections of the valve and piping. Upon application, it fills any small gaps or imperfections in the threads, creating a tight seal that prevents the escape of fluid or gas.

Torque Tool (Wrench)

Torque wrenches are used during valve manufacturing and installation to apply an accurate amount of force to bolts. The torque wrench should be used to ensure that the screws have been tightened to the manufacturer's recommended installation specifications. A *torque tool* can be either *hydraulic* to produce higher force or *manual*. In Figure 7.7, a worker uses a manual torque wrench to tighten the flange bolts installed on a piping system in an industrial yard. A worker is properly equipped with safety cloths.

Valve Tag Numbers

The valve tag number is a unique identifier assigned to each valve in a system. By using tag numbers, it is easy to identify, track, and document them throughout the project life cycle. In most cases, these numbers follow a logical format that includes information such as the type of valve (for example, BV, Ball Valve), the line or system it belongs to, and a unique sequence number. An example would be BV-1001-01, which represents the first BV installed on line number 1001. Valve tag numbers are used in project drawings, such as P&IDs, installation records, and maintenance logs. The reflected tags on the drawing must match the tags on the valve to ensure that the correct valve is installed, tested, and serviced at the appropriate location. The importance of tag numbers cannot be overstated when it comes to preventing misplacement and ensuring that every valve is correctly installed.

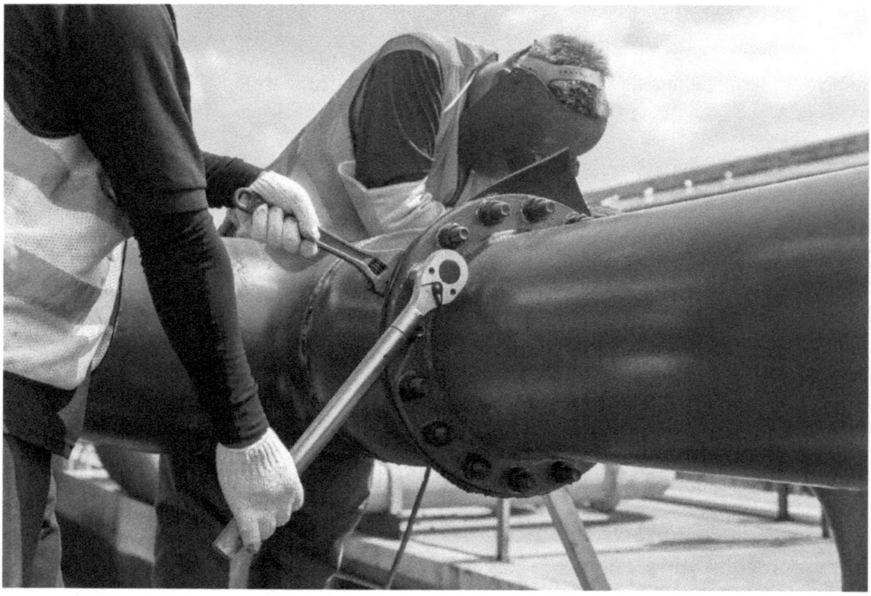

FIGURE 7.7 A worker using a torque wrench to tighten bolts of a flange on the piping system in an industrial yard. (With permission: Shutterstock.)

PART 3: SYSTEM INTEGRATION

Tests during system integration are typically considered a part of pre-commissioning. Pre-commissioning includes pressure testing such as hydrostatic test, flushing, and system integrity verification. Commissioning is performed after pre-commissioning and includes functional testing, calibration, actuator testing, and final system integration. Activities related to testing and verification in Part 4 are more related to commissioning phase.

Actuator Mounting

In system integration, actuator mounting refers to the process of attaching an actuator to a valve after the valve has been installed. It should be noted that valves are typically shipped to the yard with actuators already mounted at the valve factory. Some situations, however, necessitate actuator installation after valve installation on the pipe because of several factors, including large and bulky actuators, and space constraints during transportation or installation. The actuator controls the valve's opening and closing. More information about actuators can be found in Chapter 11. An actuator can be manually operated, electrically powered, pneumatically powered, or hydraulically driven. It is responsible for operating the valve in response to fluid pressure or control signals. Mounting is accomplished by positioning the actuator on the valve stem and fixing it in place with brackets, bolts, or flanges. API 6DX emphasizes the importance of valve–actuator integrity for smooth valve operation.

Backflow Prevention

Backflow prevention in system integration refers to measures taken to prevent the reverse flow of fluid in piping systems. As an example, backflow of fluid and gas into upstream pumps and compressors can cause significant damage to these expensive components. In order to prevent backflow, devices such as check valves, backflow preventers, and pressure relief valves are installed.

Flow Path Alignment

A flow alignment after installation during system integration involves ensuring that the valve and piping are aligned correctly so that fluid flows through the valve in the correct direction and with the least amount of resistance. Flow alignment involves ensuring that the fluid flows in the correct direction, smooth flow inside the pipes and valves, and minimizing pressure drops.

Instrument Connection

An instrument connection is the process of connecting various instrumentation devices to the piping, equipment, valves, and actuators following valve installation.

These instruments include pressure, temperature, flow, and level gauges and transmitters. Furthermore, they are connected to smart valves to provide real-time data during valve operation. A number of instruments are connected to actuators in order to provide real-time data, as well as information regarding the position of the valve and actuator as they open and close.

Joint Efficiency (Joint Integrity)

A joint's efficiency or integrity is determined by its ability to maintain its strength and tightness under pressure, temperature, and operating conditions (for example, a welded or flanged joint between pipes or valves). A joint with good joint integrity or efficiency is 1 or 100% that is properly sealed, which prevents fluid or gas leakage. Welding joint efficiency refers to the ratio between the strength of the welded joint and the strength of the base metal. Joint efficiency of 1 (or 100%) indicates that the welded joint is as strong as the base material. In contrast, if joint efficiency is less than 1, the welded joint is weaker as a result of defects in the welding process or improper welding technique. The effectiveness of welding joints can be increased to 100% by applying a radiography test to them. A high-quality, nondestructive welded joint may be assigned to increase the joint efficiency to 1 (100%), meaning it has the same strength as the pipe material itself.

As a general rule for flange connections, joint efficiency refers to the effectiveness of the flange connection sealing. To achieve 100% efficiency, the alignment of the joint, the choice of the flanges and gasket, and the correct tightening of the bolts are essential. The efficiency of a flange connection is often evaluated by performing pressure testing to ensure that the connection does not leak when the connection is in operation.

Pipe Strain

Strain refers to stress and deflection. After valve installation, pipe strain refers to the stress or deformation that occurs in the piping system. Pipe stress and deformation can be caused by a variety of factors, including improper flange alignment, thermal expansion due to high temperatures, or external forces and vibrations. Pipes can become subject to various loads if they are not correctly supported or aligned after valve installation. Loads may be tension, compression, or bending. The result can be leaks, misalignments, and damage to piping and valve connections. It is important to support piping and use expansion joints to provide flexibility to prevent pipe strain.

Piping Layout

Piping layout in system integration after valve installation refers to the arrangement and positioning of pipes, valves, and other components and equipment within a piping system. This layout ensures efficient and safe operation. Isometric drawings and P&IDs (piping and instrumentation diagrams) are essential documents used for maintaining piping layout during installation.

Pressure Equalization

When there is a high-pressure differential between the valve outlet and inlet, opening the valve immediately after installation may result in damage to the valve. In order to prevent damage or operational issues, pressure equalization of valves after installation and during system integration involves balancing the pressure on both sides of a valve to prevent damage. For valves with a large pressure difference between the inlet and outlet, pressure equalization is imperative. To accomplish this, one can open an auxiliary valve on the bypass line of the main valve to connect the inlet and outlet of the valve and equalize the pressure at the inlet and outlet. As a result of slowly opening the valve, pressure can gradually increase on both sides of the valve, resulting in a gradual equalization of pressure without causing a sudden shock. Also, control valves or pressure safety valves can be installed upstream and downstream of the valve to adjust the inlet and outlet pressures to achieve a very low-pressure differential between the inlet and outlet lines.

Safety Valve Integration

Integrating safety valves in a piping system after valve installation involves installing and configuring safety valves in order to protect against overpressure conditions.

Signal Wiring

Electrical signal wiring is used to transmit signals between components of a system, such as valves, actuators, sensors, or control systems. There are two types of signals: control signals and feedback signals. The signal wiring transmits control signals (such as 4–20 mA, 0–10 V, or digital signals) from the control system to the valve actuator. Additionally, the wiring facilitates feedback from the valve actuator to the control system, providing information on the status or position of the valve.

System Flushing

The flushing of a system refers to the removal of debris, dirt, and contaminants from the piping system after the valve has been installed. As a matter of fact, it is considered to be part of valve integration with piping system integration. In most cases, this is accomplished by opening the valve temporarily to allow high-velocity fluid (such as water or air) to flow through the pipes. By flushing, any particles or contaminants that may affect the performance of the valve or the system are removed. Flushing helps to prevent blockages or damage to sensitive valve components.

System Pressure Testing

A system pressure test after valve installation means testing the entire pipeline or piping system, including the newly installed valve and equipment. The purpose of this is to ensure that it is capable of handling the required pressure safely without leaking or failing. In most cases, the test involves pressurizing the system (piping and equipment) with a fluid (usually water or air) until a specified pressure is reached. This pressure is greater than the normal operating pressure of the system. For example, ASME B31.3 process piping code specifies that the test pressure approximately equals 1.5 times the design pressure. Pressure gauges are used throughout the test to ensure the system maintains the required pressure without any drop. A drop in pressure shown by pressure gauges during a test is indicative of the presence of leaks. In the course of the test, visual inspections and sometimes leak detection methods are used to determine whether there is any indication of leakage at valve connections, flanges, welds, or other joints. The pressure is held for a specified period of time, usually a few hours. The duration of this period is necessary to allow for adequate inspection time. The system is slowly depressurized after the test has been completed, and the results are documented.

Thermal Movement

As a result of temperature changes, pipes, valves, and other components expand and shrink in system integration after valve installation. Temperature changes can be attributed to both fluid temperature changes flowing through the piping system and the temperature changes in the environment. The expansion of materials like metals and plastics occurs when they are heated and the shrinkage occurs when they are cooled. The thermal movement may create stress on the valve, piping, and connections. This could result in leaks, misalignments, or even damage to the valve, piping, and their connections.

PART 4: TESTING AND VERIFICATION

Actuator Tests

Chapter 6 provides more information about actuator tests under the section of "Actuator Tests." Actuator tests are typically performed as a part of commissioning phase.

Body Test

After valve installation, a body test is performed to verify that the valve body is leak-free and able to withstand the pressures that it will be subjected to during normal operation. The body test can be performed as part of the hydrostatic test, which is discussed in more detail in this chapter. The valve will fail if there is any leakage from the body.

Bubble Test

After valve installation, a bubble test is performed to detect leaks in the valve body, seats, and connections. A leak detection method such as this is applicable when air or gas is used as a test medium. The test areas, such as the body and bonnet, the seat areas, and the flange joints should be sprayed or applied with a soapy water solution (bubble-forming solution). The presence of leaks is indicated by the appearance of bubbles. As part of the standard, the acceptance criteria are typically described in terms of bubbles or bubbles per minute.

Functional Test

It is necessary to perform a functional test after the valve has been installed into a piping system in the yard in order to verify that the valve performs as expected under real-life operating conditions. The purpose of this test is to ensure that the valve has been installed correctly and is operating as intended. When a functional test is conducted, proper opening and closing of the valve, as well as the duration and torque required for valve operation, are measured and compared to engineering documents such as valve specifications and datasheets. *Functional testing of valves* is typically conducted during the *commissioning phase* of a project.

Gas Test

The gas test is performed after valve installation in the yard. It involves the use of nitrogen or helium gases or mixture of both as the test medium to test the valve's resistance to pressure and leak prevention. Typically, this type of test is performed when a hydrostatic (water-based) test is not feasible or desirable. In some cases, valves are installed in high-pressure gas piping systems and testing them with high-pressure gas simulates the real operating conditions. A gas test is more hazardous than a hydrostatic test since compressed gas contains a significant amount of energy and can result in explosive failure if the valve or piping bursts. As a result, the test pressure is usually limited (e.g., 1.1× or 1.25× the design or operating pressure). Further information regarding why water cannot be used for testing can be found further in this chapter in the "Pneumatic Test" section.

Hydrostatic Test

In a hydrostatic test after valve installation, pressure is applied to the entire piping system, including the newly installed valve. It is performed to ensure there are no leaks and to ensure that the system is capable of handling the required pressure safely. It is necessary to fill the system with water (or an alternative fluid) during the test. The system is then pressurized to a pressure that is approximately 1.5 times its design pressure, as required by process piping codes such as ASME B31.3. High pressure simulates operating conditions which allows inspectors to detect leaks at valves, flanges, and joints.

Leak Detection

Chapter 6 provides more information about leak detection under the section of "Leak Detection."

Operational Test

Operational testing ensures that the valve functions correctly in the installed system under actual or simulated operating conditions after installation. As a result of this test, it is ensured that the valve performs as intended. Operational testing is quite similar to functional testing; however, there are some differences between the two. Operational tests are performed under operating conditions such as flow and pressure, as opposed to functional tests, which are conducted in an environment without flow or pressure. A function test focuses on the functionality of the actuator, the movement of the valve, and the alignment of the valve. However, operational tests evaluate the reliability of the sealing system, the performance of the seals, and the operational behavior of the system under real-world conditions. Operational test is typically done during commissioning phase of the projects.

Pneumatic Test

The pneumatic test is performed after valve installation in the yard. It involves the use of compressed air as the test medium to test the valve's resistance to pressure and leak prevention. Typically, this type of test is performed when a hydrostatic (water-based) test is not feasible or desirable. A pneumatic test is more hazardous than a hydrostatic test since compressed gas contains a significant amount of energy and can result in explosive failure if the valve or piping bursts. As a result, the test pressure is usually limited (e.g., 1.1× or 1.25× the design or operating pressure). For valve testing, air is prepared rather than water for a number of reasons. There are certain systems, especially those that involve gas or air service, that must remain dry in order to prevent corrosion or contamination. During a hydrostatic test, residual moisture may result, which is not desirable. Equipment or materials in the system that are susceptible to rusting or corrosion may not tolerate water exposure. Complex piping systems, particularly in yard installations, can be difficult to drain completely. When residual water remains after a test, it might freeze or corrode internal components. Disposing of water after a test may present environmental challenges, especially if the water has been contaminated with chemicals.

Seat Test

After valve installation, a seat test is performed to ensure that the valve seats are properly sealed and do not leak. When a valve is closed, the valve seat seals the flow path by joining the valve closure member. When the valve is closed, the seat test determines whether there is a leak between the valve seat and the valve closing element (such as the disk or plug). A valve seat pressure test involves pressurizing a valve's upstream side with a test medium (such as air, water, or nitrogen) while the valve is fully closed. In order to ensure that valve seats are sealed properly, this procedure is performed. The test pressure is increased slowly to the specified test level. Typically, the test pressure is 1.1 times the valve's working pressure. The pressure is kept for a defined amount of time, such as 1–5 minutes. This is determined by a standard such as API 6D, API 598, or ISO 5208. During this time, a visual inspection of the downstream side is performed (for liquids). Different techniques can be used to detect gases, such as using a gas detector or using soups to detect bubbles. Any leakage exceeding the limits specified in the valve specification or data sheet indicates a failure of the seat test. Normally, no leakage is permitted from soft seat valves.

Stroke Test

After valve installation, a stroke test is conducted to ensure that the valve operates smoothly between open and closed positions. By performing this test, the operator will ensure that the valve's actuator or manual mechanism is functioning correctly. Besides, this test is typically performed as part of a functional test.

System Start-Up Test

When the system is operated for the first time after valve installation in the yard, a system start-up test is conducted to verify that the valve and the entire piping system function as an integrated unit.

Tightness Test

Chapter 6 provides more information about tightness tests under the section of "Tightness Test."

Torque Test

Upon valve installation, torque testing is performed normally as a part of function test to determine the required torques for valve operation (opening, closing, or modulating) and to verify that they are within acceptable limits. Chapter 6 provides more information about torque tests under the section of "Torque Testing."

Vibration Test

After valve installation, a vibration test is conducted to determine the valve's ability to withstand and operate reliably under vibrational forces. These vibration forces may be caused by fluid pressure, mechanical equipment such as pumps or compressors, or seismic activity. This test is performed by exposing the valve to controlled vibrations. Vibration forces during this test simulate the vibration sources described above. The goal of this test is to ensure that the valve's components remain secure and functional without loosening, wearing, or degrading over time. Chapter 6 provides more information about vibration test under the section of "Vibration Test."

PART 5: DOCUMENTATIONS AND RECORDS

As-Built Documentation

The as-built documentation for valve installation in the yard represents the final conditions, configurations, and changes that were made during the installation and construction process of the valve. It provides information on how the valves and associated systems were installed, as opposed to the original design or engineering plans.

Example: Valve Replacement and Orientation Adjustment

A 10″-gate valve was specified in the engineering design for a main pipeline. To facilitate easy access, the actuator was located on top of the valve in a vertical orientation. When the valve was installed, it was discovered that the actuator would interfere with a nearby structural beam, so that it could not be installed as designed. A 10″-ball valve was chosen to replace the gate valve due to insufficient space to operate or maintain it in the intended position. The ball valve required less space for operation, but had different installation requirements. Therefore, the engineering team changed the ball valve installation direction vertically. Changing the valve and its installation direction results in a change in the as-built drawings compared to the initial valve engineering drawings.

Commissioning Report

After valve installation, a commissioning report provides a detailed record of the tests (typically function tests and actuator tests), inspections, and validations that were performed. It is necessary to perform these tests and inspections in order to ensure that the valves and associated systems are functioning as intended. A commissioning phase in a project refers to the transition from construction to full operation. A commissioning process ensures that a system, including its components and equipment, is installed, tested, and ready to use (operate).

Inspection Checklist

Inspection checklists for valve installation are systematic lists of tasks and acceptance criteria that are used prior to, during, and after the installation of the valve. This list should be used to ensure that the valves have been installed correctly and meet all design, safety, and operational requirements before, during, and after installation. Inspectors and installers can use this list to verify proper installation and identify potential problems before commissioning.

Installation, Operation, and Maintenance (IOM) Manual

IOM manuals are comprehensive documents provided by valve manufacturers that provide instruction on how to properly install, operate, and maintain valves.

Handover Checklist

A handover checklist is essential when transferring responsibility from the installation team to the operations or commissioning team. This document confirms that all aspects of the valve installation, testing, and documentation have been completed and verified. This serves as a final quality control measure to ensure that the valve and associated systems are ready for operation.

Installation Record

An installation record gives a detailed information of the valve installation process. The purpose of this report is to demonstrate that the valve has been installed according to the design specifications, industry standards, and manufacturer requirements. For example, in the installation record of a 10″-gate valve, various information is recorded, including valve information, actuator data, installation location, installation details (e.g. flange connection type and material, gasket type and material, bolts, etc.), inspection performed before installation, testing performed after installation, and adjustments or changes made during installation.

System Drawings

A valve installation system drawing is a technical diagram, a plan, and other documents. In these documents, valve layouts, placements, and connections within a piping system or process are illustrated. The purpose of these drawings and technical documents is to demonstrate where and how the valves should be installed in the piping system. Furthermore, following these documents ensures that valves are installed correctly and meet all design and manufacturing specifications. A detailed description of the design, valve components, and the entire system is provided in these documents. The type of system drawings for valve installation is provided as follows.

1. Data Sheets

Data sheets for valves provide detailed specifications for each valve, including size, type, material, pressure rating, and actuator details. Additionally, it contains valve tag numbers for ease of identification. Instrument or valve engineers prepare valve data sheets during the engineering phase for actuated valves. Typically, manual valves are covered in piping material specifications (PMS) prepared by valve or piping material engineers. For special (nonstandard) valves, such as multi-port Ball Valves, it is the responsibility of the valve or piping material engineer to prepare the data sheet.

2. General Arrangement (GA) Drawings

The manufacturer of the valve prepares this document and the buyer or the buyer and the end user approve it. All dimensions for the valve, including the face-to-face dimensions, height, and flange dimensions, are included in this document. In order to ensure that the valve fits correctly into the piping system, these drawings are essential.

Example: Using a GA Drawing of 8″ Ball Valve during Valve Installation in the Construction Yard

An GA drawing for a valve, such as an 8″-BV, provides critical structural and dimensional information that facilitates accurate installation. This document includes essential dimensional information, such as face-to-face dimensions (e.g., 203 mm), flange details (e.g., diameter, bolt size, pattern), and height and width dimensions. The GA DWGs for actuated valves also provide information concerning the actuator dimensions, type, and weight (e.g., the height and weight of pneumatic actuators). Installers can use these details to ensure that valves are installed correctly in piping. Additionally, installers can select the appropriate valve flange, gaskets, and bolts by knowing the flange end dimensions. It should be noted, however, that valve cross-sectional drawings provide more detailed information regarding the material of valve parts. Flow direction for unidirectional valves is also indicated on the drawing, which prevents the valve from being oriented incorrectly. A clearance dimension can also be provided in order to avoid interference with nearby equipment or structures. It is helpful for installers to design appropriate pipe supports by knowing the weight of the valve and actuator from GA drawings.

3. Isometric Drawings

During the engineering phase of the project, piping designers produce this document. These drawings provide a 3D representation of the piping system, including valves, fittings, and support components. Detailed dimensions, orientations, and elevation changes on these drawings assist installers in determining the exact position of valves. Furthermore, they include details such as flange connections, welds, and threading.

4. Piping and Instrumentation Diagram (P&ID)

In this document, the overall process flow is represented, as well as the location of valves, piping, and instrumentation. Moreover, it provides valve tag numbers, symbols for valve types (e.g., gate, globe, ball), and their function (manual or automated). Process engineers prepare this document during the engineering phase of a project. During the installation, operation, and maintenance phases of a project, P&IDs are essential documents for identifying and allocating valves.

5. Plot Plan

A plot plan shows the overall layout of the site, including the locations of key systems, pipelines, and valves. It is prepared by piping engineers at early engineering phase.

Warranty

For valves, a warranty represents the manufacturer's guarantee that the product will perform as specified under normal operating conditions for a specified period of time. The valve manufacturer typically provides warranty coverage for defects in materials, engineering, and manufacturing. It provides the buyer with assurance that the valve meets the required quality standards. According to the warranty, the manufacturer will repair, replace, or refund the valve if it fails due to a covered defect due to a manufacturer fault within the specified period of time (e.g., 12–24 months). The warranty is only valid if the valve is installed, operated, and maintained in accordance with the manufacturer's instructions. If the product is installed incorrectly, operated outside its design parameters, or neglected to be maintained, the warranty may be voided, resulting in the user bearing responsibility for the failure. A warranty is critical to reducing financial risks for buyers and demonstrates the manufacturer's confidence in the quality of the product.

Warranty Inspection

As part of a warranty inspection, it is determined whether the valve has been installed, operated, and maintained in accordance with the warranty terms. It involves checking whether the valve has been installed correctly (e.g., proper alignment, torque values, and orientation), has been operating within design parameters (e.g., pressure, temperature, and media compatibility), and is maintained in accordance with manufacturer IOM recommendations (e.g., lubrication, cleaning, and periodic testing). As a result of the inspection, issues such as improper handling or manufacturing defects may be identified. Improper handling of the valve by the buyer, such as mishandling during transport, incorrect installation, or misuse, is not covered under the manufacturer's warranty. However, manufacturing defects, such as issues related to materials, workmanship, or production flaws, must be compensated by the valve manufacturer in accordance with the terms and conditions of the warranty.

PART 6: PACKING AND PRESERVATION

Environmental Protection

Environmental protection refers to actions performed to protect valves during installation from environmental harmful conditions. Among these factors are moisture, dust, dirt, temperature fluctuations, UV radiation, and exposure to corrosive or abrasive elements in the surrounding environment. These can lead to valve malfunction and reduce the life of the valves.

Moisture Prevention

During valve installation, moisture prevention refers to the tasks performed to prevent the valves from being exposed to moisture. Keeping sensitive components protected from moisture is essential in order to prevent corrosion, rust, and degradation. Moisture can adversely affect the performance, lifespan, and operational reliability of valves. Valve moisture is prevented by protective wrapping, drying agents such as silica gel, and moisture-resistant materials such as plastic wrap, polyethylene covers, or volatile corrosion inhibitors (VCI).

Preservation (Long-Term versus Short-Term)

In short-term preservation, valves are protected for a limited period of time which is several weeks to a few months, before installation. It involves simple measures such as covering valve openings with caps or plugs. On the other hand, long-term preservation refers to storage periods that extend beyond several months or even years. It involves more robust measures, such as the application of thicker anticorrosion coatings, the use of VCI packaging, and the complete sealing of the valves with protective wraps.

Protective Warping

More information about protective wrapping is provided in Chapter 6, under the heading of "Wrapping (Protective Wrapping)."

Rust Prevention

Rust prevention during valve installation involves preventing corrosion caused by moisture, oxygen, and environmental factors. The application of anticorrosion coatings, such as epoxy- or zinc-based paints, is recommended. Also, it is important to seal openings with plastic caps, plugs, or plywood. Finally, valves should be stored in a dry and climate-controlled environment prior to installation. As an example, plywood (wooden-rounded components) can be used to cover and protect valve ends. The drain and vent plugs of valves must be covered with a cap or tape. Figure 7.8 illustrates a BV with plywoods completely covering both ends of the valve to prevent moisture and dirt from entering the valve. Plastic caps cover the injection ports of sealant on seats as well as the drain and vent plugs. Furthermore, the aluminum foil covers the gearbox of the valve that is filled with grease.

Rather than providing fire protection, aluminum foil is used to protect the gearbox of valves against environmental factors. Aluminum foil has several advantages over other methods of preservation such as plastic including superior moisture barrier properties, protecting against high temperature, UV protection, and fire protection. In this way, moisture, dirt, and debris are prevented from entering the gearbox. As a result, the gearbox is protected from corrosion and contamination. Furthermore, the foil protects the grease from drying out or being affected by temperature changes. Grease is filled inside the gearbox to prevent moisture ingress and air pockets into the gearbox as a part of preservation to protect against corrosion and contamination.

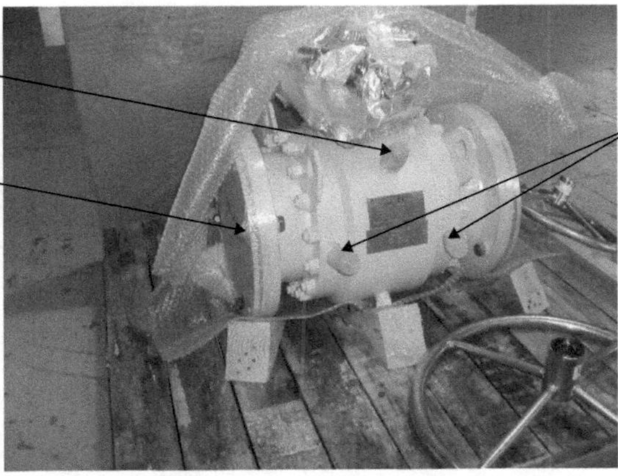

FIGURE 7.8 A ball valve with plywood covering both ends to prevent moisture and dirt ingress, plastic caps sealing the injection ports for seat sealant, and drain and vent plugs. Additionally, the gearbox, filled with grease, is protected with aluminum foil for enhanced preservation.

Temporary Storage

In the event that the piping system is not ready for valve installation, storage may be required. In relation to valve installation, temporary storage refers to the process of properly storing valves for a short period of time before they are installed in a system. It is recommended that valves be stored in a clean, dry, well-ventilated area on pallets or racks that are elevated off the ground. Additionally, it is important to preserve the valves using the methods described in this and previous chapters. In order to ensure valve quality, prevent contamination and corrosion, and avoid delays during installation, temporary storage is required. The same rules are applied to temporary storage of actuators. Figure 7.9 shows a well-organized temporary storage setup for actuators before installation. The large actuators are securely positioned on a wooden pallet. They are fastened using straps to prevent movement during storage. The actuator rests on shock-absorbing rubber pads to protect them from vibrations and potential damage. To protect the actuators from weather conditions such as rain or direct sunlight, the storage area is covered by a roof. It is a good idea to add warning label surrounding the actuators, including the phrases "Fragile" and "Handle with Care," emphasizing the need to handle it with care.

Transport Safety

In order to ensure that valves are transported safely during installation, steps must be taken to prevent damage to the valves while they are being moved to the installation site. As part of this process, the valve manufacturers typically provide strong packaging, such as crates or wooden boxes. Furthermore,

FIGURE 7.9 A temporary storage setup for hydraulic actuators before installation on the valves. The actuators are securely fastened to a wooden pallet with shock-absorbing pads.

valve manufacturers should cover valve openings with caps to prevent dirt and moisture from entering. In addition, valves should be mounted securely in order to prevent damage. Mounting valves refers to the process of securing, positioning, or fixing valves to a structure, support, or system. To avoid accidents, heavy valves should be lifted using appropriate equipment such as forklifts or slings. In order to ensure proper handling, labels with instructions such as "Handle with Care" or "This Side Up" are provided. In order to protect valves from moisture and corrosion, protective wraps or coatings may be applied. Moreover, shock-absorbing materials can prevent damage during bumps. Inspecting the valve before and after transport ensures that it arrives in good condition, ready for installation without the need for additional repairs or delays. Figure 7.10 illustrates the safe transportation of an industrial valve using the sling method in the construction yard. The large BV is securely fastened with heavy-duty slings connected to a crane hook. The slings are properly positioned to distribute the weight evenly. This prevents the valve from slipping or falling during the lift. Around the valve, the workers must wear safety gear. There must be safety signs such as "Caution" and "Heavy Load." Signs such as these emphasize the importance of adhering to safety standards during valve transportation.

FIGURE 7.10 Safe transportation of a very large and heavy industrial ball valve before installation in the yard using a sling method. The valve weight is properly distributed and safety measures are taken.

Volatile Corrosion Inhibitor

VCI is a chemical compound that prevents corrosion of metal surfaces by forming a molecular layer of protection. In packaging materials, VCI releases vapor that condenses on the metal surface, forming an invisible barrier. The barrier prevents moisture, oxygen, and other corrosive elements from interacting with the metal. During storage and transportation, this technology is particularly useful for preserving valves and their metallic components. Thus, oil-based coatings that are harmful to the environment are no longer required to be applied and removed. The VCI is nontoxic, cost-effective, and does not negatively affect the functionality of the components. This makes it an ideal choice for protecting valves and other sensitive equipment from corrosion and rust.

QUESTIONS AND ANSWERS

1. What is the first step in valve installation?
   A. Tightening the bolts
   B. Checking the valve orientation
   C. Inspecting the valve for damage
   D. Applying lubricant to the valve stem

   Answer: The correct answer is option C.

2. Which tool is most commonly used to tighten flange bolts during valve installation?
   A. Screwdriver
   B. Torque wrench
   C. Allen screw
   D. Adjustable spanner

   Answer: The correct answer is option B.

3. Why is proper unidirectional valve orientation critical during installation?
   A. To align the valve with actuator
   B. To facilitate proper lubrication of the valve
   C. To prevent corrosion
   D. To match the flow direction with system requirements

   Answer: The correct answer is option D.

4. What is the purpose of a hydrostatic test?
   A. To check for internal and external leaks
   B. To test the valve's pressure rating
   C. To ensure the valve operates smoothly
   D. To verify the flow rate

   Answer: The correct answer is option A.

5. What does a valve tag number indicate?
   A. The manufacturer's name
   B. The valve's installation date
   C. The valve's unique identification in the system
   D. The valve's warranty period

   Answer: The correct answer is option C.

6. Which standard is commonly referred to for flange dimensions in valve installation?
   A. ASME B16.5
   B. API 598
   C. ISO 9001
   D. ASTM A216

   Answer: The correct answer is option A.

7. What should you do if the valve does not match the piping alignment during installation?
   A. Apply force to align the valve
   B. Weld the valve into place
   C. Ignore the issue and proceed
   D. All three answers are wrong

   Answer: The correct answer is option D.

8. Which of the following is typically included in an as-built drawing for a valve?
   A. Valve dimensions
   B. Installation orientation
   C. Tag number and material details
   D. All of the above

   Answer: The correct answer is option D.

9. What is the main purpose of a functional test after valve installation?
   A. To check the valve's material strength
   B. To ensure the valve moves smoothly and seals properly
   C. To test the flow rate through the valve
   D. To verify the valve's flange connections

   Answer: The correct answer is option B.

10. Why is valve preservation important during storage or before installation?
    A. To ensure the valve's aesthetics remain intact
    B. To protect the valve from damage due to environmental factors
    C. To increase the valve's weight for better stability
    D. To reduce the need for testing after installation

    Answer: The correct answer is option B.

BIBLIOGRAPHY

American Petroleum Institute (API). (2021). Valve Inspection and Testing (API Standard 598). 11th edition. API: Washington, DC.

American Petroleum Institute (API). (2021). Pipeline Valves (API Standard 6D. 25th edition. API: Washington, DC.

American Petroleum Institute (API). (2018). Petroleum and Natural Gas Industries: Mechanical Integrity and Sizing of Actuators and Mounting Kits for Pipeline Valves (API 6DX). American Petroleum Institute: Washington, DC.

American Society of Mechanical Engineers (ASME). (2021). Pipe Flanges and Flanged Fittings (ASME B16.5-2021). New York, NY: ASME.

American Society of Mechanical Engineers (ASME). (2021). Butt-Welding Ends (ASME B16.25-2021). ASME: New York, NY.

American Society of Mechanical Engineers (ASME). (2020). Large Diameter Steel Flanges: NPS 26 through NPS 60 (ASME B16.47-2020). ASME: New York, NY.

American Society of Mechanical Engineers (ASME). (2020). Process Piping (ASME B31.3-2020). ASME: New York, NY.

Antaki, G. A. (2003). Piping and Pipeline Engineering: Design, Construction, Maintenance, Integrity, and Repair. CRC Press: New York, NY.

Barker, G. B. (2017). The Engineer's Guide to Plant Layout and Piping Design for the Oil and Gas Industries. Gulf Professional Publishing: Cambridge, MA.

International Organization for Standardization (ISO). (2015). Industrial Valves: Pressure Testing of Metallic Valves (ISO 5208:2015). ISO: Geneva, Switzerland.

Nesbitt, B. (2007). Handbook of Valves and Actuators: Valves Manual International. 1st edition. Elsevier: Oxford, UK.

Skousen, P. L. (2011). Valve Handbook. 3rd edition. McGraw-Hill Education: New York, NY.

Sotoodeh, K. (2019). Valves and actuators integrity and blast load calculations. Springer Nature Applied Science: Springer Switzerland, 1, 599. https://doi.org/10.1007/s42452-019-0624-z

Sotoodeh, K. (2021). Pipeline Valve Technology: A Practical Guide. CRC Press: Boca Raton, FL.

Zappe, R. W., & Smith, P. (2003). Valve Selection Handbook: Engineering Fundamentals for Selecting the Right Valve Design for Every Industrial Flow Application. 4th edition. Gulf Publishing: Oxford UK.

8 Valve Operation and Performance

A

Actuator Control System

Actuator control involves the use of external control systems, such as programmable logic controllers (PLCs) and distributed control systems (DCSs). The control systems receive information about pressure, temperature, and other process parameters and compare them with what should be, based on which they command the actuators. In these systems, actuators and valves are controlled without the need for manual intervention. As part of this process, actuators are controlled by control signals (e.g., analog 4-20 mA, air, or digital protocols). According to the requirements of the system, actuators move valves to regulate the flow of fluid or to start and stop it. In control valves, positioners installed adjacent to actuators receive signals and instruct the actuators to act. Suppose that a control valve is installed in a pipeline to regulate the flow of water. The valve is controlled by a pneumatic diaphragm actuator with a positioner. The system operates on a standard 4-20 mA signal to determine the valve's position. Upon receiving a 4-mA signal, the actuator will fully close the valve, whereas a 20-mA signal will fully open the valve. The positioner maintains the valve in half-open position if the control system sends a signal in the middle, 12 mA.

Actuator Cycle

Valve operation is performed by the actuator cycle. More information about actuator cycle is provided in Chapter 11, under the heading of "Actuator Cycle."

Actuator Remote Operation

When it comes to actuators, remote operation refers to controlling and operating a valve's actuator from a distance away from where the valve and actuator are located. In order to accomplish this, signals are sent to the actuator through communication systems. There are several types of control signals, including electrical, wireless, pneumatic, and hydraulic. Remote operation is applicable to electrical or electromechanical actuators as well as diaphragm actuators that are used to control valves. Safety is one of the main advantages of remote operation, since it allows operation from a safe distance in hazardous environments such as offshore platforms, refineries, or chemical plants. Moreover, some valves and

DOI: 10.1201/9781003640967-8

actuators are located in areas that are difficult to access. Note that control valves are equipped with positioners that interpret remote signals from DCSs) or PLCs.

Actuator Signal

As signals are transmitted from the control room to actuators, the actuators are given commands for operation and the valve is operated. Some actuators (smart actuators) are equipped with sensors that send signals to the control room informing them of the position of the valve. More information about actuator signal is provided in Chapter 11, under the heading of "Actuator Signal."

Actuator Stroke Time

Actuator stroke time refers to the amount of time it takes an actuator to move a valve from one extreme position to another. Typically, this involves moving the valve from fully open to fully closed, or vice versa, or between intermediate positions.

Actuator Type

More information about actuator type is provided in Chapter 11, under the heading of "Actuator Type."

Automation

Automation in valve operation refers to the use of actuators, controllers, and sensors to operate valves. Automated valves are operated without the need for human intervention (manual intervention). As a result of signals from a control system and/or actuator movement, automated valves may open, close, or adjust their position. Because of automation, more precise operations are achieved such as accurate flow control and pressure regulation, as well as increased system efficiency. It is common for valves to be equipped with actuators that convert control signals into mechanical motion in order to operate them. These actuators used for valve operation are electric, pneumatic, or hydraulic in most cases. For example, it is possible for pneumatic and hydraulic actuators to receive an electrical current signal from a control panel to change the position of a solenoid valve, which in turn controls the opening or closing of the actuator. Nevertheless, it should be noted that not all actuators receive control signals, and control signals are more commonly associated with control valves. For example, electrical actuators can be operated with or without a control signal, as they often include manual override options or a push button on the actuator for stop and start. Actuators used for on/off valves, such as hydraulic and pneumatic actuators, typically operate with the aid of external control signals, such as electrical inputs to solenoid valves. In complex systems, such as industrial plants and oil and gas facilities, where frequent, rapid, or remote valve adjustments are required, automation is essential to ensure operational efficiency and safety.

B

Backflow

During valve operation, backflow refers to the unintended reverse flow of fluid in a piping system. As a matter of fact, in a backflow scenario, the flow is moving in the opposite direction to the intended flow path. When there is a loss of pressure, during system shutdowns, pump failures, and pressure imbalances between different sections of the piping, this can occur. It is possible for backflow to cause contamination, damage equipment, or disrupt the operation of a system. A backflow prevention device, such as a check valve or a non-return valve, is used to prevent backflow. These components or devices allow fluid to flow in only one direction. Managing backflow requires proper installation, maintenance, and selection of appropriate valves.

Backflow Prevention

The concept of backflow prevention refers to the mechanisms and devices used to prevent the reverse flow of fluids, gases, or contaminants from a downstream system back into a clean or potable water supply. The purpose of this is to protect the quality of the water and to prevent contamination of the water. Changes in pressure, such as a drop in water supply pressure or an increase in downstream pressure, can cause backflow. In order to prevent backflow, devices such as check valves and backflow preventers are installed. Knowing the difference between a backflow preventer and a reverse flow preventer is essential. While backflow prevention is performed to prevent backflow into water systems, reverse flow prevention is a similar concept, but is referred to more broadly. The term reverse flow prevention refers to any system or method that prevents fluids from flowing in the opposite direction within a piping or pipeline system.

Backpressure

Backpressure is related to safety valves operation. More information about backpressure is provided in Chapter 10, under the heading of "Backpressure."

Blowdown Operation

The blowdown operation in valve systems refers to the controlled discharge of pressure or fluid from the system. For example, blowdown may be required before maintenance to depressurize the system or for emergency overpressure release. In this process, blowdown valves are used to rapidly discharge fluids or gases from pipelines, vessels, or equipment. It is also possible to release overpressure fluid through a blowdown valve into a flare system. A blowdown valve may be an actuated ball valve that is normally closed. In the event of an overpressure or maintenance requirement, it is opened in order to release the excess pressure or fluid.

Bypassing (Bypass Connection)

A valve bypass connection refers to a smaller and auxiliary piping system that is installed around or alongside a main valve. The bypass connection includes a bypass valve and the piping. Bypass valve and connected piping allow fluid to flow around the main valve, providing various operational and maintenance advantages. Valve bypass connections serve various purposes such as pressure equalization, maintenance and repair, start-up operation, and flow regulation. They help balance pressure on both sides of the main valve, especially in high-pressure systems, making it easier to operate or open the main valve. They allow the main valve to be isolated for inspection, maintenance, or repair while maintaining fluid flow through the bypass. They gradually introduces fluid into the system by using the bypass before fully opening the main valve, minimizing potential surges or shocks. More information about surge is provided in Chapter 9, under the heading of "Surge." Finally, they provide a way to control or limit the flow through the pipeline without operating the larger main valve. Bypass connections contain bypass pipe and valve as well as isolation valves.

Bypass valve is an important component in bypass connection. A bypass valve is a valve installed in a bypass line that allows fluid or gas to flow around a primary valve, pump, or other equipment. It provides an alternative path for the flow, enabling specific operational, maintenance, and safety functions.

Bypassing is highly connected to *pressure equalization* that is explained later in this chapter.

C

Cavitation

Cavitation is an erosion–corrosion valve and pump operational problem which occurs mainly in globe valves in liquid services. It is possible to have cavitation in other valves used for flow throttling such as butterfly. More information about cavitation is provided in Chapter 4, under the heading of "Cavitation."

Chattering

Valve chattering is the rapid and repetitive opening and closing of a valve. This operational problem usually happens in check valves or safety valves (pressure relief valves [PRV]). Vibrations, noise, and potential damage to the valve and piping are consequences of this phenomenon. This phenomenon may also shorten the valve's lifespan and reduce its effectiveness. A valve that is chattering is often a sign of improper valve sizing or pressure settings, which can compromise the valve's ability to provide safe system protection. Valve chattering can indeed be caused by process flow fluctuations, among other factors.

Choked Flow

A choked flow occurs when the fluid flow through a valve reaches its maximum velocity. As a result, increasing the valve opening or lowering the downstream pressure will not be able to increase the flow rate. Gases or steam can reach the speed of sound (also called choked flow) at the narrowest part of the valve. In this situation, the flow rate is determined by the upstream pressure and the valve design, rather than by the downstream pressure. In the event that the condition is not handled properly, it may cause noise, vibrations, or damage.

Clearance

The term "clearance" in valves refers to the intentional gap or space between specific moving and stationary components within the valve assembly. The clearance allows the valve to operate smoothly, compensates for thermal expansion, prevents binding and wear, and ensures the valve's long life. For example, clearance could be the space between the valve stem and the bonnet or packing that lets the stem move (or rotate, depending on the valve type) without friction or obstruction with packing or bonnet. Valve seat clearance is how much space between the valve disk (or ball, if it is a ball valve) and the seat allows the valve to close tightly without being too forceful. Too little clearance can cause friction, binding, or jamming, while too much clearance can cause leaks.

There is also an additional meaning to clearance in relation to the installation, operation, and maintenance of valves. Specifically, clearance refers to the physical space around the valve that is necessary for safe and effective installation, operation, or maintenance. During normal operation, sufficient space is required around the valve, especially if it has a manual operator (such as a handwheel or lever) or an electrical (motorized) actuator that requires space to move or turn by manual override. By providing this clearance, operators will be able to easily access and adjust the valve without being hindered, which is particularly important in tight piping systems or in crowded environments.

Cold Differential Set Pressure (CDSP)

1. Definition

A CDSP is the pressure at which a PRV is set to open under ambient, cold (non-operating) conditions. CDSP is established while the system is at ambient temperature, rather than at elevated or operational temperatures. Set pressure refers to the pressure at which the relief valve opens in service, while cold differential test pressure refers to the pressure at which it opens in a test stand. The CDTP can be measured and verified with a conventional relief valve since back pressure is not present during testing in the shop. In many applications, fluid expansion or thermal effects cause pressure changes in the system as a result of operating temperatures. Therefore, this makes it a baseline pressure for the valve at "cold" conditions. Once the system has reached its operating temperature, CDSP is adjusted to accommodate these factors so that the valve will open at the desired set pressure. There are several factors that affect the pressure within a system during operation, including back pressure and thermal expansion. These factors are taken into consideration when calculating CDSP. A CDSP is particularly important in systems that experience significant temperature fluctuations, such as boilers, chemical reactors, and other high-temperature applications in which thermal expansion can affect internal pressure. For example, if a valve needs to open at 10 bar during operation (e.g., at 200°C), the CDSP may be set to a lower or higher pressure than 10 bar at ambient temperature to compensate for changes in pressure at elevated temperatures. As pressure and temperature increase, the set pressure of the safety valve can be significantly impacted, particularly due to the changes in system conditions which affect fluid behavior and material properties. When the temperature of the system increases, fluids and gases expand, thereby raising the system's internal pressure. High temperatures can reduce the strength of metals, which may affect the ability of valve components (such as springs and seats) to maintain the correct set pressure. If this additional pressure is not accounted for, a safety valve could open prematurely (at a lower pressure than its set pressure). Occasionally, materials may deform or weaken, resulting in the valve opening at a lower pressure than intended. As also explained in Chapter 10, CDSP addresses safety because it is a term related to safety valve.

2. Example

Assume that the desired set pressure of the valve at 200°C is 20 bar. The valve is tested on a bench in the shop at a temperature of 20°C. As the temperature rises, the fluid expands, increasing the internal pressure. Assume that the system pressure increases by 1 bar when the temperature rises from 20°C to 200°C. Additionally, the backpressure to the valve at operating conditions is 1.5 bar. The valve is set at a lower pressure when the system is cold in order to ensure that it opens at 20 bar when the system reaches 200°C. There is no backpressure at cold condition on the test bench. Therefore, it is easier to open the safety valve without backpressure, meaning the set pressure at cold temperature must be lower than the set pressure at 200°C by amount of backpressure which is 1.5 bar. Now we have to see the effect of internal pressure on the backpressure. At the lower temperature, the fluid pressure is less, which is 1 bar lower than the fluid temperature at 200°C. Therefore, adding 1 bar to the valve set pressure at cold condition will compensate for fluid cold condition. Therefore, the CDSP equals 20 − 1.5 + 1 equals 19.5 bars. In the event that the set pressure is not adjusted for cold temperatures, it can lead to a delay in the opening of the valves both at low and high temperatures, failure of the system, a fire, or an explosion.

Condition Monitoring

Valve condition monitoring involves assessing the operation of a valve regularly to detect early signs of wear, damage, or malfunction before they lead to valve failure. In critical applications, where unexpected valve failure can disrupt operations or create hazardous situations, this proactive approach assists in ensuring that valves remain reliable, safe, and effective. In order to monitor valve condition, various sensors and diagnostic tools are typically used to track key performance indicators, such as vibration, temperature, and acoustic emissions. A number of benefits can be derived from valve monitoring, including the improvement of valve lifespan, the reduction of maintenance costs, an increase in operation efficiency, and the enhancement of safety and reliability. More information about condition monitoring and condition-based monitoring is provided in Chapter 9, under the heading of "Condition-Based Monitoring (CBM)."

Control of Fluid

Valve control refers to the regulation and control of fluid flow, pressure, level, or temperature within a system. By sending signals to the actuators, the control system adjusts the position of the valve. In order to achieve the required pressure and temperature, the position of the valve is adjusted. Actuators, such as electric, pneumatic, or hydraulic devices, are often used to control and move valves automatically. A control system (for example, a PLC or DCS) sends signals to actuators which adjust the valve opening as a result of the signals. Process parameters such as pressure, flow, or temperature are continuously monitored by sensors and the control system. Thus, control system sends commands to the actuator to adjust the valve's position. As a result, the system is maintained within the desired parameters.

Control Range

The control range of a valve refers to the range of flow rates, pressures, or other process variables that the valve can effectively control. In general, the control range is the range between the minimum and maximum values of the controlled variable (e.g., flow rate) over which the valve can maintain stable and accurate performance. The control range is typically calculated by dividing the maximum controllable parameter by the minimum controllable parameter. This is often referred to as the rangeability of the valve. For example, if a valve controls flows from 10 L/min to 1,000 L/min, its control range is 100:1. It is possible that a valve with a narrow control range will only work well under specific conditions. A flow rate outside of that range may result in inefficiency or instability. Alternatively, a valve with a wide range of operation or control ranges can easily be adjusted to meet changing process requirements without compromising accuracy. Valve type, valve design, piping system conditions such as installation, and flow conditions through the piping system affect valve control range.

Cycle

A valve cycle refers to the complete process of moving the valve from one position to another and back again. It occurs most often when a valve is moved from a fully closed position to a fully open position and then reverted to its fully closed position. A cycle represents one full operation of the valve. It is important to note that the number of cycles a valve has indicates its life expectancy. In general, valves are rated for a certain number of cycles before they require maintenance or replacement. High-cycle valves are designed to withstand repeated movement without excessive wear. The expected number of cycles plays a critical role in selecting valves for different applications. In automated or high-frequency operating systems, valves must handle repeated cycles, while those in systems that are infrequently used may not require high-cycle durability. A high-cycle valve is designed for applications that require frequent actuation, such as control valves or automated on/off valves. One important reason for using actuators for valves is to handle the frequent cycling (opening and closing) of the valve, which would be impractical or inefficient to manage manually. It is common for these valves to use more durable materials and designs in order to handle repeated cycles. Low-cycle valves are used in applications where the valve is rarely cycled, such as isolation valves that are typically kept open or closed for the majority of time.

D

Deadband

In valve operation, dead band refers to a range of input control signal changes that do not result in any movement or response from the valve. In essence, it is a zone of insensitivity within which the valve does not respond to the control signal. Deadband is commonly expressed as a percentage of the full range of the input signal. If a valve has a deadband of 1%, and the control signal changes by less than 1% (e.g., from 50% opening to 50.8% opening), the valve will not move. Only when the signal exceeds the 1% threshold (e.g., from 50% opening to 51.1%) will the valve respond.

Deadband occurs in valve operation due to several factors that create resistance or delay in the valve's response to small control signal changes. One of the major causes is friction within the valve components, such as the stem and packing, which resists minor movements. Another common cause is loose mechanical connections between the actuator and the valve body. The result is that input signal is not the cause of valve movement. In addition to wear and tear, improper maintenance, or limitations in the actuator's precision, other factors may also contribute to deadband, affecting the valve's responsiveness.

Differential Pressure

Differential pressure (ΔP) refers to the difference in pressure between a valve's inlet and outlet. A differential pressure is the force that propels the fluid through a valve. The amount of fluid that flows through a valve is determined by the differential pressure. A higher ΔP is generally associated with a higher flow rate. Differential pressure is a critical input for valve sizing calculations. This pressure drop represents energy lost during the process of the valve opening. Increasing differential pressure can result in higher energy requirements to pump the fluid, affecting the efficiency of the system. It is possible for the fluid to undergo cavitation (the formation of vapor bubbles) or flashing (permanent vaporization) when the differential pressure is too high. In order for the actuator to move the valve against the pressure differential, it must possess sufficient force or torque. A high ΔP can make it difficult for the actuator to operate the valve. Therefore, a valve actuator with a higher torque output is required when the differential pressure across the valve is high. During valve operation, pressure drop and differential pressure are effectively the same thing.

Discharge Capacity

In the context of a safety valve, discharge capacity refers to the maximum amount of fluid or gas that can be released per unit of time when the valve opens to relieve excess pressure. In the case of liquids, it is typically measured in terms of L/min and standard cubic feet per minute (SCFM) in the case of gases. By releasing enough fluid or gas to lower the system pressure to a safe level, discharge capacity is a critical specification that determines the valve's ability to release excess pressure and prevent system overpressure. The discharge capacity of a valve is affected by a variety of parameters such as the size of the valve's inlet and outlet, the type of safety valve design, the set pressure, and process conditions such as type of gas or liquid flowing through it. The larger the inlet and outlet sizes, the greater the flow, which results in a greater capacity for discharge. The pressure at which the valve opens influences discharge capacity, as higher set pressures generally force more fluid or gas through the valve. Different fluids and gases have different flow characteristics (e.g., densities and viscosities) that influence their discharge rate.

Drainage

In valve operation, drainage is the process of removing liquids that have accumulated in piping, vessels, or systems. The purpose of this is to maintain efficiency and prevent damage. Drain valves are valves (e.g., wedge gate or ball valves) used for drainage. They are typically installed at low points in the system where fluids accumulate. By using these valves, unwanted fluids can be discharged in a controlled manner in order to prevent blockages, corrosion, or freezing of the system. Drainage may be necessary for maintenance or shutdown of a system in some instances. Venting is the opposite of drainage, which is explained further in this chapter under the heading of "Venting."

Dual-Mode Operation

In valve operation, dual-mode operation refers to the functionality of valves in two different modes. Dual-mode operation is typically used to control different flow characteristics or adapt to changing operational conditions.

It is possible to define dual-mode operation as the operation of a valve that is equipped with both an electrical actuator and a manual override. This setup allows the valve to be operated both automated (using the electrical actuator) and manually (using the manual override). In automated mode, the valve is controlled by the electrical actuator. The operator can manually override the valve in manual mode by using a handwheel or lever. By using manual override, the electrical actuator or motor can be bypassed. As a result of its dual-mode functionality, the valve can still be operated manually in the event of a power failure or malfunction of the actuator.

Another example of dual-mode operation would be a control valve that can operate in both flow control mode and pressure control mode. The valve regulates the flow rate of the fluid passing through it in the flow control mode. The same valve can be operated in pressure control mode by changing the pressure rather than fluid. In pressure control mode, a pressure sensor and controller adjusts the valve's position to maintain a set pressure (not a set flow) within the system. This allows the valve to operate in two different control modes depending on the system's needs. The pressure sensor continuously monitors the system's pressure. Also, pressure sensor sends real-time pressure data to the controller. As the controller compares the actual pressure (measured by the sensor) to the desired setpoint pressure (set by the operator or control system), the pressure is adjusted accordingly by control valve operation. It is the controller's responsibility to calculate the adjustment necessary if the actual pressure deviates from the setpoint. In response to the controller's output, the control valve adjusts its position to increase or decrease the fluid pressure. In the event of excessive pressure upstream of the valve, the control valve may open more in order to release some of the pressure. By doing so, the fluid will be able to flow out and the pressure will be reduced. If the valve upstream pressure is too low, the valve may close slightly to reduce flow and allow pressure to build.

It is also important to know dual-mode operation valve. Dual-mode operating valve can perform two functions. For example, a vacuum breaker and air release valve can be considered a dual-mode operation valve, as it serves both functions—protecting the system from vacuum conditions and releasing trapped air to maintain efficient flow.

Dynamic Loads

Dynamic loads are the forces acting on valves during operation. Fluid flow, pressure and temperature changes, or the movement of valve components are the main causes of these loads. These loads can create vibrations, fluid-induced forces, and pressure surges. It is possible that these loads may affect the valve's stability, wear, and overall performance. High-pressure or high-flow systems require valves to withstand dynamic loads through proper design and material selection.

E

Emergency Actuation

The emergency actuation of an actuator occurs during an emergency to perform a critical safety function. This could involve opening, closing, or modulating a valve in order to ensure the safety of the system. Emergency actuation is often associated with the shutting down of an emergency shutdown valve (ESDV) to isolate hazardous processes, but it is not limited to this function. An emergency shutdown (ESD) can be initiated by the actuation of a blowdown valve. When an emergency occurs, blowdown valves are designed to release fluids or gases from a system as quickly as possible. As an example, a blowdown valve could be installed on a separator to release the high-pressure and high-level fluid from the separator into the flare or sub-flare line. In this process, known as blowdown, pressure is reduced and potentially hazardous substances are removed from equipment, such as piping. As a result, this minimizes the risk of fire, explosion, or structural failure due to equipment or system overload. Emergency actuations often involve fail-safe mechanisms such as spring-return actuators. As soon as power or signal is lost, the spring mechanism automatically reverts the valve to a safe position (open or closed). Emergency actuation is not limited to ESDV and blowdown valves. A control valve can indirectly adjust temperature, pressure, or liquid level by modulating the flow rate of the fluid in a system

Emergency Shutdown

ESD refers to a safety mechanism designed to isolate or stop the flow of fluids in a system during hazardous conditions. The primary goal is to protect people, the environment, and equipment by halting operations in an emergency. ESD generally applies to an entire system or process. ESD involves multiple components, including actuated valves, sensors, and control systems. During a gas leak, an ESD system may automatically close critical isolation valves to prevent a fire or explosion by stopping gas flow.

Emergency Shutdown Valve

ESDV is a critical safety valve used to isolate a section of a pipeline or process equipment in an emergency. It is imperative that this quick isolation action is taken in order to prevent fluid from flowing to the areas affected by a gas leak or fire in the event of a fire, gas leak, or equipment failure. As a primary function, it prevents the flow of hazardous substances and mitigates the escalation of an incident in order to protect personnel, equipment, and the environment. With the assistance of fast-acting actuators, ESD valves close rapidly and often within seconds of being opened.

Equal Percentage (Flow Characteristic)

More information about equal percentage flow characteristic is provided in this chapter under the heading of "Flow Characteristics (Flow Curves)."

F

Fail-Safe Position

Information about fail-safe position, including both fail-safe open and fail-safe closed, is provided in Chapter 11, under the heading of "Fail-Safe Mechanism."

Flashing

During valve operation, flashing occurs when the pressure of a liquid flowing through the valve drops. As a result, the fluid partially vaporizes into a gaseous state. This occurs when the pressure downstream of the valve drops below the liquid's vapor pressure but remains above the system's recovery pressure. A difference between cavitation and flashing is that in cavitation, bubbles collapse as pressure recovers downstream. However, in flashing, the vapor phase (bubbles) remains intact. The effects of flashing include erosion, wear, reduced flow capacity, noise, and vibration. A proper valve sizing is essential for preventing flashing.

Flow Balancing

Balanced flow in valve operation refers to the adjustment of valves in a system in order to ensure that fluid flows equally or as needed to the different parts of the system. For systems such as air-conditioning, heating, and water distribution, where each area requires the appropriate amount of flow to function properly, this is essential. It is for this purpose that special valves, known as balancing or control valves, are employed.

Flow Characteristics (Flow Curves)

The valve flow curve is a graphical representation of the relationship between the valve opening position and the flow rate through the valve. The flow rate curve illustrates how the flow rate changes as the valve moves from a closed to a fully open position. It is important to consider valve flow curves when selecting the right valve for a given application, especially when precise flow control is important.

Valve flow curve, as shown in Figure 8.1, represents the relationship between the valve's position (percent open) and the flow rate (flow coefficient C_v) through the valve. It provides insight into how the valve controls flow and is critical for

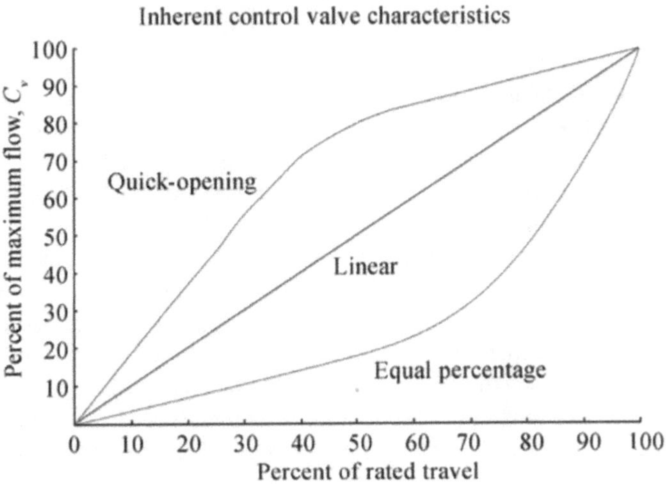

FIGURE 8.1 Valve flow curves showing the three common flow characteristics: quick-opening, linear, and equal percentage, where flow changes gradually at low openings and accelerates significantly at larger openings.

optimizing its performance in a system. Curves can be categorized into three types—the linear flow curve, the equal percentage curve, and the quick-opening curve.

- *Linear Flow Curve:* The flow rate increases in direct proportion to the position of the valve. As an example, if a valve is opened 20%, then the amount of flow that passes through the valve is 20% of the flow that passes through it in its full-open position. Flow control can be achieved with valves with linear flow characteristics, and some globe, butterfly, and V-port ball valves have linear flow characteristics.
- *Equal Percentage Curve:* Flow changes are minimal at low openings and significantly more noticeable at higher openings. Equal percentage flow characteristic valves are highly suitable for flow control and are well-suited for handling varying pressure drops. Some globe valves, V-port ball valves, butterfly valves, and diaphragm valves exhibit equal percentage flow characteristics.
- *Quick-Opening Curve:* When the valve is opened to a small opening, flow increases rapidly, making it suitable for on/off applications. Typically, ball valves have a quick-opening curve.

Flow Coefficient

For more information about flow coefficient, refer to Chapter 5, under the heading of "Flow Coefficient."

Flow Control

In valves, flow control refers to the ability to regulate the flow rate. It is possible to control flow by changing the valve's position to increase or decrease the passage area through which a fluid flows. As a result of regulation of flow rate, a fluid (liquid, gas, or slurry) passing through a valve within a pipeline or system may change its flow rate, pressure, or direction. Examples of flow control valves are globe, butterfly, and needle valves.

A flow control valve regulates the flow rate of a fluid (liquid or gas) in a system. Its primary function is to maintain the required flow rate regardless of changes in system pressure or load conditions. In accordance with the requirements of the system, these valves can either be operated manually or automatically. An example of a flow control valve would be a butterfly or globe valve that is manually operated or actuated with only purpose of flow control (not pressure, temperature, or level control). There is a difference between a flow control valve and a control valve. A control valve is a highly sophisticated valve that is used to dynamically regulate the flow, pressure, temperature, or level of a fluid within a system. To maintain desired process parameters, it operates within an automated control loop, receiving signals from a controller and adjusting its position accordingly.

Flow Directional Control

In valve operation, directional control refers to the ability of a valve to direct fluid or gas flow in a specific direction. In these valves, specific ports within the valve body are opened or closed in order to control the flow path. There are often multiple positions on directional control valves (e.g., two-way, three-way, four-way) and they can be operated manually or by an actuator. As an example, three-way valves can direct flow from one input port to one of two output ports, changing the direction of flow or combining fluids. One form of fluid directional control is *flow diverting*. More information about flow diverting is provided in this chapter under the heading of "Flow Diverting."

Flow Diverting (Flow Diversion)

It is true that flow diverting is a form of flow directional control in the operation of valves. In contrast to directional control valves, diverting valves specifically redirect the flow from one path to another within the system. In valve operation, flow diversion or diverting refers to the diversion or redirection of fluids or gases within a system. A diverting valve is typically used to accomplish this. According to operational requirements, this valve allows the operator to control and change the direction of flow. It is common for flow diversion to be used in processes in which multiple flow paths are available. Depending on certain conditions, the system may be directed to different locations or components.

The purpose of this example is to clarify the difference between flow directional control and flow diversion or diverting. Let us consider a three-way valve in operation with one inlet port A and two outlet ports B and C. For flow diversion, the fluid enters the valve through port A and moves to either port B or C. This means that, in one operation, the valve passes fluid from port A to port B, then diverts it from port A to port C. In directional control, the flow can be directed from port A to ports B and C and the inlet port can be switched. As a result, port B or port C may be used as inlet ports instead of flow diversion ports. As a result of flow diversion, ports B and C always receive flows and are always outlets.

Flow Modulation (Modulating)

In valve operation, flow modulation means an accurate control of fluid flow through a piping system by adjusting the position of the valve. In the process of flow modulation, the fluid flow rate can be varied to meet the desired value, and it does not just involve stopping or starting the flow in the piping system. There are several types of valves that are capable of modulating flow, including globe valves, butterfly valves, control valves, and V-port ball valves. In some texts and literature about valves, flow control is used as a means of controlling and changing flow rates. In spite of this, flow control has a wide range of applications since it covers both changing flow rate and stopping, starting, and directing the flow. A ball or gate valve, for example, can be used to control the flow rate in a piping system by opening and closing the valve.

Flow Rate

Flow rate in valve operation refers to the volume or mass of fluid (liquid or gas) that passes through a valve over a specific period. Volume flow rate and mass flow rate are two types of flow rate. Volume flow rate is defined as the amount of space occupied by the fluid. Volume flow rate is measured in m^3/h, L/s, or gal/min. The mass flow rate, on the other hand, is determined by the weight of the fluid. The rate is expressed in kg/s or lb/h. Units are chosen according to the application. In the case of liquids, volume flow is commonly used, while mass flow is often preferred in the case of gases. The flow rate is affected by several key parameters, including the pressure drop across the valve. The higher the pressure difference, the greater the flow. Flow rate is also affected by the size and type of valve. The larger the valve size, the larger the flow that can pass through it. As far as valve types are concerned, a full-bore ball valve in a fully open position can allow a higher flow rate than a fully open butterfly valve of the same size and pressure class. In addition, fluid properties such as viscosity, density, and compressibility have a significant impact on flow characteristics. The valve's opening percentage and flow characteristics have an effect on the flow. Higher opening percentages result in greater flow passage through the valve.

Flow Restriction

Flow restriction in valve operation refers to the intentional or unintentional limitation of fluid flow through a valve. Intentional flow restriction is a positive feature of operation, while unintentional flow restriction is an operational issue that is undesirable. Intentional flow restriction may occur when a valve is partially closed, obstructed, or designed with a narrow passageway. This results in a reduction in the amount of fluid that can pass through it. A fluid flow restriction is commonly used in applications in which it is necessary to control or regulate the flow rate, pressure, or velocity of the fluid. By restricting the flow of incoming fluid, a pressure-reducing valve (PRV), for example, lowers the pressure of incoming fluid.

It is also possible for flow restriction to occur unintentionally due to valve wear, blockages, or incorrectly set valve positions. This can have unintended consequences, such as reduced system efficiency, increased energy consumption, or uneven fluid distribution. As a result, it is necessary to manage flow restriction carefully in order to ensure that valves perform optimally and that the system operates as intended.

Flow Turbulence

Flow turbulence in valve operation occurs when fluid flow becomes chaotic and irregular. The cause of this is disturbances caused by the valve's geometry (for example, narrow areas in axial check valves, or changes in flow direction in globe valves), the valve position such as throttling the flow, or the operating conditions of the valves. In most cases, it results from throttling, sudden changes in flow direction, or velocity. It is important to install some valves such as dual-plate, axial check, and globe valves before and after certain amount of straight line (piping) to prevent fluid turbulence. Thus, incorrect valve installation near elbows or tees can lead to flow turbulence. Turbulence can lead to energy losses, pressure drops, noise, vibrations, erosion of valve components, and instability in downstream equipment. Certain valve designs, such as butterfly or ball valves, may induce less turbulence compared to globe or control valves. To minimize turbulence, it is essential to select appropriate valves, ensure proper installation with adequate straight-pipe lengths, and operate within recommended flow ranges.

Flutter

During valve operation, valve flutter refers to the rapid, oscillatory movement of a valve component. As a result of unstable flow conditions or pressure fluctuations, components such as the disk, plug, or stem may be damaged. It is common for this phenomenon to occur in partially open valves. In a partially open valve, high-velocity flow or turbulence interacts with the internal parts and causes them to vibrate. Flutter can occur in safety valves, especially if they are poorly designed or installed incorrectly. Flutter can lead to noise and increased wear on the valve components. Additionally, the valve or surrounding piping may be damaged. Knowing the difference between flutter and vibration is important. By definition, flutter involves the rapid motion of the valve's internal moving parts as a result of the flow conditions. As opposed to this, vibration involves broader oscillations of the whole valve or system, caused not only by the fluid flow, but also by external or systemic forces.

Fugitive Emission

As the name implies, fugitive emissions in valves refer to the release of gases or vapors, typically hazardous substances, into the atmosphere as a result of leaks in valve components. These emissions are commonly observed in industrial environments, particularly in the oil and gas, chemical, and petrochemical sectors, where valves handle pressurized fluids and gases. For both environmental protection and worker safety, as well as compliance with stringent regulatory requirements, fugitive emissions must be minimized. The emission of fugitive gases contributes to the accumulation of greenhouse gases and air pollution. Natural gas contains a large quantity of methane (a component that is especially potent in terms of its global warming potential). As a matter of health and safety, hazardous gases can pose serious health risks to workers if inhaled. It is possible that some emissions gases could be toxic, including hydrogen sulfide, which could kill employees on the site. The presence of fugitive emissions can also result in a risk of fire or explosion in facilities that handle flammable materials. It is common for companies to be required to comply with strict regulations pertaining to fugitive emissions, such as those set by the US Environmental Protection Agency (EPA) and the European Union. It may be necessary to pay regulatory fines and increase monitoring requirements if emissions are not controlled, adding to the operation's expense. There are different valve standards, such as API 624 and API 641 and ISO 15848, addressing fugitive emission.

G

Gearbox-Operated Valves

The gear-operated valve is a type of valve that utilizes a gear mechanism for opening, closing, or adjusting the valve. The gear mechanism, usually a worm gear or a bevel gear, reduces the amount of force necessary to open or close a valve. Consequently, it is easier for a human operator to operate large or high-pressure valves manually. Using the handwheel attached to the gearbox or gear mechanism, the valve can be operated. Using the gears, rotational inputs are converted into torques that move the valve's internal components, such as disks, balls, and gates. Gate valves, ball valves, and butterfly valves are common valves that can have handwheel and gearbox. In general, gearbox assemblies provide smoother and more efficient torque transmission by enclosing and protecting the gears.

Figure 8.2 illustrates three ball valves equipped with gearboxes and handwheel operation. These valves are installed on the sub-flare line upstream of the pressure safety valves (PSVs). The ball valves are normally kept in the open position to allow the overpressure fluid or gas to flow through the sub-flare line into the flare header. However, when a safety valve is removed from the line for maintenance or calibration, it is crucial for the operator to close the upstream ball valve to ensure safety and prevent leakage to the environment. The gearbox facilitates the operator's ability to close the valve efficiently, even under high torque requirements.

FIGURE 8.2 Ball valves with gearbox and handwheel operation on the sub-flare line upstream of pressure safety valves (PSVs). (With permission: Shutterstock.)

Once the safety valve is reinstalled, it is equally important for the operator to fully open the upstream ball valve to restore normal operation and ensure unhindered flow through the sub-flare line. Proper operation of these valves is essential for maintaining system integrity and environmental safety.

Gradual Flow

Gradual flow is a term used to describe the gradual increase or decrease in fluid flow as a valve is opened or closed. The use of this approach can prevent sudden changes in flow velocity, pressure surges, and water hammer, which can cause damage to the valve and the piping. In order to achieve gradual flow, globe and control valves that are suitable for flow control or throttling are ideal. Alternatively, the valve actuator can be operated slowly. It is also essential to have gradual flow increase or decrease in systems where precise flow control or pressure management are imperative. Equal percentage flow characteristics are suitable for flow control since they provide gradual flow increases at valve openings. With the valve slightly open, flow changes are minimal, allowing precise control and preventing sudden surges.

H

Handwheel Rotation

Handwheel rotation in valve operation involves turning a circular wheel manually to open, close, or adjust a valve. Handwheel operation is applicable to both linear and rotary stem and closure movement valves. As a result of handwheel rotation, torque is applied to the stem of the valve. By applying torque to the valve, its internal components, such as the gate, disk, or globe, will move in order to control the flow of fluid.

Handwheel Spoke and Rim

More information about handwheel spoke and rim is provided in Chapter 3, under the heading of "Handwheel."

High Pressure

In valve operation, high pressure refers to conditions in which the valve is subjected to pressures above its normal operating range. Also, valves and piping in pressure class 600 (equal to pressure nominal of 100 bar) and above are considered high-pressure piping and valves. There is a risk of leakage, deformation, or failure of components in high-pressure applications due to the intense force exerted by the fluid. High-pressure fluid service damages can be mitigated by adding sufficient wall thickness to valve components, such as the body and bonnet, as well as by selecting strong materials, such as duplex and super-duplex.

High-Pressure Fluid Dynamics

The phrase "high-pressure fluid dynamics" pertains to the behavior and effects of fluid flow within a valve under conditions of high pressure. The fluid dynamics of valves operating in high-pressure systems involve forces generated by the fluid under these conditions. First, there is a high-pressure drop which can lead to operational problems such as *cavitation*, *flashing*, *noise*, and *vibration*. There is also the possibility of *fluid turbulence* combining with *high velocity* and causing wear and erosion in the valve. Additionally, high-pressure fluid exerts *strong forces* on valve components, such as the stem, seat, and body. This necessitates the use of robust materials and design for valves for high-pressure fluids.

High Temperature

In valve operation, a high temperature refers to conditions in which the valve is subjected to elevated temperatures that are beyond its normal operating range. High-temperature applications present a number of challenges, including thermal expansion of components, degradation of sealing materials (such as elastomers or gaskets), reduction of strength, and creep or deformation of metal parts. To prevent negative effects of high temperatures, valve materials, designs, and maintenance practices must be chosen appropriately.

Hydraulic Shock

Hydraulic shock is same as water hammering. More information about hydraulic shock and water hammering is provided in this chapter under the heading of "Water Hammering."

Hysteresis

Hysteresis in valve operation refers to the failure of the valve to respond correctly to control signals. For example, a valve may act slower than it should during opening or closing. It is important to note that the valve's behavior during opening differs from its behavior during closing. There may be a variety of reasons for this, such as different stem and packing friction during opening and closing, or actuator behavior during opening and closing. As a result, there is a lag or difference between the actual position of the valve and the control signal. Because of the factors mentioned above, it is normal for a valve to behave differently during opening and closing. Hysteresis, on the other hand, refers to a situation in which the difference between opening and closing is greater than what would normally be expected. This may result in inaccurate control of flow, pressure, or temperature. Hysteresis may be caused by factors such as friction between valve stem packing and valve stem, or by abnormal actuator opening or closing behavior. Consider the case of a valve in a system where fluid flow is being controlled. A signal is given to the valve to open, and it takes 2 min for the valve to fully open. As a result, when the valve is given the same signal to close, it takes 3 min to close fully. This difference in response (opening in 2 min, closing in 3 min) is described as hysteresis. In other words, it means that the valve does not respond symmetrically to control signals, which can result in a lag or delay, which can have a negative impact on the system's accuracy and precision. In the event that this closure occurs too late, there may be damage to the system due to overpressure, loss of energy, and instability.

I

Isolation (Flow Isolation)

In valve operation, isolation is the process of completely shutting off or separating a section of pipe, pipeline, or system to prevent fluid or gas flow. In most cases, this can be achieved by using isolation valves, which when fully closed provide a tight seal. As a result of maintenance, repairs, or operational reasons, isolation valves are used to stop the flow of fluids. The most common types of isolation valves are ball valves, gate valves, and butterfly valves.

L

Laminar Flow

Laminar flow is characterized by smooth, parallel, and orderly layers of fluid, as shown in Figure 8.3. Unlike turbulent flow, each layer moves independently without disrupting the adjacent ones. This flow pattern occurs at lower velocities. In a pipe, this generally occurs when the Reynolds number, a dimensionless quantity representing the ratio of inertial to viscous forces, falls below a critical value, often around 2,000 (more precisely less than 2,300). Pipelines and piping,

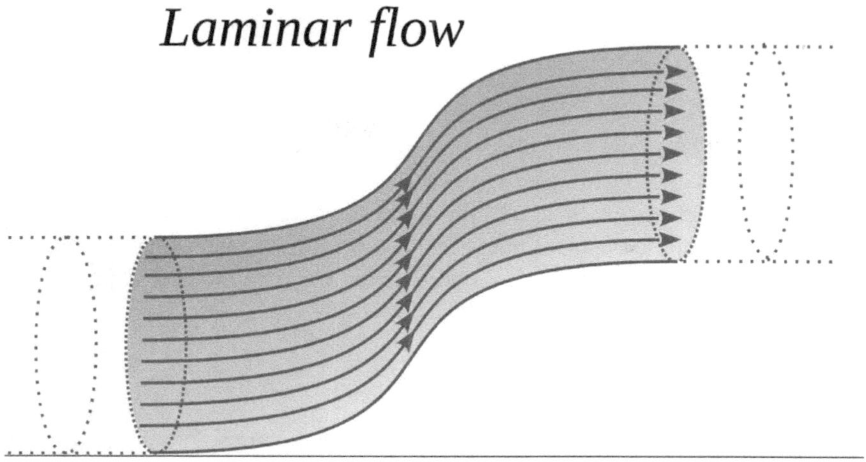

FIGURE 8.3 Laminar flow in the piping. (With permission: Shutterstock.)

medical devices, and certain industrial processes often require laminar flow for precision fluid control. That is because turbulent flow creates erosion, wear, and higher energy loss.

Leakage Class

More information about leakage class is provided in Chapter 5, under the heading of "Leakage Class."

Lever Operation

A lever-operated valve is one that is operated manually by a lever to control fluid flow by opening, closing, or throttling it. This lever is attached to the valve stem or shaft and provides a simple operation method. Rotating the lever adjusts the valve's internal components (e.g., ball, disk, or plug) to regulate flow. Lever operation allows the valve to open or close rapidly, often within 90°. In general, lever operation is used for smaller valves, while handwheels or handwheels combined with gear boxes are usually required for larger valves.

Linear Flow/linear Flow Characteristic

It refers to a flow rate that changes in direct proportion to the valve opening. For example, If a valve is opened to 50%, the flow rate will also be at 50% of the maximum flow. For a valve with a linear flow characteristic, the relationship between valve opening and flow rate is a straight line (linear) when plotted on a graph. More information about linear flow and linear flow characteristic is provided in this chapter under the heading of "Flow Characteristics (Flow Curves)."

M

Manual Operation

The manual operation of a valve, in contrast to the actuator operation, refers to the manual control of a valve's position by means of a handwheel or lever. The handwheel is typically used when the valve is larger and has a higher pressure class, so it requires greater force to operate. In some cases, a gearbox is added beside the handwheel to facilitate manual operation of the valve by increasing the force or torque applied by the operator. The operation of this type of valve is performed by human effort; therefore, it is simple and cost-effective.

Manual Override

In valve operation, manual override is related to the actuator's configuration. As the name implies, it refers to the capability of manually operating an actuated valve without the aid of an automated or a powered actuation system. In most cases, this feature is included in motorized (electrical) or pneumatic actuators. Manual override provides a backup method for opening, closing, or adjusting valves in the event of power failures, actuator malfunctions, or maintenance activities on the actuator. A manual override typically consists of a handwheel, lever, or other mechanical device that allows the operator to manually position the valve as necessary. A butterfly valve shown in Figure 8.4 is actuated by an electrical actuator and has a manual override attached to the electrical actuator.

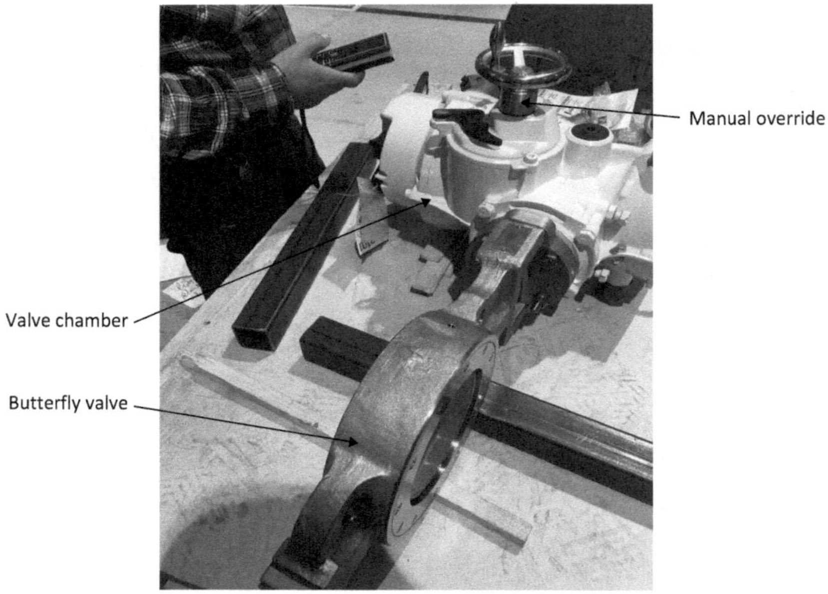

FIGURE 8.4 The butterfly valve is actuated by an electrical actuator and has a manual override attached to the electrical actuator.

Maximum Allowable Stem Torque (MAST)

More information about MAST is provided in Chapter 5, under the heading of "Maximum Allowable Stem Torque (MAST)."

N

Noise

Noise is known as a valve operational problem. In valve operation, noise occurs as a result of turbulence, pressure drops, cavitation, or mechanical vibrations. It is possible for turbulent flow to be created by high flow velocities, sudden changes in direction, or throttling. Cavitation, which occurs when vapor bubbles form and collapse, may also cause significant noise and damage to property. Furthermore, mechanical vibrations caused by moving parts or external forces can amplify the sound. The presence of excessive noise is common in high-pressure systems and during throttling. The presence of excessive noise may be indicative of inefficiencies or other potential problems such as wear and tear. It is possible to reduce noise levels in valve operation by selecting the right valve, installing anti-cavitation trims, and following good installation practices.

O

On/Off Control

An on/off valve control refers to a simple method of valve operation in which the valve is either fully opened or fully closed without any intermediate positions. The use of this type of control is typically limited to applications requiring no precise control of flow, pressure, or other process variables.

Opening Percentage

Valve opening percentage refers to how much a valve is open compared to its fully closed or fully open position. It is expressed as a percentage of the total travel range of the valve's movable element (e.g., closure member). The valve opening percentage indicates how much fluid can flow through the valve at a given position. For example, a 50% valve opening means the valve is halfway between fully closed and fully open.

Operating Envelope

A valve's operating envelope refers to the range of operating conditions within which it is capable of functioning safely, efficiently, and reliably. An example of this would be a valve in CL150 that is pressure rated for any pressure between 1 bar and 19 bar. The valve may be suitable for operating temperatures ranging from –29°C to 400°C.

Operating Pressure

The operating pressure of a valve refers to the pressure at which it is expected to function under normal, everyday conditions. It is the pressure exerted by the fluid (liquid, gas, or steam) passing through the valve during normal operation. The operating pressure is an essential parameter in the selection and design of valves, as it helps ensure that the valves are capable of handling the routine demands of the system without causing any problems.

Operating Temperature

A valve's operating temperature is the range of temperatures within which it is expected to function under normal, day-to-day operating conditions. The operating temperature of a valve refers specifically to the temperature of the fluid (liquid, gas, or steam) flowing through the valve during standard system operation. In the selection and engineering of valves, the operating temperature is a critical parameter. To prevent degradation, the valve's sealing materials and soft materials must be compatible with the operating conditions such as operating pressure and temperature. The selection of certain elastomers is based on their ability to withstand the typical temperature without softening, warping, or losing strength.

Overpressure Prevention

Overpressure prevention in valve operation refers to measures taken to prevent systems from experiencing excessive pressure. It is possible for excessive pressure to damage piping systems and equipment, to endanger the health and safety of personnel on site, and to harm the environment. This is achieved by designing and using devices such as safety valves or PRVs that automatically open in response to excessive fluid pressure. Excess pressure is released after they are opened, preventing potential equipment damage, safety hazards, or catastrophic failures.

Over-Torque

Valve over-torque occurs when excessive force is applied to the valve's operating mechanism, such as the stem and closure member. As a result, the applied load or torque exceeds the design limits for the affected components. There can be a variety of damages caused by this, including bent or broken stems, crushed or deformed seals and seats, misaligned stems, seats, or closure members, and wear of internal parts. Moreover, actuators that apply excessive torque may also malfunction as a result of damaged gears or internal components. Over-torque can be prevented by proper actuator design and sizing, following manufacturer guidelines, ensuring that proper maintenance is carried out, and applying only the required force during operation.

P

Pop Action

A pop action in a safety valve refers to the sudden, rapid opening of the valve at its set pressure. When the system pressure reaches the setpoint, the safety valve opens almost instantly, releasing excess pressure rapidly. An immediate action is critical to the operation of safety valves, as it allows for fast pressure relief in order to prevent system overpressure and potential damage to the equipment.

Positioner

More information about positioner is provided in Chapter 3, under the heading of "Positioner."

Pressure Drop

The pressure drop associated with valve operation refers to the reduction in fluid pressure during the flow of fluid through a valve. This reduction is the result of resistance created by the valve's internal components, such as the seat and disk. A number of factors affect the resistance of a valve, such as its type, size, flow rate, and degree of opening. The greater the pressure drop, the greater the energy loss within the system. Besides, higher energy loss can reduce the efficiency and performance of the piping system. Many control valves are designed to manage specific pressure drops. This allows them to regulate flow effectively, while minimizing turbulence and preventing cavitation. Choosing the right valve and sizing it properly is critical to achieve the desired pressure drop. As an example, butterfly valves have a reduced bore and therefore produce a high-pressure drop for piping that requires a full-bore valve, such as a full-bore ball valve. When it comes to valve sizing, if a 2″ valve is used for a piping system that requires a 3″ ball valve to minimize pressure drop, then such a 2″ small valve will result in a high-pressure drop.

Pressure Equalization

Pressure equalization in the context of valves refers to the process of balancing pressure on both sides of a valve. This is typically achieved using a dedicated bypass system, including bypass line and a smaller auxiliary valve. Pressure equalization is particularly important in systems with high-pressure differential to ensure safe and efficient valve operation. Large or high-pressure valves, like gate valves or ball valves, may be difficult to open or close due to the pressure difference across the valve. Equalizing the pressure reduces the force required. Abruptly opening a valve under high differential pressure can cause damage to valve components, such as seals, seats, and stems, due to the sudden fluid movement or mechanical stress due to pressure surge. Pressure equalization minimizes the risk of pipeline or system damage caused by pressure surges or shocks.

Pressure Regulation

In valve operation, pressure regulation refers to the process of maintaining or adjusting the pressure within a system to a desired level. It is common to achieve this by using pressure-regulating valves, such as control valves, and pressure-reducing valves.

Pressure Surge

A pressure surge occurs when a sudden change (sudden stop or start) in fluid velocity causes a rapid increase in pressure within a pipeline or piping system. This may occur when a valve is quickly closed or a pump is abruptly turned off. As a result of this surge, pressure spikes can occur that exceed normal operating levels. As a result, pipes, valves, and other components of the system may be damaged.

Purging

During valve operation, purging involves the removal of unwanted substances from a valve and its associated piping. Air, moisture, contaminants, or residual process fluids are examples of unwanted substances. During purging, a purge medium (e.g., inert gas or nitrogen) is introduced into the system. Purging is done for several reasons such as prevent contamination and maintain cleanliness. Besides, purging prepares equipment and components for maintenance or shutdown. In industries such as oil and gas or pharmaceuticals, even trace amounts of contaminants can compromise safety or product quality, which is why purging is essential. The purge medium is typically passed through the valve to ensure that unwanted materials are thoroughly removed without damaging the valve or the equipment connected to it. A proper purging procedure protects against corrosion, explosion risks, and system inefficiencies. As an example, nitrogen purging can be done to ensure that the piping and equipment system is free of air to prevent explosions.

Q

Quick Opening (Flow Characteristic)

More information about quick opening flow characteristic is provided in this chapter under the heading of "Flow Characteristics (Flow Curves)."

Quick Closing Valve

In critical situations, quick-closing valves are designed to close rapidly, usually within seconds, to stop the flow of fluid immediately. A quick closing valve can be actuated by means of a spring-return pneumatic or hydraulic actuator or by means of a lever that is manually operated. Despite the fact that quick closing valves appear to be similar to ESDVs, they differ in a number of ways. Typically, a quick closing valve is used to protect a single piece of equipment or a small system. In contrast, ESDV is used to protect an entire system or a section. As another difference, whereas an ESDV is always actuated, a quick closing valve can also be operated manually. Additionally, ESDV is integrated with a process safety system.

R

Rangeability

Rangeability refers to the ratio between maximum and minimum values of a process variable that a valve can control effectively while maintaining a stable and accurate performance. It reflects the valve's ability to handle variations in operating conditions, such as flow, pressure, temperature, or other operational parameters. As an example, a pressure control valve might be able to handle pressures between 10 psi and 1,000 psi, giving it a rangeability of 100:1. As an example of temperature rangeability, in the second example, a valve that controls the temperature could regulate a system from 50°C to 300°C, resulting in a temperature rangeability of 6:1. Rangeability and control range are closely related. There is more information about the control range in this chapter under the section "Control Range."

Relieving Pressure

More information about relieving pressure is provided in Chapter 10, under the heading of "Relieving Pressure."

Response Time

Response time in valve operation refers to the amount of time it takes for the actuator or positioner to move the valve from one position to another (e.g., from fully open to fully closed). It is important to note that the actuator or positioner first receives a signal from the control room. Response time includes the delay in signal processing, the mechanical movement of the actuator, and the valve's travel time. In systems that require precise control or emergency actions, fast response times are critical. In other systems, slower response times may be acceptable in order to prevent system disruptions such as water hammer.

Example: The response time of a valve system includes the duration of signal processing, actuator movement, and valve travel. For instance, the positioner interprets a command sent by the control room to close a valve, which takes about 50 ms. The actuator then moves the valve stem, which requires approximately 400 ms to engage completely. A valve with a travel distance of 50 mm and a stroke speed of 100 mm/s will travel in 500 ms. In total, the response time from signal reception to complete valve closure is approximately 950 ms (0.95 s).

Reverse Flow Prevention

In valve operation, reverse flow prevention refers to mechanisms or devices that prevent fluid from flowing backward. By doing so, the fluid will only move in the direction intended. Fluid reversal is crucial to preventing operational problems such as contamination and equipment damage. It is common to use devices such as check valves (swing, ball, or lift types) and backflow preventers.

S

Safety Relief

During valve operation, safety relief is achieved by using a safety relief valve. The purpose of this valve is to prevent overpressure in a system. This valve automatically releases excess fluid (liquid or gas) when the system pressure exceeds a predetermined level. These valves combine the features of both safety valves (used primarily for gases) and relief valves (used for liquids).

Sampling

Valve sampling is the process of extracting a small amount of fluid (liquid, gas, or slurry) from a system. A sample is taken for the purpose of testing, analysis, or quality control. In order to accomplish this task, sampling valves are specifically designed. By using them, operators are able to take a representative sample of the process fluid without disrupting the system or contaminating the sample. Depending on the pressure, sampling valves may be gate or ball valves in low-pressure applications and modular valves (double block and bleed) in high-pressure applications. It is the function of the sampling valve to open in order to obtain a sample and to close once the sample has been obtained. A sampling valve installed in a reactor, for instance, enables operators to extract a small amount of liquid. It will be possible for the operator to check the pH levels after taking a sample to ensure that the reaction is proceeding as expected.

Seal Leakage

In valve operation, seal leakage occurs when the sealing elements, such as gaskets, packing, or O-rings, fail to create an adequate seal. The result is that fluid is allowed to escape. Wear and tear, improper installation, material degradation, thermal expansion, and excessive pressure all can contribute to this problem. It is possible for seal leakage to result in decreased system efficiency, safety hazards, contamination, or equipment damage. In systems handling toxic or flammable fluids or fluids under high pressure, this problem can be particularly severe. Seal leakage can be prevented by using seals made of materials suitable for operating conditions. Additionally, it is important to install the seals correctly. Seals must be inspected regularly and damaged seals must be replaced. In addition, proper valve design and operating the valves within the operational envelop are essential.

Service Life

Based on actual, real-world performance, this is the lifespan of the valve from the time of installation to the time when it needs to be replaced or overhauled. A number of factors influence the service life of a product, including unexpected pressures, variations in maintenance practices, environmental impacts, and the specific requirements of the application. The service life of a valve, depending on how well it is maintained and the conditions it is exposed to, may be shorter or longer than its design life.

Setpoint

The setpoint in valve operation refers to the desired value or target for a specific process variable. Process variables are pressure, temperature, or flow rate, which the valve and control system aim to maintain. In a heating system, the setpoint might be 75°C, for example. By controlling the position of the valve, the control system ensures the process temperature remains as near as 75°C as possible. The setpoint could be 100 L/min in a flow control application. Responding to the control system's signals, the control valve modulates its opening in order to maintain that flow rate.

Here is an explanation of how to achieve the setpoint. The control system continuously measures the actual process variable (e.g., current flow or pressure) using sensors. The control system which gets process data from the sensors compares the measured value to the setpoint. If there is a difference (error) between the setpoint and the actual value, the control system adjusts the valve position to minimize this error.

Signal Feedback

During valve operation, signal feedback refers to the process of communicating information to the control system regarding the valve's actual position or status in order to confirm that the valve is operating as intended. As an example, a positioner on a control valve might indicate to the controller that the valve is currently open at a certain percentage. In the event that the feedback signal does not match the control signal (setpoint), the control system can make adjustments. Alternatively, an alarm may be triggered to address issues such as stiction, misalignment, or faulty actuators.

Simmer

The term simmer refers to the condition when a safety valve begins to leak or slightly discharge fluid or gas as the system pressure approaches the valve's set pressure. This phenomenon occurs just before the valve fully opens and is characterized by small, intermittent releases of fluid or gas rather than a continuous flow. This indicates that the valve is approaching the point where it will "pop" open fully. A simmer usually occurs at a pressure slightly below the valve's set pressure, usually between 90% and 95% of the set pressure. As a result of system pressure exceeding the initial spring force on the valve seat, a gradual release of fluid or gas occurs during simmer. Process conditions and how quickly the pressure reaches the valve's setpoint determine the duration of simmer.

Stem Rotation

Stem rotation in valve operation refers to rotating the valve stem by actuation or manual operator. This is typically used in rotary valves such as ball, butterfly, and plug valves. As a result of this rotational movement, usually up to 90°, the valve closure member moves. Similarly, stem travel is an important concept related to stem rotation, which is described in more detail in this chapter.

Stem Travel

In valve operation, stem travel refers to the linear movement of the valve stem. The stem travel is responsible for opening or closing the valve by moving its internal components, such as the disk, plug, or gate. The distance the stem travels determines the position of the valve, from fully closed to fully open. In most cases, stem travel is measured in millimeters or inches. The stem travel is relevant to linear motion valves such as gate valves, globe valves, and diaphragm valves.

Stiction

The term "stiction" comes from the phrase "static friction." In other words, it refers to the resistance that a valve encounters when it starts moving from a stationary position to a moving position. The problem occurs as a result of frictional forces between the moving parts of the valve (e.g., the stem and packing, or the stem and actuator). In order to initiate valve movement, the control system must overcome this resistance, which can result in delays or inconsistent operation. Imagine a valve that requires a control signal change of 3% to overcome stiction and start moving. If the process only demands small adjustments (e.g., 1% changes), the valve will remain stuck until the signal exceeds the stiction threshold, leading to delayed or abrupt movements. There is a difference between stiction and sticking. Sticking implies physical obstruction of the valve, which makes it difficult to move even with significant force. As a result of factors such as corrosion, debris, or mechanical wear and tear, this is usually a more serious problem. More information about sticking is provided in this chapter, under the heading "Valve Sticking." *Stiction* (static friction) is closely related to *deadband* in control valves. Stiction refers to the resistance that prevents a valve from moving in response to small changes in the control signal, while *deadband* represents the range within which input changes do not result in any valve movement.

Stroke Speed

The stroke speed of a valve refers to the rate at which the actuator moves the valve stem or disk between its open and closed positions. In addition, it indicates the speed at which the actuator moves from one position (e.g., fully closed) to another (fully open). Chapter 11 provides more information about actuator parameters, including stroke. Typically, it is measured in seconds per inch or seconds for a full stroke. To improve the responsiveness of actuated valves to control signals, valve and actuator manufacturers can adjust the stroke speed regarding the opening and closing times for process requirements.

T

Thermal Expansion

In valve operation, thermal expansion refers to the increase in size caused by a rise in temperature. It is possible for thermal expansion to adversely affect the performance and functionality of a valve. Depending on their thermal properties, materials expand at different rates as the temperature rises. The result could be issues such as tight valve stem movement due to friction with packing, misalignment, or binding of moving parts. It is especially important in high-temperature applications, such as steam or process systems. The reason for this is that excessive expansion can result in leaks, difficulties with valve operation, or damage to seals and seats. To mitigate the negative effects of thermal expansion, valves must be designed with specific considerations. This includes selecting materials that can withstand the operating temperature and incorporating design features such as flexible seals or adequate clearances to accommodate expansion.

Throttling

Throttling occurs when a valve is partially opened or closed in order to control (modulate) the flow of fluid. By doing so, a system's flow rate, pressure, or velocity can be regulated. In contrast to on/off operation, throttling enables precise adjustments to meet process requirements. Globe, butterfly, and V-port ball valves are well-suited to throttling. The throttling process can cause high-velocity flow and turbulence, which can lead to erosion, noise, vibration, and cavitation. A proper valve and material selection are necessary to avoid operational problems caused by throttling.

Torque Application

In valve operation, torque refers to the force required to rotate or move the valve stem and closure member in order to open, close, or throttle (fluid control) the valve. It is particularly important to consider torque when it comes to rotary valves, such as ball, butterfly, and plug valves. The actuator or handwheel is used to apply rotational force in order to overcome resistance caused by factors such as packing and stem friction, as well as friction between the valve seat and closure member. The application of torque correctly ensures the efficient operation of valves. Over-torque can cause damage to valve components, while under-torque is insufficient to shift the valve to the required position. Furthermore, torque is a very important parameter to consider when selecting an actuator or a valve gearbox. Considering a safety factor, the actuator must provide sufficient torque above what is required for the valve operator. More information regarding valves and actuator torques is provided in Chapter 11—the heading "Torque" under the "Actuator Motions and Output Types" section.

Torque is measured in units like newton-meters (Nm) or pound-feet (lb-ft). In practical terms, it is what you feel when you use a wrench to turn a bolt. Over-torque generally means better ability to perform work, especially in applications requiring rotational movement. In the valve industry, this term refers to the amount of force that can be exerted on a valve operator in order to rotate the valve closure member to be opened or closed. For example, if a ball valve hand wheel requires 200 N to be turned with one hand and the diameter of the hand wheel is 300 mm (0.3 m), the required torque for the operator to operate the valve would be $200 \times 0.30 = 60$ Nm. Various parameters influence the value of torque, including valve type, pressure and temperature, valve size, and packing and sealing type. Depending on the type of valve, different torque requirements are needed, such as gate valves, ball valves, and butterfly valves, due to their unique sealing mechanisms. Higher operating pressures and temperatures increase friction and resistance, thus requiring more torque. There is friction between the stem packing and the internal seals, which adds to the torque requirement, especially if the packing materials are tightly compressed. The larger the valve, the greater the force and torque required to operate it.

Transient Conditions

Transient conditions in valve operation refer to changes in flow, pressure, temperature, or other operating parameters during the start-up, shutdown, or change in operation of the system. If these conditions are not properly managed through appropriate design and operation procedures, pressure surges, flow instability, or cavitation can occur, which may cause damage to valves or associated equipment.

Turbulent Flow

A turbulent flow is a type of fluid movement characterized by chaotic and irregular motions. In this situation, fluid particles are moving in a random direction at varying speeds. Unlike laminar flow, which has a smooth flow pattern and parallel layers, turbulent flow is characterized by swirls, and rapid fluctuations in velocity. It typically occurs in large-diameter pipes at high flow rates. Additionally, it can occur when a fluid encounters obstacles, sharp bends, or rough surfaces. The Reynolds number has a significant impact on turbulent flow. Reynolds number is a dimensionless value that predicts the behavior of flow. In general, flow

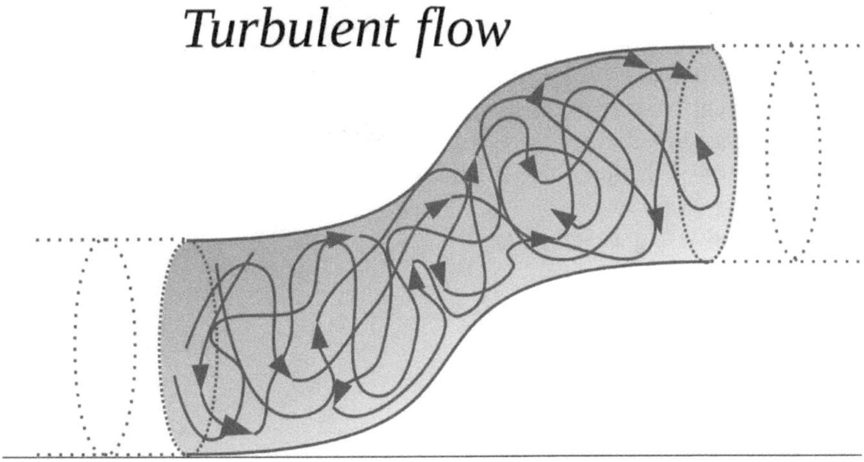

FIGURE 8.5 An illustration showing turbulent flow in a pipeline, with chaotic and irregular motion depicted by swirling arrows. (With permission: Shutterstock.)

turbulence occurs when the Reynolds number exceeds 4,000. Although turbulent flow increases energy loss due to friction and pressure drop, it also enhances mixing and heat transfer. As a result, it can be used in applications such as heat exchangers and chemical reactors. Nevertheless, it is undesirable in systems such as piping or valves. This is due to the fact that it creates excessive turbulence, erosion, noise, and a reduction in efficiency. Figure 8.5 illustrates turbulent flow inside a pipeline. Chaos and irregular motion are represented by swirling arrows pointing in different directions. As a result, fluid is behaving in an erratic manner. The overlapping arrows illustrate the presence of swirls, which are characteristics of turbulent flows. It is illustrated that the roughness of the pipe walls contributes to the development of turbulence.

V

Valve Authority

The phrase "valve authority" refers to the ability of a control valve to regulate the flow within a system effectively in relation to the entire process. This measure indicates the degree of control the valve has over the flow in comparison to other components in the system. The valve authority is typically expressed as a ratio. This is calculated by dividing the pressure drop across the valve at its design flow rate by the total pressure drop in the system, including the valve (see Equation 8.1). When the valve authority is high, it indicates that the valve has significant influence over the flow. This indicates that the valve is stable and provides precise flow control. If there is a low authority, the valve control will be dominated by other components of the system, resulting in poor performance and instabilities.

When selecting and sizing control valves, the best engineering practice is to keep N or valve authority close to 0.5, but not greater than that value. By using a control valve in this manner, an acceptable flow control can be achieved without the use of large amounts of energy consumption. If $N < 0.5$, the system's pressure drop dominates, and the valve has poor control authority, making it ineffective in regulating flow. If $N > 0.5$, the valve experiences excessive pressure drop, which can lead to energy inefficiencies.

Equation 8.1: Valve Authority

$$N = \frac{\Delta p_1}{\Delta p_1 + \Delta p_2}$$

where

N: Valve authority (dimensionless)
$\varnothing p_1$: Pressure loss in a valve in a fully open position (bara)
$\varnothing p_2$: Pressure drop across the remainder of the system (circuit)
$\Delta p_1 + \Delta p_2$: Pressure drop across the whole system (circuit)

Example: A system has a total pressure drop ($\Delta p_1 + \Delta p_2$) of 1.25 bar. If the control valve in the system has a valve authority (N) equal to 0.45, what is the pressure drop across the valve?

Answer: $N = \frac{\Delta p_1}{\Delta p_1 + \Delta p_2} \rightarrow 0.45 = \frac{\Delta p_1}{1.25} \rightarrow \Delta p_1 = 0.5625$ (pressure drop across the valve)

Valve Capacity

Generally, valve capacity refers to a valve's ability to allow fluid (gas or liquid) to flow through it under certain conditions. There is no doubt that it is one of the most important parameters when selecting a valve, as it determines how much flow the valve is capable of handling as well as ensuring that it will meet the piping system's needs to deliver the required flow rate. Typically, valve capacity is measured by flow testing or by using standardized equations for C_v or K_v valve coefficients. Using these equations, it is possible to compare valves and select the most appropriate one for specific conditions.

Valve Gain

Valve gain is mathematically expressed as the derivative of flow rate (Q) with respect to valve opening or opening percentage. This parameter is calculated from Equation 8.2. Gain should never be less than 0.5 to avoid any problem for the valve to control the flow. It should be noted that if the valve gain is below 0.5, a significant change in the position of the valve will result in only a small change in the flow rate. As a result of this lack of response, the flow control system is unable to achieve the desired flow, particularly when adjusting the flow rapidly.

A gain of 0.5 or higher ensures that the valve is responsive enough to control the flow effectively.

Gain same as the pressure drop is constant in the linear valve, whereas it increases as the valve opens in equal percentage of flow characteristics. Also, gain is high in quick-opening flow characteristic valves at a small opening.

Equation 8.2 Valve Gain

$$\text{Valve gain} = \frac{\text{Flow rate change}}{\text{Valve opening percentage change}}$$

Example: A control valve is installed in a system to regulate flow. The flow rate through the valve changes as the valve opens or closes. At 40% valve opening, the flow rate is 50 m³/h, and at 50% valve opening, the flow rate is 70 m³/h. Calculate the valve gain.

Answer: $\text{Valve gain} = \frac{\text{Flow rate change}}{\text{Valve opening percentage change}} = \frac{70-50}{0.5-0.4} = \frac{20}{0.1} = 200 \ \ \text{m}^3/\text{h}$ per valve position change

Interpretation: A high gain of 200 in this case indicates that the valve is highly sensitive. This can be useful for precise control, but may also contribute to instability. If the valve gain is too high, even small adjustments to the valve position can cause large changes in flow. This can cause flow disturbances and vibrations. The flow rate or pressure fluctuates up and down around the target value without settling. This can reduce efficiency, wear equipment, and lead to poor process control.

Valve Operation

Valve operation is performed by using a valve to regulate the flow, pressure, or direction of a fluid within a system. Generally, this involves opening or closing the valve or adjusting its position. It is possible to operate the valve using a manual operator, such as a lever and handwheel, or with an automated system, such as an actuator. In accordance with the requirements of the application, valve operation can be categorized as on/off, throttling (modulating flow), overpressure prevention, and avoiding return of flow to the valve.

Valve Operation Sequence

An overview of the valve operation sequence, including both the opening and closing sequences of the valve.

1. Opening Sequence

The opening sequence describes the steps involved in transitioning the valve from a closed position to an open position and allowing fluid to flow through it. Initially, the actuator or lever or handwheel is moved. The valve operator (actuator or manual operator) transfers force to the valve stem. The valve stem is responsible for moving the valve closure member. It is necessary to overcome the valve's stem sealing force (packing friction) and friction between the valve seat and closure member in order to move the stem and closure member. The flow path gradually expands as the valve is opened, resulting in an increase in the flow of fluid. Once the valve has been fully opened with maximum flow capacity, the opening sequence ends. Depending on their design and function, different valve types (e.g., gate, globe, or ball valves) have unique flow curves during opening. According to the opening percentage of the valve, the flow curve indicates the amount of flow passing through the valve.

2. Closing Sequence

The closing sequence in valve operation involves transitioning from an open to a closed position. As a result of this action, the flow of fluid through the valve is stopped. In order to begin the process, the actuator or a manual operator must move the stem. The stem then moves the closure member toward the closing position. During the travel of the closure member to the closed position, the flow rate is reduced and the flow path becomes narrower. The flow decreases significantly as the valve approaches full closure. As soon as the valve is fully closed, a tight seal between the valve closure member and seats is ensured to prevent leaks. Depending on the type of valve, the exact characteristics of the closing sequence will vary. Ball valves, for example, have a quick closing and opening mechanism.

Valve Position

In valve operation, valve position refers to the specific degree of openness or closure of a valve. The position of the valve determines how much fluid flows through it. In a valve, the position can range from fully open (allowing maximum flow) to fully closed (blocking all flow). Moreover, intermediate positions are used for regulating the flow by throttling or modulating it. Valve positions are generally controlled manually (by handwheels or levers) or automatically (by actuators and positioners). In order to maintain process conditions such as flow rate, pressure, and temperature, valve positioning must be accurate.

Valve Sticking

Valve sticking is a type of valve operational issue. The sticking of a valve occurs when the valve becomes difficult or impossible to operate. There are several reasons for this, including dirt or debris buildup, corrosion, a lack of lubrication, extreme temperatures, and mechanical wear. It can affect the valve's ability to open, close, or modulate flow, resulting in reduced system performance and potential safety hazards. Fouling from contaminants, thermal expansion, overtightened packing, and malfunctioning actuators are some of the most common causes of failure. The prevention of valve sticking requires regular maintenance, proper lubrication, corrosion-resistant materials, and appropriate operating conditions.

Valve Synchronizing

Synchronization of valves refers to the coordination of multiple valves. It is done in order to ensure that valves are operating in accordance with the sequence and harmony required. In process systems, in which valves control the flow together or in a specific sequence, this is essential. An automated control system such as a PLC or DCS is often used to create synchronization. A water treatment plant, for example, might have multiple pipelines that supply water to different parts of the system. Parallel pipes are working in such a way that they must provide the same amount of flow to different parts of the system. In order to maintain consistent flow rates and pressures throughout the pipeline, valves must be synchronized so that they supply the same flow rate. As an example, let's suppose that a pump pumps water into two parallel piping or pipeline systems. For the flow to be distributed equally, both pipelines' valves must be opened simultaneously. Synchronizing the valves would prevent pipelines from being either overpressurized or underutilized. As a result, the system remains in balance. The synchronization of valves through interlocks is a great example of valve synchronization. As a result, a valve's operation is mechanically or electronically linked to another valve's operation. It is essential that valves connected by interlocks or locking devices operate according to a specific sequence or condition.

ValveWatch

ValveWatches are used to verify that critical valves in a system are functioning correctly and are ready to be operated when needed, particularly in safety-critical environments such as oil and gas facilities, chemical plants, and power plants. Various sensors can be installed on or around critical areas of a valve, including its cavity, stem, actuator, and other parts, to continuously monitor its performance and condition. It is important to measure the force or torque applied to the stem in order to identify issues such as excessive friction, wear, or resistance within the stem and packing. In order to detect flow anomalies that could indicate valve problems, flow sensors are installed on either the upstream or downstream side of the valve. When fluid leaks through a closed valve or a damaged seal, acoustic emission sensors are used to detect the high-frequency sound waves that are emitted. They are used to detect even very small leaks that might not be visible to the naked eye or measurable with other methods.

A ValveWatch is a critical component for ensuring the reliability of key valves (such as ESDVs, PRVs, and control valves) that play crucial roles in process safety and operations. A ValveWatch program involves continuous or periodic monitoring of valves for signs of malfunction, leakage, or wear that could compromise their reliability. It is possible for valves under a ValveWatch to undergo routine testing in order to ensure that they are functioning correctly. This may include stroke testing (confirming that the valve can fully open or close) and leak testing to ensure a proper seal. ValveWatches are useful for identifying valves that may need to be maintained before they fail. Preventing unexpected shutdowns and ensuring that safety-critical valves remain operational can be achieved by proactive repair or replacement of parts. ValveWatch is a critical tool that bridges the gap between *safety* and *operation*. More information about ValveWatch is provided in Chapter 10, under the heading of "ValveWatch."

Venting

As the name implies, a vent in a valve is a small opening or mechanism designed to allow air, gas, or fluid to escape from a certain area of the valve, such as the upper body cavity or piping system. Many valve designs incorporate vents to ensure safe, efficient, and effective operation under a variety of conditions.

Venting in valve operation refers to the release of gases or air from a piping system, valve, or a vessel. The purpose of venting is to maintain proper pressure, or remove trapped air. In addition, it is done to ensure the safety of the operation. For example, air trapped in a liquid system can create air pockets. These pockets can restrict fluid flow and cause localized pressure drops. These pockets can restrict fluid flow, leading to localized pressure drops. A proper venting system is essential to prevent pressure drops, overpressure, or air pockets. It is possible that all of these conditions may disrupt fluid flow, cause inefficiency, or damage equipment. There are two types of valves commonly used for venting—vent valves and air release valves. Typically, they are installed at high points in the system where gases accumulate naturally. Drainage is opposite of venting that is explained in more detail in this chapter under the heading "Drainage."

Vibration

Vibration is known as a valve operational problem. Vibration in valve operation refers to the shaking of a valve and its connected piping caused by dynamic forces acting on the valve during operation. Vibrations can result from various factors, including turbulent flow, pressure fluctuations, mechanical issues, or external forces.

Vibration Impact

Vibration impact occurs when a valve is subjected to mechanical vibrations due to equipment such as pumps and compressors or fluid-induced vibrations during operation. As a result of vibrations, valve components may wear, fatigue, or even fail, as well as cause noise and instability in the system. To mitigate these effects, proper support, dampening mechanisms, and vibration-resistant designs are imperative.

Vortex Formation

Vortex formation occurs when fluid flow is disrupted by changes in velocity, direction, or pressure. As the fluid passes through the valve or piping, this can occur. Flow disturbances, such as high-velocity flow passing through partially open valves, cause this swirling or rotational motion. It is possible for vortex formation to be exacerbated by sharp bends or improper installation. Vortices can cause energy loss, noise, vibration, and erosion. Additionally, it may result in flow instabilities, particularly in high-pressure or turbulent flow systems. Typically, it is found in throttling valves, such as globe valves and butterfly valves. In the event that pressure drops significantly, vortex formation may also contribute to cavitation. This can be mitigated by selecting valve designs that have streamlined shapes that result in a reduced pressure drop, while maintaining proper installation with adequate straight-pipe lengths before the valve. *Fluid turbulence* that is explained earlier in this chapter is a broader term, and *vortex flow* is a specific manifestation or type of turbulence.

W

Water Hammering

In water hammering, also referred to as hydraulic shock, fluid flow changes suddenly due to the rapid closure of a valve. A sudden closure of the valve results in a sudden interruption of fluid flow, which has a high amount of kinetic energy. This kinetic energy will be transferred to the walls of the pipes in which the fluid is flowing, as there is no way for the fluid to move forward. Furthermore, shockwaves travel back and forth through the system due to the movement of the fluid. When the fluid is stopped suddenly, its pressure increases, which is called a pressure surge. This is the actual wave of energy that travels through the fluid and the pipeline as a result of the pressure surge. Noise, vibrations, and potential damage to pipes, valves, and other equipment may result from these pressure surges. To mitigate water hammer, gradual valve operation such as equipping the swing check valves with dashpot is recommended which lead to check valve disk closure slowly.

Wedging Effect

More information about wedging effect is provided in Chapter 5, under the heading of "Wedging Effect."

QUESTIONS AND ANSWERS

1. What is the typical characteristic of a valve that suffers from "flutter" in operation?
   A. It opens and closes too quickly.
   B. It exhibits uncontrolled vibrations or oscillations.
   C. It is resistant to pressure changes.
   D. It leaks fluid through the seal.

   Answer: The correct answer is B.

2. What does "flow diversion" mean in valve operation?
   A. Redirecting the flow of fluid to a different path
   B. Increasing the flow rate in a pipeline
   C. Preventing fluid from flowing through the valve
   D. Reducing the pressure in the system

   Answer: The correct answer is A.

3. What does "hysteresis" in valve operation refer to?
   A. The valve's inability to fully open or close
   B. The delay in response to changes in input signals during opening and closing
   C. The wear of valve components over time
   D. The valve's resistance to pressure changes

   Answer: The correct answer is B.

4. How does cavitation affect valve performance, and which valve type is most susceptible to it?
   A. Cavitation improves valve performance by increasing flow efficiency.
   B. Cavitation can cause damage to valve seats and increase noise, and is most likely in globe valves.
   C. Cavitation decreases valve seating pressure, making the valve easier to close.
   D. Cavitation improves the sealing ability of the valve and reduces leakage.

   Answer: The correct answer is B.

5. What does valve rangeability refer to?
   A. The maximum flow rate a valve can handle
   B. The range of temperatures a valve can operate in
   C. The pressure drop across the valve
   D. The ratio of maximum to minimum controllable flow rates

   Answer: The correct answer is D.

6. What is deadband in valve operation?
   A. A range of signal changes that cause immediate valve movement
   B. A type of valve packing material
   C. The range of signal changes where the valve does not respond
   D. The noise produced by the valve during operation

   Answer: The correct answer is C.

7. What can excessive pressure drop across a valve cause?
   A. Increased system efficiency
   B. Reduced flow turbulence
   C. Cavitation and erosion
   D. Improved valve sealing

   Answer: The correct answer is C.

8. What is stiction in valve operation?
   A. The sticking of the valve due to material build-up
   B. The static friction that resists valve movement
   C. The vibration of the valve during operation
   D. The pressure drop across the valve

   Answer: The correct answer is B.

9. Which type of valve has a quick-opening flow characteristic?
   A. Globe valve
   B. Gate valve
   C. Butterfly valve
   D. Ball valve

   Answer: The correct answer is D.

10. What happens if a control valve operates continuously at a very low valve opening percentage (e.g., less than 10%) for an extended period?
    A. The valve will maintain precise control with minimal wear.
    B. The valve's rangeability will increase due to enhanced precision at low openings.
    C. The valve may experience erosion, noise, and instability due to high velocities and turbulence.
    D. The pressure drop across the valve will decrease, reducing the system's energy loss.

    Answer: The correct answer is C. When a valve is operated at very low openings, high fluid velocities and turbulence occur due to the flow being squeezed through a small area. As a result, the valve internals may be eroded, noise may increase, and control instability may result. Typically, it is recommended to size the control valve so that it operates within an optimal range of openings, typically between 20% and 80%. In cases of high openings over 80%, the valve is not able to regulate flow further due to the fact that it is nearly fully open. In addition, at near-full openings, fluid flows at very high velocities through the valve. This can cause erosion of internal components.

BIBLIOGRAPHY

American Petroleum Institute. (2014). Type testing of rising stem valves equipped with graphite packing for fugitive emissions (API Standard 624). Washington, DC.

American Petroleum Institute. (2016). Type testing of quarter-turn valves for fugitive emissions (API Standard 641). Washington, DC.

American Society of Mechanical Engineers (ASME). (2012). Design and Fabrication of Pressure Vessels. Boiler and Pressure Vessel Code. ASME Section VIII Div.02. New York, NY.

Dickenson, T. C. (1999). Valves, Piping, and Pipelines Handbook. 3rd edition. Elsevier.

Fisher. (2005). Control Valve Handbook. 4th edition. Fisher.

Hellemans, M. (2009). The Safety Relief Valve Handbook: Design and Use of Process Safety Valves to ASME and International Codes and Standards. 1st edition. Butterworth-Heinemann: Oxford, UK.

International Organization for Standardization. (2015). ISO 15848-1: Industrial Valves—Measurement, Test and Qualification Procedures for Fugitive Emissions—Part 1: Classification System and Qualification Procedures for Type Testing of Valves. ISO: Geneva, Switzerland.

Nesbitt, B. (2007). Handbook of Valves and Actuators: Valves Manual International. 1st edition. Elsevier: Oxford, UK.

Skousen, P. L. (2011). Valve Handbook. 3rd edition. McGraw-Hill Professional: New York, NY.

Smith, P., & Zappe, R. W. (2004). Valve Selection Handbook: Engineering Fundamentals for Selecting the Right Valve Design for Every Industrial Flow Application. 5th edition. Elsevier: Amsterdam, Netherlands.

Sotoodeh, K. (2019). Noise and acoustic fatigue analysis in valves (case study of noise analysis and reduction for a 12″ × 10″ pressure safety valve). Journal of Failure Analysis and Prevention, 19, 838–843. https://doi.org/10.1007/s11668-019-00665-3

Sotoodeh, K. (2020). Development of a numeric method to validate the reliability improvement of safety critical valves during operation through online monitoring implementation. Journal of Sensing and Imaging, 21, 57. https://doi.org/10.1007/s11220-020-00323-1

Sotoodeh, K. (2021). Analysis and failure prevention of nozzle check valves used for protection of rotating equipment due to wear and tear in the oil and gas industry. Journal of Failure Analysis and Prevention, 21, 1231–1239. https://doi.org/10.1007/s11668-021-01162-2

Sotoodeh, K. (2021). A Practical Guide to Piping and Valves for the Oil and Gas Industry. 1st edition. Gulf Professional Publishing: Houston, TX.

Sotoodeh, K. (2023). Industrial Valves: Calculations for Design, Manufacturing, Operation and Safety Decisions. 1st edition. Wiley: New York, NY.

Sotoodeh, K. (2023). The design of the pressure safety and relief valves for overpressure protection: Essential considerations. Transactions of the Indian National Academy of Engineering, 8, 273–287. hpps://doi.org/10.1007/s41403-023-00396-w

9 Valve Inspection, Maintenance, and Troubleshooting

INTRODUCTION

In this chapter, seven key sections address critical aspects of maintaining, inspecting, and troubleshooting valves. The purpose of this is to maintain valves optimal performance, safety, and reliability. In this chapter, common failure modes, maintenance strategies, and troubleshooting techniques are discussed. Thus, it provides engineers with the tools and best practices to enhance valve service life and reduce downtime. These seven headings are listed as follows:

1. **Common Valves Failure Modes**
   This section identifies typical failure valve modes such as leakage, valve seat wear, actuator malfunctions, and environmental damage. It explores the operational and environmental factors that contribute to these issues. Thus, it provides a foundation for understanding valve reliability challenges.
2. **Inspection Techniques**
   It highlights advanced techniques such as nondestructive testing (NDT), pressure testing, and performance monitoring to ensure comprehensive assessments of valve health. While previously discussed in Chapter 6, this section reviews inspection in the context of maintenance and troubleshooting.
3. **Maintenance Practices**
   This section is about various maintenance practices such as preventive and proactive maintenance. It details essential activities such as regular inspections, lubrication, part replacements, and scheduled maintenance strategies. All of these are done to prevent unexpected failures and extend valve service life.
4. **Troubleshooting Operational Issues**
   It is the purpose of this section to provide a practical troubleshooting guide by addressing common operational problems. Operational problems may include improper sealing, flow irregularities, actuator faults, and noise. To minimize downtime, this guide provides step-by-step solutions and best practices.
5. **Reliability and Safety Practices**
   This section focuses on reliability-centered maintenance (RCM) and failure mode and effect analysis (FMEA). These are essential strategies to improve valve quality and reliability. It also discusses safe operation through safety measures such as safety integrity level (SIL) standards.

DOI: 10.1201/9781003640967-9

6. **Operational Challenges**
   Valve operation presents operational challenges due to the complexity of the fluid flow passing through the piping and valves. The challenges include high pressures and temperatures, corrosive or abrasive fluids, and many others. The most common issues include sticking, cavitation, excessive vibration, and improper sealing. These issues may compromise the performance of the valve and the efficiency of the system. Often, these challenges are caused by factors such as incorrect valve selection, harsh operating conditions, or inadequate maintenance.
7. **Best Practices for Maintenance and Troubleshooting**
   The final section summarizes best practices for valve maintenance and problem-solving. Strategies include creating effective maintenance checklists, training, and documentation for consistent results.

PART 1: COMMON VALVE FAILURE MODES

Actuator Malfunction

A valve actuator malfunction occurs when the device that operates the valve does not function as intended. As a result, the actuator loses its ability to open, close, or regulate the flow inside the valve. Typical causes include power supply failures, signal loss, mechanical wear, improper calibration, and environmental factors such as corrosion.

It is possible that these malfunctions may result in improper valve positioning, delayed response times, and system inefficiencies. All of these may result in safety risks or operational interruptions.

Keeping actuators in good working condition requires regular maintenance, including inspecting connections, calibrating settings, and replacing worn components. For a particular application, it is essential to select the appropriate actuator type (e.g., electric, pneumatic, or hydraulic). It would be incorrect, for example, to select an electrical actuator for an emergency shutdown valve (ESDV) because the actuator is not capable of closing the valve rapidly. Troubleshooting techniques, such as verifying power inputs and checking for blockages, can be used to identify and resolve problems quickly.

Cavitation

More information about cavitation is provided in Chapter 4, under the heading of "Cavitation."

Corrosion

More information about cavitation is provided in Chapter 4, under the heading of "Corrosion." Proper material selection is the best way to mitigate the risk of corrosion.

Erosion

More information about cavitation is provided in Chapter 4, under the heading of "Erosion." Using harder materials or hard-facing materials with hard alloys such as Stellite or Tungsten carbide are essential to prevent erosion.

Fatigue

As materials undergo cyclic or repetitive loading over time, fatigue occurs as a progressive and localized structural damage. It is important to note that fatigue is destructive even if the applied stresses are below the material's ultimate tensile strength. The process begins with the formation of microscopic cracks, usually in stress concentrator areas such as welds or areas of surface imperfection. With each loading cycle, it will continue to grow. Eventually, these cracks propagate and cause sudden and catastrophic failures. There are several factors that influence fatigue, including the magnitude of the stress, the number of cycles, the material properties, the environment (e.g., corrosion), and the finish or condition of the surface. Engineers use methods such as reducing stress concentrations in order to prevent fatigue. It is also important to apply surface treatments (e.g., shot peening) as well as choose fatigue-resistant materials, such as austenitic stainless steels and nickel alloys. It is also possible to design for lower stress ranges or loads. A regular inspection and maintenance program can also assist in detecting early signs of fatigue failure.

Flashing

Flashing occurs when a liquid passes through a valve used for throttling, such as a globe valve. When liquid passes through the valve, it experiences a pressure drop below its vapor pressure. As a result, part of the liquid vaporizes, creating a two-phase flow of liquid and vapor. Typically, this phenomenon occurs at the outlet of the valve and remains in the system as vapor.

Due to high-velocity vapor and liquid droplets, flashing may cause erosion of valve components. The effects of erosion are greater on materials that are softer or less resistant to erosion. Furthermore, it may result in flow inefficiencies and long-term damage to the valve's internal components.

Flashing can be prevented by selecting erosion-resistant materials, such as hardened alloys. Furthermore, the valve should be designed to handle pressure drops effectively, such as using a multistage trim design as explained in Chapter 5. Managing flashing and maintaining reliable system performance require proper valve operation, including avoiding excessive throttling. It is also possible to mitigate flashing and cavitation risks by using Y-type globe valves, which provide less pressure drop than straight globes. Flashing damage can be detected and mitigated through regular inspections and maintenance. Note that all the approaches outlined in this paragraph are suitable for both preventing and mitigating cavitation.

Galling

Galling occurs when two metal surfaces slide against each other, resulting in adhesive wear. Under pressure, the galling effect can be intensified, resulting in surface damage and potential component breakdown. The problem commonly occurs in fasteners, valves, and sliding parts made of metallic materials. Aside from high pressure, galling is also exacerbated by high loads, poor lubrication, rough surfaces, and repetitive motions. Lubrication and the use of materials with low friction coefficients are two methods of preventing friction. Additionally, it is important to reduce surface roughness by polishing, to use hard linings, such as nitriding, and to select dissimilar metals with hardness difference in order to minimize adhesion. Mitigating galling requires proper design and maintenance practices.

Jamming

A jammed valve is an operational failure mode in which the valve becomes stuck and is unable to move to its intended position. Although jamming and sticking are very similar, they are not exactly the same. Sticking refers to partial obstructions where the valve is difficult to operate and requires extra effort to move, but is still operable with effort. Jamming occurs when a valve becomes completely immobile and cannot be moved to its intended location. It is possible for this position to be partially open or fully closed, which may result in a disruption of flow control. There are several factors that can contribute to this issue (jamming):

1. Debris or foreign particles obstruct valve movement.
2. Parts become stuck as a result of internal corrosion. In the event of corrosion on the valve stem, seats, or internal surfaces, rough, uneven areas may be created that increase friction. By-products of corrosion, such as rust (iron oxide) or other deposits, can accumulate in tight spaces, such as between the stem and packing or around the seat. It is also possible for corrosion to alter the shape of the component (deform) causing friction. The corrosion-induced swelling of the valve stem is an example of deformation resulting from corrosion that leads to valve jamming. As a result of corrosion, oxides and other by-products accumulate on the surface of the stem, increasing its diameter as a result. A swelling (enlargement) of the stem can cause it to become too large to move freely through the packing or guides, resulting in jamming of the device.
3. Excessive friction from inadequate lubrication.
4. Mechanical deformation of components like the stem as explained earlier or seat.
5. Over-torque during operation, thermal expansion, or misalignment can also contribute to valve jamming. Over-torque can lead to stem bending and increase of wear and friction between the seat and valve closure member.

Jamming can lead to significant operational disruptions, safety hazards, or process inefficiencies. Preventive measures include regular cleaning, proper lubrication, and material selection for corrosion resistance. In addition, setting maintenance schedules is essential to minimize the risk of this failure mode.

Leakage

The leakage of valves is a common issue that can occur internally or externally. An internal leak occurs when the valve fails to seal properly, allowing fluid to pass through the seat even when the valve is closed. An external leak occurs when fluid escapes from the valve body and bonnet connection or from the valve stem packing.

The leakage prevention strategies include regular maintenance, proper material selection, proper alignment between the valve and the mating flange, and condition monitoring. The most effective method of troubleshooting involves inspecting the valve seats, tightening, or replacing packing and gaskets. It is important to maintain proactive maintenance, which is described in greater detail in this chapter. In fact, proactive maintenance minimizes downtime and ensures safe, efficient operation.

Misalignment

Misalignment in valves occurs when the valve components, such as the stem, seat, or closure element, are not properly aligned. This issue can happen due to several reasons such as improper installation, wear and tear, or excessive mechanical stress during operation.

Misalignment can lead to problems such as incomplete sealing, increased internal or external leakage, wear on the valve seat and stem, higher torque requirements, or even failure to operate. In automated valves, misalignment can also cause actuator inefficiencies or damage.

Preventing misalignment involves precise installation, regular valve component inspection, and proper actuator calibration. Addressing misalignment promptly by realigning or replacing damaged parts helps maintain optimal valve performance and extends service life.

Seat Wear

Valve seat wear is a common problem. As a result of this issue, sealing of the valve seat is disturbed, resulting in leakage or operational inefficiencies. As a result of repeated contact between the valve seat and the closure element (e.g., ball, disk, or plug), it occurs. Abrasive or corrosive media, frequent cycling, or high-pressure or high-temperature operation can intensify this constant contact.

Seat wear is primarily caused by erosion, corrosion, cavitation, and debris accumulation. During the course of time, seat wear can lead to internal leaks, increased or reduced torque requirements for valve operation, and a reduction in valve performance.

Seat wear is prevented by durable, corrosion-resistant materials. Furthermore, correct seat placement in valves during valve manufacturing is essential. Performing regular inspections and maintenance on the seats during valve operation can reduce the chance of seat excessive wear. It is necessary to replace or refurbish valve seats in cases of significant wear in order to achieve an optimal performance and prevent further damage.

Stem Damage

In valves, stem damage refers to deformation, wear, or fracture of the valve stem. As a result, valve operation may be impaired, leading to leaks or failures. It may be caused by excessive torque, misalignment, corrosion, galling, improper material selection, or inadequate maintenance. Damage to stems can also be caused by high operating pressures, frequent cycling, and exposure to corrosive or abrasive media. Corrosion-resistant materials should be used in order to prevent corrosion. Furthermore, since the stem is subject to high operating loads, it is advisable to choose a high strength material for the stem at the time of valve design. Furthermore, the valve stem must be aligned properly during installation and excessive operating forces should be avoided. It is essential to maintain the actuator regularly and properly set the actuator settings in order to avoid applying excessive force or torque to the stem beyond the maximum allowable stem torque (MAST). Stem guides and hard coatings can further minimize the risk of stem damage. More information about MAST is provided in Chapter 5, under the heading of "Maximum Allowable Stem Torque (MAST)."

Sticking

When a valve experiences resistance to movement, it sticks. As a result, it is difficult to open or close the valve even with an extra force. Typically, this issue is caused by corrosion, debris or foreign particles, inadequate lubrication, or wear. These problems can occur on critical components, such as the stem, seat, or closure member. Misalignment or thermal expansion can also result in sticking as they increase friction between moving parts. Besides impairing valve performance, sticking can also increase the torque required for operation. Consequently, actuators or stems may be damaged. The prevention of valve sticking requires regular maintenance. In addition, proper lubrication, cleaning, and material selection can prevent this problem.

Wear

A common cause of valve failure is wear, which occurs as a result of the gradual loss of material from valve components. During operation, wear can occur as a result of friction, abrasion, or erosion. A valve's stem, seat, and closure member are typically affected by wear. As a result, leakage occurs, sealing performance is reduced, and operating torque is increased. As an example, wear can cause a rough surface on the valve stem, increasing friction with the packing and thereby increasing the torque required to operate the valve. High-cycle frequencies, abrasive media, improper lubrication, and misalignment of moving parts contribute to wear. Over time, wear can degrade the valve's ability to maintain a tight seal and may cause operational issues such as leakage. Wear can be prevented by selecting resistant materials (e.g., applying Stellite 6 overlay on seat and closure member), proper alignment, and lubricating the valve components. For valves to perform reliably, regular inspection and timely replacement of worn components are essential.

PART 2: INSPECTION

Dye Penetration Testing (DT)

More information about DT is provided in Chapter 6, under the heading of "Nondestructive Testing."

Magnetic Particle Testing (MT)

More information about MT is provided in Chapter 6, under the heading of "Nondestructive Testing."

Nondestructive Testing (NDT)

More information about NDT is provided in Chapter 6, under the heading of "Nondestructive Testing."

Pressure Testing

More information about pressure testing is provided in Chapter 6, under the heading of "Pressure Test."

Proof Test

More information about proof test is provided in Chapter 10, under the heading of "Proof Test."

Radiography Testing (RT)

More information about RT is provided in Chapter 6, under the heading of "Nondestructive Testing."

Stroke Testing

Stroke testing is a critical test used to verify the operational readiness of control valves, safety valves, and ESDVs by ensuring they can fully or partially open and close as required. This testing is essential for maintaining process safety, reliability, and regulatory compliance in industries such as oil and gas, and chemical processing. Stroke testing is performed during operation by operating the valve and measuring its response, travel time, and accuracy. Besides, potential operational issues such as sticking or excessive friction can be observed during the stroke test. The results help to determine whether the valve operation is acceptable or requires maintenance. Stroke test is divided into two categories: partial stroke testing (PST) and full stroke testing (FST).

PST is a method used primarily for safety-critical valves, such as ESDVs and pressure relief valves. Safety and ESDVs are designed to fully close during normal operation. Fully stroking them during normal operations would disrupt production and could pose operational risks. In this method, the operation suitability of the valves is evaluated without fully interrupting the process (typically 10–30% of full stroke). The main PST advantage is that it allows testing without disrupting normal operations. PST also minimizes operational downtime and maintenance costs. However, PST has limitations, as it only tests a portion of the valve's movement, which may not reveal issues that occur during full travel, such as misalignment.

FST, on the other hand, involves moving the valve through its entire range of motion from fully open to fully closed (or vice versa) to validate complete functionality. This test is typically performed during planned maintenance shutdowns. FST provides comprehensive validation by verifying the valve's full operational capability. FST has limitations, including operational downtime and potential process interruptions (in case it is done during operation), and higher costs.

Torque Testing

More information about torque testing is provided in Chapter 6, under the heading of "Torque Testing."

Ultrasonic Testing (UT)

More information about UT is provided in Chapter 6, under the heading of "Nondestructive Testing."

Visual Inspection

More information about visual inspection is provided in Chapter 6, under the heading of "Nondestructive Testing."

PART 3: MAINTENANCE PRACTICES

Bolt Tightening

It is necessary to tighten bolts in valve maintenance in order to achieve a tight seal. The bolts can be used to attach valve components such as the body and bonnet, or the valve flanges to the piping mating flanges. Bolts may loosen over time as a result of vibration, thermal expansion, pressure cycling, or corrosion. This may result in leaks and emissions from the valve. The correct torque values specified by the manufacturer must be used during tightening. Typically, valve manufacturers provide recommended bolt torques in a document referred to as the installation, operation, and maintenance manual (IOM). Maintenance personnel can use these values to ensure uniform stress distribution and avoid overtightening. Bolts and gaskets can be damaged if they are overtightened. Furthermore, under-tightening can cause leakage through bolted joints. In order to apply precise force, tools such as torque wrenches or hydraulic tensioners are often used. In order to achieve a correct result, sequential tightening in a cross-pattern is commonly used. A regular inspection and retightening of bolts is an essential part of preventive maintenance. In this way, leaks are minimized, environmental pollution is reduced, and safe operation is maintained.

Cleaning

Cleaning is an essential component of valve maintenance. It may be performed as a single task or in conjunction with the replacement of a worn or malfunctioning component, such as a seat or seal. In order to maintain reliable performance of the valve after maintenance, it is necessary to clean it. Over time, valves can accumulate debris, scale, or deposits from the fluid flowing through them. Particles such as these are undesirable since they may obstruct flow, damage internal components, or impair sealing efficiency. For safety reasons, it is necessary to remove the valve from service (isolate) and to depressurize it before cleaning. To access the internal components of the valve, it is also necessary to disassemble the valve. Depending on the nature of the contaminants, the cleaning method will vary. From mechanical cleaning with brushes and tools to chemical cleaning, there is a wide range of cleaning methods available. A great deal of attention is paid to the cleaning of sensitive components, such as seals, seats, and trim, in order to ensure that no damage is caused. The valve is then reassembled, inspected for proper operation, and returned to service after it has been cleaned. The regular cleaning of equipment prevents operational problems, enhances efficiency, and reduces the risk of costly downtime.

Corrective Maintenance

Following the identification of a fault or issue, corrective maintenance is performed to restore the equipment to a proper working condition. If the problem is detected early during inspections or monitoring, it may be possible to plan for it. The failure may, however, occur unexpectedly if it is unplanned. To minimize downtime and ensure equipment reliability, this type of maintenance addresses specific faults. In addition, corrective maintenance can also be performed prior to failure by identifying minor issues or early signs of deterioration.

Lubrication

Lubrication is an essential component of preventive maintenance when it comes to valve repair and maintenance. In addition to facilitating smooth operation, lubrication reduces wear and prevents galling of valve components. This process involves lubricating moving parts such as stems, bearings, and seats with suitable lubricants. The purpose of this is to minimize friction and facilitate ease of operation. A properly lubricated valve will also maintain a reliable seal, reduce the risk of galling and seizing (sticking), and prolong its lifespan. Technicians inspect the lubricants as part of preventive maintenance. Also, it is important to clean the components before lubrication, and to apply the right type and amount of lubricant. The type and amount of valve lubricant is determined by the operating conditions and the compatibility of the materials. Lubrication ensures optimal valve performance and prevents costly failures.

Maintenance

Maintenance of valves refers to the routine inspection and servicing of valves. This is done to ensure that valves function reliably, efficiently, and safely. As a result of regular maintenance, issues such as leaks and unexpected failures can be prevented or detected at an early stage. As a result, the operational cost and downtime are reduced and the reliability of the valves improved. The type of valve, operating conditions, and specific industry requirements determine the type and interval of maintenance tasks.

Maintenance Check List

It is the responsibility of the technicians to ensure that all necessary steps are followed during maintenance by using checklists. There are specific tasks including inspections and maintenance, such as checking seals for leak detection. Performing maintenance by using checklist ensures consistency and thoroughness and reducing the possibility of missing a step. More information about maintenance checklist is provided in Part 7 of this chapter under the heading of "Checklist."

Maintenance Logbook

It serves as a record of all maintenance activities performed on a valve. It includes dates, technicians involved, and specific tasks completed. The logbook provides a history of all maintenance work, including inspections, repairs, and replacements. However, maintenance documentation and logbooks are not the same, even though they are closely related. They are often used together to maintain comprehensive records of a valve's maintenance history. Logbooks are a type of maintenance documentation designed to provide a detailed overview of maintenance activities. They focus on what was done, when, and by whom. In contrast, maintenance documentation is broad. Maintenance documents contain detailed procedures and technical references to guide and standardize maintenance work.

Packing Adjustment

Valve packing adjustment is a crucial maintenance task. The importance of this is due to the fact that leakage through packing directly affects the environment. In the case of toxic and flammable fluids and gases, this could be particularly critical. It is necessary to adjust the packing as part of a maintenance program in order to ensure that there are no leaks around the stem as well as stem smooth movement. A valve packing material can wear out or degrade over time as a result of repeated valve operation, temperature and pressure, and contact with fluid. As a result, leaks or excessive friction can occur in the stem. To adjust the packing, the gland bolts and nuts must be tightened evenly. In this way, the packing is compressed and its sealing ability is restored. To avoid overtightening the gland bolt and nuts, it is important to follow the valve manufacturer's recommendations. The overtightening of the gland bolt and nut could damage the packing or restrict the movement of the stem. It may be necessary to replace the packing if the adjustment does not work. During valve maintenance, packing should be inspected and adjusted regularly to maintain the integrity of the seal. As a result of packing adjustment, leaks of fluid or gas are prevented that could compromise the efficiency or safety of the system.

Predictive Maintenance (PdM)

PdM is a proactive maintenance strategy that involves the use of advanced monitoring techniques and data analysis. In this way, this type of maintenance is able to predict when equipment is likely to fail. As a result, maintenance can be performed in a timely manner in order to prevent unexpected breakdowns. Technology such as vibration analysis, thermal imaging, oil analysis, and Internet of Things sensors are used to continuously monitor the condition of assets such as valves.

As PdM becomes more connected to machine learning (ML), it becomes more accurate and efficient. In order to identify patterns and predict failures, ML algorithms analyze vast amounts of historical and real-time data collected from equipment sensors. The use of vibration analysis, oil analysis, and thermal imaging are not forming of ML, but they are techniques for monitoring and assessing the health of equipment that can be enhanced by ML. Vibration analysis involves monitoring and assessing the condition of rotating or moving machinery, such as motors, pumps, bearings, and turbines. A thermal imaging system utilizes infrared cameras to detect temperature variations on the surface of equipment or systems. An oil analysis method evaluates the condition of lubricants used in machinery to detect wear, contamination, and chemical degradation.

PdM allows organizations to perform maintenance only when necessary by analyzing real-time data and identifying patterns or anomalies. As a result, downtime and maintenance costs are reduced. By utilizing this approach, efficiency is increased and equipment life is extended.

Preventive Maintenance (PM)

PM refers to the proactive approach to maintaining equipment and systems. The purpose of this is to prevent unexpected failures and downtime by performing regular inspections, servicing, and repairs. The maintenance program involves scheduled activities such as cleaning, lubrication, adjusting seals or other parts, and replacing parts as necessary. As a result, assets are operated efficiently and reliably. A PdM program differs from a reactive maintenance program. In reactive maintenance, issues are addressed after they have arisen. Essentially, preventive maintenance seeks to identify and mitigate potential problems before they result in costly breakdowns or disruptions.

Among the benefits of preventive maintenance are the extension of equipment's lifespan, the reduction of repair costs, and the improvement of safety. Identifying wear and tear or other issues early allows organizations to avoid sudden and costly failures. It is important to plan carefully, adhere to schedules, and keep detailed records in order to implement an effective PM program. The purpose of this is to track the condition of assets and optimize maintenance activities.

Proactive Maintenance

The purpose of proactive maintenance is to address the root causes of equipment failures in order to prevent them from occurring in the first place. As an example, vibration dampeners can be installed on a pump to reduce the stress placed on the bearings and prevent failure over time. A proactive maintenance program includes a combination of preventative measures (e.g., regular lubrication and inspections) and a deeper root cause analysis (RCA) to eliminate recurring problems.

Reactive Maintenance

A reactive maintenance program involves repairing or servicing equipment only after it has failed. It is also known as "run-to-failure" maintenance. It is because the equipment is allowed to operate until it fails before the problem is addressed. Since maintenance activities are performed in response to unexpected equipment failures, they are not scheduled. The costs of reactive maintenance are often higher as a result of emergency repairs and downtime. Corrective maintenance may appear similar to reactive maintenance, but there are differences between the two. The process of reactive maintenance is always conducted after a failure, whereas the process of corrective maintenance may occur before or after a failure. Furthermore, in reactive maintenance, no maintenance is performed before valve failure, in contrast to corrective maintenance.

Seal Replacement

The replacement of seals is an important part of valve maintenance. It is necessary to replace the seal in order for the valve to function properly and be safe. In the course of time, seals can degrade due to wear and erosion, corrosion, high temperatures, or exposure to damaging media. As a result, leaks develop and valve performance is reduced. During seal replacement, the valve is usually isolated from the piping system. It is also essential to depressurize the valve during seal replacement for safety reasons. Additionally, some components of the valve are disassembled as needed so that the maintenance personnel can access the seals. During the design and selection of valve materials, it is essential to select seals materials with extra care. Suitable materials for seals are those that are compatible with the operating conditions of the valve. Generally, fluid service, pressure, and temperature affect the selection of valve seal materials. It is essential to perform proper cleaning, inspection of seals, and accurate new seal installation during seal replacement to avoid future leakage from the valve. In order to prevent unplanned shutdowns and maintain the integrity of the system, it is essential that seals are regularly replaced. Seal replacement should be done as a preventive maintenance program rather than a reactive one.

Seat Replacement

It is imperative that valve seats be replaced as part of regular valve maintenance. The reason for this is that the contact between the valve seat and closure member prevents internal leakage of the valve when it is closed. However, valve seats can wear out over time due to erosion, corrosion, pressure drop and cavitation, high pressure and deformation, misalignment, or repeated cycles of operation. Therefore, valve seats are replaced in order to restore seat sealing. Additionally, it is crucial to ensure that the valve seat is properly sealed in order to achieve reliable operation. As with replacing the seal, replacing the seat typically involves isolating and depressurizing the valve. To gain access to the seat, it is necessary to disassemble the valve parts, such as the stem and bonnet. Before installing a new seat, it is essential to clean the seat pocket thoroughly. During the installation of a new seat, the seat should be aligned properly with the body and closure member. Material selection plays a significant role in determining the life of the seat during operation. Seat materials should be compatible with service conditions, including temperature, pressure, and type of fluid. As part of a maintenance program, regular seat replacement prevents leaks. Consequently, this improves valve performance and extends its service life.

Spare Parts

Spare valve parts are replacement components kept in stock to enable maintenance, repair, and overhaul of valves. In industrial settings, having spare parts readily available is crucial, as it enables operators to quickly replace worn or damaged parts, thereby minimizing downtime and ensuring that the valve functions reliably. Typical spare parts include small components such as seals and gaskets as well as large components such as actuators and valve closure members.

Spare Parts Inventory

It is important to have a spare parts inventory for valves. This is done to store necessary components for valve maintenance and repair. It has been demonstrated that storing spare parts can facilitate the repair and replacement of valve parts. This will minimize downtime and operational disruptions. A typical inventory of critical items includes stems, seats, gaskets, packing, seals, actuators, and fasteners. In the course of operation, these items may wear out, corrode, or fail. A well-organized inventory is essential to ensure having enough spare parts for quick maintenance of the valve. Managing inventory properly involves tracking usage and implementing inventory control systems. In addition, it is imperative to evaluate stock periodically in order to prevent shortages and overstocking.

Stem Greasing

To maintain smooth operation, stem greasing is an essential maintenance activity. As a result of this task, friction is reduced and wear on the valve stem is prevented. During the course of time, valve stems may be damaged or corroded if they come into contact with corrosive fluid or external environments. Temperature and high pressure also contribute to the degradation of valve stems. Grease applied to the stem during maintenance facilitates smooth movement of the stem during opening and closing operations. As a result, the stem is protected from corrosion and wear. These should be taken into account that wear and erosion can be combined with corrosion in valve components and intensify the effects of wear, erosion, and corrosion when they occur together. Grease should be compatible with both valve material and service conditions, such as temperature and fluid type. In addition to maintaining valve efficiency and reducing operational failures, regular stem lubrication is an integral part of a maintenance schedule.

PART 4: TROUBLESHOOTING

Actuator Fault

Troubleshooting valves requires an in-depth understanding of actuator faults. This is due to the fact that the actuator controls the movement of the valve in order to regulate fluid flow. The most common actuator faults include loss of power, incorrect air or hydraulic pressure, mechanical wear, misalignment, and improper calibration. The most obvious symptom of an actuator fault is that the valve does not open, close, or reach its intended position. In addition, it is possible that the valve is completely disconnected from the actuator.

The first step in troubleshooting actuator faults is to determine whether there is a problem with the power or pressure supply. It is also important to inspect the mechanical linkages to the valve. In addition, it is important to ensure that the actuators are aligned and that control signals are being transmitted and received correctly. Electric actuators must have their motors and wiring connections checked in order to function properly. In the case of pneumatic or hydraulic actuators, it is essential to examine pressure gauges and seals for leaks. When actuator faults are identified and addressed promptly, the valve will operate as intended. By performing regular maintenance, such as lubrication, calibration, and timely replacement of worn components, actuator-related problems can be prevented.

Calibration Issues

In valve troubleshooting and maintenance, calibration issues refer to a problematic condition in which valve performance is different from its intended performance. As a result, the valve cannot reach its setpoint and adjust the process parameters as it should. Incorrect alignment or settings of control components, such as actuators, positioners, or sensors, lead to calibration issues. A poor calibration can lead to inaccurate flow control, failure to achieve full opening or closing, and an ineffective response to control signals. Inefficiency or instability can result as a result of this. The first step in troubleshooting calibration issues is to verify the valve's setpoints in order to understand how the valve is intended to operate and move. Another step is to check the actuator performance, especially its response to control inputs. Finally, the operator must ensure that the positioner accurately place the valve in its actual or intended position. Depending on the manufacturer's specifications or the process requirements, it may also be necessary to recalibrate components. As part of preventive maintenance, regular calibration checks and adjustments are essential. In the process of calibration and adjustment, the operation efficiency is improved, and optimal performance for the valve and control system is maintained.

Flow Irregularities

Flow irregularities in valves are disruptions or inconsistencies in the flow of fluid. Turbulence and obstructions are responsible for this phenomenon. An obstruction caused by the sudden closure of a valve may indeed result in water hammer, a phenomenon where a sudden change in flow velocity causes pressure shock and irregular flow waves. Cavitation can also cause flow irregularities. The reason for this is that cavitation is associated with pressure fluctuations, which are characterized by a reduction in pressure followed by an increase in pressure. As a result of these irregularities, there can be a reduction in efficiency, noise and vibration, and in some cases, the valve and its surrounding piping may be damaged. In order to resolve flow irregularities, it is necessary to inspect the valve for damage and clean any debris that may have accumulated. Furthermore, it is important to ensure that the valve is suitable for the flow conditions. Choosing the appropriate valve type and size based on the process parameters (flow rate, pressure, temperature, and fluid properties) will ensure that the valve is suitable for flow conditions and minimize the risk of valve failure due to flow irregularities. The choice of materials should be compatible with the fluid. The design should prevent issues such as cavitation or turbulence. In addition, valve performance should be verified using standards, simulations, or testing. To prevent flow irregularities and maintain smooth operation, regular maintenance is essential, such as cleaning, lubrication, and replacement of worn parts.

Improper Sealing

In valve troubleshooting, improper sealing refers to when the valve does not create a tight seal. Leaks and reduced operational efficiency are the obvious consequences of improper sealing. Leakage through a closed valve, for example, reduces its efficiency, especially if it is designed to provide zero leakage or a tight shutoff. The cause of this issue can be attributed to a number of factors. As an example, damaged or worn seals cannot provide sufficient sealing. A stem or seat that has become eroded or corroded due to material incompatibility with the fluid service, particularly in the area where seals are located, can lead to improper sealing and leakage of this fluid. Incorrect alignment of the closure member or stem may also contribute to improper sealing. It is possible for debris or contaminants to damage or obstruct the sealing surface, which may result in poor sealing. Improper sealing can cause leaks and fluid loss. In addition, improper sealing causes pressure drops and compromises the efficiency of the operating system. The first step in troubleshooting involves inspecting the sealing surfaces for wear and damage and removing any debris that may be present. The proper alignment of valve components, particularly the seat, closure member, and stem, should also be checked and verified. It is essential to replace worn or degraded seals with materials that are appropriate for the operating conditions. The prevention of improper sealing can be achieved through regular maintenance and proper installation.

Noise

When troubleshooting valves, noise is an important indicator. In many cases, it can provide valuable insight into the underlying issues within the valve. The presence of unusual sounds, such as hissing, whistling, banging, or vibration-related noises, can indicate a number of operational problems. These problems may be caused by cavitation, flashing, fluid turbulence, or mechanical malfunctions. It may be indicative of a leaking seal, for example, if you hear a hissing noise, while a banging sound may indicate water hammer or loose internal components. An increase in vibration-induced noise may indicate improper flow conditions or wear on critical components, such as valve seats or trims. Diagnosing issues and determining corrective measures require an understanding of the source and nature of the noise. Monitoring noise-related problems regularly and addressing them early can prevent further damage and enhance valve performance.

Over-Torque

Over-torque is a common troubleshooting issue when it comes to valve maintenance. It can damage the bolts and bolted components, including the valve body and bonnet, as well as break the gaskets. When it comes to gland bolts and nuts, they can damage the packing and increase friction between the valve stem and packing. Moreover, over-torque can lead to difficulties operating the valve, leaks, or even a complete failure of the valve. When troubleshooting, it is necessary to look for signs of damage, such as damaged bolts, cracked components, or stem deformation. Earlier in this chapter, we discussed the importance of tightening bolts with the torque values recommended by manufacturers. The failure to do so can result in the bolt being over torqued. To prevent over-torque, it is important to adhere to the torque specifications recommended by the manufacturer. For the purpose of fastening the bolts and nuts, calibrated and accurate tools must be used. In addition, operators should be trained to avoid applying unnecessary force. By addressing over-torque promptly, valve functionality can be restored and further damage can be prevented. It should be noted that over-torque could be due to high torque applied by actuator that can lead to valve internals damages. Opposite to over-torque is under-torque that is explained further in this chapter.

Positioner Fault

A positioner fault in valve troubleshooting and maintenance occurs when the positioner fails to accurately control the valve's position in response to input signals.

There are several possible causes of this fault. The first is an improper calibration of the positioner. Positioners must be calibrated to match the input signal (e.g., 4–20 mA) with the valve's mechanical movement. If the positioner is not calibrated properly, the signal is not accurately translated into the correct valve position. The second reason is the mechanical wear and damage of the internal components of the positioner. As a result, it is important to inspect the positioner regularly and replace any worn components. The third reason is due to a problem with the air supply applicable to pneumatic positioners. In order for pneumatic positioners to function, a steady supply of pressurized air is required. An air supply line leak, a blockage, or a malfunctioning compressor can interrupt the flow of air. Next, there is a problem with the positioner's electrical system. Positioners that utilize electropneumatic technology use electrical signals to control pneumatic actuators. Damaged wiring or loose electrical connections can disrupt the flow of signals.

Symptoms of a positioner fault include the valve not reaching its desired position or remaining stuck. During troubleshooting, the positioner's calibration and air supply must be verified. Furthermore, it is necessary to inspect for loose or damaged connections, as well as confirm that the positioner is receiving and interpreting the input signal correctly. It is possible to prevent most positioner-related problems with proper maintenance, such as regular calibration and cleaning.

Pressure Drop

When troubleshooting valves and systems, pressure drop is an important parameter to consider. When a valve is in operation, it displays the difference in pressure between the inlet and outlet. A high-pressure drop during valve operation can indicate a number of problems, such as blockages and wear. Components like the seat and trim could wear out or become misaligned, leading to unintended flow restrictions. There could be debris obstructing the valve, or damaged internal components like seats or trim causing excessive pressure drop. An unusually low-pressure drop when the valve is closed could mean the valve is not closing properly or there is a leak. By measuring and analyzing pressure drop, you can figure out what is causing performance issues or potential failures.

Root Cause Analysis

In maintenance and troubleshooting, RCA involves identifying the underlying reasons behind failures and malfunctions of equipment or components, for example, valves. It is not only important to identify and address the immediate failure issue, but also to eliminate the root causes of the failure in order to prevent similar failures from occurring in the future. This process involves the collection of data regarding the failure and the analysis of contributing factors. Through methods such as fault tree analyses or cause-and-effect diagrams, it identifies the primary cause of the problem or failure. As an example, RCA might reveal that the root cause of a leaky valve is improper installation, incorrect material selection, or insufficient maintenance. Operators can improve valve reliability and reduce downtime by identifying, addressing, and eliminating the root cause instead of just focusing on the faults.

Signal Loss

In valve troubleshooting and maintenance, signal loss occurs when the control signal sent to the actuator is not interrupted, weakened, or not received. As a result, the valve does not function as intended. It is possible for pneumatic, hydraulic, and electrically operated valves to suffer signal loss.

The following paragraph discusses some of the main causes of signal loss. The loss of signal can be caused by a damaged wiring connection between the valve, or actuator, or positioner and the control system. If the wiring is damaged, the communication between the control system and the valve or actuator will be interrupted. This will result in the valve not responding or responding incorrectly. For their basic operation, pneumatic and hydraulic actuators do not require electrical wiring. This is due to the fact that their primary power source is compressed air (pneumatic) or pressurized fluid (hydraulic). It is possible, however, that they may still have electrical wiring in certain components. It is possible, for instance, to use pneumatic actuators in conjunction with a poisoner for valve control. In many cases, pneumatic actuators used for valve control are equipped with electropneumatic positioners that use electrical signals (4–20 mA or digital signals) to control the position of the actuator. It is necessary to install wiring for the signal input on these devices. Alternatively, limit switches or position feedback devices may be used to transmit the actuator's position to the control system, requiring electrical wiring. Furthermore, solenoid valves, which are electrically operated, control the flow of air to pneumatic or hydraulic actuators and require wiring to operate. In addition, there may be a problem with the control system since it is the control system that transmits signals to the valve or actuator. Besides, loose connections, such as wires, may lead to signal loss. There are also other factors that contribute to valve malfunction and not reaching its setpoint during operation.

In order to determine the cause of signal loss, it is necessary to systematically examine the signal source (e.g., control system or PLC), verify continuity of electrical wiring, and inspect actuator electrical components such as limit switches and solenoid valves. Additionally, it is important to ensure that communication protocols are functioning properly. Regular inspections of connections are part of preventive maintenance.

STICKING

In valve troubleshooting and maintenance, sticking occurs when the valve fails to open, close, or move smoothly as a result of mechanical resistance. In contrast to jamming, where the valve is locked in its position, sticking permits the valve to be moved with additional force.

There are several reasons why sticking can occur, which are outlined in this paragraph. When both the stem and packing are not lubricated, there is a greater likelihood of sticking due to the unsteady movement of the stem in relation to the packing. In addition to corrosion, erosion, and wear, enlarging a component or making a rough surface can result in sticking. As a result of corrosion, the stem can bind with the packing, thereby increasing friction and resistance during operation. In the same way, corrosion can alter the geometry of the seat, resulting in misalignment or increased contact with the closure member. Therefore, it is more difficult to open or close the valve smoothly. The misalignment of the seat or closure member may also result in increased friction between these components, which may hinder smooth operation. By misaligning valve components such as the seat, stem, and closure member, vibration can also cause sticking. Finally, the accumulation of particles around the seat and stem results in their sticking to the closure member and packing, respectively.

When a valve sticks, it can lead to operational inefficiencies, flow disruptions, or even system shutdowns. If a valve sticks, it may not close tightly, resulting in leakage. In processes involving hazardous or expensive fluids, leaks may cause the safety system to shut down the line where the valve is located. Troubleshooting involves identifying the root cause by inspecting the internal components of the valve. Cleaning any buildup particles and checking for wear, corrosion, or misalignment are essential. In order to facilitate smooth operation, the next step is to replace a part or apply some lubricant. In order to prevent sticking, regular inspections, cleaning, and lubrication should be performed as part of a routine maintenance program.

Surge

Surge, also known as water hammer or hydraulic shock, occurs when fluid flows through pipes suddenly undergo rapid changes in flow velocity. Consequently, pipes, valves, and other equipment may be subjected to a significant amount of mechanical stress and damage. Surges can occur in a variety of circumstances. When a valve is closed too quickly, a pressure wave is reflected back through the piping as a result. Alternatively, when a pump suddenly stops, it results in a rapid deceleration of fluid and a subsequent spike in pressure (surge). Surge can also occur if a valve with fluid accumulated behind it is opened too quickly. This results in a sudden surge of fluid that generates excessive force within the system as a result. Other scenarios include check valves slamming due to reverse flow, which primarily occurs with swing check valves. Surging can also be caused by trapped air pockets in the fluid that suddenly compress and expand.

Troubleshooting

During maintenance, the process of identifying and resolving operational problems with the valve (see Figure 9.1) is called troubleshooting. Valve troubleshooting

FIGURE 9.1 Operator inspecting and troubleshooting a ball valve for smooth and reliable operation. (With permission: Shutterstock.)

involves identifying, diagnosing, and resolving issues or malfunctions in valves to ensure they operate correctly and reliably. Among the most common valve issues are corrosion and material failure, erosion and wear, leaks, sticking and jamming, noises, and incorrect flow control.

Under-Torque

Under-torque can also be an issue when troubleshooting valves. This is because insufficient torque may result in leaks in valves and a failure to provide an effective seal. When bolts, packing glands, or actuator settings are not tightened to the recommended specifications, under-torque occurs. There may be problems such as leaks and improper sealing as a result of this. Under-torqued gland bolts and nuts, for example, may fail to adequately compress the packing. The result is stem leakage, while under-torqued flange bolts can result in flange misalignment and fluid leakage.

In order to troubleshoot under-torque, it is necessary to identify the areas where torque is insufficient. It is essential to use calibrated tools to measure the torque applied and compare it with the manufacturer's specifications. In order to ensure proper sealing, it is important to address under-torque promptly by tightening the bolts and nuts to the proper torque levels. The regular inspection and maintenance of bolting components can assist in the detection and prevention of under-torque issues before they become serious.

Vibration

It is important to consider vibration when troubleshooting valves. This is due to the fact that excessive vibration can indicate problems with the valve's performance or integrity. Fluid turbulence, cavitation, and flow-induced forces all cause vibration. There are more reasons for vibration such as looseness of valve components as well as misalignment. Vibration can accelerate the wear of internal parts and damage valve components. Additionally, vibration can result in leakage, fatigue, as well as damage to adjacent piping. When troubleshooting vibrations, it is necessary to identify the root cause of the vibration. That is done by inspecting the valve's internal components and operating conditions. The reduction of flow velocity or preventing cavitation by changing the differential pressure across the valve can mitigate vibration. Regular monitoring and maintenance for vibration can reduce vibration failures. In addition, it can enhance valve reliability and extend valve life.

PART 5: RELIABILITY PRACTICES

Asset Management

In valve maintenance, asset management involves the strategic planning, monitoring, and maintenance of valves. As part of the process, the valves are expected to operate efficiently and reliably throughout their useful lives. The objective of this approach is to maximize the performance and lifespan of valves while minimizing downtime and maintenance costs. It involves activities such as tracking valve performance, implementing condition-based and predictive maintenance, and providing timely replacement or refurbishment of valves. An effective asset management program ensures that valves remain in optimal condition, supporting the overall goals of productivity and sustainability.

Condition-Based Monitoring (CBM)

As part of condition monitoring, a variety of tools and techniques are used to monitor the health and performance of equipment and components like valves. A number of sensors can be used to detect leaks through valve seats, vibration analysis, thermal monitoring, and oil analysis which can be used to determine the effect of lubrication on equipment. As an example, acoustic sensors that detect leaks are considered part of condition monitoring and ValveWatch system. Identifying changes or anomalies in operating conditions that might indicate wear, degradation, or impending failure is the objective of condition monitoring. This is a general term that refers to both PdM and condition-based maintenance strategies. CBM involves monitoring equipment regularly and performing maintenance only when problems arise. A variety of sensors and tools are used to monitor the condition of parts such as valves, checking for issues such as wear, vibration, temperature changes, or leaks.

Example: Advantages of Condition-Based Monitoring

Here is an example of how condition monitoring or real-time data can lead to both predictive and condition-based valve maintenance. The inlet pressure of a control valve in a chemical processing plant is monitored by a pressure sensor. The sensor is part of a real-time monitoring system. During the course of several days, the sensor detects a gradual increase in pressure beyond the expected operating range. According to the data analysis, the valve is having difficulty regulating flow due to partial obstruction, which is likely the result of fouling or debris accumulation. Monitoring systems are triggered by abnormal pressure trends. CBM is scheduled to clean or inspect the valve only if sensor data indicates that a problem is developing. The maintenance is performed at the appropriate time to prevent complete valve failure or production interruption. In PdM, sensors perform more than simply display real-time data. The historical data from the pressure sensor combined with the flow rate and operating history. In this way, it is possible to determine when the valve will fail completely if no action is taken. It is predicted that the obstruction will worsen within two weeks, potentially resulting in a production shutdown. In PdM, real-time or historical data is used to determine when a valve (or any equipment) is likely to fail. In this way, the maintenance team is able to avoid reacting hurriedly or unplanned: Instead, with a clear timeline for when maintenance should be performed. As an example, PdM can be used to estimate how long a valve will continue to function reliably even if its sensor detects abnormal vibration or pressure trends caused by wear or partial blockage. It is possible to schedule maintenance during a convenient low-production period within 30 days if the analysis predicts that the valve will fail in 30 days, thus avoiding the need for urgent and disruptive repairs.

Failure Mode and Effect Analysis

FMEA in valve maintenance is a systematic method for identifying potential failure modes in valves and evaluating their impact on system performance and reliability. As part of the FMEA process, it is common to analyze each component of the valve, such as the stem and seals. It is possible in each analysis to identify and assess how failures such as leakage or wear might occur for each valve component. It is then possible to quantify the risk score for each failure mode by assessing its severity and likelihood. Failures are more severe when the severity and likelihood are higher. The severity of each failure can be multiplied by the likelihood of the failure in order to determine its risk score. When the risk score is higher, there is a greater risk involved, and therefore a greater priority is placed on addressing and resolving the risk score so that the risk can be reduced. By using this information, maintenance efforts can be prioritized and risk mitigation strategies can be developed.

Mean Time between Failures (MTBF)

More information about MTBF is provided in Chapter 10, under the heading of "Mean Time between Failures (MTBF)."

Mean Time to Failures (MTTF)

More information about MTTF is provided in Chapter 10, under the heading of "Mean Time to Failures (MTTF)."

Mean Time to Repair (MTTR)

More information about MTTR is provided in Chapter 10, under the heading of "Mean Time to Repair (MTTR)."

Reliability-Centered Maintenance

The RCM approach focuses on the reliability and functionality of equipment and components, such as valves. According to this method, the focus of inspection and maintenance is on those systems and components that are more critical, and their lack of reliability has a greater impact on overall system performance. RCM, for example, prioritizes maintenance on safety-critical valves, such as pressure relief valves and isolation valves, over less critical components. It defines the functions of an asset, the ways in which it can fail, and the consequences of its failure. By doing so, maintenance efforts can be directed to the areas in need of the most attention.

In reliability-based maintenance (RBM), the primary goal is to prevent failures from happening by combining RCA and FMEA. To optimize maintenance and minimize downtime, this strategy combines PM, PdM, and condition-based maintenance. In PdM, real-time monitoring and analytics are used to predict when components, such as valves, will fail, allowing maintenance to be performed only when necessary. Condition-based maintenance focuses on the current state of the equipment, triggering actions based on measurable parameters such as temperature, pressure, vibration, or leakage.

Safety Integrity Level

More information about SIL is provided in Chapter 10, under the heading of "Safety Integrity Level (SIL)."

PART 6: OPERATIONAL CHALLENGES

Abrasive Media

In valve operation, abrasive media refers to fluids containing solid particles, such as sand, slurry, or other particulates, which can wear down valve components by mechanical *erosion*. Due to wear and tear, this poses a significant operational challenge and increases maintenance costs. It can also cause sticking and accumulation of particles in critical areas of the valve, such as the seat. The selection of appropriate materials is the most important step in mitigating this challenge.

Corrosive Environment

When it comes to valve operation challenges, corrosive environments refer to conditions in which the valve is exposed to corrosive fluids or an external corrosive environment. Fluids that are corrosive are those that contain carbon dioxide, hydrogen sulfide, chloride, or other corrosive compounds. The valves may be exposed to corrosive environments offshore or subsea as a result of contact with the seawater. The chloride content of seawater can cause pitting and stress cracking corrosion (SSC) of metals. The challenges associated with corrosive environments include deterioration of valve components (such as the body, stem, seat, or seals). Further, there is an increase in maintenance requirements, a reduction in lifespan, and the possibility of leaks or failure. The valves must be made from corrosion-resistant materials in order to mitigate these issues.

Dust Buildup

Valves and actuators are particularly susceptible to dust buildup in dusty environments. As dust accumulates, it can obstruct moving parts, such as the stem, or actuator, which can lead to sticking, wear, or inefficient operation. Additionally, dust can contaminate seals, reducing their effectiveness and causing leaks. Dust buildup can cause valves with electronic or pneumatic actuators to malfunction by blocking airways or interfering with electrical components. In order to mitigate dust buildup, it is essential to maintain the equipment by regularly cleaning, inspecting, and replacing worn seals. Additionally, enclosures must be equipped with adequate ingress protection (IP) ratings. This chapter provides a more detailed explanation of IP.

Emergency Shutdown (ESD)

Due to the fact that ESDs are designed to stop the plant's operations entirely in response to safety-critical events, they pose a significant operational challenge. The sudden shutdown process can result in significant costs in terms of downtime, lost production, and potential equipment wear due to sudden stoppage of fluid and water hammering. More information about ESD is provided in Chapter 8, under the heading of "Emergency Shutdown (ESD)."

High Pressure

More information about high pressure is provided in Chapter 8, under the heading of "High Pressure."

High Temperature

More information about high temperature is provided in Chapter 8, under the heading of "High Temperature."

Ingress Protection (IP)

It is important to consider IP when maintaining and operating valves. This is particularly important in environments where dust, water, or other contaminants present. These contaminants must be prevented from entering the valve, actuator, and valve gearboxes. According to the IP rating system, valves, actuators, or gearboxes are classified according to the degree of protection they provide against the ingress of solids (such as dust or particles) and liquids (such as water or oil). It is important to note that inadequate ingress protection can lead to several operational problems. These are internal contamination, corrosion, electrical malfunctions in actuators, or even additional wear on mechanical components in harsh environments. There are significant operational challenges associated with this situation. It is important to select valves with suitable IP ratings for the operating conditions. In addition, it is important to ensure that the protective features are routinely checked to prevent failures due to contamination.

An IP rating, such as IP68, indicates the level of protection a device or enclosure provides against the entry of solids and liquids. It was the author's experience to have IP 68 valve gearboxes on one offshore project. The first digit (number 6) indicates the level of protection against solid objects such as dust or debris. A 6 rating indicates the enclosure is completely dust-tight, offering the highest level of protection. By doing so, dust is prevented from entering the gearbox. The second number indicates the level of protection against liquids such as water. A rating of 8 indicates that the device is designed to withstand continuous immersion in water under conditions specified by the manufacturer. The conditions associated with number 8 usually include a greater depth (often more than 1 m) and a longer duration (such as more than 30 min).

Seal Extrusion

The process of seal extrusion occurs when the seal material is forced out of its designated pocket or groove. It is possible for seals to extrude for a variety of reasons, including excessive pressure, high temperatures, or mechanical stress. Typically, this occurs when the seal material is too soft and the operating conditions exceed the seal's design limits. There are other reasons for seal extrusion, including very tight tolerances between the seal and the sealed component, such as the stem or seat, causing additional friction resulting in the seal being forced out of its pocket. It is possible that seal extrusion may result in leakage, loss of sealing integrity, and accelerated wear of both the seal and its surrounding components. It is important to select seals made from materials that are compatible with the operating conditions (e.g., pressure, temperature, and chemical exposure) in order to prevent seal extrusion. Additionally, it is essential to ensure that proper tolerances are maintained between the seal and the sealed component, such as the stem, during valve assembly. There is a challenge in that tight tolerances between the seal and stem can cause friction and seal extrusion, while a lack of tight tolerances may result in leaks. It is also beneficial to use backup seals or rings. It is possible to seal the ball valve stem with three rubber O-rings instead of one rubber O-ring. It is particularly important to use triple O-rings when the valve will be operated at high pressures and temperatures. Inspection and maintenance are essential for detecting and addressing seal extrusion before it leads to the failure of the valve and the connected piping.

Thermal Expansion

More information about thermal expansion is provided in Chapter 8, under the heading of "Thermal Expansion."

Water Hammering

More information about water hammering is provided in Chapter 8, under the heading of "Water Hammering."

PART 7: BEST MAINTENANCE PRACTICES

Checklists

Maintenance checklists are a simple but highly effective method of practice. All necessary tasks can be accomplished in an organized and systematic manner by using checklist. It serves as a guide for technicians and maintenance personnel. A checklist lists essential maintenance activities such as inspections, lubrication, cleaning, and even testing after maintenance. Maintenance teams can ensure that the job has been performed correctly by following a checklist. Furthermore, checklists help standardize the maintenance program so that all maintenance personnel follow the same procedure rather than their own methods and experiences. Moreover, checklists serve as a means of documenting the completion of tasks and supporting audits. In order to improve efficiency, reduce downtime, and enhance the performance of valves, well-designed and regularly updated checklists are essential.

Continual Improvement

Continual improvement in maintenance refers to the continuous process of improving maintenance practices and outcomes. The purpose of this is to enhance the reliability and cost-effectiveness of the system. It involves regularly evaluating maintenance performance based on key metrics related to reliability and maintenance and trying to make improvements. These metrics include downtime, mean time between failures (MTBF), and mean time to repair (MTTR). It is important to employ techniques such as RCA, feedback from technicians, and adopting new technologies such as PdM tools and data-driven analytics. Data-driven analytics is the process of combining ML with data (e.g., real-time data or condition monitoring data) in order to predict the future of the valve regarding failure or required maintenance.

Data Analysis

The analysis of data is an essential part of maintenance best practices. It involves analyzing historical and real-time data in order to identify trends, identify valuable observations, and predict potential failures. The use of advanced techniques such as predictive analytics and ML enables accurate forecasting of valve health and remaining useful life (RUL).

Documentation

Effective maintenance practices require proper documentation. The process involves recording all maintenance activities and the history of the equipment. It is important to maintain a detailed record of maintenance activities such as repairs and replacements for each equipment. Having clear and documented procedures ensures that all maintenance tasks are performed in a consistent manner and in accordance with best practices. Records of maintenance and diagnostics assist in identifying root causes of problems. Records of maintenance are often required to comply with industry standards such as ISO 9001 as well as government regulations.

Monitoring

Monitoring is a critical component of maintenance best practices. Monitoring the performance and condition of valves is carried out continuously or periodically. Detecting early signs of operational problems, such as wear, degradation, or failure, is carried out in this manner. Among the tools used in modern monitoring are sensors, real-time data systems, and condition monitoring technologies. A condition monitoring system may include vibration analysis, thermal imaging, and acoustic emission. Through these techniques, we can obtain a great deal of information from inside the valves that we are not able to see during operation. Maintenance teams can identify problems through monitoring before they escalate into costly failures.

Proactive Maintenance

In proactive maintenance, potential problems are identified through condition monitoring, including real-time sensors, and addressed before they lead to equipment failure or downtime. Instead of reacting to problems, it emphasizes preventing them from occurring in the first place. In this strategy, the root causes of failures are monitored, analyzed, and improved. Thus, it extends the life of equipment, enhances its reliability, and reduces the cost of long-term maintenance.

Risk Mitigation

Mitigating risks is a vital component of maintenance best practices. In this process, potential risks are identified, assessed, and minimized. The purpose of this is to ensure that equipment and systems are operated safely and efficiently. In order to prevent failures and associated hazards, proactive strategies such as regular inspections, condition monitoring, preventive maintenance, and even more are employed.

Training

In the field of maintenance, training is a critical best practice. Training provides personnel with the knowledge and skills needed to perform maintenance tasks efficiently and safely. Well-trained technicians can diagnose problems accurately and apply appropriate repair techniques. This can result in a reduction of valve downtime and an increase in the reliability. Training programs should include technical knowledge related to valves, maintenance procedures specific to valves, and troubleshooting techniques. A regular training program is designed to ensure that staff members stay up-to-date on the latest technologies, tools, and best practices.

QUESTIONS AND ANSWERS

1. Which type of valve maintenance focuses on early fault detection through real-time monitoring?
   A. Preventive maintenance
   B. Corrective maintenance
   C. Condition-based maintenance
   D. Reactive maintenance

   Answer: The correct answer is C.

2. Which of the following conditions is most likely to cause valve seat damage?
   A. Frequent throttling at low pressure
   B. Abrasive particles in the fluid
   C. High ambient temperature
   D. Valve being fully open during operation

   Answer: The correct answer is B.

3. In a ball valve, what is the most common cause of excessive torque during operation?
   A. Misalignment of the actuator
   B. Over-greasing of the stem
   C. Improper flow direction
   D. Corrosion or debris in the ball and seat area

   Answer: The correct answer is D.

4. Which of the following is a key principle of maintenance best practices?
   A. Performing maintenance only after equipment fails
   B. Regularly monitoring equipment and addressing issues proactively
   C. Avoiding documentation to save time
   D. Scheduling maintenance based solely on assumptions

   Answer: The correct answer is B.

5. Why is it important to maintain a detailed maintenance log as a best practice?
   A. To eliminate the need for preventive maintenance schedules
   B. To reduce the cost of spare parts inventory
   C. To provide a record of activities, identify recurring issues, and track technician performance
   D. To replace the need for training documentation

   Answer: The correct answer is C.

6. What is the most common cause of a control valve losing its signal in a pneumatic system?
   A. Corroded valve stem
   B. Faulty actuator calibration
   C. Air supply issues or damaged signal wiring
   D. Incorrect flow direction through the valve

   Answer: The correct answer is C.

7. Which of the following is a likely cause of a positioner malfunction in a control valve?
   A. Properly calibrated actuator
   B. Clean and dry air supply
   C. Tight packing gland around the valve stem
   D. Mechanical wear, improper calibration, or electrical signal disruption

   Answer: The correct answer is D.

8. Which of the following is a key component of reliability-based maintenance practices?
   A. Ignoring historical failure data to focus on reactive repairs
   B. Using predictive analytics and condition monitoring to prevent failures
   C. Performing maintenance only during equipment failure
   D. Scheduling maintenance tasks randomly without data analysis

   Answer: The correct answer is B.

9. Why is reliability-centered maintenance (RCM) considered an effective approach?
   A. It prioritizes maintenance based on the criticality of equipment and its failure impact.
   B. It focuses on repairing all equipment, regardless of importance.
   C. It eliminates the need for preventive maintenance entirely.
   D. It relies solely on reactive maintenance practices.

   Answer: The correct answer is A.

10. Which of the following correctly matches the type of maintenance with its description?
    A. Corrective maintenance: Maintenance tasks performed before or after failure happens.
    B. Predictive maintenance: Uses real-time data and analytics to predict when maintenance is needed.
    C. Preventive maintenance: Maintenance performed only after equipment failure occurs.
    D. Reactive maintenance: Addresses the root cause of failures to prevent them from occurring.
    E. Proactive maintenance: Maintenance performed with no regard to equipment condition.

    Answer: The correct answer is A.

BIBLIOGRAPHY

Al Ruqaishi, M. A., & Al Kalbani, F. K. (2024, November). Revolutionizing Leak Detection and Cost Savings in Valve Predictive Maintenance. Paper presented at the ADIPEC Conference. https://doi.org/10.2118/222620-MS

FieldInsight. (2022). Valve Maintenance Guide: Preventive Measures and Scheduling Best Practices. Retrieved from: https://www.fieldinsight.com/blog/valve-maintenance/?utm_source=chatgpt.com (accessed January 14, 2025).

Hettegger, M., & Altmann, C. (2023). Valve inspection with acoustic emission: What are the boundaries, what are new opportunities for predictive maintenance? e-Journal of Nondestructive Testing, 28(1). doi: 10.58286/27642

Lindberg Process Equipment. (2024). Troubleshooting Common Valve Problems: A Practical Guide. Retrieved from: https://lindbergprocess.com/2024/10/troubleshooting-common-valve-problems-a-practical-guide/ (accessed January 14, 2025).

Nesbitt, B. (2007). Handbook of Valves and Actuators: Valves Manual International. 1st edition. Elsevier: Oxford, UK.

Sotoodeh, K. (2020). Failure mode and effect analysis (FMEA) of pipeline ball valves in the offshore industry. Journal of Failure Analysis and Prevention, 20, 1175–1183. doi: 10.1007/s11668-020-00924-8

Sotoodeh, K. (2023). Industrial Valves: Calculations for Design, Manufacturing, Operation and Safety Decisions. 1st edition. Wiley: New York, NY.

Sotoodeh, K. (2023). The mathematical analysis and review of water hammering in check valves in offshore industry. Journal of the Institution of Engineers (India): Series C, 104, 879–885. doi: https://doi.org/10.1007/s40032-023-00965-6

Weldon Valves. (2025). Maintenance and Troubleshooting of Ball Valves. Retrieved from: https://www.weldonvalves.com/news/maintenance-and-troubleshooting-of-ball-valves.html?utm_source=chatgpt.com (accessed January 14, 2025).

10 Valve Safety, Reliability, and Compliance (SIL, SAR)

A

Absolute Pressure

The absolute pressure is the total pressure measured relative to a perfect vacuum (or absolute zero pressure). In addition to the pressure exerted by the atmosphere, any pressure within a system is considered. As a result, this measurement is often used in industries such as the valve industry, engineering, and science where precise pressure readings are required, particularly in applications that involve vacuums or variable atmospheric conditions. When compared to gauge pressure, which measures pressure in relation to the surrounding atmospheric pressure, absolute pressure measures pressure in relation to a perfect vacuum. You can calculate absolute pressure by adding gauge pressure (the pressure within a system relative to atmospheric pressure) and atmospheric pressure.

$$\text{Absolute pressure} = \text{Gauge pressure} + \text{Atmospheric pressure}$$

If the gauge pressure of a system is 2 bars and the atmospheric pressure is 1 bar, the absolute pressure would be as follows:

$$2 + 1 = 3\,\text{bar}\,(\text{absolute})$$

There are several ways of measuring absolute pressure, including psia (pounds per square inch absolute), bara (absolute), or kPa (absolute), in which the "a" indicates the measurement is absolute.

Absolute Temperature

According to the valve industry, absolute temperature refers to the temperature measured from absolute zero, the theoretical point at which molecular motion ceases (0 K or −273.15°C). The most commonly used absolute temperature unit in the valve industry is Kelvin (K), though Rankine (°R) is also sometimes used, especially in certain engineering contexts in the United States. The absolute temperature is important in fields such as gas flow and pressure control in valves where precision of temperature is required for accurate calculations.

DOI: 10.1201/9781003640967-10

A system's absolute pressure is directly affected by its absolute temperature. When the temperature of a closed system increases, the pressure within the system also increases. Designing safety features for pressure relief or setting pressure ratings on valves requires an understanding of this relationship. Cryogenic (extremely low temperature) and superheated steam (extremely high temperature) valves require absolute temperatures for setting operational limits. Materials behave very differently at extreme temperatures.

Acoustic Emission (AE) Sensors

AE sensors are used in ValveWatch Systems in order to detect high-frequency sound waves generated by dynamic events within the valve, such as leaks, cracks, or structural problems. By detecting faults that are not visible or detectable through other monitoring methods, these sensors contribute to early fault detection, which makes them particularly valuable in safety-critical and high-performance applications. Installing it on the cavity is an effective method of detecting leakage through the valve, particularly when the valve is closed.

Actuator Fail-Safe Mechanism

An *actuator fail-safe mechanism* is a critical feature in valve automation designed to move the valve to a predetermined safe position (open or closed) in the event of a system failure, such as power loss, signal disruption, overpressure, or actuator malfunction. This mechanism ensures that the valve transitions to a state that minimizes potential hazards, protects equipment, and ensures personnel safety. The actuator fail-safe mechanism plays a vital role in ensuring operational safety by preventing uncontrolled fluid flow, reducing risks during failures, and ensuring compliance with industry safety standards like API 6A, API 6D, or IEC 61508 for safety integrity levels (SILs).

Anti-Surge Valve

An *anti-surge valve* is a critical safety component in compressor systems, designed to prevent surge conditions that can severely damage the compressor and jeopardize the safety of the process. Surge occurs when the compressor's flow drops below a critical level, causing high-pressure fluctuations and reverse flow, which can lead to mechanical failure or hazardous situations in systems handling flammable or toxic gases. The anti-surge valve rapidly opens to recycle gas from the compressor discharge back to the suction side or to a safe location, maintaining the minimum flow required to avoid surge. Typically, anti-surge valves are fast-acting control valves, such as globe valves or ball valves, equipped with actuators and positioners for precise and rapid operation. These valves are integrated into a surge control system, ensuring real-time response to changes in pressure and flow to protect equipment and ensure safe operation.

ATEX

ATEX stands for *atmosphères explosibles*, which is French for *explosive atmospheres*. It refers to the European Union directives that regulate the design, manufacturing, and use of equipment in environments where explosive atmospheres may occur due to flammable gases, vapors, dust, or fibers. These directives aim to ensure the safety of workers and equipment in hazardous environments. It comprises two main directives: Directive 2014/34/EU, which governs the design, manufacturing, and certification of equipment and protective systems used in explosive atmospheres, and Directive 1999/92/EC, which focuses on worker safety by regulating the classification of hazardous zones, equipment selection, and implementation of preventive measures. The directive requires equipment to meet stringent standards for explosion protection and to be marked with specific ATEX certifications, ensuring suitability for hazardous zones. Widely applied in industries like oil and gas, chemical processing, and mining, the ATEX Directive aims to minimize explosion risks, protect workers, and ensure compliance with safety standards across the EU.

B

Backpressure

1. Definition

Backpressure is the pressure that exists on the outlet side of a safety valve (downstream side) when the valve is open and closed. As a result, this pressure may have a substantial impact on the performance of the safety valve and must be taken into account when designing, selecting, and operating the valve.

2. Effect on Valve Operation

Backpressure refers to the pressure exerted on the outlet side of a valve. It can impact the performance of safety valves, particularly in systems where the backpressure is high or fluctuates. Safety valves are designed to open and close at specific pressure thresholds, and *high backpressure* can prevent them from functioning correctly by interfering with the valve's opening force. This can cause safety valves to fail to open when they are needed to release excess pressure, which could lead to dangerous situations, such as equipment failure, overpressure, or system rupture. A safety valve that discharges into the atmosphere is not subject to backpressure. However, backpressure can cause *chattering* (rapid opening and closing) or *improper closing* of the valve, compromising the safety function of the valve.

3. Backpressure Types

It is possible for the backpressure value to be *constant* or *variable*. Constant backpressure occurs when the pressure exerted on the outlet side of the valve remains relatively stable and does not fluctuate significantly during operation.

There are several reasons why constant backpressure occurs, including the design of the system, such as long discharge pipes or a pressured collection system that the safety valve discharges. The opening pressure of the valve is affected by constant backpressure. It is possible that the valve could open at a pressure greater than its set pressure if this factor is not considered. Safety valves designed for constant backpressure are less likely to experience issues like valve chattering.

Variable backpressure can be caused by dynamic system conditions, such as other equipment venting into the same discharge system or changes in system operation conditions such as pressure or temperature. In some cases, variable backpressure can impact the valve's opening pressure and closing behavior and cause problems like *chattering* and *premature valve closing*. An excessive backpressure may prevent the valve from properly relieving pressure, resulting in system overpressure. Premature closing refers to a situation where a valve closes before it is supposed to, often leading to unintended consequences such as pressure buildup, system instability, or equipment failure.

4. Backpressure Effect Mitigation

Balanced safety valves are designed specifically to handle *backpressure* variations. The valve's internal *balanced trim* ensures that it can function correctly even when backpressure is fluctuating, making it an ideal solution for applications with changing outlet pressures. Balanced safety valves are explained further in this chapter.

5. SIL and Compliance

The design and operation of valves in relation to backpressure are evaluated for *SIL* in safety-critical systems. High *backpressure* conditions are considered during SIL analysis to assess whether the valve will meet safety standards and perform reliably. The performance of safety valves under *backpressure* is critical for compliance with safety regulations and industry standards such as *API 520, ISO 4126*, and *ASME Section VIII.*

6. Testing and Monitoring

In safety systems, *backpressure testing* is crucial to verify that the valve will operate correctly even under varying or high-backpressure conditions. Monitoring and maintenance practices ensure that valves can handle these conditions without compromising safety.

In conclusion, backpressure can have a significant impact on the performance of a safety valve, including its ability to open and relieve pressure at the correct set point. Managing backpressure through design considerations, such as the use of balanced bellows or the proper sizing of the valve and piping, is crucial to maintaining safety and efficient system operation. Figure 10.1 illustrates the effect of the backpressure to the safety valve in the outline pipe that is connected to the safety valve.

FIGURE 10.1 Backpressure effect on a safety valve.

Balanced Safety Valve

1. Definition and Purpose

Balanced safety valves are designed to maintain consistent performance regardless of fluctuations in backpressure (the pressure at the safety valve's outlet). A feature such as this is essential in systems where there may be variations in backpressure as a result of piping configurations, downstream equipment, or other process conditions. In industries such as oil and gas, chemical processing, and power generation, balanced safety valves are commonly used to prevent overpressure without affecting the safety valve performance due to changes in safety valve outlet pressure. High backpressure can affect the ability of standard safety valves to open at the set pressure and compromise their relief capacity when high backpressure pushes against the valve seat. Balanced safety valves are equipped with secondary pistons or bellows that counteract the effects of backpressure on the valve disk, allowing them to open and close consistently at the set pressure regardless of fluctuations in backpressure. In order to balance the forces acting on the valve disk, a bellows or piston is installed within the valve. As a result of the bellows, the process fluid is isolated from the spring chamber, allowing the valve to operate independent of backpressure. Like standard safety valves, a balanced safety valve could have a spring mechanism. However, this mechanism remains unaffected by variations in backpressure.

2. Applications

Balanced safety valves are widely used in industries where pressure systems experience fluctuating or varying backpressure conditions. These valves are essential in *oil and gas*, *chemical processing*, *refining*, and *power generation* industries, where maintaining stable pressure is critical for operational safety.

3. Codes and Standards

API 520, API 526, ISO 4126, and ASME Boiler and Pressure Vessel Code provide guidelines for the design and use of balanced safety valves, ensuring their safe and effective use in industrial processes.

Bellows Safety Valve

A *bellows safety valve* is a specialized type of safety valve designed to address challenges associated with *backpressure* and ensure reliable operation in safety-critical systems. The key feature of this valve is the *bellows element*, which isolates the valve's spring mechanism from the process fluid. Bellows compensates for variable backpressure, ensuring consistent set pressure and opening characteristics, even in systems where backpressure might otherwise impair the valve's performance. In addition, bellows prevent exposure of valve spring and internals to corrosive, toxic, or high-temperature media. Bellows safety valves are widely used in industries such as *chemical processing, power generation*, and *oil and gas*. Leak prevention and reliable overpressure protection are critical for safety and environmental compliance in industries mentioned above. By maintaining stability under fluctuating conditions, these valves enhance the safety integrity of pressurized systems and mitigate risks of hazardous failures.

Bellows Seal Bonnet

In valves, a bellows seal bonnet provides a highly reliable, leak-tight seal around the valve stem, effectively isolating the valve's internal process fluid from the external environment. Typically, this design is used in systems handling hazardous, toxic, or highly volatile substances, in which fluid leakage to the environment needs to be minimized. By expanding and contracting as the valve stem moves, the bellows, which are typically made of stainless steel or other corrosion-resistant alloys, form a seal around the valve stem. In this manner, no process fluid can escape from the stem, resulting in a zero-leakage seal. It is especially important when it comes to systems that require high levels of purity, safety, or environmental protection.

Blowdown Valve

1. Definition

The blowdown valve is a type of valve used to remove or discharge excess fluids, pressure, or sediment from equipment, typically boilers, pipelines, and pressure vessels.

2. Safety Function

Generally, a blowdown valve is designed to maintain optimal operating conditions, protect equipment, and enhance system efficiency by removing contaminants, reducing pressure, or removing liquid buildup. Normally, the valve is closed and opens if there is an overpressure situation in the piping or equipment. As an example, blowdown valves are safety critical valves that are designed to rapidly depressurize equipment by venting gases to flare stacks. The blowdown valves are configured as fail-safe open valves.

3. Type of Blowdown Valves

There are two types of blowdown valves—continuous and intermittent. In the context of a blowdown valve designed to release overpressure from a separator to the flare line, both continuous and intermittent blowdown serve different purposes. Continuous blowdown involves the steady, controlled release of gas or liquid from the separator. Intermittent blowdown involves the periodic, high-flow release of pressure from the separator when pressure builds up beyond normal operating levels.

4. Compliance and Reliability

Blowdown valves must be designed and maintained according to safety standards such as *ASME*, *API*, and *ISO*. These standards ensure that the valve will operate correctly under extreme pressure conditions, providing a reliable safety function. As part of a *safety instrumented system* (SIS), blowdown valves are evaluated for their *SIL* to ensure that they are capable of performing their function without failure in high-risk situations.

5. Connection to Safety Systems

Blowdown valves are often used in *pressure relief* or *emergency venting systems*, which are directly tied to the safety and protection of personnel, equipment, and the environment. These valves are often part of *pressure relief systems*, where they perform safety-critical functions. In this context, they contribute directly to *system safety* and *reliability*.

C

Chattering

1. Definition and Application

Valve chattering is the rapid and repetitive opening and closing of a valve, usually occurring in safety valves or pressure relief valves (PRVs) as well as check valves. Vibrations, noise, and potential damage to the valve and its surrounding piping can occur as a result of this phenomenon. This phenomenon may also shorten the valve's lifespan and reduce its effectiveness. A valve that is chattering is often a sign of improper valve sizing or pressure settings, which can compromise the valve's ability to provide safe system protection.

2. Causes

Improper valve sizing (oversized valve), fluctuating backpressure, or inadequate flow conditions (low flow conditions).

3. Prevention and Mitigation

Proper sizing, maintenance, and monitoring of system conditions; using balanced safety valve.

Cold Differential Set Pressure (CDSP)

1. Definition and Application

CDSP is the pressure at which a PRV is set to open under ambient, cold (nonoperating) conditions. CDSP is established while the system is at ambient temperature, rather than at elevated or operational temperatures. Set pressure refers to the pressure at which the relief valve opens in service, while cold differential test pressure refers to the pressure at which it opens in a test stand at normal ambient temperature. If the set pressure is too high, the valve may fail to relieve pressure during emergencies. Conversely, if it is too low, it could open prematurely, leading to unnecessary system depressurization

2. Measurement and Verification

The CDTP can be measured and verified with a conventional relief valve since backpressure is not present during testing in the shop. In many applications, fluid expansion or thermal effects cause pressure changes in the system as a result of operating temperatures. Therefore, this makes it a baseline pressure for the valve at "cold" conditions. Once the system has reached its operating temperature, CDSP is adjusted to accommodate these factors so that the valve will open at the desired set pressure.

3. Evaluation and Calculation

There are several factors that affect the pressure within a system during operation, including backpressure, thermal expansion, and compressibility. These factors are taken into consideration when calculating CDSP. A CDSP is particularly important in systems that experience significant temperature fluctuations, such as boilers, chemical reactors, and other high-temperature applications in which thermal expansion can affect internal pressure. For example, if a valve needs to open at 10 bar during operation (e.g., at 200°C), the CDSP may be set to a lower or higher pressure than 10 bar at ambient temperature to compensate for changes in pressure at elevated temperatures. As pressure and temperature increase, the set pressure of the safety valve can be significantly impacted, particularly due to the changes in system conditions which affect fluid behavior and material properties. When the temperature of the system increases, fluids and gases expand, thereby raising the system's internal pressure. High temperatures can reduce the strength of metals, which may affect the ability of valve components (such as springs and seats) to maintain the correct set pressure. If this additional pressure is not accounted for, a safety valve could open prematurely. Occasionally, materials may deform or weaken, resulting in the valve opening at a lower pressure than intended.

4. Case Study

Case Study: Cold Differential Set Pressure for Pressure Relief Valve in a High-Temperature System

Background

In a *chemical processing system*, a PRV is used to protect the system from overpressure. The desired *set pressure* of the valve is designed for operation at *200°C*, where the internal pressure should be 20 bar. However, during testing at a much lower temperature of *20°C*, the pressure characteristics change, leading to the need for adjustments in the valve's *CDSP*.

Scenario

Assume that the PRV is intended to open at *20 bar* when the system operates at *200°C*. However, when tested at *20°C*, the system pressure is lower. As the temperature increases from 20°C to 200°C, the fluid inside the system expands, which causes the pressure to increase. In this case, the pressure increases by *1 bar* as the temperature rises by *180°C* (from 20°C to 200°C). Additionally, the system experiences a *backpressure* of *1.5 bar* during normal operation at the desired set pressure.

Challenge

In order to ensure that the valve opens at the correct pressure of *20 bar* when the system reaches *200°C*, the *CDSP* must be set lower to compensate for the pressure increase due to the temperature rise. Without proper adjustment for *CDSP*, the valve could open either *too early* (at a lower pressure) or *too late* (at a higher pressure), risking *overpressure conditions*.

Solution

At *low temperature* (*cold conditions*), backpressure is *zero* because it is measured under *ambient conditions* on the test bench. However, at *high temperature* (200°C), the *backpressure* increases to *1.5 bar* due to operational conditions. Since backpressure opposes the opening of the valve and does not exist in the cold temperature, it should be deducted from the desired set pressure when calculating the CDSP. The thermal expansion of the fluid increases the system pressure as the temperature rises. This is an opening force in high temperature that assists in the valve opening and reducing the set pressure when the temperature increases from 20°C to 200°C. Since we do not have this opening force in ambient or cold condition, it should be added to the CDSP.

$$\begin{aligned}\text{CDSP} &= \text{Desired set pressure} + \text{Thermal pressure increased} - \text{Backpressure}\\ &= 20\,\text{bar} + 1\,\text{bar} - 1.5\,\text{bar} = 19.5\,\text{bar}\end{aligned}$$

Conclusion

This case study demonstrates the importance of properly setting the *CDSP* in systems where temperature changes can affect internal pressure. By adjusting the valve's *set pressure* at *cold conditions* to account for the pressure increase at higher temperatures, engineers can ensure the valve opens at the correct pressure under all operating conditions, preventing system failure and ensuring the safety of the plant.

Constant Backpressure

A constant backpressure in a safety valve is the presence of an unchanging, constant pressure on the outlet side of the valve, often caused by an existing downstream pressure source. It is important to note that this backpressure remains constant regardless of whether the safety valve is open or closed. Constant backpressure is commonly found in systems that discharge into a pressurized header, vent system, or other closed piping system that maintains a constant level of pressure.

Cryogenic Safety

Cryogenic safety for valves refers to the specialized design, materials, and operational considerations required for valves that operate in extremely low-temperature environments, typically below −150°C (−238°F). These valves are critical in handling cryogenic fluids such as liquid nitrogen, oxygen, hydrogen, helium, and liquefied natural gas (LNG). Proper cryogenic safety measures ensure the reliable operation of valves under these extreme conditions while preventing failures that could lead to leaks, equipment damage, or safety hazards. Many parameters are important in design of cryogenic valves like material, leak tight, extended stem.

D

Dangerous Detected Failure (λ_{DD})

A dangerous detected failure is a type of failure in a safety-critical component, such as a valve, where the component fails in a way that could lead to a hazardous condition but the diagnostic mechanisms of the system detect it before it can cause harm. When high pressure is detected in an oil pipeline, an ESDV is designed to close automatically. Assume that the valve actuator malfunctions and prevents the valve from closing (a potentially dangerous situation), but that the system's diagnostics detect the malfunction.

Dangerous Failure (λ_D)

A dangerous failure of a valve is one that can lead to hazardous conditions when it fails to perform its intended safety function. This failure mode poses risks to personnel, equipment, and the environment, as it prevents the system from responding correctly to a potentially hazardous situation. A dangerous failure occurs in an ESDV when it is supposed to close in an emergency but does not. This can allow hazardous materials, such as toxic gases or flammable liquids, to continue flowing, leading to fires, explosions, or exposure to harmful substances. There are some applications where a valve may have to be opened in order to relieve pressure or redirect fluid flow to a safer location. It is possible for excessive pressure to build up or damage the system if the valve fails to open. The other scenario of dangerous failure is when a valve will provide a tight seal when closed, thereby preventing any leakage of hazardous materials. In the event that the valve fails to seal completely, dangerous fluids or gases could leak, creating the risk of fire, environmental contamination, or toxic exposure. The dangerous failure is equal to the sum of dangerous detected and undetected failures.

$$\lambda_D = \lambda_{DD} + \lambda_{DU}$$

Dangerous Undetected Failure (λ_{DU})

An undetected dangerous failure occurs when a safety-critical component, such as a valve, fails in a way that may result in a hazardous condition, and the system's diagnostic mechanisms fail to detect it. A valve may become stuck in an open position in a shutdown system due to internal corrosion, debris, or mechanical failure. During an emergency, the valve will fail to close, which will allow hazardous materials to continue to flow.

Demand Mode

In the context of SIL, demand mode refers to how often a safety instrumented function (SIF) is expected to operate or respond in order to prevent a hazardous event. The safety function in demand mode is only activated under certain conditions, such as high pressure, temperature limits, or emergency shutdown (ESD) situations. In demand mode, a valve may only need to close in response to a high-pressure event rather than remaining closed continuously. As a result of low demand, the safety function is rarely needed, typically less than once per year or once per operational cycle. In low-demand mode, an ESDV operates only in rare circumstances, such as in the case of an overpressure scenario. During high-demand periods, the safety function is often required more than once a year. The probability of failure per hour (PFH) is calculated for high-demand scenarios, whereas the probability of failure on demand (PFD) in used for low-demand scenarios.

Design Pressure

Design pressure in the context of valve safety refers to the maximum pressure a valve and its components are engineered to withstand under normal operating conditions without compromising its structural integrity or functionality. It is a critical parameter that directly impacts the valve's selection, application, and safety. Design pressure for a valve is typically is affected by industry standards such as ASME B16.34 (for pressure–temperature ratings) or API 6D (for pipeline valves). Valves are classified into pressure classes (e.g., ASME 150, 300, 600) that correlate with their pressure–temperature ratings. In most cases, design pressure should be less than or equal to the pressure rating of the selected pressure class. If the *design pressure* of a system is *40 bar at 200°C*, the selected pressure class should have a rating greater than 40 bar at 200°C.

Design Temperature

Design temperature in relation to valve safety refers to the maximum and minimum temperatures a valve is designed to withstand while maintaining its functionality, structural integrity, and leak-tightness under operating conditions. It is a critical parameter that, along with design pressure, defines the safe operating limits of the valve.

Diagnostic Coverage (DC)

DC is a safety engineering metric used to assess the effectiveness of a system's diagnostics in detecting failures. The percentage of potentially dangerous failures can be identified by the diagnostic mechanisms of the system. SIS rely heavily on DC, which impacts the SIL, since higher DC values enhance system reliability and reduce PFD. DC can be defined as the ratio between the number of discovered dangerous failures and the number of all dangerous failures (both detected and undetected). In most cases, it is expressed as a percentage. When DC values are higher, there are fewer undetected dangerous failures, which lowers the PFD and enables the system to meet higher SIL requirements.

Example: A safety system has 200 dangerous failures, of which 150 are detected. What is its DC?

Answer: $DC = \frac{150}{200} = 75\%$

Double-Block and Bleed (DBB)

DBB is a valve configuration designed to enhance safety by providing two independent sealing mechanisms (blocks) and a bleed (vent) system located between them. This setup ensures complete isolation of the upstream and downstream flow, allowing for the safe removal of fluid or gas trapped between the two blocking valves. DBB valves are critical in applications requiring high safety standards, such as the oil and gas, petrochemical, and chemical industries.

Double Isolation and Bleed (DIB)

DIB is a valve configuration designed to provide an enhanced level of safety by incorporating two independent sealing mechanisms with a bleed (vent) system in between. This design ensures complete isolation of the process fluid, prevents leakage, and provides a method for verifying the integrity of the seals. DIB valves are widely used in high-risk applications where leak-tightness and safety are critical, such as in oil and gas pipelines, chemical processing plants, and offshore platforms.

DIB design types are used in ball and through conduit gate valves. They could be DIB1 including two double-piston effect (DPE) seats or DIB2 including one DPE and one self-relieving (SR) seat. Definitions of DPE and SR seats are provided in Chapter 5.

Dynamic Leak Measurement

1. Introduction and Importance

A measurement of the leakage from the valve packing when the valve stem is moving. When it comes to fugitive emission testing, dynamic valve leak measurement is the process of measuring leaks from a valve when it is operating under simulated or real-world conditions. Unlike static testing (which is conducted when the valve is stationary), dynamic testing measures leaks while the valve is in operation.

2. Maintenance and Testing

Dynamic leak measurement must be done as a part of regular safety testing and maintenance protocols to ensure that safety valves and other critical components are functioning correctly and preventing leaks. This method mimics real operating conditions by opening and closing the valve, cycling the stem, or subjecting it to fluctuating pressures and temperatures. Standards include ISO 15848 (for valves under fugitive emission conditions) and API 622 (for valve packing). This test is important since it ensures that valves comply with strict environmental regulations, such as those imposed by the US Environmental Protection Agency (EPA) and leak detection and repair (LDAR) program.

3. Safety and Compliance Focus

Dynamic leak measurement is highly relevant to *safety and compliance*, as it ensures that the valve system is *leak-free* and operating under safe conditions, particularly in *high-pressure or hazardous fluid systems*. Leaks can lead to *equipment failure*, *system inefficiencies*, or even *dangerous accidents*. Therefore, this type of measurement can directly contribute to maintaining *system reliability*, ensuring that the valve is functioning as designed and preventing *safety hazards*.

E

Emergency Shutdown

ESD refers to a safety system that rapidly isolates a process or an equipment to prevent or mitigate hazardous situations, such as equipment failure, leaks, fires, or explosions.

Emergency Shutdown Valve

ESDV is a type of safety-critical valve that is designed to automatically isolate or shut down the flow of fluids or gases in a system during critical conditions. These valves are key components of SIS and are used in industries such as oil and gas, petrochemicals, power generation, and chemical processing. It is the primary function of an ESDV to prevent the escalation of potentially hazardous events in order to protect personnel, equipment, and the environment.

Emergency Venting

Emergency venting refers to the rapid and controlled release of excess pressure, vapor, or gas from a system to prevent overpressure conditions that could lead to equipment failure, explosions, or other hazardous events.

Emergency Venting Valve

Emergency venting valves are designed to rapidly release excess pressure, vapor, or gas from a system to prevent overpressure conditions that could lead to equipment failure, explosions, or environmental harm. Emergency venting valves come in various types to address specific safety needs across industries. PRVs and *pressure safety valves* (PSV) are widely used for releasing excess pressure in vessels and pipelines, while *pilot-operated relief valves* provide precise pressure control in high-capacity systems. *Vacuum relief valves* and *breather valves* protect tanks from underpressure or vacuum conditions that could lead to structural failure.

EPA Method 21

EPA Method 21 is a standardized procedure established by the US EPA for detecting and measuring volatile organic compound (VOC) leaks from process equipment. It is widely used in industries such as oil and gas, petrochemicals, and manufacturing to identify fugitive emissions from components like valves, flanges, pumps, and compressors.

Explosion Proof Construction

Explosion-proof construction for valves refers to designing and manufacturing valves to safely operate in environments where flammable gases, vapors, or dust are present, minimizing the risk of ignition and explosion. These valves are critical in industries like oil and gas, chemical processing, and mining, where explosive atmospheres can create severe safety hazards. Explosion-proof valves are designed to safely operate in environments with flammable gases, vapors, or dust by incorporating spark-free materials, sealed actuators, and flame arrestors to prevent ignition. They feature robust enclosures to contain internal sparks or explosions, fail-safe mechanisms to move valves to a safe position during emergencies, and material compatibility with high temperatures and corrosive conditions. These valves comply with stringent certifications like ATEX, ensuring safe operation in hazardous zones. Regular maintenance and proper installation are essential to maintaining their explosion-proof integrity and protecting personnel, equipment, and the environment.

F

Factory Acceptance Test (FAT)

Performing a FAT on valves is a quality control procedure carried out at the manufacturer's facility prior to shipment to the client. As a result of this test, it is ensured that the valve meets the technical specifications, safety standards, and functional requirements outlined in the purchase order or project specifications. For industries such as oil and gas, power generation, and water treatment where valve performance and reliability are important, FATs are vital for quality assurance. The FAT includes visual inspection, pressure testing, dimensional checks, and functional testing. Additionally, the valve inspector reviews relevant certificates, such as material certificates, during the testing process. The manufacturer must pass the FAT in order to obtain approval from the client to ship the valves. Passing FAT can minimize the risk of problems during installation or commissioning. FAT of a large size butterfly valve with an electrical actuator on a test bench is shown in Figure 10.2.

Fail-As-Is

A fail-as-is actuator (also known as a fail-in-place actuator or a fail-last actuator) holds a valve in its current position in the event of power, control signal, or pressure failure. Unlike fail-safe actuators, which are designed to move to a predefined position upon failure, a fail-as-is actuator holds the valve in its last position, whether it is open, closed, or partially open. By default, double-acting actuators are considered fail-as-is. When a double-acting actuator is in use, both opening and closing movements rely on an external power source (such as air, hydraulic fluid, or electric power), rather than relying on a spring to return to a safe position in the event of a power failure.

Fail-Safe

The fail-safe position of an actuated valve is the predetermined position to which the valve moves if power or control signal is lost. Having this feature ensures safety in systems. A fail-safe position can be achieved by using a single-acting or spring-return actuator. Depending on the application, these actuators may be pneumatic or hydraulic.

FIGURE 10.2 Performing FAT on an electrically actuated butterfly valve on a test bench.

Fail-Safe Close (Closed)

Fail-safe closed actuators are designed to automatically close a valve in the event of a loss of power or control signal. Typically, this type of actuator is used in systems where shutting off fluid flow in the event of an emergency or a failure is critical for safety or operation integrity. These actuators use a spring to close the valve once power is removed. For example, in a pneumatic or hydraulic actuator, pressure is applied to open the valve, and when the pressure source is lost, the spring forces the valve back to its closed position. In the event of a power failure, some electric actuators incorporate battery backups or mechanical springs to close the valve.

Fail-Safe Open

In the event of a power or control signal loss, a fail-safe open actuator is designed to open the valve automatically. These actuators have a spring mechanism that opens the valve when power or pressure is lost. In a pneumatic or hydraulic actuator, pressure is applied to keep the valve closed, and when the pressure source is removed, the spring automatically moves the valve to the open position. In case of a power failure, some electric actuators are powered by backup batteries or mechanical springs that force the valve open. In case of fire, emergency water supply valves should remain open to facilitate the movement of water that will be used to extinguish the fire. Blowdown valves are critical safety valves in flare systems that are designed to rapidly depressurize equipment by venting gases to flare stacks. The blowdown valves are configured as fail-safe open valves.

Fail-Safe Position

The position in which a valve defaults to in the event of failure, such as fail-closed or fail-open.

Fail-Safe State

Fail-safe states are designed conditions in which a system or component reverts to a safe or nonhazardous state following a failure. In safety-critical systems, specifically those in industrial settings such as SIS, the fail-safe state is essential for preventing accidents, injuries, and damage when unexpected issues occur, such as loss of power, equipment malfunction, or communication failure.

Fail-to-Close

Same as fail-safe close.

Fail-to-Open

Same as fail-safe open.

Failure Analysis Report

If a valve fails, a failure analysis report identifies the root cause of the failure and describes the corrective actions taken to prevent recurrence. As a result of these reports, maintenance practices can be improved and similar problems can be avoided in the future.

Failure Rate (λ)

When calculating SILs, failure rate is a critical metric that quantifies the likelihood of a component or system failing over time. The number of failures per unit of time (e.g., failures per hour) is typically used to assess the reliability and safety of components in safety-critical systems, such as SIS. Different failure rates are classified as safe, dangerous, detectable, and undetectable.

Final Control Element

Final control elements in SIS act directly on processes to bring them to a safe state when a potentially dangerous condition is detected. At the final stage of the SIS response chain, it receives signals from the logic solver (a part of control system) and executes the necessary physical actions to mitigate or prevent a hazardous event. An example of a final control element would be an ESDV or a PSV.

Fire Safe (Design)

A fire-safe valve is specially designed and tested to maintain its functionality and prevent catastrophic leakage in the event of a fire. These valves are designed to contain flammable or hazardous fluids, thus preventing leakage even under extreme temperatures caused by a fire, thereby preventing fire escalation and enhancing safety. As part of the design of a fire-safe valve, there are several key features to consider, including primary and secondary seals, high-temperature resistance sealing, antistatic springs, blowdown proof stems, and external leakage protection. Double-sealing typically includes a primary seal (soft materials such as PTFE) and a secondary metal-to-metal seal that activates in the event of a fire. Secondary seals prevent leakage when the primary seals burn or degrade. Antistatic devices are installed between the stem and closure member to provide a path for static electricity to be discharged. When pressure is applied, a blowout proof stem prevents the stem from being ejected.

Fire Test

1. Purpose

In order to minimize the risk of leaks, blowouts, or other failures that could exacerbate the fire or cause additional hazards, a valve must be able to continue performing its safety function or maintain a certain degree of sealing under extreme heat and pressure. Fire testing verifies that valves, especially safety relief valves, shutoff valves, and check valves, maintain their integrity and functionality when exposed to extreme temperatures.

2. Procedure

In the process, the valve is heated to 750–1,382°C (1,382–1,832°F) for a specified period of time, often reaching temperatures between 750°C and 1,000°C. The test duration and temperature vary according to standards and intended application, but usually last between 15 min and 30 min. As part of the fire test, the valve is also subjected to internal pressure, simulating operational conditions. This pressure ensures that the valve's sealing elements are tested under realistic fire conditions. The test standard may specify either the standard operating pressure or a fraction of it. A cooling phase is conducted after the fire exposure phase to assess whether the valve can reseal or maintain its integrity after it has been exposed to fire and then cooled. During the test, any leakage through the valve (whether through its stem, body, or seats) is measured. For the valve to pass the test, it must remain within acceptable leakage limits in order to maintain a seal under fire conditions.

3. Standards

Fire tests can be conducted in accordance with various standards, such as API 6FA or ISO 10497 or API 607. As long as the valve passes the test, it means that the seals are in good condition and it has a good structural integrity when exposed to fire. Finally, passing a fire test indicates that a valve can be operated in the event of a fire. A valve fire test is illustrated in Figure 10.3.

Flare

Flares are commonly used in industrial facilities, such as oil refineries, petrochemical plants, and natural gas processing plants, for the purpose of burning off excess gases that cannot be processed or stored. It is imperative that flares are used in order to ensure the safety of operations and to prevent the uncontrolled release of flammable or toxic gases into the atmosphere. Gas flaring in an offshore platform is shown in Figure 10.4.

FIGURE 10.3 Performing a valve fire test.

FIGURE 10.4 Gas flaring in an offshore platform. (With permission: Shutterstock.)

Functional Safety

As part of overall safety, functional safety ensures that a system or a piece of equipment operates correctly in response to specific inputs or conditions, reducing or eliminating the risk of accidents, harm to people, and damage to equipment or the environment. It is particularly important to ensure functional safety in safety-critical systems, such as those used in industries such as oil and gas, automotive, rail, and manufacturing, where malfunctioning equipment could pose a risk to public safety.

H

Hardware Fault Tolerance (HFT)

The concept of HFT is used in safety engineering to refer to the ability of a system to continue performing its safety function despite one or more failures of its hardware components. HFT refers to the number of faults a system can withstand without losing its ability to achieve or maintain a safe state. The ability of hardware to tolerate faults is essential in SIS and directly impacts a system's SIL. HFT = 0 indicates that the system does not have fault tolerance; a single failure of the hardware can result in a loss of the safety mechanism. The HFT value is 1 if the system is capable of coping with one hardware failure while still performing its safety function. System HFT = 2 indicates that the safety function of the system may not be compromised by two hardware failures. Higher HFT levels provide greater reliability, as the system can continue functioning safely despite multiple failures. Higher HFT is typically achieved by redundancy that means combining multiple independent components to perform the same function in order to increase fault tolerance. For a SIL 2 requirement, a single component with HFT = 0 may be sufficient if frequent diagnostics and proof testing are implemented to ensure reliability. However, to achieve SIL 3, the system would typically require an HFT of at least 1 and using redundancy, such as a 1oo2 (one-out-of-two) configuration, to meet the more stringent reliability and fault tolerance criteria.

Health, Safety, and the Environment (HSE)

HSE refers to a methodology that is used to ensure occupational health and safety in addition to protecting the environment. A concern of many companies, especially those in the oil and gas industry, is the implementation of HSE guidelines and procedures in order to prevent problems such as oil spills, fugitive emissions, and leaks.

High-Integrity Pressure Protection System (HIPPS)

1. Definition

A *HIPPS* is a specialized SIS designed to prevent overpressure in pipelines, vessels, or equipment by shutting off the source of pressure before it exceeds safe operating limits. Unlike conventional relief systems that release pressure into the environment, HIPPS acts as a "last line of defense" by isolating the pressurized system, ensuring minimal environmental impact and compliance with strict safety and emission standards. HIPPS systems are commonly used in industries such as oil and gas, petrochemicals, and energy, where high-pressure processes are routine. They typically consist of sensors, logic solvers, actuators, and final elements, such as HIPPS valves, which all work together to detect and respond to overpressure scenarios. It is important to know that HIPPS has several advantages over ESD such as faster closing action (response time).

2. Valves

HIPPS valves are the critical final elements in the system, designed to shut off fluid flow rapidly and reliably in response to detected overpressure conditions. These valves must meet stringent safety and reliability requirements such as SIL 3 or SIL 4 as outlined in standards such as *IEC 61508* and *IEC 61511*. Safety concerns surrounding HIPPS valves include their ability to perform quickly under extreme conditions, such as high pressure, high temperature, and corrosive environments.

I

Inherently Safe Design

1. Definition

Inherently safe design refers to engineering principles that minimize or eliminate hazards directly through the design of equipment, such as valves, rather than relying on additional safety systems or procedures. In the context of valves, inherently safe design focuses on preventing risks like leaks, operational failures, or catastrophic incidents by addressing potential issues at the design stage. This proactive approach enhances reliability, and ensures long-term safety in industrial systems.

2. Case Study: Valve Failure Due to Poor Seal Material Selection

Background

A gas processing plant in the Middle East experienced a near-miss incident due to a valve failure in a high-pressure gas line. The system operated at 120 bar and 250°C, handling flammable and toxic gases, including sour gas containing hydrogen sulfide (H_2S). The failed valve was a critical isolation valve responsible for controlling the flow of gas between processing units. The failure resulted in a significant leak, creating a severe safety hazard. This incident prompted the plant's management to prioritize a redesign of the valve system, focusing on incorporating key principles of safe valve design.

Incident

The failure analysis revealed that the valve's sealing material was not suitable for high-temperature applications, leading to gasket degradation and leakage. Additionally, the valve's actuator lacked a fail-safe mechanism, causing the valve to become stuck in a partially open position during a power failure. This allowed gas to escape uncontrolled. Internal erosion caused by cavitation due to rapid pressure fluctuations further weakened the valve body, contributing to the incident. The investigation highlighted inadequate consideration of operating conditions and an overreliance on external safety systems rather than addressing inherent design flaws in the valve.

Action Taken

To address the issues, the plant implemented a comprehensive redesign of the valve system based on the principles of safe valve design. Leak-tight construction was achieved by selecting double-sealing or double-isolation and bleed ball valves with fire-safe certification and high-integrity sealing materials, such as graphite gaskets, to withstand high temperatures and pressure. Stainless steel valve bodies with Inconel trim compatible with NACE MR0175/ISO 15156 suitable for sour service (hydrogen sulfide containing) were chosen to resist corrosion and erosion, while anti-cavitation measures were introduced to prevent internal damage. Fail-safe mechanisms, including spring-return actuators with backup pneumatic systems, were installed to ensure automatic valve closure during power failure.

Outcome

The redesigned valve system was successfully tested and commissioned. It provided enhanced safety by eliminating leakage risks and incorporating fail-safe and fire-safe features. The upgraded materials and simplified designs resulted in improved operational reliability and reduced maintenance requirements. The plant's valve system now complies with industry safety standards, including *API*

6D and *ASME B16.34*, ensuring the system's integrity under high-pressure and high-temperature conditions.

L

Leak/Leakage

In a valve, leak/leakage occurs when fluid (liquid or gas) escapes through areas that should be sealed off, either internally or externally. There may be issues with valve components, wear, improper sealing, or damage resulting in this unwanted flow. The presence of leaks in valves can compromise the performance of the system, pose safety hazards, increase operational costs, and pose an environmental threat, especially when dealing with hazardous fluids. Internal leakage is also known as seat leakage, and occurs when fluid flows through the valve despite it being fully closed. Generally, this occurs as a result of worn or damaged seats, erosion, or buildup of debris, which prevents a tight seal. External leakage occurs when fluid escapes from the valve stem, packing, or body joints to the environment.

Leak Detection and Repair

1. Introduction

LDAR in valves refers to a program or process aimed at identifying and repairing leaks of gases or liquids from valves, pipelines, and other equipment in industrial systems. LDAR is particularly important in industries such as oil and gas, petrochemicals, and power generation, where leaks can lead to environmental harm, safety hazards, and financial losses. Figure 10.5 illustrates the LDAR process for a valve in an industrial plant. The image shows a technician using a portable gas analyzer to check for leaks around the valve, transporting toxic, and flammable hydrogen sulfide.

2. Leakage Risk Mitigation

The primary objective of LDAR programs is to prevent leaks of VOCs, hazardous gases, and other substances that can pose serious health, safety, and environmental risks. It is closely related to preventing fire hazards, explosions, or toxic exposure, which are key concerns in industrial safety.

3. Leakage Risk Mitigation

As part of a facility's safety and risk management practices, many safety and environmental regulations (e.g., Occupational Safety and Health Administration [OSHA], EPA) require LDAR programs. By maintaining compliance, the workplace and surrounding areas remain safe for workers and the general public.

4. Environmental Safety

While LDAR also addresses environmental concerns, such as reducing emissions, its safety implications (e.g., minimizing leaks that could lead to accidents) are significant.

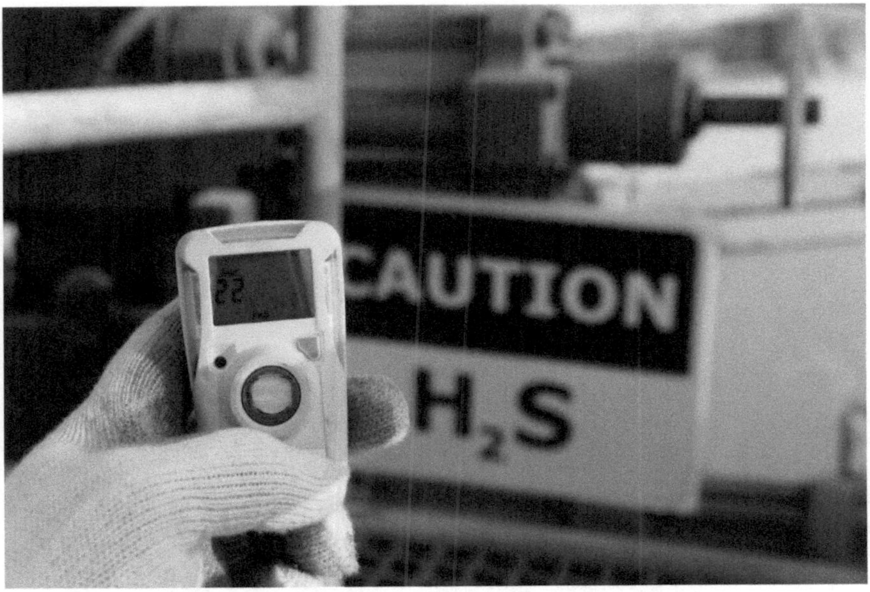

FIGURE 10.5 LDAR for a valve in the plant. (With permission: Shutterstock.)

Leakage Prevention

Leakage prevention in valves is a critical aspect of safety in industrial systems, as even minor leaks can lead to significant risks, including equipment failure, environmental harm, and hazards to human health. Effective leakage prevention begins with proper valve design and material selection. Valves must be engineered to meet the specific requirements of the system, including pressure, temperature, and fluid characteristics. Standards such as API 598 and ISO 5208 establish acceptable leakage limits and provide guidelines for valve testing. Regular *inspection*, *testing*, and *maintenance* are essential to ensure valves remain leak-tight over their operational lifespan.

Locking Device

1. Definitions

Locking devices for valves are mechanical or electromechanical tools designed to secure a valve in a fixed position (open, closed, or throttled) to prevent unauthorized or accidental operation. These devices are critical for ensuring safety during maintenance, servicing, or system isolation, as they help eliminate risks associated with unintended valve movement. Locking devices are commonly used in industries like oil and gas, chemical processing, and power generation, to maintain operational and worker safety. Figure 10.6 shows locking device installed on the handwheel of the valve to prevent unauthorized operation of the valve.

FIGURE 10.6 Locking device is installed on the handwheel of a valve to prevent unauthorized valve operation. (With permission: Shutterstock.)

2. Types

Locking devices for valves are indispensable tools for ensuring operational and worker safety in industrial systems. These devices, including padlock mechanisms, valve lockouts, chains, and electromechanical locks, secure valves in fixed positions to prevent unauthorized or accidental operation. They are a critical part of lockout/tagout (LOTO) procedures, offering robust protection during maintenance or system isolation. By immobilizing valve handles or actuators, locking devices enhance safety, prevent tampering, and ensure compliance with safety regulations in high-risk environments such as oil and gas processing, chemical plants, and power-generation facilities.

Lockout/Tagout

1. Definition

LOTO is a safety procedure used in industrial and maintenance settings to ensure that dangerous machinery, equipment, or systems such as valves are completely de-energized and isolated from their energy sources before maintenance or servicing begins. It should be noted that LOTO is also related to installation of pressurized equipment and components like valves. LOTO prevents accidental start-up or the release of stored energy, protecting workers from hazards such as electrical shocks, mechanical movements, or the release of hazardous fluids or gases.

2. Key Features

Lockout devices physically secure energy control points, such as valve handles, circuit breakers, or switches, using padlocks or specialized mechanisms to prevent accidental re-energization. *Tagout devices* provide visible warnings with tags attached to the lockout points, displaying critical information such as the reason for isolation, the name of the responsible person, and the date.

3. Example

Imagine a maintenance team is working on a high-pressure steam valve in a power plant. To ensure safety, they implement a *LOTO* procedure. First, they identify all energy sources connected to the valve, including the steam supply line and control systems. The steam line is isolated by closing the valve, and a lockout device is applied to the valve to secure it in the closed position. A padlock is placed on the valve handle, and a tag is attached, indicating that the valve is locked out for maintenance, along with the name of the responsible technician and the date.

Logic Solver

The logic solver of a SIS is the central processing unit which receives input from sensors, analyzes conditions against safety criteria, and decides whether to activate the final control elements to bring the process to a safe state. Logic solver is a part of control system.

M

Maximum Allowable Accumulated Pressure (MAAP)

1. Definition

MAAP is the maximum pressure a system is allowed to accumulate during an overpressure event, including normal operating pressure and pressure accumulation due to relief valve activation. It includes all pressure sources during upset conditions but must remain within safe limits for the equipment. It should be noted that exceeding MAAP leads to valve deformation, leakage, and rupture.

2. Relationship to MAWP (Maximum Allowable Working Pressure) and Calculation

MAWP is the maximum pressure for normal operating conditions, whereas *MAAP* accounts for pressure excursions during emergency situations. MAAP is typically higher than MAWP because it considers temporary overpressure during relief events. Codes like ASME (American Society of Mechanical Engineers) allow a specified percentage of pressure accumulation over MAWP during relief events. In general, 10% accumulation is allowed for pressure relief systems, while 21%

accumulation is allowed for fire-related scenarios (in accordance with ASME Section VIII).

MAAP = MAWP + Allowable accumulation.

If MAWP = 100 psi and the allowable accumulation is 10%, the MAAP = 110 psi.

3. Regulatory Compliance

It is defined by safety standards such as ASME Boiler and Pressure Vessel Code, API 520/521 (Pressure-Relieving and Depressurizing Systems), and OSHA.

4. Industrial Applications

Pressure vessel (boilers, storage tanks, heat exchangers), piping systems, and relief devices such as safety valves and rupture disks.

Maximum Allowable Working Pressure

1. Definition and Safety Implications

MAWP defines the maximum pressure that a vessel or piping system can safely withstand during operation. In valves, the MAWP refers to the highest pressure that a valve is capable of handling safely under normal operating conditions. Exceeding this limit poses serious risks, such as rupture or explosion, making it a critical safety parameter. It ties directly to operational safety measures, like PRVs and overpressure protection systems. MAWP decreases as operating temperature increases because most materials weaken at higher temperatures. The MAWP of a valve with a MAWP of 10 bar at 20°C may only be 8 bar at 200°C. Safety valves and relief systems are set to prevent pressure from exceeding the MAWP, with a small allowable margin (typically 10%) refer to MAAP. Operating within the permitted 10% MAAP margin should not cause immediate damage to the valve, provided the exposure duration is limited.

2. Key Factors Affecting MAWP

Material properties especially strength of material, corrosion allowance, design codes and standards, wall thickness of the component, and weld joint efficiency.

3. Relation to Other Pressure Terms

To ensure safety, the operating pressure must always be lower than the MAWP during normal operation. As a general rule, the design pressure is the intended operating pressure of the system, often slightly below the MAWP in order to allow for a safety margin. Burst pressure is the maximum pressure at which the valve will fail catastrophically. A burst pressure exceeds MAWP, acting as the ultimate safety limit. A set pressure is the pressure at which a safety valve begins to open in order to relieve excess pressure. In order to prevent overpressure, set pressure is typically below the MAWP. In addition, MAWP is higher than the design pressure.

4. Case Study: Understanding Pressure Parameters in a Pressure Vessel System

System Description

This case study examines a pressure vessel equipped with a safety valve, operating in an industrial process, such as a refinery or chemical plant. The vessel contains a pressurized fluid and must maintain safe operations under varying conditions, including pressure surges and emergencies.

Key Parameters:

- Design pressure: 50 bar
- MAWP: 55 bar
- Set pressure: 52 bar
- Burst pressure: 120 bar

In this example:

- *Design Pressure:* The pressure the system is designed to operate at under normal conditions
- *MAWP:* The maximum pressure the valve can safely withstand during normal operation
- *Set Pressure:* The pressure at which the safety valve opens to relieve excess pressure
- *Burst Pressure:* The ultimate pressure limit beyond which the valve would fail catastrophically

Process Upset Scenarios:

0–50 Bar (Normal Operating Range): During normal operations, the pressure vessel operates at a steady *50 bar*, well within the system's design parameters. Due to a malfunction in an upstream compressor, the pressure begins to rise. Let us analyze how the system responds at each critical pressure point:

50–52 Bar (Safety Valve Activation): The system exceeds the *design pressure*. While this does not immediately cause failure, it signals that the safety valve needs to activate to prevent further pressure buildup.

52 Bar (Safety Valve Opening): At this point, the *safety valve opens*, releasing pressure to protect the system and keep it below the *MAWP (55 bar)*.

52–55 Bar (Pressure Relief Zone and Less Than MAWP): Pressure Relief Zone where the safety valve is actively relieving pressure, ensuring the system does not exceed the MAWP.

Beyond 55 Bar (Catastrophic Danger Zone): The vessel's integrity could be compromised if the safety valve does not operate properly or is undersized, resulting in leakage, damage, and failure if pressure exceeds the MAWP.

Consider the case in which the compressor fails or the outlet pipe connected to the safety valve is blocked, resulting in a buildup of pressure within the system. As a result of a few emergency cases mentioned above, we are now able to include MAAP in this study. First of all, the MAAP must be calculated at a level that is 10% higher than MAWP in non-fire cases, resulting in MAAP at a level of 60 bars. It is now possible to add a new pressure zone between 55 bar and 60 bar, which is referred to as the emergency accumulation zone.

55–60 Bar (Emergency Accumulation Zone): This represents the *MAAP* zone, which accounts for extreme emergency conditions where pressure exceeds the *MAWP* of 55 bar. This range typically occurs during rare and severe scenarios, such as fires (thermal expansion), process failures, or equipment malfunctions, where the safety valve is fully open and relieving pressure but cannot immediately reduce it back to normal levels. Operating within the permitted 10% MAAP margin should not cause immediate damage to the valve, provided the exposure duration is limited.

Mean Time between Failure (MTBF)

MTBF measures the average time between failures of a repairable system or component. In contrast to mean time to failure (MTTF), which applies to nonrepairable components that are replaced upon failure, MTBF is used primarily for systems that can be repaired and returned to service after a failure. MTBF is calculated as the total operational time divided by the number of failures observed during that time. As an example, if a pump is operated for 10,000 h and fails four times, its MTBF would be 2,500 h.

Mean Time to Failure

MTTF is a reliability metric that estimates how long a nonrepairable system or component will last before it fails. In the context of SIL, MTTF is a crucial measure in understanding the lifespan and reliability of components in SIS, particularly when evaluating PFD or PFH. The MTTF of a component is calculated by dividing the total operating time of the component by the number of failures observed. A higher MTTF indicates greater reliability, meaning the component will last longer before failure. For safety-critical applications, components with high MTTF are preferred to ensure fewer failures over time. As part of higher SILs (e.g., SIL 3 or SIL 4), components with higher MTTFs are often required to reduce the likelihood of failure and to achieve stricter targets for PFD and PFH.

Example: Consider the case of a safety-critical valve that has been operating for 10,000 h and has failed twice during that time. Calculate the MTTF.

Answer: $\text{MTTF} = \dfrac{\text{Total operational time}}{\text{Number of failures}} = \dfrac{10,000}{2} = 500 \text{ h}$

Mean Time to Repair (MTTR)

In maintenance, MTTR represents the average amount of time required to repair a failed component or system and restore it to full operational capability. MTTR typically includes the time spent diagnosing the issue, gathering tools and parts, repairing or replacing the component, testing it to ensure proper function, and bringing the system back online. There is an inverse relationship between MTTR and availability. A lower MTTR value indicates that the system can be restored quickly, resulting in a higher level of availability. MTTR is an important parameter in the calculation and determination of SIL. A lower MTTR value reduces the amount of time a system remains in a failed state, thereby reducing the PFD and enhancing the safety of the system. In reliability engineering and asset management, the MTTR is widely used to determine how quickly equipment can be repaired following a failure, which is crucial for minimizing downtime and improving the availability of systems. The MTTR is calculated by dividing the total downtime spent on repairs by the number of events that need to be repaired. As an example, if there are three valve repairs that require repair times of 4, 5, and 6 h, the MTTR would be equal to the average of the repair times that is 5 h.

O

Operating Pressure

The operating pressure of a valve refers to the pressure at which it is expected to function under normal, everyday conditions. It is the pressure exerted by the fluid (liquid, gas, or steam) passing through the valve during normal operation. The operating pressure is an essential parameter in the selection and design of valves, as it helps ensure that the valves are capable of handling the routine demands of the system without causing any problems.

Operating Temperature

1. Definition

A valve's operating temperature is the range of temperatures within which it is expected to function under normal, day-to-day operating conditions. The operating temperature of a valve refers specifically to the temperature of the fluid (liquid, gas, or steam) flowing through the valve during standard system operation. In the selection and engineering of valves, the operating temperature is a critical parameter. To prevent degradation, the valve's sealing materials and soft materials must be compatible with the operating conditions such as operating pressure and temperature. The selection of certain elastomers is based on their ability to withstand the typical temperature without softening, warping, or losing strength.

2. Case Study: Failure Analysis of a Valve Due to Mismatch between Design and Operating Conditions

Background

A refinery in the Middle East processes crude oil in a high-temperature and high-pressure environment. The system includes a pressure-reducing control valve installed in a line transferring hot crude oil from a high-pressure separator to the distillation column. The valve was designed to handle specific pressure and temperature conditions while maintaining leak-tightness and durability. However, frequent pressure and temperature fluctuations during operation resulted in unexpected valve failures, causing production disruptions.

Design and Operating Conditions

The valve was originally designed with a *design pressure of 50 bar* and a *design temperature of 400°C*, using a carbon steel body with stainless steel trim and PTFE seals. The system operated at *normal conditions of 40 bar pressure* and *380°C temperature*, but occasional pressure spikes of up to 10–15% above normal occurred during start-up and shutdown phases.

Incident Description

The valve started showing signs of failure within six months of installation. Symptoms included external leakage around the bonnet, cracks on the valve body near the flange, and deformed seals leading to internal leakage. The valve failure caused an unplanned shutdown of the crude oil transfer line, resulting in significant production losses and emergency maintenance costs.

Investigation Finding

The root cause analysis revealed that the PTFE soft seals were not rated for temperature spikes exceeding 350°C, leading to deformation and leakage. Additionally, the carbon steel body exhibited thermal fatigue due to prolonged exposure to temperatures near its upper limit. During pressure surges, the valve's design pressure was exceeded, creating cumulative stress on the body and flange. These issues were compounded by inadequate consideration of transient operating conditions during the valve selection process and insufficient insulation around the valve.

Corrective Actions

To address the issues, the refinery replaced the valve with a redesigned model rated for *60 bar pressure* and *450°C temperature*, providing a safety margin for transient conditions. The new valve featured a chrome–molybdenum steel body for enhanced thermal resistance and graphite seals compatible with high temperatures and pressure surges. Pressure dampeners were installed to mitigate spikes, and additional insulation was applied around the valve to stabilize its temperature environment. Real-time monitoring sensors for pressure and temperature were also implemented, and routine inspections were introduced to detect early signs of wear or failure.

Lesson Learnt

This incident highlighted the importance of aligning design and operating conditions for critical valves. Valves should be selected with adequate safety margins for both pressure and temperature, considering potential transient conditions. Material compatibility must be carefully evaluated, particularly for high-temperature or corrosive applications. Dynamic operating conditions such as pressure spikes and temperature fluctuations should be accounted for during the design process. Finally, real-time monitoring and proactive maintenance are essential to ensure long-term valve reliability and safety.

Outcome

The redesigned valve performed reliably under operating conditions for over three years without leakage or failure, significantly reducing downtime and maintenance costs. The refinery adopted a new standard requiring a 20% safety margin for valve pressure and temperature ratings in critical systems. This case demonstrated the importance of considering both design and operational dynamics to ensure valve integrity and system safety.

OSHA

OSHA is a federal agency of the US Department of Labor, established in 1970 under the Occupational Safety and Health Act. OSHA is responsible for ensuring safe and healthy working conditions for employees by setting and enforcing standards and providing training, outreach, education, and assistance.

Overpressure

An overpressure refers to an increase in pressure above the set pressure at which a safety valve begins to open. In order for a valve to perform its intended function of relieving excess pressure from a system safely and efficiently, it is expressed as a percentage of the set pressure. In the case where a valve set pressure is 10 bars and the system pressure reaches to 11 bars, the overpressure is 1 bar, or 10% of the set pressure. Overpressure makes sure that the safety valve opens fully by allowing sufficient flow to relieve excessive pressure in the system. For most applications, ASME Section VIII allows an overpressure of 10%.

$$\text{Overpressure} = \frac{\text{Relieving pressure} - \text{Set pressure}}{\text{Set pressure}} \times 100\%$$

Example: A safety valve is set to open at *100 psi*. If the pressure rises to *110 psi* before the valve fully opens, calculate the overpressure.

$$\text{Overpressure} = \frac{\text{Relieving pressure} - \text{Set pressure}}{\text{Set pressure}} \times 100\% = \frac{110 - 100}{100} \times 100\% = 10\%$$

Overpressure Protection

Overpressure protection refers to a system or method designed to prevent pressure in a process, piping system, or equipment from exceeding a safe limit, which could lead to equipment damage, leaks, explosions, or other hazardous events.

P

Personnel Protection

Personnel protection in valve safety involves implementing design features and operational safeguards to minimize safety hazard risks for personnel on site associated with valve operation and maintenance. Measures such as isolation mechanisms, ergonomic designs, splash guards, and automation protect workers from exposure to high pressure, extreme temperatures, or hazardous fluids. *Ergonomic designs* focus on making valve handles, actuators, and controls easy to operate, reducing strain and injury risks, especially in high-use or hard-to-reach areas. *Splash guards* shield workers from exposure to hazardous fluids, such as corrosive chemicals or high-temperature liquids, by containing leaks and preventing splashes.

Pilot-Operated Safety Valve (POSV)

POSV uses a smaller pilot valve to control the opening and closing of the main safety valve. Using this design allows the safety valve to provide more precise pressure relief and handle higher set pressures. In addition, this design has other advantages such as reduced leakage and improved stability compared to conventional spring-loaded safety valves. The pilot valve responds to system pressure changes, ensuring the main valve opens only when the set pressure is exceeded.

Pop Action

A pop action in a safety valve refers to the sudden, rapid opening of the valve at its set pressure. When the system pressure reaches the set point, the safety valve opens almost instantly, releasing excess pressure rapidly. An immediate action is critical to the operation of safety valves, as it allows for fast pressure relief in order to prevent system overpressure and potential damage to the equipment. If the valve fails to pop open properly, the system could remain overpressurized, leading to potential failure. Pop action is a characteristic of safety valves typically used for gas services.

Pressure Relief

Pressure relief refers to the mechanism or system designed to prevent excessive pressure buildup in a system, such as a piping system or pressure vessel, to ensure safety and prevent equipment failure, explosions, or other hazards. Pressure relief is a critical aspect of industrial safety. It ensures that systems operate within their designed pressure limits, thereby protecting equipment, preventing accidents, and safeguarding human lives and the environment.

Pressure Relief Valve

PRVs are designed to relieve excess pressure in a controlled manner in order to protect equipment and piping systems. In the event that the pressure within the system rises above the set pressure, it gradually opens (without pop action), releasing fluid and maintaining a safe pressure level. As pressure increases, PRVs open gradually and close gradually as pressure decreases, making them proportional in their operation. PRVs are commonly used in liquid systems (such as water or oil lines) where pressure increases slowly and a gradual pressure release is sufficient to protect the system. From a safety perspective, PRVs form part of a layered defense strategy, working alongside pressure monitoring systems and alarms. Regulatory standards such as API 520 and ASME Boiler and Pressure Vessel Code provide guidelines for PRV sizing, installation, and maintenance. Regular testing and calibration are essential to ensure these valves operate effectively when needed. While PRVs are not designed for extreme emergency relief, their precise operation helps mitigate the risks of escalating pressure fluctuations, protecting both equipment and personnel.

Pressure Safety Valve

A PSV is also designed to release excess pressure, but it is typically used in emergency situations in high-pressure systems where a quick response is necessary. As soon as the set pressure is reached, the pop action valves open fully and immediately, which is known as the pop action valve. In safety-critical applications, PSVs are governed by strict regulatory standards, such as ASME Section VIII, which mandates their use for pressure vessels operating above specific limits. Proper calibration of the set pressure and regular maintenance are crucial to ensure reliable performance.

Pressure Surge

1. Definition

Pressure surge and water hammering are critical safety considerations in valve systems, as they can lead to severe operational and structural challenges. A *pressure surge* occurs when the velocity of a fluid changes rapidly, causing a sudden increase in pressure. This phenomenon often results from quick valve closures, abrupt pump start-ups or shutdowns, or flow obstructions.

2. Mitigation

To manage the risks associated with pressure surge, proper valve selection and system design are essential. Slow-closing valves, such as non-slam check valves or check valves with dashpot or anti-surge valves, help prevent sudden flow interruptions, reducing the intensity of pressure surges.

Pressure–Temperature Rating

It can be called "pressure and temperature rating" indicating the pressure that a valve is able to withstand often decreases as the temperature increases. A valve has a pressure–temperature rating that indicates the maximum pressure that can

TABLE 10.1
Carbon Steel Pressure–Temperature Rating

Temperature (°C)	Maximum Allowable Pressure (bar)
−29 to 38	40
100	35
200	30
300	25
400	20
425	18

be applied at a given temperature, including the design temperature. Let us take a carbon steel valve with a pressure–temperature rating as per ASME standards in Table 10.1.

When the temperature ranges from −29°C to 38°C, the valve can withstand a maximum pressure of 40 bar. It is important to note that at 300°C, the maximum allowable pressure drops to 25 bar as a result of the effects of high temperatures on the material strength of the valve. At 425°C, the maximum pressure rating of the valve is reduced further to 18 bar due to the effects of high temperatures on the valve material strength.

Pressure–Temperature Relationship

Operating temperature often affects the pressure the valve can withstand. Higher temperatures generally lower the valve's pressure rating, so it is essential to know the operating temperature when determining pressure compatibility.

Pressure Test

In valves, a pressure test is performed to determine whether the valve can withstand a specified internal pressure without leaking. Valve integrity, durability, and the ability to maintain a proper seal under operating conditions must be tested under pressure. Chapter 6 provides more information about valve tests performed by valve manufacturers.

Probability of Failure on Demand

PFD refers to the likelihood that a safety system or component, such as a valve, will malfunction when it is needed. PFD, specifically in the context of SIS, indicates the likelihood that a system will fail to perform its safety function when called upon in potentially hazardous situations. For example, if an ESDV has a PFD of 0.01, this means that there is a 1% chance that it will not close when needed in an emergency. A component's PFD is an important factor in determining its SIL. There is a specific range of acceptable PFD values associated with each SIL level:

- *SIL 1:* PFD of 0.1 to 0.01 (10% to 1% chance of failure on demand)
- *SIL 2:* PFD of 0.01 to 0.001 (1% to 0.1% chance of failure on demand)
- *SIL 3:* PFD of 0.001 to 0.0001 (0.1% to 0.01% chance of failure on demand)
- *SIL 4:* PFD of 0.0001 to 0.00001 (0.01% to 0.001% chance of failure on demand)

The PFD of a safety component like a valve is influenced by its testing frequency. Regular testing and maintenance reduce the PFD. In low-demand systems (where safety functions are rarely needed), PFD is generally calculated as follows:

$$\text{PFD} = \frac{\lambda}{T}$$

where

- λ is the failure rate of the component (failures per hour),
- T is the test interval (time between tests).

When systems are more complex, with diagnostics and redundancy, the PFD calculation can become more complex:

$$\text{PFD} = \frac{\lambda \times (1 - \text{DC}) \times T}{2} + \frac{\lambda \times \text{MTTR}}{2}$$

where
DC: diagnosis coverage;
MTTR: mean time to repair.

There is another formula for PFD calculation. In fact, the *PFD* of a safety valve is influenced by two key factors—*MTTF* and *MTTR*. A longer MTTR means the system is exposed to failure for a greater duration, increasing the overall PFD:

$$\text{PFD} = \frac{\text{MTTR}}{\text{MTTF}}$$

Example: The failure rate of a valve is 1 out of $100{,}000\,\text{h}\,(1 \times 10^{-5}$ failure per hour). The test interval for detecting the failure is one year (8,760 h). There is a 70% DC and an 8-h mean time to repair. Determine the SIL by calculating the PFD.

Answer: $\text{PFD} = \frac{1 \times 10^{-5} \times (1 - 0.7) \times 8{,}760}{2} + \frac{1 \times 10^{-5} \times 8}{2} = 13.18$

In this example, the calculated PFD of 13.18 falls outside typical SIL thresholds, indicating that the valve does not comply with the SIL requirements. Therefore, it would be necessary to make adjustments, such as increasing testing frequency, adding redundancy, or reducing mean time to repair in order to improve safety reliability.

Probability of Failure Per Hour (PFH)

This is often used in calculating SIL for continuous operations as it is the likelihood of a system failing in a continuous or high-demand mode. In contrast to PFD, which is used for low-demand systems with occasional safety demands, PFH is relevant for systems with steady operation or frequent safety demands. In SIL assessment, specific PFH ranges correspond to each SIL level, with lower PFH values required for higher SILs. The lower the PFH, the more reliable the system, and the higher the SIL level it can achieve. Here is a general breakdown:

- *SIL 1:* PFH between 10^{-5} and 10^{-6} failures per hour
- *SIL 2:* PFH between 10^{-6} and 10^{-7} failures per hour
- *SIL 3:* PFH between 10^{-7} and 10^{-8} failures per hour
- *SIL 4:* PFH between 10^{-8} and 10^{-9} failures per hour

For a single component with a failure rate λ if diagnostics and repair time are considered, PFH is calculated as follows:

$$\text{PFH} = \lambda \times (1 - \text{DC}) \times \text{MTTR}$$

Proof Test

Proof tests are scheduled, comprehensive tests used to verify that a valve in a SIS is capable of performing its safety function. An important objective of the proof test is to detect any hidden failures—especially dangerous undetected failures—that could prevent the valve from functioning correctly in an emergency. In order to maintain the reliability and integrity of safety-critical valves over time, proof testing is essential.

Proof Test Coverage

The percentage of dangerous failures that can be detected by a proof test. The higher the proof test coverage, the greater the level of reliability and the more likely it is to maintain SIL requirements.

Proof Test Interval

PTI is the scheduled period between proof tests of a safety-critical system or component, particularly in SIS.

R

Redundancy

1. Definition and Application

In terms of SIL, redundancy refers to the practice of incorporating multiple, independent safety components or systems in order to increase reliability and reduce the likelihood of dangerous failures. When it comes to SIS, redundancy is an important concept, since it ensures that even if one component fails, other components can carry out the safety function. In redundant systems, one or more components may fail without compromising the safety of the system. As a result of this capability, known as HFT, higher SIL levels can be achieved. There are different types of redundancy such as 1oo2 (1 out of 2) and 2oo3 (2 out of 3). *1oo2 (1 out of 2)* means two identical components are used, but only one needs to operate correctly for the safety function to succeed. If one fails, the other can still perform the function. *2oo3 (2 out of 3)* means three components are used, and at least two must operate correctly to fulfill the safety requirement.

2. Case Study: Impact of Redundancy on SIL and PFD

Background

A chemical processing plant operates a high-pressure reactor that requires overpressure protection to prevent catastrophic failure. The safety system includes a *PSV* and an *ESD system*. The plant evaluates the existing system to meet a required *SIL 2* level (PFD range of 1×10^{-2} to 1×10^{-3}) as per IEC 61508/61511 standards. The current configuration includes a single PSV, which does not meet the SIL 2 requirement due to its high PFD.

To improve system reliability, the plant adds redundancy by installing a second PSV in a *1oo2 (one out of two)* voting configuration. The engineering team conducts a SIL verification to assess the impact of redundancy on the overall system's PFD and SIL.

System Configurations and Analysis

Original Configuration: Single PSV:

- Device PFD: 0.02 (within SIL 1 range).
- The single PSV provides basic overpressure protection but lacks sufficient reliability for high-demand scenarios.
- Failure of this single PSV during an overpressure event would lead to a process failure, as no backup exists.

Improved Configuration: Redundant PSVs in 1oo2:

- *Voting Logic:* The system activates if at least one of the two PSVs operates correctly.
- *Calculation of Combined PFD:*

- For two independent PSVs, each with PFD = 0.02:
  - Combined PFD for 1oo2: PFD1oo2=PFD1 × PFD2 = PFD1 × PFD2
  - PFD 1oo2 = 0.02 × 0.02 = 0.0004 (within SIL 2 range).
- The redundancy reduces the overall PFD to a level that meets the SIL 2 requirement, significantly improving system reliability.

Conclusion: The original system (SIL 1) could not ensure reliable protection for high-demand scenarios. Adding redundancy upgraded the system to *SIL 2*, providing higher confidence in overpressure protection. Redundancy in the 1oo2 configuration reduced the PFD by 98% (from 0.02 to 0.0004). This decrease greatly reduces the likelihood of failure during an overpressure event.

Relieving Pressure

As a result of overpressure conditions, the relieving pressure is the maximum pressure for which the safety valve is designed to function effectively. It indicates the pressure at which the valve is fully open and discharging enough fluid to prevent a further rise in the system's pressure. Relieving pressure is equal to set pressure plus overpressure.

$$\text{Relieving pressure} = \text{Set pressure} + \text{Overpressure}$$

Example: What is the value of relieving pressure if the set pressure is 10 bar and the overpressure is 10% of the set pressure?

$$\text{Relieving pressure} = \text{Set pressure} + \text{Overpressure} = 10 + 10 \times 10\% = 11\,\text{bar}$$

Reseat

Reseat or reseat pressure is the system pressure at which a safety valve fully closes after it has opened to relieve excess pressure. It is a pressure below the set pressure at which the valve closes tightly after discharging the overpressure. Standard safety valves typically reseat at *90–95%* of the set pressure. Reseat pressure is a critical parameter in safety valve operation, ensuring the valve closes reliably after relieving excess pressure. Proper reseating is vital for system safety, as a valve that fails to reseat may lead to continuous venting, fluid loss, or instability in downstream systems.

Reseat Pressure

"Reseat pressure" is sometimes referred to as "reset pressure," but they mean the same thing in most contexts.

Risk Reduction Factor (RRF)

RRF measures how effectively a SIF reduces risk. In particular, it quantifies the degree to which a safety function, such as an ESDV, and reduces the probability of a hazardous event occurring. When assessing SIL assessments, the RRF is particularly relevant, as higher SIL levels correspond to higher RRFs, which means a greater reduction in risk. The RRF is calculated as the inverse of the PFD for systems operating in low demand mode or the PFH for systems operating in continuous mode.

$$\text{RRF} = \frac{1}{\text{PFD}}\left(\text{Low} - \text{demand mode}\right)\text{RRF} = \frac{1}{\text{PFH}}\left(\text{High demand mode}\right)$$

S

Safe Failure (λ_S)

Safe failures are those that do not result in hazardous conditions and instead cause the system to revert to or remain in a safe state. An example of a safe failure would be the failure of an emergency shutoff valve. If the valve fails, it automatically closes, thus stopping the flow of potentially hazardous fluids or gases.

Safe Failure Fraction (SFF)

In reliability and safety engineering, SFF is a key metric used to evaluate the reliability of safety-critical components, such as those in SIS. An indicator of safety measures the proportion of failures (total failures) that are either "safe" (don't result in a hazardous condition) or "detected" by diagnostics, allowing preventive measures to be taken. The formula for SFF is given as follows:

$$\text{SFF} = \frac{\text{Safe failures} + \text{detected dangerous failures}}{\text{Total failures}} = \left(\frac{\lambda_S + \lambda_{DD}}{\lambda_S + \lambda_D}\right) = \left(\frac{\lambda_{TOTAL} - \lambda_{DU}}{\lambda_{TOTAL}}\right)$$

SFF is critical for determining SIL requirements and HFT, since it indicates how effectively a system will be able to avoid or mitigate dangerous failures. Having a high SFF value indicates that most failures are either safe or detected, reducing the likelihood of undetected dangerous failures resulting in hazardous events. A SFF of 90% or higher is considered high.

Example: Suppose a SIS system has 100 failures. There are 60 safe failures, 30 dangerous detected failures, and 10 dangerous nondetectable failures. Determine the SFF.

$$\text{SFF} = \frac{\text{Safe failures} + \text{Detected dangerous failures}}{\text{Total failures}} = \frac{60 + 30}{100} = 90\%$$

Safety Analysis Report

SAR provides a detailed assessment of the safety of a facility, system, or process. It is typically prepared for facilities dealing with hazardous materials or high-risk operations, such as nuclear power plants, chemical processing facilities, or industrial systems, in order to demonstrate compliance with safety standards and regulatory requirements. In the SAR, SIL analyses are provided for the safety-critical valves.

Safety Critical Valves

1. Definition

Safety-critical valves play an important role in protecting personnel, equipment, and the environment by preventing hazardous events such as leaks, explosions, and uncontrolled flow. Designed to perform specific safety functions under normal and emergency conditions, these valves are potentially catastrophic if they fail. Among the safety critical valves are ESDVs, PSVs, control valves, and blowdown valves. SIL analysis should be applied to these valves. In a SAR, the SIL analysis for these valves is documented.

2. Case Study: Safety Critical Valves in Offshore Oil and Gas Production

Background

In offshore oil and gas production, safety-critical valves are essential for ensuring the safe operation of high-pressure systems that handle hazardous fluids such as hydrocarbons. A major offshore platform in the North Sea experienced an incident involving the malfunction of two *safety critical valves* in the gas compression system. This case highlights the importance of proper selection, testing, and maintenance of these valves.

Incident Description

The platform's gas compression system relied on a set of *PSVs* and ESDVs to manage overpressure and isolate the system during emergencies. During a scheduled start-up after routine maintenance, the gas compressor experienced a sudden surge in pressure due to sudden start of the fluid in the piping. The PSV installed to protect the system failed to open at its set pressure of *250 bar*, allowing pressure to build up to *270 bar*, close to the vessel's MAWP of *280 bar*.

Simultaneously, the ESDV intended to isolate the compressor failed to close fully, allowing a significant quantity of high-pressure gas to escape into the flare header, leading to an unexpected flare system overload. These combined failures resulted in a process shutdown and posed a risk of catastrophic failure.

Investigation Findings

The investigation revealed that the primary cause of the incident was the failure of two safety-critical valves and a lack of redundancy in the system. The PSV failed to open at its set pressure of 250 bar due to a sticking valve disk caused by hydrocarbon deposits that had solidified over time. PSV failure was a result of insufficient testing and maintenance exceeding the recommended proof test interval of two years. Simultaneously, the ESDV failed to close fully because of degraded seals exposed to corrosive substances in the gas stream, which had not been replaced despite being beyond their recommended service life. Additionally, the system's reliance on a single PSV and ESDV without redundancy further compounded the risk, as no backup was available to mitigate the failures. These issues highlighted gaps in the platform's maintenance program and safety system design.

Consequences

The incident resulted in a significant operational impact, including a three-day unplanned shutdown of the offshore platform, leading to substantial production losses. Financially, the repairs, valve replacements, and downtime costs amounted to approximately *$2.5 million*, highlighting the economic repercussions of safety system failures. Although no injuries occurred, the event underscored the potential for catastrophic failure, as the system pressure had approached the vessel's MAWP of 280 bar. Furthermore, the flare system overload raised concerns about the platform's ability to handle emergency venting scenarios, demonstrating the broader safety risks posed by inadequate safety valve performance.

Lesson Learnt

This incident emphasizes the importance of regular testing, inspection, and maintenance of safety-critical valves to ensure their reliability, as recommended by industry standards like API 576 and API 598. Material selection and seal integrity must be carefully considered for valves exposed to corrosive fluids, with timely replacement of aging components to prevent failures. Additionally, incorporating redundancy into the safety system, such as dual PSVs in a *1oo2 configuration*, could significantly enhance reliability and reduce risks. Finally, a comprehensive risk assessment should evaluate worst-case scenarios, addressing potential single-point failures and ensuring maintenance schedules are optimized to minimize vulnerabilities in safety-critical equipment.

Safety Instrumented Systems

SIS provides a critical layer of protection in industrial processes by automatically executing predefined safety functions in order to detect and prevent dangerous situations. In high-risk industries such as oil and gas, petrochemical, chemical processing, and power generation, these systems are commonly used to protect people, the environment, and assets from hazardous events. A SIS is governed by standards such as IEC 61511 and IEC 61508 and is part of the broader functional safety framework. The main components of SIS are the sensor, the logic solver, and the final control element. The most common sensors found in a SIS are pressure sensors, temperature sensors, and flow meters. It is the logic solver (often a programmable logic controller, or PLC) which receives the signals from the sensors, analyzes them, and determines if a safety action is necessary. In the event that an unsafe condition is detected, the logic solver will send a signal to activate the final control elements (e.g., actuated valve). These components act on the process as a result of the signals received from the logic solver. A safety shutdown valve (e.g., an ESDV) and a relief valve are common final control elements.

Safety Integrity Level

SIL is a measure of the safety and reliability of systems and components, including valves, used in critical processes within industries such as oil and gas, petrochemical, and manufacturing. In hazardous situations, it represents the probability that a system or component (such as a valve) will perform its intended safety function without failure. Functional safety standards, primarily the IEC 61508 and IEC 61511 standards, establish guidelines for designing and evaluating safety-critical systems. Each SIL level represents a quantitative measure of risk reduction and the likelihood of a system failure, ranging from SIL 1 (lowest) to SIL 4 (highest). For valves, SIL is associated with a risk reduction factor (RRF). Accordingly, a SIL 1 valve offers a lower risk reduction than a SIL 3 valve. SIL relates to the PFD, which represents the probability that a valve will fail when required to perform its function. When the PFD is lower, the valve is more reliable, resulting in a higher SIL level. SIL-rated valves must meet specific performance criteria, such as reliability, DC, and failure rates, to ensure they perform as intended in critical conditions. Redundancy and fail-safe mechanisms may be required for higher SIL levels. SIL is applicable to safety-critical valves, such as ESDVs and SIS.

Safety Shutoff

1. Definition

Safety shutoff refers to a system or mechanism designed to automatically stop the flow of a fluid (liquid or gas) in a piping system or process when unsafe conditions are detected. Safety shutoff devices are critical for preventing accidents, protecting equipment, and ensuring the safety of personnel and the environment.

2. Key Features

Safety shutoff systems are designed to automatically halt fluid flow in a process when unsafe conditions are detected, ensuring rapid response and minimizing risks. Key features include automatic operation triggered by sensors monitoring parameters like pressure, temperature, or flow, and fail-safe designs that move the valve to a safe position (open or closed) during power loss or malfunctions. These systems could be equipped with actuators for quick action and are tailored to specific applications, such as handling hazardous fluids or extreme high-pressure and high-temperature conditions. Their fast response time and robust construction make them essential for protecting lives, equipment, and the environment while ensuring regulatory compliance.

Safety Valve

Safety valves are used to release excessive pressure from pressurized equipment during operation. Pressure is automatically released when it exceeds a pressure limit, preventing potential hazards, such as explosions or equipment failure. As a matter of fact, a safety valve is a normally closed valve that opens to relieve excessive pressure under abnormal conditions such as high pressure and/or high temperature. Protecting assets, personnel, and the environment is the main purpose of using this valve. As it pertains to asset protection, it is important to note that overpressure scenarios in piping and equipment may result in fire and explosion that can cause significant damage to the facility and components. The purpose of safety valves is to prevent such damages, to prolong the life of equipment and facilities, and to reduce their repair costs by opening when the pressure reaches a certain degree to release the overpressure. Furthermore, safety valves can contribute to the safety and health of personnel by preventing fires and explosions. The final benefit of safety valves is that by controlling the release of gases and liquids, they prevent harmful substances from escaping into the environment, minimizing pollution and environmental damage. It is common to find safety valves in a variety of applications, such as boilers, pressure vessels, and industrial equipment. The role they play in ensuring the safety of the public and compliance with regulatory requirements is critical.

Sensor

They detect abnormal conditions (e.g., high pressure, high temperature, or toxic gas leaks). SIS sensors include pressure sensors, temperature sensors, and flow meters. Process control systems rely heavily on sensors.

Set Pressure

Essentially, safety valve set pressure refers to the pressure at which the safety valve opens and relieves excess pressure. Setting the set pressure ensures that the valve will remain open during overpressure situations so that equipment failure and safety hazards can be avoided.

Simmer

1. Definition

Simmer refers to the slight leakage or partial opening of a safety valve as system pressure approaches the set pressure. This indicates that the valve is approaching the point where it will "pop" open fully. A simmer usually occurs at a pressure slightly below the valve's set pressure, usually between 90% and 95% of the set pressure While simmering is a normal behavior in safety valves, excessive or prolonged simmering can compromise system safety by causing fluid loss, pressure instability, or confusion for operators. In critical applications, improper simmering may indicate valve malfunction or incorrect calibration, potentially delaying proper valve activation and opening during an overpressure event. To mitigate risks associated with simmering, industry standards such as ASME BPVC and API 526/527 define acceptable simmer ranges and behavior for safety valves. Regular testing and maintenance are essential to address simmer-related issues and maintain the integrity of safety-critical systems.

2. Case Study

Example (Fluid Loss): In a natural gas pipeline, a safety valve with a set pressure of *100 bar* begins to simmer at *95 bar.* If the simmering lasts for several hours due to fluctuating pressure, a significant volume of gas may be lost, reducing efficiency and increasing operational costs.

Example (Pressure Instability): In a steam system, a safety valve set at *10 bar* begins to simmer at *9.5 bar* but does not transition to full opening. The small, repeated leaks lower the system pressure intermittently, causing steam supply to downstream turbines to fluctuate. This can reduce turbine efficiency and lead to uneven power generation, disrupting operations.

Example (Operator Confusion): On a refinery unit, a safety valve set at *150 psi* starts to simmer at *145 psi.* An operator, hearing the slight hissing sound from the valve, believes there is something wrong in the system and shuts down the process unnecessarily. Operators are often trained to take conservative actions when a potential safety issue arises. Hearing the valve hiss sound may have triggered the operator for an overreaction to avoid a worst-case scenario such as fire and explosion. However, the system was still operating within safe limits. This leads to production downtime and financial loss, even though the valve had not reached its set pressure or required a shutdown.

Example (Potential Malfunction): In a chemical reactor, a safety valve set to *200 psi* begins to simmer at *170 psi*, far below its designed simmer range of *180–195 psi.* Upon inspection, engineers discover a weakened spring inside the valve, reducing its set pressure accuracy. This malfunction compromises system safety, as the valve may fail to respond correctly during an actual overpressure event.

Smart Valve

A smart valve is an advanced type of valve equipped with sensors, actuators, and communication technology. Smart valve allows real-time monitoring, control, and diagnostics. Valves equipped with ValveWatch systems can be considered smart valves because they incorporate sensors, advanced monitoring, diagnostics, and data communication features. Smart valves play a critical role in enhancing safety by combining real-time monitoring, diagnostics, and automated response capabilities. Integrated with SIS, smart valves provide precise control and rapid response in safety-critical applications, such as ESD and overpressure protection. Their ability to diagnose issues, such as actuator wear or misalignment, and report these conditions to the control room in real time ensures timely maintenance, reducing the risk of failure during emergencies. For instance, a smart ESDV can detect and report degraded seal performance, ensuring corrective actions are taken before the valve's reliability is compromised.

Spring-Loaded Safety Valve

Spring-loaded safety valves are essential components of overpressure protection systems, designed to protect pressure vessels and equipment from exceeding safe pressure limits. These valves rely on a spring mechanism to remain closed during normal operation and open automatically when system pressure exceeds the set pressure. By rapidly venting excess pressure, spring-loaded valves prevent catastrophic failures such as ruptures, explosions, or fires. A key feature of spring-loaded safety valves is their simplicity and reliability, making them widely applicable across industries, including power generation, oil and gas, and chemical processing. These valves are designed to comply with industry standards such as ASME BPVC and API 520, ensuring consistent performance and safety.

Static Leak Measurement

1. Definition

During a static leak measurement, the leakage rate of a valve or other equipment is evaluated while it is stationary or nonoperating. Static leak measurement is performed when the valve's stem is stationary, and the internal components (e.g., the plug, seat, or disk) are not moving. As opposed to dynamic leak measurement, which tests the valve under operating conditions (e.g., opening, closing, cycling), static leak measurement checks the valve for leakage when it is fixed in a closed position or another stable state.

2. Case Study: Static and Dynamic Leakage Measurement for Valves in a Chemical Processing Plant

Introduction

In a chemical processing plant, control valves play a vital role in regulating the flow of hazardous chemicals under varying pressures and temperatures. Any leakage from these valves can compromise operational efficiency, pose safety hazards, and violate environmental regulations. This case study explores the performance of a globe control valve tested for leakage under *static conditions* (valve closed) and *dynamic conditions* (valve operating under flow and pressure). The aim was to assess the valve's performance, identify potential issues, and recommend measures for improvement.

Objective

The primary objective of this study was to evaluate the valve's sealing performance in both static and dynamic scenarios. The testing aimed to identify potential weaknesses in the valve seat and packing, ensuring compliance with industry standards like *API 598* for static leakage and *ANSI/FCI 70-2* for dynamic leakage.

Test Setup

The tested valve was a *globe control valve*, commonly used in high-pressure chemical applications. The testing setup included:

- *Fluid:* Water.
- *Pressure Range:* 10–50 bar.
- *Temperature:* 25°C (ambient conditions). Leakage measurements were performed using calibrated flowmeters and sensors, ensuring high accuracy. Static and dynamic leakage were evaluated against established industry standards to quantify performance.

Static Leakage Test

The valve was fully closed, and upstream pressure was applied while monitoring for leakage downstream. The test measured the ability of the valve seat to prevent flow under no-motion conditions. Results revealed minor leakage of *0.5 ml/min*, exceeding the *Class IV leakage limit* specified by *ANSI/FCI 70-2*. Upon inspection, surface irregularities caused by wear and corrosion were found on the valve seat, reducing its sealing performance.

Dynamic Leakage Test

Dynamic testing assessed the valve's performance under flow conditions with varying pressures and flow rates. Leakage was measured through the valve seat (internal leakage) and around the valve stem packing (external leakage). The dynamic leakage rate reached *3 ml/min* at 50 bar, far exceeding acceptable limits. Investigation revealed that the packing material around the valve stem was worn, failing to maintain a proper seal under fluctuating operating conditions.

Consequences

The findings indicated that while static leakage was relatively minor, dynamic leakage presented a more significant issue. The increased leakage under dynamic conditions could lead to the following:

- Process inefficiencies due to lost fluids
- Potential safety hazards if hazardous chemicals were to escape
- Environmental noncompliance and increased operational costs due to repairs and downtime

Recommendations

- Replace the worn valve seat with a corrosion-resistant material to eliminate static leakage.

- Upgrade the valve packing material to a high-performance grade to improve dynamic sealing capabilities.
- Establish a more frequent and rigorous testing schedule that includes both static and dynamic conditions.

Stroke Testing

1. Introduction

Stroke testing refers to the process of verifying the operational functionality of a valve by moving it from one position to another within its travel range. This testing is critical for ensuring that valves operate reliably in process systems, particularly for safety-critical applications. This test is done during the valve operation. Stroke testing can be categorized into *partial stroke testing* and *full stroke testing*, depending on the extent of valve movement.

2. Types and Comparison

Partial stroke testing involves moving the valve a small percentage of its full travel range, typically 10–25%, without completely opening or closing it. This test is designed to check the valve's responsiveness and detect potential issues, such as sticking or excessive friction, that may impede its operation. Partial stroke testing is commonly used in safety-critical systems, like ESD valves, to ensure they can operate when needed without disrupting the process. It is nonintrusive, can be performed frequently, and helps identify mechanical or operational issues early. However, it does not provide assurance of full valve functionality or sealing capability.

Full stroke testing involves moving the valve through its entire travel range, from fully open to fully closed (or vice versa). This test confirms the valve's complete functionality, including its ability to seal effectively at both ends of its travel range. Full stroke testing is typically conducted periodically to ensure compliance with safety and regulatory requirements. While it provides comprehensive validation of valve performance, it often requires process shutdown or bypassing, making it more time-consuming than partial stroke testing. Full stroke testing typically causes more production disturbance compared to partial stroke testing. Frequent full stroke tests may also contribute to wear and tear on the valve. A comparison between partial stroke testing and full stroke testing methods is summarized in Table 10.2.

V

Valve Integrity

1. Definition

Valve integrity refers to the reliability, functionality, and durability of a valve throughout its operational life, ensuring that it performs its intended function without failure under specified conditions. Maintaining valve integrity is critical for safety, efficiency, and the protection of equipment and the environment.

TABLE 10.2
Comparison between Partial Stroke and Full Stroke Testing

Aspect	Partial Stroke Testing (PST)	Full Stroke Testing (FST)
Movement	Partial (10–25% of full travel)	Full (100% open to 100% closed)
Impact on process	Minimal	Significant (requires process interruption)
Frequency	Frequent (e.g., monthly)	Less frequent (e.g., annually)
Scope	Verifies functionality partially	Verifies full functionality and sealing
Testing time	Short	Longer

2. Key Features

Valve integrity ensures a valve's reliable performance, leak-tightness, and structural durability throughout its operational life. It involves maintaining the valve's ability to withstand specified pressures, temperatures, and environmental conditions while preventing internal and external leaks. Key aspects include using corrosion-resistant and compatible materials, adhering to industry standards, ensuring proper functionality (opening, closing, or throttling), and maintaining actuator precision in automated systems. Regular inspections, testing, and maintenance, such as lubrication and seal replacement, are essential for long-term integrity, as is compliance with fire-safe and pressure–temperature ratings to enhance safety and reliability in critical applications.

Valve Lock

See the "Locking Devices" section in this chapter.

ValveWatch

1. Definition

ValveWatch is a sophisticated monitoring system designed to provide real-time diagnostic data on the condition and performance of critical valves in industrial processes. As part of a safety framework, ValveWatch plays a vital role in ensuring the reliability and functionality of safety-critical valves, such as ESDVs, PSVs, and PRVs. By continuously monitoring these valves, ValveWatch helps prevent failures that could compromise safety, ensuring the system is always prepared to respond during emergency scenarios.

2. Importance

One of ValveWatch's key contributions to safety is its ability to detect issues such as valve sticking, seat leakage, actuator malfunctions, or improper valve position. This real-time diagnostic capability allows operators to identify and address potential problems before they escalate into safety-critical incidents. For example, in an oil and gas processing plant, ValveWatch can alert operators if an ESDV fails to fully close during a shutdown, enabling immediate corrective action to prevent hazardous conditions. Additionally, its integration with SIS ensures seamless communication and validation of valve performance as part of a SIF, providing an extra layer of assurance in high-risk operations.

ValveWatch also enhances condition-based and predictive maintenances, ensuring valves remain operational and compliant with safety standards such as IEC 61508/61511. By providing a clear view of valve performance trends, the system helps extend valve life, reduce unplanned downtime, and maintain the integrity of overpressure protection systems. This proactive approach to monitoring and maintenance not only improves plant safety but also minimizes the costs associated with reactive repairs and shutdowns. Overall, ValveWatch is a critical component of modern safety systems, offering the reliability and insights necessary to manage risks effectively in complex industrial environments.

3. Case Study

Background

An oil refinery processing 200,000 barrels per day faced recurring reliability issues with its ESDVs and PSVs, resulting in unplanned downtime, safety risks, and regulatory compliance challenges. Failures, including valve sticking, actuator malfunctions, and improper closure, frequently disrupted operations and posed significant hazards in handling high-pressure hydrocarbons. To address these issues, the refinery implemented *ValveWatch*, a real-time valve monitoring system integrated into its SIS. ValveWatch continuously tracked valve position, actuator performance, and potential leakage while providing predictive diagnostics for early detection of wear or degradation.

Challenges before ValveWatch

Before implementing ValveWatch, the refinery faced several challenges. Unplanned downtime caused significant production losses, as unreliable valves led to unexpected shutdowns. Safety risks were heightened due to valves failing to respond during shutdowns or emergencies, especially when handling flammable hydrocarbons. Maintenance relied heavily on periodic manual inspections, which often failed to detect hidden or intermittent issues, and the plant struggled to meet safety standards like IEC 61508/61511, risking penalties and operational restrictions. These challenges highlighted the need for a proactive and reliable valve-monitoring solution.

Implementation of ValveWatch: Results and Improvements

Over the first year, ValveWatch identified 15 cases of valve sticking and 10 instances of actuator performance issues, allowing targeted maintenance and avoiding costly unplanned shutdowns. This proactive approach reduced downtime by 40%, saving approximately *$4 million annually*, and improved compliance with safety standards. For example, during an ESD test, ValveWatch detected and flagged an actuator misalignment on an ESDV, enabling corrections before a critical failure could occur. ValveWatch's real-time diagnostics ensured that valves operated reliably during emergencies, enhancing both safety and operational efficiency.

ValveWatch's integration with the refinery's SIS also improved predictive maintenance and regulatory compliance. By continuously monitoring valve performance, the system reduced dependency on routine manual inspections and provided operators with early warnings of potential issues. This approach minimized emergency repairs, extended valve life, and ensured that all safety-critical valves performed as required under dynamic operating conditions. The clear audit trail and data provided by ValveWatch also helped the refinery meet regulatory requirements, avoiding penalties and building confidence in its safety systems.

Conclusion

This case study demonstrates how advanced valve monitoring systems like ValveWatch can significantly enhance the reliability and safety of critical valves. By addressing safety risks, reducing downtime, and ensuring compliance with industry standards, ValveWatch provided measurable benefits, underscoring its importance in high-risk industrial environments such as oil refineries.

Variable Backpressure

Variable backpressure is a critical factor in the operation of safety valves, defined as the fluctuating pressure on the downstream side of a valve during or after activation. These fluctuations can significantly impact the valve's performance, altering its effective set pressure, reducing its relieving capacity, or causing operational instability such as chattering. Inadequate management of variable backpressure can compromise the safety system, increasing the risk of overpressure and potential equipment failure.

To address these challenges, safety valves must be carefully selected and designed to handle variable backpressure conditions. Balanced safety valves, equipped with bellows or pistons, are effective in neutralizing backpressure effects, ensuring stable operation. Alternatively, pilot-operated safety valves offer precise control, unaffected by downstream pressure variations. Industry standards such as API 520/521 emphasize the importance of analyzing backpressure during the design phase to ensure proper valve sizing and system safety. By mitigating the effects of variable backpressure, safety systems can maintain their reliability, even in dynamic operating environments.

W

Water Hammering

Water hammering is a specific type of pressure surge that occurs in liquid systems when the flow of liquid is suddenly stopped or changes direction, creating a shockwave that travels through the pipeline. This phenomenon is often caused by rapid valve closures or opens, pump start-ups or shutdowns, or flow obstructions.

Water hammering can result in loud noises, vibrations, and significant pressure spikes, leading to damage to valves, pipes, and other system components. To mitigate water hammering, slow-closing valves, such as non-slam check valves, can be used to reduce sudden flow disruptions.

QUESTIONS AND ANSWERS

1. MTTR is primarily used to measure:
   A. The time between failures of a system
   B. The average time required to repair a failed component or system
   C. The total time a system remains operational
   D. The time taken to design a safety system

   Answer: Option B is the correct answer.

2. How does a shorter MTTR affect the probability of failure on demand (PFD) in a safety instrumented system (SIS)?
   A. Increases PFD by reducing repair times
   B. Has no effect on PFD
   C. Decreases PFD by reducing the time the system remains in a failed state
   D. Causes PFD to become unpredictable

   Answer: Option C is the correct answer.

3. If a critical safety valve in a high-SIL application has a long MTTR, what impact could this have on its PFD and compliance with SIL requirements?
   A. It would lower the PFD and help maintain the SIL.
   B. It would increase the PFD, possibly compromising the SIL requirements.
   C. It would have no impact on PFD or SIL compliance.
   D. It would eliminate the need for diagnostic testing.

   Answer: Option B is the correct answer. A long MTTR for a critical safety valve in a high-SIL application increases the PFD, potentially causing the system to exceed the allowable PFD for its required SIL. This can compromise safety, regulatory compliance, and operational reliability. To mitigate these risks, it is essential to minimize MTTR, enhance valve reliability, and, where necessary, incorporate redundancy or improve system design to maintain compliance with SIL requirements.

4. Which of the following best describes the relationship between DC and undetected dangerous failures?
   A. Higher DC results in more undetected dangerous failures.
   B. DC has no relationship with undetected dangerous failures.
   C. Lower DC increases safe failures.
   D. Higher DC decreases undetected dangerous failures.

   Answer: Option D is the correct answer.

5. Which of the following is a primary purpose of lockout/tagout (LOTO)?
   A. To allow workers to bypass safety procedures for faster maintenance
   B. To prevent the accidental release of hazardous energy during maintenance
   C. To mark equipment as needing replacement
   D. To identify equipment that is under warranty

   Answer: Option B is the correct answer.

6. Which of the following is a key feature of smart valves?
   A. They are exclusively operated manually.
   B. They cannot be integrated with control systems.
   C. They include sensors and communication systems for real-time monitoring.
   D. They are only used in residential plumbing systems.

   Answer: Option C is the correct answer.

7. Which of the following is a key feature of personnel protection in valve safety?
   A. Using ergonomic designs to reduce operator stress and discomfort
   B. Allowing unauthorized operation of critical valves
   C. Disabling all safety mechanisms for faster operation
   D. Replacing mechanical seals with electrical components

   Answer: Option A is the correct answer.

8. What is a critical safety consideration for valves used in cryogenic applications?
   A. Use of materials prone to brittleness at low temperatures
   B. Elimination of extended stem designs
   C. Use of materials resistant to extreme cold and thermal stress
   D. Avoidance of pressure relief mechanisms

   Answer: Option C is the correct answer.

9. Which of the following best describes the purpose of valve locking devices?
   A. To ensure valves can be operated remotely without manual intervention
   B. To secure valves in a specific position and prevent unauthorized or accidental operation
   C. To speed up valve operation during emergencies
   D. To allow continuous operation of valves without maintenance

   Answer: Option B is the correct answer.

10. What is the primary purpose of a high-integrity pressure protection system (HIPPS) and its associated valves?
    A. To regulate the flow rate of fluids in pipelines
    B. To prevent overpressure by isolating the source of pressure before exceeding safe limits
    C. To release excess pressure into the environment during overpressure conditions
    D. To monitor and control the temperature of pressurized systems

    Answer: Option B is the correct answer.

BIBLIOGRAPHY

American National Standards Institute (ANSI)/Fluid Controls Institute (FCI). (2016). ANSI/FCI 70-2: Control Valve Seat Leakage. Fluid Controls Institute: Cleveland, OH.

American Petroleum Institute. (2002). API 526: Flanged Steel Pressure Relief Valves. API Publishing Services: Washington, DC.

American Petroleum Institute. (2003). API 520: Sizing, Selection, and Installation of Pressure-Relieving Devices in Refineries. Part II: Installation. API Publishing Services: Washington, DC.

American Petroleum Institute. (2008). API 520: Sizing, Selection, and Installation of Pressure-Relieving Devices in Refineries. Part I: Sizing and Selection. API Publishing Services: Washington, DC.

American Petroleum Institute. (2014). API 521: Pressure-Relieving and Depressuring Systems. API Publishing Services: Washington, DC.

American Petroleum Institute. (2016). API 598: Valve Inspection and Testing. 10th edition. API Publishing Services: Washington, DC.

American Petroleum Institute. (2019). API 6FA: Fire Test for Valves. 4th edition. API Publishing Services: Washington DC.

American Petroleum Institute. (2021). API 527: Seat Tightness of Pressure Relief Valves. 5th edition. API Publishing Services: Washington, DC.

American Petroleum Institute. (2021). API 607: Fire Test for Quarter-Turn Valves and Valves Equipped with Non-Metallic Seats. 8th edition. API Publishing Services: Washington, DC, USA.

American Petroleum Institute. (2021). API 6A: Specification for Wellhead and Christmas Tree Equipment. 21st edition. API Publishing Services: Washington, DC.

American Petroleum Institute. (2021). API 6D: Specification for Pipeline and Piping Valves. 25th edition. API Publishing Services: Washington DC.

American Society of Mechanical Engineers. (2012). Design and Fabrication of Pressure Vessels. Boiler and Pressure Vessel Code. ASME Section VIII Div. 02. ASME: New York, NY.

American Society of Mechanical Engineers. (2019). ASME Boiler and Pressure Vessel Code, Section VIII: Rules for Construction of Pressure Vessels, Division 1. ASME: New York, NY.

American Society of Mechanical Engineers. (2021). ASME B16.34: Valves—Flanged, Threaded, and Welded End. ASME: New York, NY.

Caputo, R. S. (2021). Valve Monitoring and Predictive Maintenance. 1st edition. CRC Press. Boca Raton, FL.

Crosby Valve Inc. (1997). Crosby Pressure Relief Valve Engineering Handbook. Technical document No. TP-V300.

European Union. (2014). Directive 2014/34/EU: Equipment and Protective Systems Intended for Use in Potentially Explosive Atmospheres (ATEX). GOV.UK: Brussels, Belgium.

Gruhn, P., & Cheddie, H. L. (2006). Safety Instrumented Systems: Design, Analysis, and Justification. 2nd edition. ISA (International Society of Automation): Research Triangle Park, NC.

Hellemans M. (2009). The Safety Relief Valve Handbook: Design and Use of Process Safety Valves to ASME and International Codes and Standards. 1st edition. Butterworth-Heinemann. Oxford, UK.

Hunt, T. (1997). Safety Valve Design. 1st edition. Mechanical Engineering Publications: London, UK.

International Electrotechnical Commission. (2010). IEC 61508: Functional Safety of Electrical/Electronic/Programmable Electronic Safety-Related Systems. IEC: Geneva, Switzerland.

International Electrotechnical Commission. (2016). IEC 61511: Functional Safety: Safety Instrumented Systems for the Process Industry Sector. IEC: Geneva, Switzerland.

International Organization for Standardization. (2015). ISO 5208: Industrial Valves—Pressure Testing of Metallic Valves. 3rd edition. ISO: Geneva, Switzerland.

International Organization for Standardization. (2020). ISO 15156: Petroleum and Natural Gas Industries—Materials for Use in H_2S-Containing Environments in Oil and Gas Production. Parts 1–3. 3rd edition. ISO: Geneva, Switzerland.

International Organization for Standardization. (2022). ISO 10497: Testing of Valves—Fire Type-Testing Requirements. 4th edition. ISO: Geneva, Switzerland.

Knapp, E. D., & Langill, J. (2014). Industrial Automation and Control System Security Principles. 1st edition. Syngress: Waltham, MA.

Macdonald, D. (2003). Practical Industrial Safety, Risk Assessment, and Shutdown Systems for Industry. 1st edition. Newnes: Oxford, UK.

National Association of Corrosion Engineers. (2015). NACE MR0175/ISO 15156: Materials for Use in H_2S-Containing Environments in Oil and Gas Production. NACE: Houston, TX.

Nesbitt, B. (2007). Handbook of Valves and Actuators: Valves Manual International. 1st edition. Elsevier: Oxford, UK.

Smith, D. J. (2010). Functional Safety: A Straightforward Guide to IEC 61508 and IEC 61511. 2nd edition. Butterworth-Heinemann: Oxford, UK.

11 Actuators and Valve Automation

A

Actuator

Actuators are mechanical or electrical devices that convert air, hydraulic oil, or electricity into mechanical motion for opening, closing, or controlling valves. In a variety of industries and environments, actuators are crucial for valve automation and accurate fluid control.

Actuator Alignment

Alignment refers to ensuring that the actuator is properly positioned and aligned with the valve to which it is attached. Integrating valves and actuators correctly is so critical that a standard, API 6DX, was developed to address this issue. As an example, if there is no connection between the actuator stem and the stem of the valve, the actuator force will not be transferred to the valve, and the valve will not move. A second disintegration between valve and actuator could be the difference between the position of the valve and value recorded by the position sensor installed on the actuator. In addition, the coupling between the valve and actuator must be strong so that it can withstand the loads and weight of the actuator.

Actuator Components

Cylinder: The cylinder of an actuator is an important part that contains the piston and acts as the chamber where fluid energy is converted into mechanical motion. By sealing the actuator's internal components, such as the piston and hydraulic fluid or compressed air, it keeps the outside environment out of the actuator's internal components.

Gearbox: The gearbox is an integral component of some types of actuators (e.g., electrical or electromechanical actuators) that transmits torque, speed, and motion from the actuator to the valve. Gearboxes ensure that the valve receives the appropriate amount of torque and speed, as well as torque and force direction, needed for operation. In addition to increasing torque, this component reduces speed for smoother valve movement, and transforms or redirects motion, such as converting rotational motion into linear movement or changing the axis of rotation. As mentioned, gearboxes are primarily used in electromechanical actuators,

DOI: 10.1201/9781003640967-11

which consist of electric motors combined with mechanical parts, such as gears, in order to achieve accurate control and adjust force. When it comes to quarter-turn valves (balls or butterfly valves) or multi-turn valves (gates or globe valves), where high torque or multiple rotations are required, gearboxes are crucial. Using gearboxes to control electromechanical actuators provides many advantages, such as an increase in the efficiency of the actuator's output force or torque amount, as well as direction for more effective valve control. Consequently, they are not typically used in pneumatic or hydraulic actuators, where torque intensification is achieved through different mechanisms, such as fluid pressure or springs.

Housing: An actuator housing is the outer casing that encloses and protects the actuator's internal components.

Motor: Motors in actuators are the components that convert electrical energy into mechanical motion, enabling the actuator to move the valve it is connected to. With electromechanical actuators, the motor provides rotary motion, which may be used directly for rotary valves or converted into linear motion using mechanisms such as lead screws or gearboxes. According to the application and precision requirements, motors may be AC, DC, or servo motors. Electrical and electromechanical actuators use motors to move and control valves, respectively, since both types of actuators rely on electrical energy.

Piston: The piston is an essential component of pneumatic and hydraulic actuators, since it converts pressurized fluid energy into mechanical motion to operate valves. As a result of fluid pressure application, the piston moves inside the cylinder to generate the necessary force to drive the valve stem or another moving component. In gate and globe valves, pistons provide linear motion. However, quarter-turn valves like ball and butterfly valves use components such as scotch-yoke or rack-and-pinion to convert linear motion to rotary motion. To prevent fluid leakage and maintain pressure, the piston is in contact with seals. A piston rod (or stem) transmits the piston's motion to the valve.

Positioner: For more details about *positioner*, please refer to the "Positioner" section in this chapter.

Rack and Pinion: For more details about *rack and piston*, please refer to the "Rack and Pinion Actuators" section in this chapter.

Sealing: In actuators, seals prevent leakage of fluids (e.g., air, gas, hydraulic fluid) flowing inside the cylinder and protect internal components from contamination from external elements such as dust, moisture, and chemicals.

Spring: In pneumatic and hydraulic actuators with spring returns, a spring provides force to close or open the valve connected to the actuator. By incorporating a spring, the actuator will move to the defined fail-safe position (either open or closed) in the event that the primary driving force (compressed air, hydraulic pressure, or electrical power) is lost.

Providing this function is essential for the safety and reliability of industrial applications involving valves and actuators.

Stem/Piston Rod: In most actuators, the actuator stem, also known as the piston rod, transmits force or torque from the actuator to the valve stem in order to facilitate valve operation. Linear actuators can connect the piston rod directly to the valve stem, whereas rotary actuators can use couplings or linkages to connect the piston rod to the valve stem. It should be noted that some actuators, such as scotch–yoke actuators, do not have an actuator stem, but instead use the scotch and yoke mechanism to convert the actuator's linear force into rotary motion for the valve stem. Using this mechanism, the scotch–yoke transmits torque generated by the actuator's power source (e.g., pneumatic or hydraulic pressure) directly to the valve stem without requiring a piston rod. Actuators with stems or piston rods are generally linear actuators that are used to operate multi-turn valves, such as gate valves and globe valves, where the stem of the actuator is critical to transferring linear motion to the valve stem. As a general rule, rotary actuators and scotch–yoke systems do not require an actuator stem. Because piston rods must be able to withstand high amounts of pressure, particularly in hydraulic actuators, due to high hydraulic oil pressure and high cyclic loads, it is important that they be made from strong alloys such as super duplex stainless steels or martensitic stainless steels.

The actuator stem is often called the piston rod in many types of actuators, particularly pneumatic and hydraulic linear actuators. It is possible to use the terms "actuator stem" and "piston rod" interchangeably, but their meanings are not the same. The piston rod is usually the internal component of linear actuators, whereas the actuator stem is the extension that extends from the actuator body and is used to connect to the valve stem. As opposed to the piston rod, the actuator stem is in direct contact with the valve stem and transmits motion to the valve's closure mechanism.

In Figure 11.1a, a hydraulic fluid enters a piston and cylinder in a linear actuator (see actuator mechanisms for more information) and pushes the piston upward. During this process, the hydraulic energy is converted to mechanical motion as the piston moves inside the cylinder. The piston's motion is transmitted to the piston rod (or actuator stem), which is in turn responsible for driving the valve stem. In Figure 11.1b, hydraulic fluid enters from the top and moves the piston downward. As seen in this picture, a double-acting actuator uses hydraulic oil to drive both sides of the piston in order to facilitate bidirectional movement of the piston and piston rod.

Torque Arm: An actuator torque arm is a mechanical component designed to transfer torque (rotational force) generated by the actuator to the valve or device it controls. In valve actuators, torque arms are sometimes used

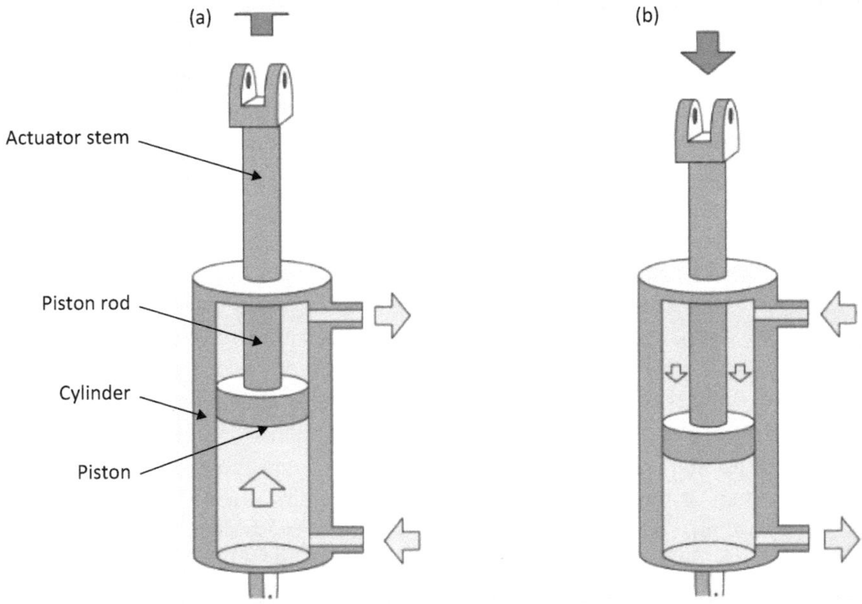

FIGURE 11.1 Illustration of a double-acting hydraulic actuator (a) hydraulic enters the cylinder from bottom and pushes the piston upward. Therefore, hydraulic fluid exits the upper chamber. (b) hydraulic enters the cylinder from top and pushes the piston downward. Therefore, hydraulic fluid exits the lower chamber. (With permission: Shutterstock.)

in *linkage mechanisms* to transmit force to the valve stem. It is important to note that the torque arm functions in the same way as the actuator stem in that it drives the valve stem. It is important that this component provides smooth motion transmission to the valve stem. In rotary actuators, torque arms are often used to operate large valves with quarter-turn stem rotation (e.g., butterfly valves, ball valves, or plug valves) and/or high-pressure classes (e.g., ASME pressure class 600 and above).

Actuator Cycle

During an actuator cycle, a valve is moved from its initial position (e.g., fully closed) to the opposite position (e.g., fully open) and then returned to its original position.

Actuator Housing

In an actuator, an outer casing or structure is used to enclose the internal components of the actuator.

Actuator Mechanisms

There are two rotary actuator mechanisms—*rotary* and *linear*.

1. Rotary Actuator Mechanism

It is common to use these mechanisms to operate quarter-turn valves (e.g., ball, butterfly, and plug valves) since they produce a rotational motion.

Common Rotary Mechanisms

- *Rack-and-Pinion:* Utilizes rack and pinion components to convert linear motion of a piston into rotary motion.
- *Scotch and Yoke:* Utilizes a yoke to convert linear motion into rotary motion, resulting in high torque being produced by the actuator during valve operation.
- *Direct Drive (Electrical Motor):* It is powered by a motor in the actuator and provides rotary motion directly to the valve stem without the need for any mechanical interface.
- *Gearbox-Based Rotary Mechanisms (Electromechanical Mechanism):* It transmits torque efficiently by using gears such as worm or bevel gears in conjunction with a motor.

2. Linear Actuator Mechanism

These mechanisms produce linear motion (movement), which is used for valves with linear stem and closure member movement such as gate and globe valves.

Common Linear Actuators

- *Piston and Cylinder:* It is common for pneumatic or hydraulic actuators to move the piston through fluid pressure to produce linear motion.
- *Lead Screw/Ball Screw:* Provides accurate linear motion to electromechanical actuators by converting rotary motion into linear motion.
- *Rack Drive:* A rack component used for some linear actuators to create linear motion.

Actuator Motions and Output Types

1. Acceleration

The acceleration of an actuator refers to the rate at which the output velocity changes over time during the movement of the actuator. For applications that require good control over actuator movement, this parameter is considered to be an important parameter. It should be noted that excessive acceleration is undesirable since it may cause vibration as well as additional stress to the system. The acceleration of an actuator depends on the type of actuator (e.g., electric, hydraulic, or pneumatic), the load it moves (the type, size, and pressure class of the valve), and the input control signal from the control room.

2. Displacement

An actuator's displacement is the movement it produces in response to a control signal, pressure of air or hydraulic oil, or electrical power. In this measurement, the actuator's output (piston or piston rod) is measured in terms of the distance (or angle) that it moves to reach the desired position or shift the connected component (such as a valve stem).

3. Force

In mechanical terms, actuator force refers to the force generated by an actuator when it is used to move or control a load (for example, a valve stem). In most cases, this parameter depends upon the type and pressure or load of input energy (hydraulic, pneumatic, or electric) as well as the design of the actuator (e.g., the size of the piston or the amount of torque available).

4. Modulation

Modulation in an actuator refers to the ability to adjust its output position, speed, or force continuously and accurately in response to a control signal, rather than operating in a simple on/off or fully open/close manner. It is common for modulating actuators to respond to variable input signals (e.g., 4–20 mA or 0–10 V) in order to achieve accurate positioning. Rather than simply controlling flow, pressure, or other process parameters statically, modulation allows for dynamic (continuous) control. The use of modulation is crucial in applications requiring high accuracy and frequent changes in the position of the actuator and valve.

5. Opening/Closing

As the name implies, opening/closing refers to the action of actuators in operating (moving) a valve from a fully closed position (blocking flow in the piping system) to a fully open position (allowing maximum flow passage), and vice versa. It is possible for a valve to move linearly (such as a globe or diaphragm valve) or rotationally (such as a ball or butterfly valve).

6. Stroke

When it comes to actuators, stroke refers to the total distance that the actuator's output (such as a piston) moves between its fully retracted and fully extended positions. This measurement describes the range of motion the actuator is capable of, and is typically expressed in millimeters (mm) or inches. In a linear actuator, the stroke is the distance that can be pushed or pulled, whereas in a rotary actuator, it may refer to the angle (e.g., degrees). When selecting an actuator, it is important to take into consideration the range and length of the stroke movement. In most cases, the actuator stroke and the valve stroke occur simultaneously since the actuator is directly responsible for driving the valve's movement.

7. Torque

A torque is a force applied at a distance from the axis of rotation that causes the valve stem and closure member to rotate. In actuators, it refers to the force applied by the actuator to rotate the valve stem and disk. Various torques are used for valve operation, including break torques (the torque required to start the valve from a full-open or full-close state, which is usually the largest torque for valve operation due to static friction). A running torque is the torque required to move a valve between the open and closed positions, while a seating/unseating torque is the torque needed to fully open or close the valve, eliminating sealing friction. The actuator torque is typically higher than the valve torque in order to ensure the valve can be opened and closed easily. An actuator may be required to produce two times the torque required for the valve to open, indicating a safety factor of 2. An actuator torque measures the rotational force generated by an actuator to move a valve stem in order to open or close a quarter-turn valve. This is a critical parameter in selecting an actuator for a specific application, and is usually measured in Newton-meters (Nm) or pound-feet (lb-ft). The output of linear actuators is typically measured in terms of force rather than torque due to the linear nature of the motion (straight line rather than rotational).

Different torque parameters (applicable to both valves and actuators) depending on the valve closure member positions are listed as follows:

Break to Close (BTC) Torque: BTC torque refers to the amount of torque required to initiate the movement of a valve from its fully open position to the start of the closing position. Like the break to open (BTO) torque, it is an important parameter in valve operation, but it applies specifically to closing the valve.

Break to Open Torque: BTO torque is a term used in valve engineering to describe the amount of torque required to initially move a valve from its fully closed position to the start of the opening position. It is one of the most critical torque values for valve operation, as breaking the valve free from its sealed position often requires more effort than moving it once it is partially open.

End to Close (ETC) Torque: The ETC torque refers to the torque required for a valve to fully close and achieve a tight seal at the very end of its closing motion. As a result of the additional forces required to compress seals, overcome resistance, and ensure a proper shutoff, this torque is often higher than the torque needed during the intermediate stages of valve closure.

End to Open (ETO) Torque: The ETO torque is the torque required to completely open a valve, moving it from a nearly closed position to the fully open position. This torque is typically less than the BTO torque.

Running Torque: Torque needed to keep the valve moving during operation approximately in the middle positions between open and closed positions. As soon as the valve begins to move, the required torque generally decreases. In general, running torque is less than breakaway torque because it is required to keep the valve moving throughout its range. Generally, running torque is defined as the torque required to keep the closure member (ball, disk, or gate) moving after the valve has already started, covering its intermediate positions from slightly open to fully open. All positions between partially open (half-open) and fully open are included in this category.

Breakaway Torque: In the case of a valve, the breakaway torque is the initial torque required to initiate the movement from its fully closed (or sometimes fully open) position to its fully open position. A valve's torque is defined as the rotational force required to open or close it, usually applied to the valve stem or actuator. The torque specification is an essential requirement for the operation of valves, especially for automated or motor-operated valves, where the actuator must deliver sufficient force to operate the valve reliably. Thus, *breakaway torque* is essentially the same as *BTO torque* or *BTC torque*, depending on the context (opening or closing the valve).

Seating Torque: The torque required to fully seat the valve in its closed position to ensure a tight seal. It should be noted that *seating torque* and *ETC torque* are related but not necessarily the same. The seating torque is calculated based on the amount of force or torque required to compress and deform the soft seat or achieving tight metal-to-metal seal between the valve metal seat and closure member. The focus of this torque is to create a tight closure of the valve. However, the focus of ETC torque is to complete the valve closure.

Unseating Torque: The torque required to unseat the valve from its closed position, typically equal to or slightly higher than BTO torque.

8. Velocity/Speed

When an actuator is in operation, velocity refers to the speed at which it moves (extends or retracts). In this chapter, we have already discussed actuators as devices that convert energy (hydraulic, pneumatic, or electric) into mechanical motion, and velocity denotes the speed at which they can do so. There is a small difference between the speed and velocity of the actuator. It is important to note that the speed does not take into account the direction of movement and is always positive. In spite of this, velocity can be either minus or plus depending on the direction of movement. As an example, if the actuator moves backward with a speed of 5 m/s, the velocity would be –5 m/s.

Actuator Signal

A valve actuator signal consists of a command or input sent to the valve actuator that initiates, stops, or modulates the movement of the valve. As a result of this signal, communication takes place between the control system and the actuator. Many actuators are equipped with sensors that provide feedback to the control room regarding the valve's status and operation.

Actuator Types

Manual Actuators: The manual actuators are operated by hand and utilize levers, hand wheels, or gearboxes in order to open and close the valves. In addition to being simple and cost-effective, they are suitable for applications in which automation is not required. They are not common since they are unable to provide automation, which is the main purpose of actuators.

Diaphragm Actuators: It is common for control valves to utilize diaphragm actuators to provide precise and reliable valve positioning. The mechanism is based on a flexible diaphragm that moves when air or hydraulic pressure is applied to one side. In turn, this movement results in linear movement of the valve stem, which allows for accurate flow control. In response to a control signal, a diaphragm actuator controls the position of a valve stem using a flexible diaphragm and a spring mechanism. To open, close, or modulate the valve, the actuator converts the control signal value into linear motion. Diaphragm actuators can be either *direct acting* or *reverse acting*. The pressure signal is applied to the top of the diaphragm in a direct-acting diaphragm actuator. In response to an increasing signal value, the diaphragm is pushed more forcefully downward, resulting in a compression of the spring and a downward movement of the valve stem. As the signal value decreases, the spring pushes the diaphragm and valve stem upward to their original position. In most globe valves with a fail-open mechanism, a direct-acting actuator means higher air pressure pushes the actuator stem downward, which closes the valve. In a reverse-acting diaphragm actuator, the pressure is applied beneath the diaphragm. As the signal pressure increases, it pushes the diaphragm upward, compressing the spring and moving the valve stem upward. When the pressure decreases, the spring forces the diaphragm and valve stem back downward. Reverse-acting actuators could be used where increasing pressure opens the valve (e.g., a control valve with a "fail-close" action).

Pneumatic Actuators: In pneumatic actuators, compressed air is used to produce motion. For quick and repetitive valve operations, they are reliable. Due to the fact that air is used as a source of actuation, these could be considered as a preferred choice of actuation due to their environment-friendly nature. There are two types of pneumatic actuators available—single-acting (spring-return) and double-acting. Both are explained further in this chapter. As a result of the possibility of air volume changes, they may not be accurate; however, because of the air's unflammability, they are suitable for use in hazardous conditions.

Hydraulic Actuators: A hydraulic actuator uses a pressurized liquid, such as hydraulic oil, in order to deliver high forces that are much greater than those delivered by pneumatic actuators when operating valves. This is due to the fact that hydraulic oil pressure is much higher than compressed air pressure. These actuators are suitable for heavy-duty applications requiring greater force for valve operation, such as subsea or high-pressure valves of large size. The valves are extremely accurate, but to prevent leakage of hydraulic oil into the environment, they require reliable sealing systems. This is especially true for valves that are installed underwater.

Gas Actuators: A gas actuator is a type of actuator that is powered by pressurized gas. Actuators powered by gas are used in industrial applications where natural gas or other types of gases are available as a power source. In remote areas without access to electricity or other power sources, they are very popular. For example, in a pipeline transporting natural gas, a gas-over-oil actuator could be used to obtain pressurized gas from the pipeline and use this gas to pressurize hydraulic fluid in order to generate the force needed to operate a large valve.

Electrical Actuators: Electric actuators are driven by electric motors. In addition to being environment-friendly due to the use of electricity, these actuators are also cost-effective and compact. Electrical actuators can be operated remotely, making them well-suited for remote operations. However, their performance may be less effective in environments with high forces or explosive hazards. An electric actuator is shown in Figure 11.2, which is used to automate the valves in industrial plants. A cylindrical housing or shell contains a motor that generates torque in order to operate a valve, while an attached gearbox increases torque and adjusts speed for efficient operation. It is possible to open, close, and stop the valve by pressing the buttons on the control panel or screen on the front. Furthermore, buttons can be used to switch between remote, local, and off modes. There is a digital display that indicates the valve's current position (e.g., "100" for fully open). On the bottom of the actuator, there is a flange that connects it to the valve. An electric actuator's handwheel

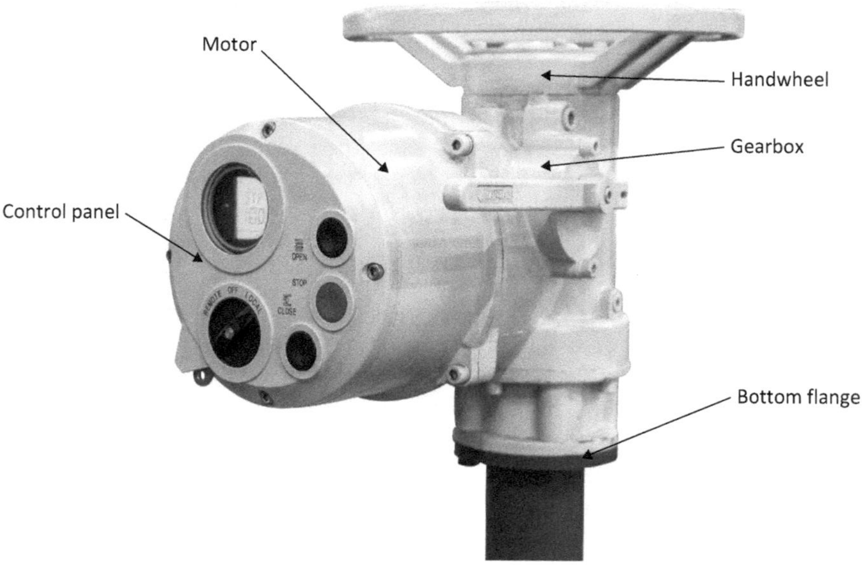

FIGURE 11.2 An electrical actuator with control panel and mounting flange. (With permission: Shutterstock.)

serves as a manual override, which means that the valve can be operated manually when the actuator cannot function automatically due to power failure, system failure, or maintenance requirements.

Electrohydraulic Actuators (EHAs): EHAs combine hydraulic power and electric control to deliver high force with excellent precision. These actuators use an electrically driven pump to generate hydraulic pressure, which is then directed to a cylinder or motor to produce the mechanical motion needed to operate the valve. They are ideal for applications requiring hydraulic fluids in areas lacking hydraulic infrastructure. These actuators are versatile but more complex and costly compared to stand-alone hydraulic or electrical actuators. EHAs have fewer hydraulic components compared to traditional hydraulic actuators because they use electrically driven pumps instead of relying on a centralized hydraulic system.

Smart Actuators: Smart actuators integrate sensors, IoT, and communication technologies for real-time monitoring and control. They enhance system efficiency by providing data on valve performance, enabling accurate monitoring, predictive maintenance, and remote adjustments. Ideal for modern automated systems, they reduce downtime and optimize operational reliability.

Thermal Actuators: Thermal actuators rely on the expansion or contraction of materials in response to temperature changes to operate valves. They are used in temperature-sensitive systems, such as thermal relief or safety applications. While simple and reliable, they are limited to specific environments and functions.

Electromechanical Actuators: Electromechanical actuators convert electrical energy into mechanical motion using linkages or gears. They are commonly employed for precise linear or rotary valve control in automation systems. These actuators are compact, efficient, and versatile, but may not handle very high forces. What is the difference between electrical and electromechanical actuators? Electrical actuators convert electrical energy directly into mechanical motion, often using simpler mechanisms like electric motors. These actuators are best suited for light-duty (low force) actuation. In contrast, electromechanical actuators combine electrical components (like motors) with mechanical systems (such as gears, screws, or levers) to amplify force or torque to handle heavier loads. While electrical actuators are simpler and faster, electromechanical actuators are more complex but ideal for applications requiring greater force or torque.

Solenoid Actuators: Solenoid actuators use electromagnetic force to create linear motion, often in smaller valves for on–off control. They are fast, compact, and ideal for simple applications where quick response and high cycle valves are needed. However, they are limited in terms of force output and are typically used in low-pressure systems that require low force for valve operation.

Servo Actuators: A *servo actuator* is an electrical actuator using servo motor. It provides high-precision, fast response, and high torque. It is driven by a *servo motor* combined with a feedback system to achieve highly accurate and controlled motion.

Automation

Automation regarding actuation refers to the integration of actuators into automated systems to perform mechanical movements of valves or control processes without the need for direct human intervention.

C

Calibration

The calibration of an actuator refers to the adjustment of each component's operation so that the actuator will function according to its design and operation requirements. Calibration is done in order to provide the movement, force, and torque that are required for valve operation. As an example, the actuator of an ESD valve was unable to provide the required force to return the valve to a fail-safe position after years of operation. Considering that an ESD valve could not perform its critical function, this situation was considered to be very unsafe regarding process control. The problem was caused by a weak spring in the actuator. As a result, the spring was calibrated since many actuators have a mechanism (e.g., screws) for adjusting the spring's initial tension or compression.

Closed-Loop Control

For more details about *closed-loop control*, please refer to the "Feedback System" section in this chapter.

Compensation System: Pressure Compensation

A *compensation system* in subsea actuators ensures reliable operation in harsh underwater environments by equalizing internal and external pressures, preventing structural damage, contamination, and fluid loss. It uses mechanisms like bladder, bellows, or capsule compensators to maintain equilibrium by adjusting to hydrostatic pressure at depth while keeping internal hydraulic fluids sealed from seawater. Without compensation, the internal actuator housing will experience a significant pressure differential as external hydrostatic pressure increases with depth. This can cause the actuator casing or seals to deform, crack, or collapse under the extreme pressure. Figure 11.3 illustrates a compensation capsule (system) integrated into a subsea actuator to equalize hydrostatic pressure resulting from the actuator's installation at significant depths.

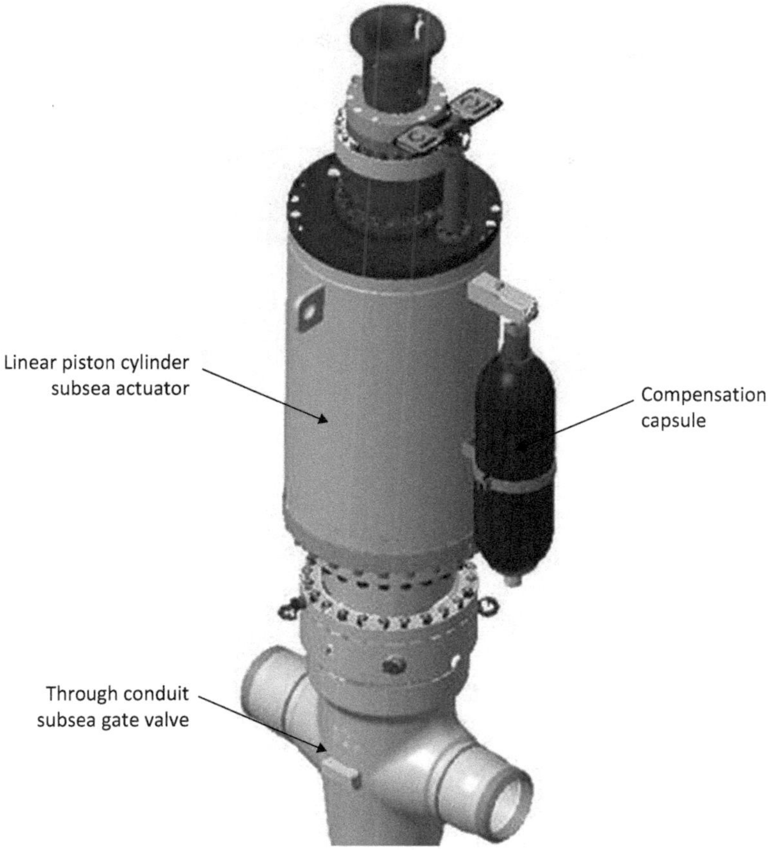

FIGURE 11.3 Compensation capsule installed on a subsea actuator to equalize hydrostatic pressure applied by deep-water deployment.

Condition-Based Monitoring

Condition-based monitoring for an actuator involves analyzing real-time data, such as pressure and temperature, and historical performance data, such as torque and valve position (open, close, half-open, etc.), to assess its health and detect early signs of wear or malfunction. This proactive approach enables the early detection of potential malfunctions in valves and actuators, ensuring that maintenance is performed only when necessary. As a result, downtime is minimized, and the operational lifespan of the valves and actuators is significantly extended.

Control Panel

The actuator control panel is an assembly that controls the flow of hydraulic oil and air to the actuator. In most cases, the panel includes components such as tubes for transporting fluid, valves such as ball and solenoid valves for controlling fluid

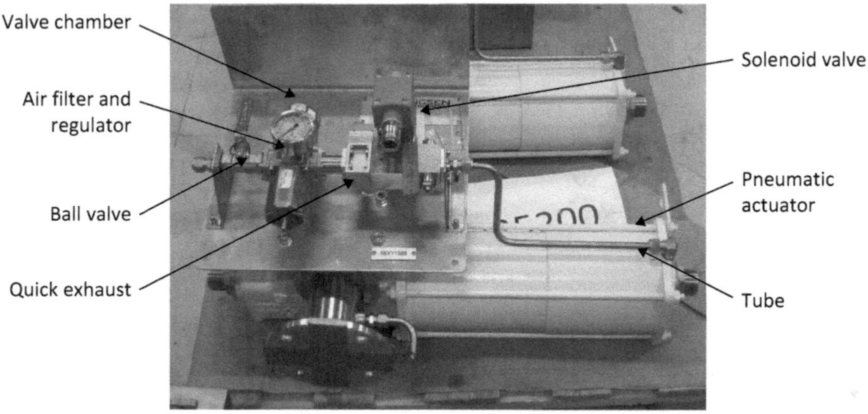

FIGURE 11.4 A pneumatic actuator control panel is used to regulate airflow to the actuator and to ensure smooth operation.

flow, and an air filter and pressure regulator. Control panels may be equipped with programmable logic controllers (PLCs) or other electronic components that process signals received from sensors and transmit commands to actuators. Modern systems are capable of containing advanced capabilities such as remote control and signal feedback. Figure 11.4 illustrates a control panel mounted on a pneumatic actuator. Located on the right side of the control panel is a ball valve that allows air to enter the panel. As the ball valve is operated to stop and start the flow of air into the actuator, the air filter and regulator perform two functions. A filter removes particles from the air in order to prevent damage to the actuator, while a regulator adjusts the air pressure in order to achieve the pressure that the actuator is designed for (e.g., 5 bars). In pneumatic control systems, the quick exhaust (valve) is a component that allows compressed air to be vented from the actuator quickly, which allows the actuator to return to its original position faster. Control panels contain solenoid valves, which are electromechanical valves used to control the flow of air or hydraulic oil. There is a tube at the end of the control panel that connects it to the actuator's cylinder.

Control Signal

Control signals are input signals that instruct actuators how to operate. For example, it directs an actuator to move to a specific position, open or close a valve by a specific percentage to start, stop, or regulate flow. PLCs, DCSs (distributed control systems), and other automation controllers provide the signal which instructs the actuator to move in a particular direction. It is possible to communicate digitally or analogically. Examples of analogue signals are 4–20 mA current loops, and 0–10 V signals. Alternatively, digital communication protocols such as HART, Profibus, and Foundation Fieldbus could be used as control signals. The purpose of an analogue signal is to control continuous variables, such as the position of a control valve. A 4-mA control signal represents 0% valve position (fully closed), while 20-mA signal corresponds to 100% valve position (fully open). Any value within the 4–20 mA range indicates a proportional valve opening (e.g., 12 mA = 50% open).

In contrast, a digital signal is used for binary control, such as opening and closing a valve. For instance, a digital "1" can signal the valve to open, while a digital "0" commands it to close.

Modern systems are increasingly adopting wireless communication to remotely monitor and control valve operations, enhancing efficiency and reducing the need for physical intervention.

Corrosion Resistance

In actuators, corrosion resistance refers to the ability to withstand degradation caused by corrosive environments. These environments include marine atmospheres, both offshore and subsea, as well as moisture, humidity, and chemicals. Without proper corrosion resistance, these factors can corrode actuator materials and reduce their lifespan. There are several strategies available for combating corrosion. These include the use of corrosion-resistant materials, such as stainless steel, instead of carbon steel, or protective coatings, such as epoxy-based coatings. Using proper sealing systems can prevent corrosive substances from entering the actuator internals. Viton rubber O-rings are a good example of robust seals. It is important to ensure corrosion resistance in corrosive environment conditions. These include offshore and subsea marine, chemical plants, where harsh conditions can lead to corrosion.

Cylinder

For more details about *cylinder*, please refer to the "Actuator Components" section in this chapter.

D

Diaphragm Actuators

For more details about *diaphragm actuators*, please refer to the "Actuator Types" section in this chapter.

Digital Twin

The virtual representation or model of the actuator that illustrates its performance in real time and the conditions in which it operates. Through the use of digital twins, actuator performance can be simulated, tested, and optimized through data analysis without affecting (damaging) the physical actuators.

Diagnostics

In actuators, diagnostics refers to the monitoring, analysis, and identification of the actuator's operational health and performance. Diagnostics assist in detecting issues such as wear, malfunctions, power consumption, pressure, and temperature in actuators. It is possible to perform diagnostics on actuators by using sensors, either independently or in conjunction with a condition monitoring system. It is also possible to monitor actuators by integrating actuators performance with an appropriate control system, such as a PLC. A diagnostic procedure may enable early maintenance to be performed before the problem worsens. As an example, diagnostics may detect increased torque being produced by an actuator, which can cause damage to the valve stem and wear in the seat and disk of the valve upon closing.

Direct Drive Actuators

Direct drive actuators are rotary actuators in which the motor, which drives the actuator, is directly connected to the valve stem or output shaft without the use of intermediate mechanical components such as gears or links. The direct connection of the actuator provides an efficient and simple method of transferring energy for the actuator motor to the valve stem. Due to its simple design, this actuator is lighter and more space-efficient as it eliminates the need for additional components such as gearboxes.

Double-Acting Actuator

As the name implies, a double-acting actuator is a mechanism in which an actuator is opened and closed by a fluid (hydraulic oil or air) (no spring action). Unlike single-acting or spring-return actuators with a fail-safe close or open function, if the fluid supply is interrupted, the double-acting actuator will remain in its last location, a condition known as *fail-as-is*. See "Single-Acting Actuator" in this chapter.

E

Electrical Actuators

For more details about *electrical actuators*, please refer to the "Actuator Types" section in this chapter.

Electrohydraulic Actuators

For more details about EHAs, please refer to the "Actuator Types" section in this chapter.

Electromechanical Actuators

For more details about *electromechanical actuators*, please refer to the "Actuator Types" section in this chapter.

Emergency Shutdown (ESD) Valve Actuator

ESD actuators are devices or components that operate emergency shutdown valves (ESDVs). ESDVs are safety-critical valves that must be closed very rapidly in the event of an emergency, for example, when a sensor detects an extremely high pressure in the piping system. In fact, the ESDV moves to a safe position (typically fully closed) during an emergency situation. ESDV actuators are equipped with spring-return, fail-safe mechanisms to ensure the valve closes automatically in the event of a power or control signal failure. Typically, pneumatic and hydraulic actuators (rather than electrical and electromechanical ones) are preferred in emergency situations due to their speed. This is due to the fact that electric motors are unable to provide fast operation of the valve during emergency conditions, as opposed to actuators operated by air and hydraulic fluids.

Energy Efficiency

The energy efficiency of an actuator relates to its ability to convert energy (electrical, pneumatic, or hydraulic) into mechanical movement while minimizing energy losses. In a pneumatic actuator, for example, friction is generated between the piston and the cylinder as a result of the friction of the seal. The resistance increases the amount of compressed air and pneumatic energy required for the piston to move. When the seal friction is too high, the actuator may require 15% more air pressure than that considered by the design engineer, resulting in increased energy consumption and increasing operating costs.

Environment Compliance

Achieving environmental compliance with actuators refers to designing, manufacturing, and operating actuators in a manner that complies with environmental regulations and minimizes ecological impact, which is especially important in the oil and gas industry. For instance, pneumatic or electrical actuators, which are more environment-friendly, can be selected during the actuator design and selection process as alternatives to hydraulic actuators to minimize environmental impact.

F

Fail-Safe Mechanism

The fail-safe mechanism in valves and actuators ensures the system moves to a predetermined safe position in the event of a power or pressure loss in order to protect equipment and maintain safety. In a fail-safe closed actuator, the valve closes when the power supply is interrupted (e.g., pneumatic or hydraulic), preventing leaks or hazards. The fail-safe open mechanism, on the other hand, opens when power is lost, ensuring that fluid flows in the connected piping system to avoid failure resulting from overheating or overpressure, for example. In all cases, springs are used inside the actuator to provide the fail-safe action.

Feedback System

A feedback system in actuators is a control mechanism that monitors the actuator's output (such as position, speed, or torque) and adjusts its operation to ensure it aligns with the desired setpoint. It typically consists of a sensor (e.g., potentiometers or encoders) to measure performance, a controller to compare the feedback with the setpoint, and an actuation mechanism to correct any deviation. In a closed-loop feedback system, the actuator continuously receives real-time adjustments, while an open-loop system operates without feedback system, making it less accurate. In an open-loop control system, the controller does not know what is happening in the process because there is no feedback from sensors. Instead, the controller sends commands to the valve based on predefined conditions such as time or manual operation. For example, the controller is programmed to open the valve for 30 min daily. Feedback systems enhance accuracy, reliability, and efficiency, compensating for variations due to load or environmental changes. The lack of feedback can also mean that there is no sensor on the valve itself to report its real condition to the control system. For more information regarding lack of feedback from the valve position, let us consider the following example.

Example: Lack of Feedback System Issues

In an open-loop control system without feedback in the oil and gas industry, a common problem could occur during the operation of a control valve used for flow regulation in a gas pipeline. Suppose the control system sends a signal to open the valve to 50% to regulate the gas flow rate. Without a feedback mechanism, the system assumes the valve has reached the 50% open position. However, due to external factors such as valve stem friction, pressure variations, or mechanical wear, the valve may only open to 30% or overtravel to 70%.

This lack of precise positioning can lead to severe consequences:

1. *Flow Imbalance:* The incorrect valve position could result in insufficient or excessive gas flow, disrupting downstream processes like compression or separation.
2. *Operational Safety Risks:* Over-delivery of gas might exceed pressure limits, risking a pipeline rupture, while under-delivery could starve critical equipment like turbines or heaters.
3. *System Downtime:* Without feedback, operators may not detect the issue until it causes a process failure, resulting in unplanned shutdowns and maintenance.

This highlights the importance of a *feedback system* in ensuring precise valve operation and mitigating risks in critical processes.

Fluid Power

Fluid power in an actuator refers to the force produced by pressurized fluids such as instrument air or hydraulic oil to move the connected valve. For hydraulic and pneumatic actuators, fluid power is necessary for motion and force generation.

Force-to-Weight Ratio

The force-to-weight ratio is a key performance metric in actuator selection, indicating how much force an actuator generates relative to its weight. A higher force-to-weight ratio means the actuator is lightweight but powerful, which is beneficial in aerospace and subsea. It is calculated in accordance with Equation 11.1. For example, a hydraulic actuator produces high force but it is heavy.

Equation 11.1 Force-to-Weight Calculation for Actuators

$$\text{Force-to-weight ratio} = \frac{\text{Output force of actuator (N or Lbf)}}{\text{Actuator weight (kg, lb)}}$$

Fugitive Emission

Fugitive emissions are typically associated with gas actuators, since they involve the unintended release of gaseous substances. Both pneumatic and electrical actuators are environment-friendly. The hydraulic actuator is not environment-friendly since it contains hydraulic oil, which can contaminate the environment. As a matter of fact, fugitive emissions occur when pressurized gases are emitted from the actuator's components, especially the seals. It is the seals that are the most important component of an actuator when it comes to preventing emissions, especially with gas actuators.

G

Gas Actuators

For more details about *gas actuators*, please refer to the "Actuator Types" section in this chapter.

Gas-over-Oil Actuator

A gas-over-oil actuator is a type of gas-hydraulic actuator used to operate large pipeline valves, commonly in oil and gas industries. It is common that the valve with this type of actuator is installed on the gas pipeline. The actuator uses high-pressure gas (typically from the pipeline itself) as the power source. Instead of directly using gas to move the actuator, the gas pressurizes hydraulic oil in a separate chamber. The pressurized oil moves a piston, which in turn opens or closes the valve smoothly. Many gas-over-oil actuators include a spring-return mechanism or gas pressure release system to automatically close the valve in case of gas supply failure. Gas-over-oil actuators eliminate the need for hydraulic piping and pumps because they use high-pressure gas from the pipeline itself to pressurize the hydraulic oil inside the actuator.

Gearbox (Actuator Component)

For more details about *gearbox*, please refer to the "Actuator Components" section in this chapter.

Gearbox-Based (Actuator Component)

Gearbox-based mechanisms in valve actuation use a gearbox along with an electric motor to control rotary valves, such as ball, butterfly, and plug valves, as well as multi-turn valves like gate and globe valves. However, it is important to note that a gearbox only produces rotary motion. In the case of linear valves like gate and globe valves, the rotary motion from the gearbox is converted into linear motion through a threaded valve stem. The gearbox enhances torque, reduces speed, and optimizes control.

H

Hydraulic Actuators

For more details about *hydraulic actuators*, please refer to the "Actuator Types" section in this chapter.

I

IoT (Internet of Things)

IoT for an actuator involves embedding sensors, control systems, and communication capabilities into the actuator to connect it to a network. This setup allows valves and actuators to interact with other devices, systems, and operators over the internet or local networks. This connectivity transforms the actuator from a stand-alone mechanical device into an intelligent, data-driven component of a larger automated system.

L

Lead Screw/Ball Screw

A *lead screw actuator* uses a threaded screw and a mating nut to create linear motion. When the screw is rotated by a motor power, the nut moves along its axis, providing linear motion. A *ball screw actuator* features a threaded screw, a nut with recirculating ball bearings, a ball return system, and a motor. The ball screw also has helical threads, but these threads are shaped to hold recirculating ball bearings between the screw and the nut. The ball bearings roll between the threads, minimizing friction and enabling smooth motion.

In Figure 11.5, the *lead screw* on the right demonstrates a simple threaded screw design where the nut directly engages the threads on the screw. This

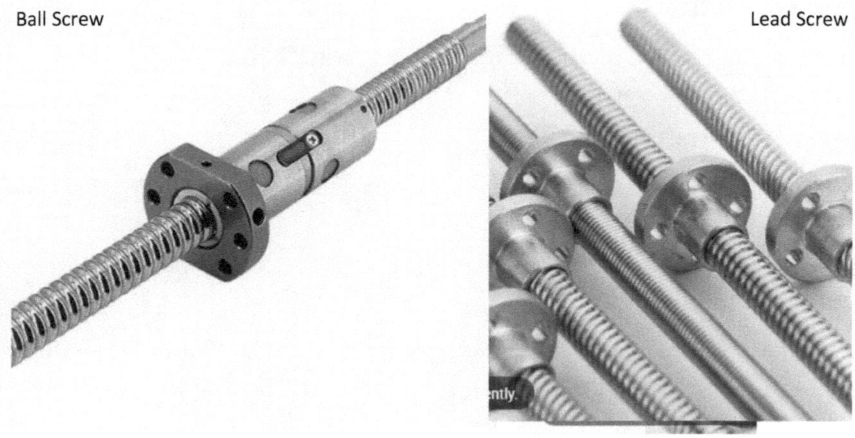

FIGURE 11.5 Comparison of lead screw (*right*) and ball screw (*left*) mechanisms.

mechanism relies on sliding friction between the screw and the nut to convert rotary motion into linear motion. The nut moves along the threaded shaft as the screw rotates, and the system inherently resists back-driving, providing self-locking capability. On the left, the *ball screw* illustrates a more advanced design with recirculating ball bearings placed between the threads of the screw and the nut. These ball bearings reduce friction by rolling instead of sliding, enabling smoother and more efficient motion. The nut moves linearly along the shaft with minimal resistance, making ball screws ideal for high-precision and high-speed applications.

Leak Detection

An actuator leak detection system identifies and monitors leaks, primarily in gas, pneumatic, and hydraulic actuators, which may compromise the safety and reliability of the actuator. Additionally, gas and hydraulic oil leaks can result in safety hazards and environmental contamination. Electric actuators are not associated with leaks due to the fact that they use electricity. Several factors can cause a leak, including worn seals, damaged components such as cylinders and housings, and loose tubes and control panel components. The methods of detection include visual inspection for fluid leakage, pressure testing, which indicates leakage, and ultrasonic sensors for nondestructive testing to detect possible defects. Performing regular inspections and maintenance is crucial to preventing leaks or minimizing the effects of leaks.

Limit Switch

In order to ensure that the valve is moved sufficiently without overtravel, a limit switch is installed on the actuators.

Limit Switch Box

Limit switch installed in a box known as limit switch box.

Linear Actuators

For more details about *linear actuators*, please refer to the “Actuator Mechanisms” section in this chapter.

Linear Motion

For more details about *linear motion*, please refer to the “Valve Stem Motion Type” section in this chapter.

Load Capacity

Load capacity is the maximum *torque* (for rotary actuators) or *thrust* (for linear actuators) that an actuator can produce.

M

Manual Actuators

For more details about *manual actuators*, please refer to the “Actuator Types” section in this chapter.

Manual Override

For more details about *manual override*, please refer to the “Manual Override” section in Chapter 8.

Modulating Actuator

A *modulating actuator* is a type of actuator designed to precisely control the position of a valve or similar device in response to a variable control signal. This actuator allows for fine adjustments rather than just fully open or fully closed positions. Modulating actuators typically receive an *analogue control signal*—commonly 4–20 mA, 0–10 V, or digital signals like those in communication protocols (e.g., HART or Profibus). The actuator translates these signals into proportional movement, enabling the valve to respond dynamically to changes in the control system’s setpoint. These actuators are widely used in applications requiring precise process control.

Motor

For more details about *motor*, please refer to the “Actuator Components” section in this chapter.

O

Open-Loop Control

For more details about open-loop control, please refer to the "Feedback System" section in this chapter.

P

PID Controller

The PID controller stands for proportional–integral–derivative controller. It refers to a control mechanism used to maintain a desired output, such as position, torque, or speed, continuously during the operation of the actuator. It is possible for actuators to change their positions, as well as other parameters, dynamically through the implementation of this system.

Piston

For more details about *piston*, please refer to the "Actuator Components" section in this chapter.

Piston and Cylinder Actuator

A *piston and cylinder actuator* is a type of mechanical actuator that uses a *piston* moving within a *cylinder* to generate and transmit linear motion. It operates by applying pressure to a fluid (either compressed air in pneumatic systems or hydraulic fluid in hydraulic systems) to move the piston, which is connected to an external component like a valve stem. Figure 11.1 shows a piston and cylinder linear actuator.

Working Mechanism: Fluid pressure is introduced into the cylinder through an inlet port. This pressure acts on the piston, causing it to move linearly within the cylinder. The piston divides the cylinder into two chambers (upper and lower), with the pressure differential driving its movement. The movement of the piston translates the air or hydraulic force into linear mechanical motion. The piston is connected to a *piston rod* that extends out of the cylinder. This rod transmits the linear motion to an external mechanism, such as a valve stem or a linkage system. The direction of piston motion is controlled by regulating the fluid flow into and out of the cylinder. In *double-acting actuators*, fluid is alternately supplied to both sides of the piston for bidirectional motion. In *single-acting actuators*, fluid pressure moves the piston in one direction, with a spring returning it to its original position.

Key components are ports, piston, cylinder, piston rods, and seals. Seals ensure no fluid leakage between the chambers or outside the cylinder.

Piston Rod

For more details about *piston rod*, please refer to the "Actuator Components" section in this chapter.

Pneumatic Actuators

For more details about *pneumatic actuators*, please refer to the "Actuator Types" section in this chapter.

Positioner

A positioner is a device used in control valves to ensure the valve stem or actuator is accurately positioned according to the control signal received from a process control system. It acts as an intermediary between the controller and the actuator, comparing the desired setpoint signal (typically 4–20 mA) with the actual valve position and adjusting the actuator to minimize any deviation. A positioner adjusts the actuator by controlling the air pressure supplied to the actuator. Positioners are commonly used in applications requiring precise control, especially in throttling valves like globe valves or butterfly valves. They are available in various types, including pneumatic, electropneumatic, and digital (smart) positioners, which can offer additional features like remote calibration, diagnostics, and improved control accuracy. By ensuring accurate positioning, positioners enhance process efficiency, reduce wear, and improve valve performance in complex process systems.

Position Indicator

As its name implies, it displays the position of the actuator (valve). In pneumatic and hydraulic actuators, a position indicator can be installed to indicate whether the actuator is in the on or off position. It is possible to view the position of electromechanical or electrical actuators on the screen integrated with the actuator. A position indicator is shown in Figure 11.6 on a pneumatic actuator, indicating that the valve is in the open position.

Position Sensor

The position sensors are components that collect information about valve position and transmit it to the control room or remote locations used to control the process. It is important to note that position sensors are special sensors that are not automatically included in the standard design of actuators. In actuator systems, proximity switches are commonly used as position sensors.

Predictive Maintenance

Predictive maintenance in actuators involves using real-time data, historical performance trends, and advanced analytics (often aided by machine learning)

FIGURE 11.6 It is a position indicator mounted on top of a pneumatic actuator that indicates the valve is closed.

to predict when an actuator is likely to fail or require servicing. This approach allows maintenance to be scheduled proactively, reducing unplanned downtime and extending the actuator's lifespan while optimizing operational efficiency.

Proportional Control

In actuators, proportional control means the actuator's output (e.g., position, speed, torque) is adjusted in proportion to the signal sent to it. A 4–20 MA signal, for example, closes the valve completely if 4 MA is sent, and fully opens it if 20 MA is sent. Now, if the actuator gets a signal of 12 MA, it keeps the valve half-open.

R

Rack and Pinion

For more details about *rack and pinion*, please refer to the "Rack and Pinion Actuators" section in this chapter.

Rack and Pinion Actuators

A rack and pinion actuator is a type of mechanical actuator that converts linear motion into rotary motion, or vice versa, using a rack (a straight, toothed bar) and a pinion (a circular gear). This design is commonly used in quarter-turn valve applications such as ball valves or butterfly valves, where precise and controlled rotary motion is required. Figure 11.7 shows the rack and pinion. The rack moves linearly, driven by a piston (in pneumatic or hydraulic actuators) or another force source. The rack's teeth engage with the teeth on the pinion gear, causing the pinion to rotate. The pinion's rotation is then transmitted to the valve stem, opening or closing the valve.

Real-Time Monitoring

Real-time monitoring for actuators involves continuously tracking key parameters such as position, torque, temperature, and power consumption by using sensors and communication system between the actuator and the operator/control room to ensure optimal performance and detect anomalies as they occur. Compared to condition-based monitoring, real-time monitoring could be considered very similar. An example of how they differ can be seen in the following paragraph.

FIGURE 11.7 Illustration of a rack and pinion mechanism. (With permission: Shutterstock.)

Example: Monitoring an Actuator in a Valve Control System

- *Real-Time Monitoring:* Continuously tracks the actuator's parameters, such as its position and torque. If the actuator fails to reach the commanded position within a specific time or exceeds a torque threshold, an immediate alert is triggered. For example, if the actuator is supposed to open the valve to 50% but stops at 30%, the system flags an error in real time.
- *Condition-Based Monitoring:* Analyzes trends in the actuator's performance over time, such as a gradual increase in the torque required to move the valve. This trend might indicate wear, friction buildup, or misalignment. Instead of triggering an immediate alert, condition-based monitoring schedules maintenance only when the data suggests a likely failure, such as when torque consistently exceeds normal levels for similar operations.

Redundancy

Redundancy in actuators refers to adding an additional component as a backup in actuators or control systems in order to improve safety and reliability. As a result, if one component fails, the other can be replaced without any downtime. Redundant control systems, such as secondary electronics or sensors, ensure that operations continue uninterrupted. Another example is the use of an additional solenoid valve on the control panel for ESD valves. A solenoid valve on the control panel is responsible for transferring air or hydraulic oil between the control panel and the actuator, so that in the event that one fails, the backup valve can transfer the fluid to the actuator.

Remote Control

The phrase "remote control" refers to the ability to operate, monitor, and adjust an actuator from a distance without having to physically interact with it. In cases where actuators are located in hazardous areas, remote locations, or locations that are difficult to access, this parameter is essential.

Rotary Actuators

For more details about *rotary actuators*, please refer to the "Actuator Mechanisms" section in this chapter.

Rotary Motion

For more details about *rotary motion*, please refer to the "Valve Stem Motion Type" section in this chapter.

S

Safety-Critical Actuators

As the name implies, safety-critical actuators are those that operate safety-critical valves. The safety-critical valves and actuators that work as a system are those whose functionality is necessary to ensure the safety of the process and piping systems. For example, ESD valves and associated actuators are always considered safety-critical since if they fail to close quickly in case of overpressure or any other alarming condition, catastrophic conditions such as fire and explosion and death will result.

Safety Factor

When designing and sizing actuators, the safety factor is calculated by comparing the torque or force of the actuator against the equivalent torque or force of the valve. For example, if an actuator produces 400 Nm torque during the closing of the valve while the valve requires only 200 Nm torque to close, the safety factor is 2. In actuator design and sizing, a safety factor refers to a margin of additional force or torque built into the actuator's capacity to ensure reliable operation even in unexpected conditions. The size of the actuator must take into account the valve's operating torque, as well as friction, process pressures, temperature variations, and dynamic conditions such as pressure surges or changes in fluid flow. The safety factor for actuators is typically 1.25–2 times the maximum operating torque or thrust of the valve. This is to ensure that the actuator is capable of operating the valve even in severe conditions.

Scotch and Yoke Actuators

A *scotch and yoke actuator* is a mechanical actuator that converts *linear motion* into *rotary motion* using a *scotch and yoke mechanism*. It is commonly used in *quarter-turn valves* like ball, butterfly, or plug valves, and is favored for its ability to provide high torque, particularly at the start and end of the valve stroke. This torque profile makes it ideal for applications requiring significant force during these critical moments. Figure 11.8 shows a scotch and yoke actuator. In a pneumatic or hydraulic scotch and yoke actuator, the linear motion of a piston (driven by pressurized air or fluid) moves back and forth within a cylinder. The piston rod is connected to a *yoke* mechanism. As the piston moves linearly as well as piston rod (piston extension), the yoke follows the scotch guide (a slotted track or arm), converting the linear motion into *rotary motion*. The rotary motion is transferred to the valve stem to open or close the valve, typically over a 90° or 180° rotation.

Sealing (Mechanism)

For more details about *sealing*, please refer to the "Actuator Components" section in this chapter.

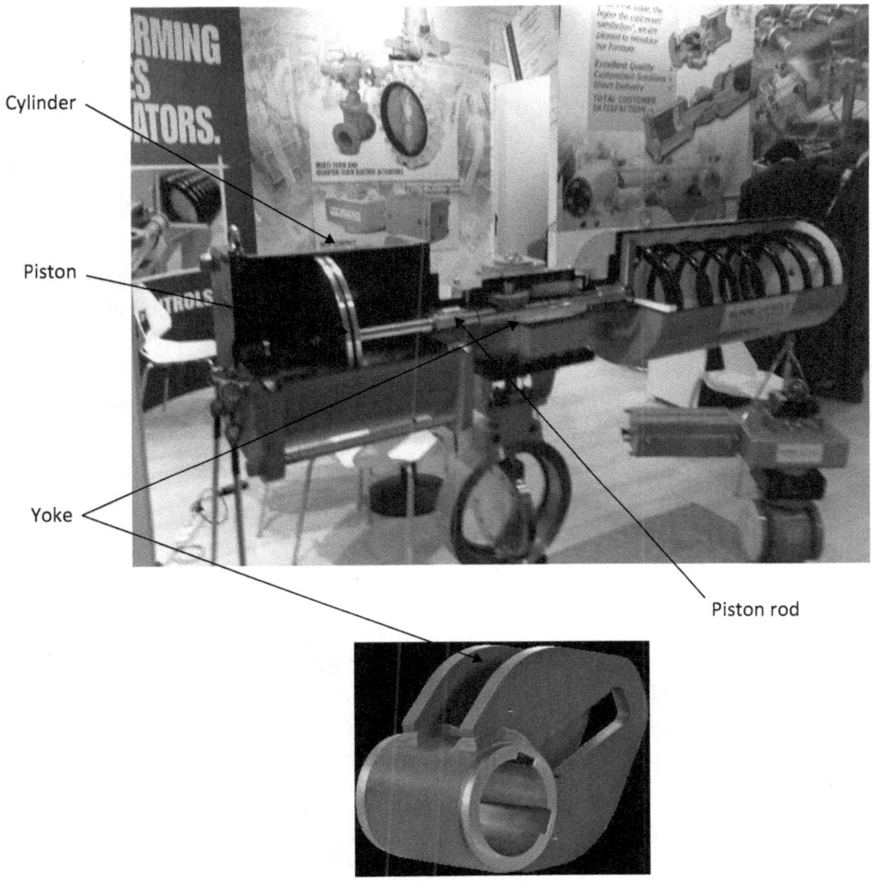

FIGURE 11.8 Illustration of a scotch and yoke actuator.

Servo Actuators

For more details about *servo actuators*, please refer to the "Actuator Types" section in this chapter.

Servo Motors

Servo motors are high-precision motors designed for applications requiring accurate control of position, speed, and torque. Servo motors are operating within a closed-loop system with real-time feedback from sensors like encoders. Unlike typical motors, which usually function in an open-loop system with constant or simple control, servo motors can dynamically adjust to achieve precise movement and positioning, making them ideal for automation. While servo motors are more expensive and require advanced control systems, they excel in tasks requiring precision, whereas typical motors are better suited for constant-speed, general-purpose operations.

Setpoint

The *setpoint* in an actuator refers to the *desired position*, *pressure*, or *other parameter* that the actuator and valve are instructed to achieve or maintain during its operation. It acts as the target value in a control system, against which the actual performance of the actuator is continuously measured and adjusted. In a linear actuator, the setpoint may be the specific distance the actuator's rod needs to extend or retract (e.g., 50 mm or fully extended position). For a rotary actuator, the setpoint is the desired angular position or degree of rotation (e.g., 90°, 180°, or fully open/closed for a valve). The setpoint for a pneumatic or hydraulic actuator could be a specific pressure or force that the actuator must achieve by considering a specific air or hydraulic pressure such as 5.5–9 bar for air pressure or 160–180 bar for the hydraulic pressure.

Signal Input

An actuator's signal input is the control signal sent to control and guide the actuator operation. As a result of this input, the actuator will move and maintain a specific position, torque (force) to implement required control of the system. This signal can be sent either directly to the actuator or to a positioner, which then adjusts the actuator to achieve precise positioning and response.

Single-Acting Actuators

Single-acting actuators are characterized by the failure-safe function, which returns the actuators to the fail-safe open or closed position. An actuator with fail-safe capability is designed with a spring as part of its design. In contrast to double-acting actuators, which open and close simultaneously with fluid force, single-acting actuators either open or close with fluid force, while springs carry out the other action. As should be noted, the fail-safe position of a fail-safe actuator is achieved by the spring force. For instance, a fail-safe actuator opens by means of a spring and closes by means of a fluid force.

Smart Actuators

For more details about *smart actuators*, please refer to the "Actuator Types" section in this chapter.

Solenoid Actuators

For more details about *solenoid actuators*, please refer to the "Actuator Types" section in this chapter.

Solenoid Valves

On control panels, these valves are used. A detailed description of the control panel and solenoid valves can be found in this chapter. In addition, Chapter 2 provides a detailed explanation of solenoid valves.

Spring-Return Actuator

An actuator with *spring-return mechanism.*

Spring-Return Mechanism

The *spring-return mechanism* in actuators is a design feature that ensures the actuator returns to a default or fail-safe position by the spring force when the external power source (such as air, hydraulic pressure, or electricity) is removed or fails. This mechanism is commonly used in valve actuators to ensure safety and reliability in critical systems, such as oil and gas, and chemical processing. When the actuator is powered (e.g., supplied with compressed air or hydraulic fluid, or receiving electrical input), it moves the valve to the required position (open or closed). During this time, the power source compresses a spring (or springs) within the actuator. The spring stores potential energy as it is compressed by the actuator's power source. If the power source is removed or fails, the stored energy in the compressed spring is released. The spring forces the actuator to return to its default or fail-safe position (e.g., fully open or fully closed).

Stem (Actuator Component)

For more details about *stem*, please refer to the "Actuator Components" section in this chapter.

T

Thermal Actuators

For more details about *thermal actuators*, please refer to the "Actuator Types" section in this chapter.

Thermal Expansion

The phrase thermal expansion is used to describe the change in size or volume of the actuator's components caused by changes in temperature. The temperature could be affected by operating pressures and temperatures or it could be affected by the environment (e.g., places that have high-temperature fluctuations). For more information, as a simple physics principle, when materials are heated, they expand, while when cooled, they contract. A negative effect of expansion or contraction of actuator components is wear and leakage, which may affect the effectiveness of seals and actuator performance. It is especially critical for actuators that require a high level of accuracy in their operation. This problem can be mitigated through a variety of strategies, including selecting materials with low thermal expansion or contraction, installing insulation around the actuator, or changing the design of the actuator.

Torque Arm

For more details about *torque arm*, please refer to the "*Actuator Components*" section in this chapter.

V

Valve Control

Valve control in terms of actuators refers to the process of managing the position and operation of a valve using an actuator to regulate the flow of fluids or gases in a system. The actuator provides the necessary force to move the valve to its desired position—open, closed, or partially open—based on signals from a control system or an operator input.

Valve Stem Motion Type

A valve stem's motion can be divided into *linear motion* and *rotary motion*, which determines how the valve controls a medium's flow through both stem and closure member movements. In linear motion valves, the stem moves in a straight line to open, close, or throttle the valve, such as in gate and globe valves. It is necessary to use linear actuators in order to produce linear motion for a valve. The conversion of rotary motion to linear motion can also be accomplished using mechanisms such as lead screws. In contrast, rotary motion valves feature a stem that rotates 90° in order to open or close the valve. The most common types of valves are ball valves, butterfly valves, and plug valves, which are controlled by rotary actuators or electric motors. Identifying the type of stem motion is crucial for selecting the most suitable actuator that should be compatible with the valve in terms of stem motion. In Table 11.1, the linear and rotary motions of stems for various types of valves are summarized.

Variable Frequency Drives (VFD)

VFD are used to control and adjust the speed, torque, and position of an electric actuator by controlling the frequency and voltage of the electrical power supplied to its motor. There is a significant difference in the level of control and efficiency between electrical actuators with and without variable frequency drives. By

TABLE 11.1
Stem Linear or Rotary Motion for Different Valves

Category	Linear Motion	Rotary Motion
Valve types	Gate valve, globe valve, diaphragm valve	Ball valve, butterfly valve, plug valve
On/off operation	Gate valve	Ball valve, butterfly valve, plug valve
Throttling operation	Globe valve, diaphragm valve	Butterfly valve, plug valve
Actuator type	Linear actuator	Rotary actuator

adjusting the frequency and voltage of the motor's power supply, VFD-equipped actuators can vary speed, torque, and position dynamically. It should be noted that servo motors and actuators are significantly more precise than electrical actuators with VFD.

QUESTIONS AND ANSWERS

1. Which component or item in a fail-safe closed actuator provides closure of the valve and actuator?
   A. Spring
   B. Electrical motor
   C. Air
   D. Hydraulic oil

   Answer: Option A is the correct answer.

2. Which of the following does not constitute a primary function of an actuator control system?
   A. Controlling the actuator's speed and position
   B. Constantly monitoring and adjusting output torque or force
   C. Converting electrical energy to mechanical energy
   D. Providing feedback to ensure accurate control

   Answer: Option C is the correct answer. The actuator itself is responsible for converting energy (for example, electrical to mechanical) and not the control system.

3. Identify the actuator component that is primarily responsible for preventing leaks in the actuators.
   A. Seal
   B. Piston
   C. Piston rod
   D. Motor

   Answer: Option A is the correct answer.

4. What type of actuator is capable of emitting fugitive emissions?
   A. Pneumatic actuators
   B. Electrical actuators
   C. Hydraulic actuators
   D. Gas actuators

   Answer: Option D is the correct answer. It is only gas actuators that are capable of producing fugitive emissions because they use pressurized gases. These gases can escape through seals and connections, causing environmental damage and safety risks to the environment.

5. What is the main consequence of miss-alignment between an actuator and the valve during operation?
   A. Energy efficiency reduction
   B. High risk of fugitive emission
   C. Wear of actuator components
   D. Material corrosion

   Answer: Option A is the correct answer. An actuator that is misaligned will have a reduced force transmission to the valve, resulting in a higher level of energy consumption by the actuator.

6. What is the primary purpose of using diagnostics in actuators?
   A. To reduce the actuator energy consumption
   B. To minor the health of actuators during operation
   C. To increase the actuator speed of operation
   D. To improve the actuator materials corrosion resistance

   Answer: Option B is the correct answer.

7. From the control room, a 4–20 mA signal is received by a proportional control actuator. The value of 4 mA corresponds to a fully closed valve (0%) and the value of 20 mA corresponds to a fully open valve (100%). What is the valve position when the actuator receives a signal of 12 mA?
   A. 25% open
   B. 50% open (half-open)
   C. 75% open
   D. Fully open

   Answer: Option B is the correct answer.

8. What is a set-point in actuator?
   A. Maximum possible force produced by an actuator
   B. The speed of actuator
   C. The position, speed, or torque of an actuator during operation
   D. The calibrated point of an actuator

   Answer: Option C is the correct answer.

9. Which sealing material is suitable for an actuator in corrosive fluid and high-temperature application?
   A. Viton
   B. Nitrile rubber
   C. Carbon steel
   D. Aluminum

   Answer: Option A is the correct answer.

10. Which of the following is considered an advantage of electrical actuators over hydraulic and pneumatic actuators? (Choose more than one answer)
   A. Maintain the valve position between openings and closings
   B. Higher force output
   C. Faster operation
   D. More compact design

   Answer: Options A and D are correct answers.

BIBLIOGRAPHY

American Petroleum Institute. (2020). Standard for Actuators and Mounting Kits for Valves (API Specification 6DX). 2nd edition, with Addendum 1 and Errata 1. API Publishing Services: Washington, DC.

Forouzanfar, M., Gagliardi, G., Tedesco, F., & Casavola, A. (2024). Integrated model-based control allocation strategies oriented to predictive maintenance of saturated actuators. IEEE Transactions on Automation Science and Engineering, 21(1), 123–135. https://doi.org/10.1109/TASE.2024.3358912

Janocha, H. (Ed.). (2004). Actuators: Basics and Applications. Springer: Berlin.

Nesbitt, B. (2007). Handbook of Valves and Actuators: Valves Manual International. 1st edition. Elsevier: Oxford, UK.

Pawlak, A. M. (2017). Sensors and Actuators in Mechatronics: Design and Applications. CRC Press: Boca Raton, FL.

Šimic, M., & Herakovič, N. (2024). Investigation of innovative high-response piezoelectric actuator used as smart actuator–sensor system. Applied Sciences, 14(18), 8523. doi: https://doi.org/10.3390/app14188523

Sotoodeh, K. (2019). Valves and actuators integrity and blast load calculations. Springer Nature Applied Science, 1, 599. https://doi.org/10.1007/s42452-019-0624-z

Sotoodeh, K. (2019). Actuator sizing and selection. Springer Nature Applied Science, 1, 1207. https://doi.org/10.1007/s42452-019-1248-z

Sotoodeh, K. (2021). Prevention of Actuator Emissions in the Oil and Gas Industry. Gulf Publishing Company: Houston, TX. doi: https://doi.org/10.1016/C2020-0-03511-6

Sotoodeh, K. (2021). Subsea Valves and Actuators for the Oil and Gas Industry. Gulf Publishing Company: Houston, TX. doi: https://doi.org/10.1016/B978-0-12-820261-0.00001-1

12 Valves in the Energy Industry

From Conventional to Emerging Applications: Hydrogen and Carbon Capture

A

Additive Manufacturing (3D Printing)

Additive manufacturing (AM), also known as 3D printing, is an advanced production method that builds objects layer by layer from digital designs. It uses techniques like laser melting, with materials such as metals, plastics, and ceramics. Unlike traditional manufacturing, which cuts material away, AM only adds material where needed, reducing waste and allowing faster production. This technology is widely used in industries like aerospace, healthcare, and automotive for creating lightweight, custom, and complex designs. In valve manufacturing, AM helps produce high-performance and custom parts, but traditional methods are still more cost-effective for mass production.

Figure 12.1 shows a metal 3D printing process, where a laser-based system is being used to manufacture a mechanical component. The print head, equipped with multiple nozzles and a laser, is depositing metal powder layer by layer to build up the part. The cylindrical shape of the component suggests it could be used for industrial applications such as valve bodies or seats, or custom mechanical parts. The setup includes laser machine, laser beams, and the component.

Example: AM Valve Component

A titanium valve body for use in the aerospace industry was produced using selective laser melting (SLM). It featured optimized internal flow channels, reducing weight by 30% while maintaining strength. The same valve body, made by casting and machining, required multiple components to be assembled, resulting in a heavier part with less-efficient flow paths.

DOI: 10.1201/9781003640967-12

FIGURE 12.1 Metal additive manufacturing (3D printing) process using a high-precision laser machine to manufacture industrial components. (With permission: Shutterstock.)

Air Vent Valve

Air vent valves (also called air release valves) are automatic devices designed to remove trapped air or noncondensable gases from fluid. Removal of trapped air maintains efficient operation, preventing issues like airpockets, corrosion, or reduced flow. Air vent valve mechanism typically involves a float-operated system (see Figure 12.2) or a thermostatic element. In float-operated valves, trapped air lowers the liquid level in the valve chamber, causing the float to drop and open the vent, allowing air to escape; once air is purged, the float rises with the liquid, sealing the vent. These valves work automatically to maintain system efficiency, protect equipment such as pumps, and reduce maintenance requirements in piping and equipment.

Anti-Surge Valve

1. Definition

An anti-surge valve is a specialized control valve (e.g., globe or C-type ball valve) designed to protect compressors from surge conditions. Surge occurs when the flow through the compressor drops below a critical threshold, causing unstable operation and potential mechanical damage. The valve quickly opens to bypass excess gas back to the compressor inlet or another safe location, maintaining the minimum flow required for stable operation. Anti-surge valves use fast-acting actuators and advanced control systems to quickly adjust to pressure or flow

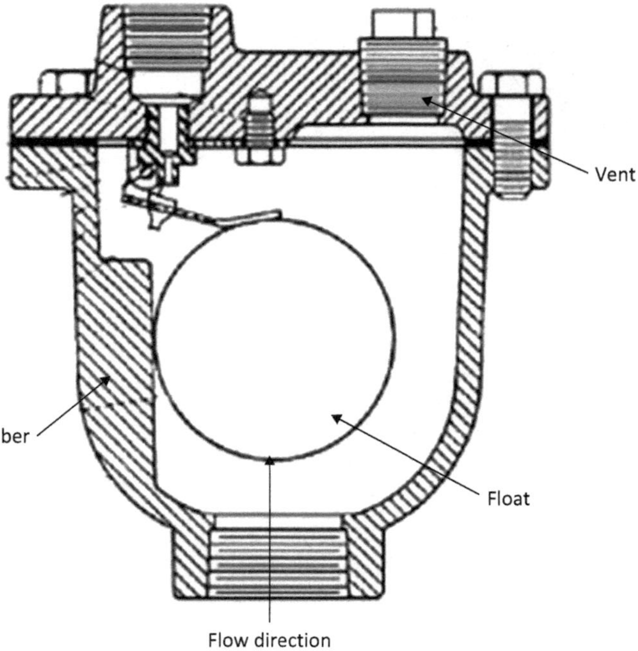

FIGURE 12.2 Mechanism of a float-operated air vent valve. The float lowers to open the vent for releasing trapped air and rises to seal it as liquid refills the chamber.

fluctuations, preventing compressor surge. This rapid response ensures the reliability, efficiency, and longevity of compressors in critical industrial processes.

2. Applications

These valves are used in various plants such as carbon capture, hydrogen, and geothermal. In carbon capture plants, anti-surge valves are essential in the CO_2 compression unit, where they prevent surges during the pressurization of CO_2 for transport or storage. In hydrogen production and storage, anti-surge valves are used to maintain stable compressor operation. Geothermal energy systems use these valves to stabilize compressors managing geothermal steam or working fluids.

3. Example

Anti-Surge Valve in Carbon Capture

Scenario: CO_2 Compression in a Carbon Capture Plant In a carbon capture plant, after CO_2 is separated in the regeneration unit, it is sent to a compression unit for pressurization to a supercritical state (100–150 bar) for transport or storage. A large, multistage centrifugal compressor is typically used for this purpose. Compressors operate within a specific flow range to remain stable; if the flow rate drops too low, a surge can occur, causing vibrations, backflow, and potential damage to the compressor.

Role of the Anti-Surge Valve The anti-surge valve is installed on a bypass line that connects the compressor outlet back to its inlet or another safe location. Its function is to maintain the minimum flow through the compressor. When the compressor detects a drop-in flow rate or pressure below the surge threshold, the anti-surge valve opens automatically and quickly. This allows the excess CO_2 gas to bypass the downstream system and flow back to the compressor inlet, ensuring the compressor maintains its minimum flow and stays stable.

Suppose the compressor's surge threshold is 500 kg/h of CO_2, and the flow unexpectedly drops to 450 kg/h. The anti-surge valve, controlled by a surge control system, senses this condition and opens partially or fully to recirculate a portion of the CO_2. This action restores the flow to above 500 kg/h, preventing the surge from occurring. Once conditions stabilize, the valve closes again to return the system to normal operation.

B

Biomass Energy Valves

1. Biomass Energy Concept

Biomass energy is a type of renewable energy that comes from natural materials such as plants, wood, and organic waste. It is called biomass because it refers to natural materials derived from living organisms, such as plants, animals, and organic waste. The term combines "bio" (life) and "mass" (material), highlighting its origin from biological sources. These materials, known as biomass, store energy from the sun. Biomass materials use the sun's energy through the process of photosynthesis, where plants absorb sunlight to convert carbon dioxide and water into sugars for growth, while also releasing oxygen into the air. When biomass is used for energy, it is either burned or converted into fuel to produce heat, electricity, or other forms of power. Biomass, like trees and crops, is called a renewable resource because new plants can be grown to replace the ones that are used. Even though the burned or used biomass does not grow back, new trees and crops can be planted, making it a continuous cycle.

The stored energy in biomass comes from the sun. Plants absorb sunlight through photosynthesis and convert it into chemical energy, which is stored in the bonds of glucose molecules. When these sugars are later broken down—either by the plant itself, animals that eat the plant, or when biomass is burned—the stored energy is released and can be used as fuel. The stored energy in biomass can be used in various ways as follows:

1. *Burning (Combustion):* Burning biomass, like wood or crop waste, to generate heat, which boils water to produce steam. This steam is then used to drive a turbine connected to a generator, converting the mechanical energy into electricity.
2. *Heating without Burning:* Pyrolysis is a process where biomass is heated without oxygen, so it doesn't burn. Instead, it breaks down into bio-oil,

biochar, and syngas. Biochar helps improve soil. Bio-oil can be used for energy. Syngas is a mix of gases (hydrogen, carbon monoxide, and carbon dioxide) that can also be used for energy or industry. The advantage of pyrolysis is that it does not involve burning, making it more environment-friendly than combustion. It produces fewer emissions while still turning biomass into useful energy products.

3. *Turning into Gas (Gasification):* Gasification is a process that converts biomass into a combustible gas called syngas by heating it at high temperatures with very little oxygen. This syngas can be used to generate electricity, power industrial processes, or serve as a raw material for producing chemicals and fuels.
4. *Using Microorganism (Fermentation):* Fermentation uses microorganisms, like bacteria and yeast, to break down sugars and starches in biomass materials such as corn, wheat, and sugarcane. During this process, the microorganisms convert the sugars into alcohol, mainly ethanol, which can be used as a biofuel to replace gasoline.
5. *Anaerobic Digestion:* Anaerobic digestion is a process where bacteria break down organic waste, like food scraps, manure, and crop leftovers, without oxygen. This process is done inside sealed tanks called anaerobic digesters. Without oxygen, the waste produces biogas, mainly methane and carbon dioxide, which can be used for electricity, heat, or fuel. It also creates digestate, a nutrient-rich material that works as a fertilizer for farming. Anaerobic digestion is an eco-friendly way to manage waste while producing renewable energy and useful materials.

2. Biomass Plant Process Units

A biomass plant comprises several key facilities and equipment designed to efficiently convert biomass materials into energy. Each part of the system has an important job in handling, converting, and producing energy. Different types of valves are used to control the flow of materials, steam, and gases safely and efficiently.

1. *Biomass Storage and Handling Unit:* This unit stores raw biomass, like wood chips, crop waste, or organic waste, and prepares it for processing. It includes hoppers, conveyors, feeders, and shredders to move and size the biomass efficiently. A hopper is a large funnel-shaped container that holds bulk materials and controls their flow into processing equipment. A conveyor is a system with belts, rollers, or chains that moves biomass or other materials from one place to another. A feeder controls the flow of biomass from storage units, like hoppers, to processing machines, ensuring a steady supply. A shredder cuts large pieces of biomass into smaller sizes for easier processing.
2. *Combustion or Boiler Unit:* In the combustion unit, biomass is burned in a boiler to produce heat. The heat is used to convert water into high-pressure steam that drives a turbine.

3. *Gasification Unit:* In a gasification unit, biomass is heated in a controlled, low-oxygen environment to produce syngas (a mix of hydrogen, carbon monoxide, and carbon dioxide). Key equipment includes gasifiers, cyclones, and gas cleaning systems. A gasifier turns solid biomass, like wood or agricultural waste, into syngas by heating it with a small amount of oxygen or steam, which breaks it down into gases instead of burning it. A cyclone separator removes dust and solid particles from the gas. Gas cleaning systems, with filters or scrubbers, remove impurities like tar and sulfur to make the syngas cleaner for use.
4. *Anaerobic Digestion Unit:* This unit processes organic materials like food waste or animal manure in sealed tanks called digesters. Bacteria break down the materials in the absence of oxygen, producing biogas (methane and carbon dioxide) and digestate. Key equipment includes digesters, mixers, and gas storage tanks. A mixer keeps the organic materials, bacteria, and nutrients evenly spread inside the digester. This helps the digestion process work smoothly and prevents floating layers or stagnant spots. A gas storage tank safely holds the biogas produced during anaerobic digestion until it is needed for use.
5. *Steam Turbine and Generator Unit:* In the turbine unit, high-pressure steam from the boiler drives a steam turbine connected to a generator, producing electricity.
6. *Flue Gas Cleaning and Emission Control Unit:* Flue gas is the smoke and gases released when biomass burns. It contains dust, ash, and harmful chemicals. The flue gas cleaning and emission control unit removes these pollutants to keep the air clean. Filters trap dust and ash. Scrubbers wash out harmful chemicals like sulfur dioxide. Electrostatic precipitators use electric charges to remove fine particles. This system helps reduce air pollution and follows environmental rules to protect the air and environment.
7. *By-Product Handling Unit:* This unit stores and manages by-products like ash, biochar, and digestate. Ash is the leftover powder after biomass burns. Biochar is a charcoal-like material made by heating biomass without oxygen, used to improve soil. Digestate is the nutrient-rich liquid or solid left after anaerobic digestion, often used as fertilizer.

   The unit contains different equipment, such as ash conveyors, storage silos, and pumps for liquid digestate. Ash conveyors move ash from the combustion area to storage or disposal. Storage silos are tall containers that hold processed biomass like wood chips, pellets, or ash. Processed biomass is biomass that has been treated or prepared for use in energy production. Pumps for liquid digestate transfer the nutrient-rich liquid left from anaerobic digestion to storage or fields for use as fertilizer.

Figure 12.3a shows a biomass plant adjacent to a farm, surrounded by green fields and yellow crops. The facility includes large anaerobic digestion tanks (the large domed structures), storage silos (cylindrical storage units next to storage

FIGURE 12.3 (a) A biomass plant integrated with a farm, utilizing organic waste to produce biogas. (b) A biomass plant from burning the biomass till electricity generation. (With permission: Shutterstock).

tanks) to keep wood chips or fermented feedstock, and processing units that convert organic waste into bioenergy. Several industrial vehicles and transport trucks are visible, indicating the movement of raw materials and processed biofuel.

Figure 12.4 shows how biomass fuel is used to generate electricity. Biomass sources, such as crops, wood, animal waste, sewage, and municipal waste, are burned in a burner to produce heat. This heat turns water into steam in a boiler, which moves through pipes to spin a turbine. The spinning turbine powers a generator that produces electricity. The electricity then passes through a power transformer to change the voltage and is sent to homes and buildings. A chimney releases steam and some gases after filtering to reduce pollution. This process converts biomass energy into usable electricity in an eco-friendly way.

3. Valve Types

Biomass Storage and Handling Unit

a. *Knife Gate Valves:* Knife gate valves control the flow of solid biomass and can cut through bulky or fibrous materials. They are also used for slurry and larger materials like wood chips, food waste, and animal manure. Slurry materials are thick, wet mixtures of solids and liquids. They can flow like a liquid but also contain solid particles. (For more details, see the section "Knife Gate Valves" in this chapter.)
b. *Slide Valves:* Slide valves control the release of biomass from storage containers or hoppers. They are similar to knife gate valves but work better for dry, fine materials and powders, such as ashes, wheat, fine wood particles, and small agricultural waste, rather than slurries. (More detailed information can be found in this chapter under the section "Slide Valves.")

Combustion or Boiler Unit

Valves are important for controlling flow and ensuring safety in a biomass plant's combustion or boiler unit.

- Control valves (globe and butterfly valves) adjust steam, fuel, and airflow for efficient combustion.
- Safety valves (spring-loaded and pilot-operated) release excess pressure to protect the system.
- Shutoff valves (gate and ball valves) stop the flow of fuel, steam, or water during maintenance or emergencies.
- Feedwater valves (angle and globe valves) keep the boiler water level steady. (More details are provided in the "Feedwater Valves" section.)
- Blowdown valves (globe valves) remove impurities to prevent scaling inside the boiler.
- Air vent valves release trapped air when starting the system.
- Fuel control valves ensure steady fuel flow. Rotary valves handle solid biomass, while ball and globe valves control liquid or gas fuels. (More details are provided in the "Fuel Control Rotary Valves" section.)
- Steam valves (pressure-reducing and globe valves) control steam pressure and flow for downstream use.

Gasification Unit

In a biomass gasification unit, different valves control the flow of gases, biomass, and by-products. Rotary valves feed biomass into the gasifier while preventing gas backflow. Check valves ensure that syngas flows in one direction and prevents backpressure. Pressure relief valves release excess pressure to keep the system safe. Ball and butterfly valves provide on/off control for gas and air, while globe valves precisely regulate the flow of air and oxygen into the gasification chamber.

Anaerobic Digestion Unit

In an anaerobic digestion unit, check valves are used to ensure unidirectional flow of biogas from the digester to storage or utilization systems, preventing backflow. Pressure relief valves are critical for releasing excess gas pressure to maintain system safety. Additionally, ball valves are commonly used for on/off control in pipelines carrying biogas or slurry, ensuring efficient operation and maintenance flexibility.

Steam Turbines and Generator Units

In steam turbines and generator units of a biomass plant, control valves regulate the steam flow to the turbine, optimizing efficiency and power output. Safety relief valves are essential for releasing excess steam pressure to protect the system from overpressure conditions.

Flue Gas Cleaning and Emission Control Unit

In the flue gas cleaning and emission control unit of a biomass plant, butterfly valves regulate the flow of flue gases through scrubbers, filters, and other cleaning systems. Check valves prevent backflow of gases into critical equipment, while control valves precisely manage the injection of cleaning agents, such as water or chemicals, to reduce pollutants.

By-Product Handling Unit

In the by-product handling unit of a biomass plant, rotary valves are commonly used to control the flow of solid by-products, such as ash or char. Slide gate valves provide on/off control for isolating sections of the system during maintenance or operation. Additionally, butterfly valves may be used for handling and regulating bulk flow of materials in pneumatic or gravity-fed systems.

C

Carbon Captures Valves

1. Carbon Capture Process

There are different methods of carbon capture. The post-combustion carbon capture method, which is the most common, is explained here. The carbon capture process involves extracting CO_2 from industrial emissions or power plant flue gases to reduce greenhouse gas emissions. The process begins with flue gas treatment, where gases are cleaned of particulates and impurities. The cleaned gas then enters an absorption column, where a chemical solvent, such as amine, selectively absorbs CO_2 from the mixture. The CO_2-rich solvent is then heated in a regeneration unit to release pure CO_2, allowing the solvent to be recycled.

FIGURE 12.4 Carbon capture facility with insulated pipelines and industrial process units. (With permission: Shutterstock.)

The released CO_2 is compressed in a compression unit, transforming it into a dense or supercritical state for efficient transport. In some systems, CO_2 may be cooled in cryogenic units to a liquid state. Finally, the captured CO_2 is transported via pipelines or stored in geological formations, completing the process. Each stage is tightly controlled with specialized valves to maintain safe, efficient, and reliable operation.

Figure 12.4 shows a carbon capture plant and large insulated pipelines connected to the plant. Industrial process columns to capture CO_2, and storage facilities set in a remote landscape. The pipelines transport captured CO_2. The facility highlights the role of valves in regulating flow of carbon dioxide transportation as well as process units.

2. Valves in Process Units

In a carbon capture plant, various types of valves play critical roles in different process units to ensure efficiency and safety. Control valves regulate the flow of flue gases into the absorption columns in post-combustion systems. Isolation valves, such as ball or gate valves, are used in pipelines and absorber column units to isolate sections during maintenance or emergencies. Check valves prevent backflow in CO_2 transport pipelines, ensuring unidirectional flow to compressors or storage tanks. Pressure relief valves are critical in compressors and transportation pipelines, protecting the system from overpressure scenarios. In post-combustion carbon capture, anti-surge valves prevent pressure surges in the

CO_2 compression stages. Additionally, cryogenic valves are used if the process involves cooling CO_2 to a liquid state for transportation.

3. Type of Valves

a. **Control Valves**
- Regulate the flow of gases during the capture process.
- Essential for maintaining proper pressure and flow rates
- Example: Pneumatic or electric control valves.

b. **Isolation Valves**
- They are used for safely isolating parts of the system during maintenance or emergencies.
- Common types: Gate valves, ball valves, and butterfly valves.

c. **Pressure Relief Valves**
- Ensure system safety by releasing excess pressure.
- Critical for handling high-pressure CO_2 streams.

d. **Check Valves**
- Prevent backflow in CO_2 pipelines.
- Ensure one-way flow to protect equipment and processes.

e. **Special Valves**
- **Anti-surge Valves:** Prevent pressure surges in compressors.
- **Multi-port Valves:** Direct CO_2 streams to different parts of the system.

4. Materials and Design Requirements

Valves used in carbon capture systems must be designed to withstand high pressures, temperatures, and the corrosive nature of CO_2,. Materials such as stainless steel, Inconel, Hastelloy, and other corrosion-resistant alloys are essential to prevent degradation from carbonic acid, which forms when CO_2 mixes with moisture. Valves must also have robust seals to ensure tight shutoff and prevent fugitive emissions of CO_2, complying with environmental regulations. The design should accommodate high-cycle operations, precise flow control, and resistance to erosion caused by particulates in flue gases. For cryogenic processes, valves must perform efficiently at extremely low temperatures, while maintaining structural integrity and functionality under rapid temperature changes.

5. Challenges for Valves

Valves in carbon capture systems face challenges such as exposure to corrosive fluids, high operating pressures, and frequent temperature fluctuations, which can lead to material fatigue and seal failures. The risk of CO_2 leaks poses a significant safety and environmental concern, requiring advanced sealing technologies and stringent compliance with emission standards like ISO 15848. Additionally, maintaining precise flow control in dynamic processes like CO_2 absorption and compression demands high-performance actuation and control systems. Another challenge is ensuring long-term durability in systems with high cycling rates, as wear and tear can reduce operational efficiency and increase maintenance costs. Balancing these technical requirements with cost-effectiveness remains a critical challenge in valve design for carbon capture applications.

6. Example

Multi-port Valve in a Carbon Capture Plant

In a carbon capture plant integrated with hydrogen production, a multi-port valve can be used to manage the flow of captured CO_2-based hydrogen gas after compression. For example, a three-way multi-port valve can be installed downstream of the hydrogen compressor to divert the hydrogen between a storage tank for onsite use and a pipeline for transport to an external facility. This system allows the plant to adjust carbon dioxide use based on demand. Extra CO_2 can be stored when demand is low or sent through pipelines for selling when demand is high. A multi-port valve makes the system simpler by replacing multiple valves, reducing piping cost, and ensuring precise control of carbon dioxide flow.

Christmas Tree Valves

A subsea Christmas tree (see Figure 12.5) is an assembly of valves, pipe spools, fittings, and steel structures installed on a subsea wellhead to control the flow of oil or gas from the well. A Christmas tree is a key part of subsea production systems. It controls the flow of fluids, allows well intervention, and ensures safety by isolating the well when needed. The main valves used are choke valves and gate valves. Choke valves regulate flow rate and pressure to keep production

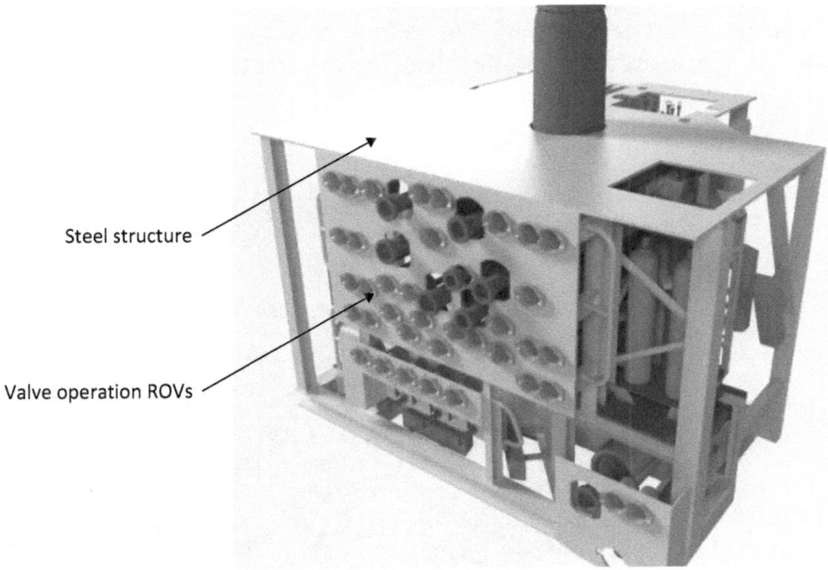

FIGURE 12.5 Subsea tree.

stable and protect downstream equipment located after the well. Gate valves are used for isolation, allowing the well to be opened or shutoff during production or maintenance. The two main types of gate valves for Christmas trees are slab gate valves and expanding gate valves. Slab gate valves use a flat gate to block or allow flow. Expanding gate valves create a tight seal through mechanical expansion, making them ideal for high-pressure environments. Both are through conduit gate valves, designed to keep the flow path smooth and minimize pressure drop. More information about slab and expanding gate valves is provided in Chapter 2. The figure shows a ROV (remotely operated vehicle), working on a subsea valve. The ROV holds a torque tool in its manipulator arm, which fits into the valve's torque receptacle. The torque tool rotates the valve to open or close it. This setup allows remote operation of valves in deep-sea environments where humans cannot go.

Subsea and onshore Christmas trees include various valves, each serving specific roles in well control and production. Gate valves on an onshore Christmas tree are illustrated in Figure 12.6. The *master valve* is the primary isolation valve located at the bottom of the tree, controlling the main flow of oil or gas from the well. It is typically a *gate valve* designed to handle high pressures and temperatures. Above the master valve is the *swab valve*, which provides access for well intervention operations, such as running tools or wireline into the wellbore. Well intervention refers to any operation performed or completed to maintain, repair, and enhance production. The *production wing valve* (PWV) is located on the horizontal outlet of the tree and controls the flow of production fluids into the pipeline and production facilities such as separator. The *kill wing valve* (KWV), on the opposite side, allows the injection of fluids into the well, such as for performing maintenance. Additionally, the *choke valve* is used to regulate flow rates and manage downstream pressure. The choke valve is typically located after the PWV in a Christmas tree or subsea tree. Together, these valves ensure safety, control, and efficient operation of the well.

FIGURE 12.6 Christmas tree TCG valves. (With permission: Shutterstock.)

Cryogenic Valves

1. Definition and Applications

Cryogenic temperatures are extremely low, typically below −150°C (−238°F). At these temperatures, materials undergo significant changes such as certain alloys become more brittle and prone to cracking. One of the primary applications of cryogenics is in the liquefied natural gas (LNG) industry. Figure 12.7 illustrates an LNG plant in Barcelona, highlighting various process units, particularly storage tanks. Cooling natural gas to a liquid state drastically reduces its volume, making storage and transportation more efficient.

The hydrogen industry also depends on cryogenic conditions for storing and transporting liquid hydrogen, which is essential for fuel cells and clean energy applications. Hydrogen is liquefied primarily to facilitate storage and transportation, as its gaseous form occupies a large volume, making it inefficient to store and move in bulk. Liquefied hydrogen also poses a lower risk of fire compared to gaseous hydrogen. Additionally, liquid hydrogen does not cause corrosion, unlike hydrogen gas, which can lead to hydrogen embrittlement corrosion, weakening metals over time. These factors make liquefied hydrogen crucial for applications in fuel cells, aerospace, and clean energy solutions.

FIGURE 12.7 Schematic of an LNG plant in Barcelona showing gas pretreatment, cryogenic heat exchangers, liquefaction trains, storage tanks, and export facilities. The role of cryogenic valves is highlighted in managing low-temperature gas and liquid flows. (With permission: Shutterstock.)

2. Valve Types

Cryogenic applications include ball valves, globe valves, gate valves, check valves, butterfly valves, and safety valves. Ball valves provide excellent sealing and quick on/off actuation, while globe valves are ideal for precise flow control in cryogenic systems. Gate valves, known for their low-pressure drop, are commonly used in applications requiring full flow or complete shutoff, such as LNG and liquid hydrogen transfer. Check valves prevent backflow, protecting equipment and maintaining pipeline integrity. Butterfly valves, being lightweight and compact, are suitable for low-pressure cryogenic systems. Safety valves, a crucial component in cryogenic systems, are designed to protect against overpressure by automatically venting excess pressure when it exceeds safe limits. This is vital in cryogenic applications where thermal expansion of liquids can rapidly increase pressure.

3. Essential Valve Design Considerations

Designing cryogenic valves involves several essential considerations to ensure safe and reliable operation in extremely low-temperature environments. Extended stems (see Figure 12.8 showing an extended stem butterfly valve) or bonnets are critical to minimize extremely cold transfer from the valve body to the actuator or handle, and packing. As a result, this extended distance reduces the risk of

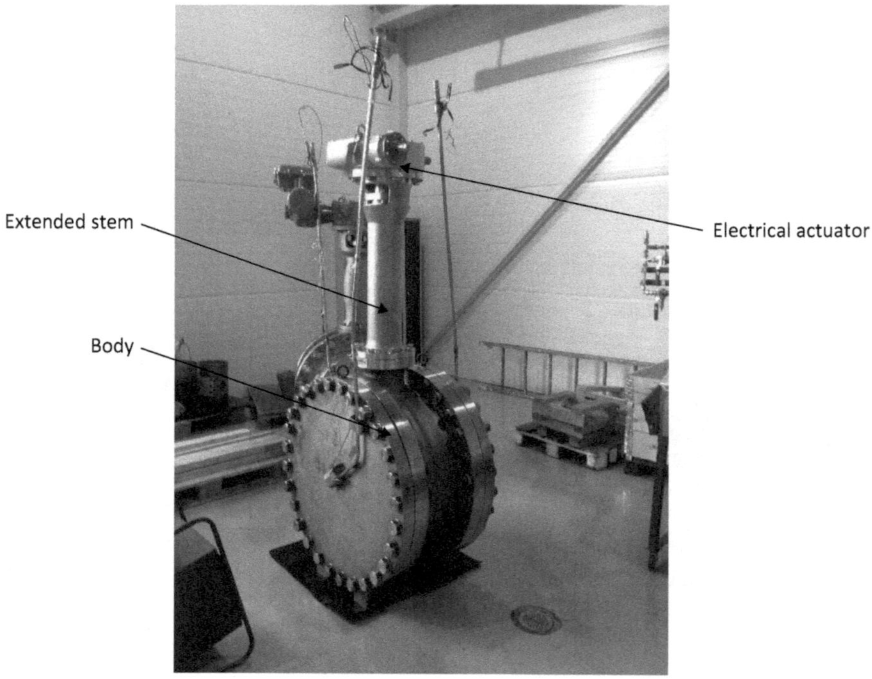

FIGURE 12.8 An extended stem butterfly valve for LNG plant.

valve packing freezing and maintaining proper functionality. Cryogenic materials, such as stainless steel (SS316) and nickel alloys, are commonly used due to their excellent toughness and resistance to embrittlement at cryogenic temperatures. Low thermal conductivity materials are also considered to reduce heat leakage. Pressure relief features are vital to accommodate the expansion of cryogenic liquids as they warm, preventing overpressure. Valves must have bubble-tight sealing to ensure no leakage of cryogenic fluids, often achieved through lip seal or similar materials for seats and seals. Cryogenic temperature testing and helium leak detection are used typically for cryogenic valves. A fire-safe design is essential for applications like LNG, which becomes flammable upon vaporization, to prevent leakage during external fire scenarios. It should be noted that cryogenic LNG itself does not pose a fire risk in its liquid state. However, in case of a leakage from the valve or exposure to an external fire, it can vaporize, and the resulting natural gas may become flammable if mixed with air in the right concentration. The internal surfaces of cryogenic valves should be polished and free from contaminants to prevent freezing or clogging. Vacuum-insulated jackets can be added to reduce external heat gain. Additionally, valves should be tested at cryogenic temperatures to verify their performance under actual operating conditions. These considerations together maintain durability, safety, and efficiency in cryogenic systems. Besides, antistatic devices in the valve design prevent the buildup of static charges that can ignite flammable cryogenic fluids like LNG or hydrogen.

The low thermal conductivity requirement in cryogenic valves is crucial to minimize heat transfer between the extremely cold cryogenic fluid and the surrounding environment. Cryogenic systems operate at temperatures as low as –150°C (–238°F) or lower, where even small amounts of heat ingress can cause vaporization of cryogenic liquids. The heat ingress inside the cryogenic LNG valve results in energy loss, increased pressure, and ice and frost formation on the valve that can damage the valve and cause safety hazards. Insulation jackets are protective enclosures used around cryogenic valves and piping to minimize heat transfer and maintain the extremely low temperatures of cryogenic fluids. These jackets create a vacuum layer, which acts as a barrier to heat transfer, significantly reducing thermal ingress.

4. Valve Standards

Standards addressing cryogenic valves include BS 6364, which focuses specifically on valves for cryogenic service, covering design, testing, and marking requirements; ISO 28921, which provides guidelines for floating and trunnion-mounted ball valves used in cryogenic applications; and API 600 and API 602, which outline requirements for gate valves suitable for industrial use, including cryogenic environments. Additionally, ISO 15848-1 addresses fugitive emissions performance in valves, crucial for handling volatile cryogenic fluids, and API 607 ensures fire-safe design in valves used in flammable cryogenic systems like LNG. These standards collectively ensure safety, reliability, and operational integrity in cryogenic valve applications.

D

Depth Rating

The maximum water depth at which a subsea valve can reliably operate, typically measured in meters or feet, such as 3,000 meters or 10,000 feet.

Digitalization

Digitalization in the valve industry means using smart technologies, data analytics, and automation. It helps improve performance, monitoring, and maintenance of valves. Sensors, IoT (Internet of Things), and advanced software enable real-time monitoring. They also support predictive maintenance and better operational efficiency. Three-dimensional printing, or additive manufacturing, is part of digitalization. It uses digital technology to improve traditional manufacturing processes.

Digital Twin

1. Introduction

A digital twin is a virtual model of a physical system, process, or component. It uses real-time data like pressure, temperature, and flow to mirror real-world performance. Using smart valves with integrated sensors and modern communication is part of a digital twin. It allows operators to monitor the valve in a 2D or 3D model that is integrated with process data. In valve applications, digital twins help monitor, predict, and optimize performance. They are used in oil and gas, carbon capture, and renewable energy industries. Sensor data and advanced analytics enable predictive maintenance. They help detect issues like wear, leakage, or blockages before they occur. This reduces downtime and repair costs. Advanced analytics also models the real valve and simulates outcomes. Digital twins improve reliability, extend valve life, and support better decision-making.

2. Comparing Digital Twin with Digitalization and Machine Learning

Digital twin and digitalization are related but not the same. A digital twin is a virtual model of a physical asset, system, or process. It uses real-time data to monitor, simulate, and optimize performance. Digitalization, on the other hand, is the broader process of using digital technologies to improve efficiency and decision-making. Digitalization transforms business operations by making them data-driven and automated. A digital twin is a specific application of digitalization, providing insights into individual components or systems. Simply put, a digital twin focuses on virtual modeling and real-time analysis, while digitalization improves overall workflows and processes.

In valves, digitalization means using digital technologies like IoT sensors and cloud systems to collect and analyze data. This includes flow rate, pressure, and temperature, helping with better monitoring and maintenance. A digital twin goes further by creating a virtual replica of the valve, simulating its real-time behavior.

For example:

- *Digitalization* lets operators track a valve's condition (e.g., pressure, temperature, flow) from the control room.
- A *digital twin* can do more than just monitoring. It can identify simple patterns in operational data and predict failures and optimize processes during disturbances. By analyzing real-time and historical data, a digital twin enhances system performance and reliability. To maximize its potential, a digital twin should be integrated with machine learning. Machine learning enables it to detect anomalies, predict failures, and recommend predictive maintenance before issues arise. A digital twin alone can provide some level of predictive maintenance, but its accuracy and effectiveness significantly improve when integrated with machine learning. This helps reduce downtime, extend equipment life, and improve overall efficiency.

  Digital twin and machine learning both help predict and optimize system performance, but they work differently.
- A *digital twin* is a virtual model of a real valve. It uses real-time data and simulations to mirror its behavior and predict how it will respond to changes.
- *Machine learning* analyzes past and current data to find patterns and make predictions. It learns from trends to improve over time.

Both technologies predict future performance, but they use different methods. A digital twin relies on physical models and real-time data, while machine learning finds hidden patterns from past data.

3. Example to Compare Digital Twin with Machine Learning

Imagine monitoring a subsea valve in an offshore pipeline. A digital twin creates a virtual model of the valve using design data (materials, dimensions) and real-time sensor data (flow rate, pressure, temperature). It simulates the valve's behavior and predicts what will happen if conditions change. For example, if the flow rate increases by 20%, the digital twin might predict cavitation or stress.

Machine learning works differently. It analyzes past data (historical trends, failures) and current data (like increasing actuator torque) to find patterns. For example, it might detect that a valve could fail in 30 days based on similar past cases.

Both predict future performance, but they use different methods. A digital twin relies on real-time data and simulations, while machine learning finds patterns in historical and current data. Together, they provide better insights:

Conclusion: The digital twin predicts how the valve will behave under different conditions. Machine learning *predicts* why and when it might fail.

Directional Control Valves

A directional control valve directs the flow of fluids or gases in hydraulic or pneumatic systems. It works by opening, closing, or switching flow paths between multiple ports. The valve uses a spool, poppet, or rotary mechanism to route fluid to actuators like cylinders or motors. The valve gets a signal from the system (electric, hydraulic, or mechanical). This signal tells the valve which way to direct the flow.

Directional control valves are classified by the following:

- Number of ports (e.g., two-way, three-way, and four-way)
- Number of switching positions
- Actuation method (e.g., manual, solenoid, pneumatic)

Directional control valves manage dynamic flow in hydraulic or pneumatic circuits. They control actuators or motors, like in wind turbine braking systems. These valves work under dynamic conditions, meaning they shift between different positions constantly in response to constant changes in process control. However, multi-port valves that used for flow diversion and directional change work under static condition, so they do not need to operate frequently. Directional control valves focus on routing flow to different path like multi-port valves. However, proportional control valves precisely modulate flow and pressure in a dynamic and demanding process condition. Directional control valves often use an electro-solenoid actuator. A solenoid shifts the spool to direct hydraulic fluid or air. Proportional control valves use proportional solenoids to adjust the spool position smoothly, allowing fine control of flow rate and pressure.

E

Emerging Technologies for Valves

Emerging technologies in valves are revolutionizing their performance and reliability to meet modern industrial demands. With advancements in energy systems such as hydrogen, LNG, and renewables, valves are being designed to handle extreme operating conditions, including cryogenic temperatures, high pressures, and corrosive environments.

Smart valves with IoT sensors and 3D printed components (additive manufacturing) are transforming the industry. They allow real-time monitoring, enhanced valve manufacturing techniques, and improved efficiency in critical applications. Digitalization is also changing valve monitoring and maintenance. With IoT sensors and real-time data analytics, industries can now use predictive maintenance, which helps reduce downtime and improve system reliability. Additionally, digital twin technology enables virtual simulation and performance analysis before a valve is physically installed. This helps optimize design and functionality.

Other innovations include nanolinings, advanced coatings, enhanced sealing mechanisms, and hydrogen-compatible materials. These advancements are improving valve performance in energy, aerospace, and chemical industries, helping build a smarter and more sustainable future.

F

Feedwater Valves

Feedwater valves are critical components in boiler systems in biomass plants, regulating the flow of water into the boiler to maintain proper water levels and ensure efficient steam generation. These valves must provide precise control to match the rate of water input with steam demand, preventing issues like low- or high-water levels in the boiler.

Common feedwater valves include globe valves for precise flow control and angle valves for space-saving designs. In modern biomass plants, these valves often have automated controls and actuators, allowing them to adjust dynamically to system needs, improving performance and reducing manual effort.

Fuel Control Rotary Valves

A fuel control valve, like a rotary valve, regulates biomass fuel flow in a biomass boiler. It ensures efficient and stable combustion by controlling fuel input. The rotary valve has a rotating rotor with pockets or vanes. These carry measured amounts of fuel from the feed system into the boiler.

A rotary valve consists of a circular rotor inside a cylindrical housing. The rotor, positioned at the center, has a shaft that allows it to spin. Around the rotor, multiple vanes (or blades) are evenly spaced, creating compartments for material flow.

Material, such as biomass fuel, enters from the top inlet and fills the spaces between the vanes. As the rotor rotates, the vanes carry the material downward to the outlet, where it is discharged into the boiler or process system.

The arrangement of vanes maintains a seal, preventing backflow of air or gases. This function, known as an airlock, ensures a steady and controlled material flow, making rotary valves ideal for biomass fuel feeding and industrial applications.

G

Geothermal Plant and Valves

A geothermal power plant makes electricity by using heat from deep inside the Earth. It works by bringing up hot water or steam, which spins a turbine to produce power. Thereafter, the cooled water is sent back underground to keep the process going.

Figure 12.9 shows how a geothermal power plant works. On the left side of the picture, hot water from deep underground is brought up through a well. The heat from the water converts it into steam, which spins a turbine. The turbine then drives a generator, producing electricity. The electricity generated by the

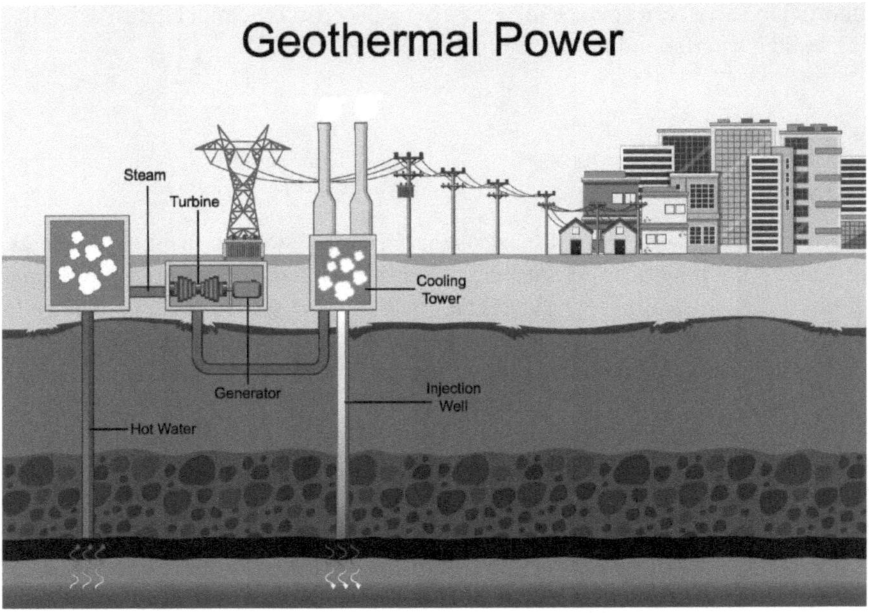

FIGURE 12.9 Geothermal power plant process. The figure shows how a geothermal power plant generates electricity. (With permission: Shutterstock.)

generator is sent through power lines to the residential houses and buildings in the background. Once the generator produces electricity, it goes to a transformer, which adjusts the voltage to make it suitable for long-distance transmission. The electricity then travels through high-voltage power lines to reach cities, homes, and businesses.

The steam cools down in a cooling tower and turns back into water. If the steam remained in gas form, it would escape into the air, and it would not be possible to inject it back into the reservoir. The cooled water is then sent back underground through an injection well so that the process can continue. This system helps generate clean energy without burning fuel, making it a good source of renewable power.

Valves are very important in a geothermal power plant because they control the flow of hot water, steam, and cold water. At the wellhead, choke valves control the pressure and flow of hot water or steam from underground, while gate valves are used to start or stop the flow when needed. Inside the plant, steam control valves (like globe or butterfly valves) regulate how much steam goes to the turbine, keeping it running smoothly and preventing high-pressure damage. In the cooling system, cold water valves (ball, butterfly, or globe valves) control the flow of cooling water to the condenser and cooling tower. This helps cool the steam and turn it back into water for reuse in the system. Check valves prevent water from flowing backward, keeping everything moving in the right direction. Safety valves (PSVs) protect the system by releasing extra pressure to prevent damage.

H

High-Integrity Pressure Protection System (HIPPS) Valves

1. Introduction

HIPPS is a safety system used in subsea production to stop overpressure in pipelines and equipment. It can be installed in different locations. One critical application is when it is placed after the wellhead and before the manifold to protect flowlines, risers, and other downstream sections from high pressure. HIPPS works by quickly detecting pressure spikes and shutting off the flow to prevent damage, ensuring safety and environmental protection.

2. HIPPS Valves and Types

A HIPPS has three main parts—pressure sensors, a logic solver, and HIPPS valves. The valves are usually fast-closing ball or gate valves. These valves are built to shut quickly and seal tightly under extreme pressure conditions. They shut off the flow to protect subsea equipment downstream. Subsea HIPPS reduces the need for traditional pressure relief systems, which means less process fluid is released into the environment. It also allows high-pressure wells to operate safely without upgrading the entire piping or pipeline. Since HIPPS valves isolate the flow, the piping after HIPPS does not need to handle extreme high pressure in case of a wellhead failure.

3. Example

Figure 12.10 illustrates a field development schematic for a HIPPS module that could be placed between the wellhead and manifold; this would result in the use of lower pressure rated piping and the possibility of using smaller sized piping

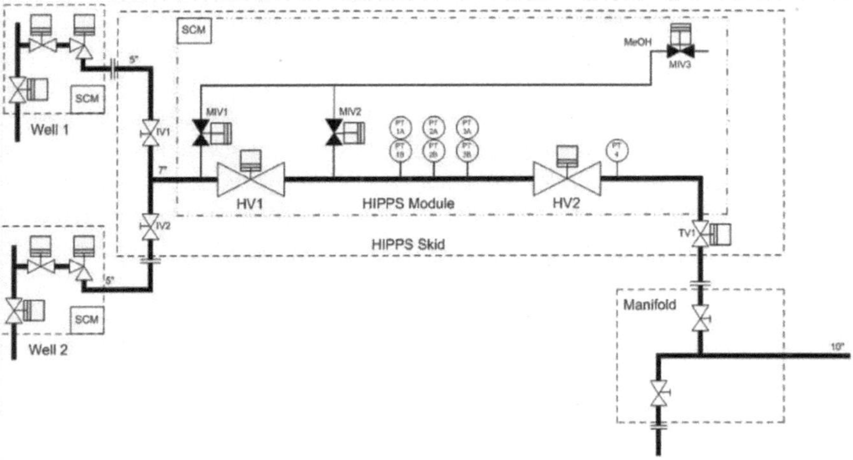

FIGURE 12.10 HIPPS module between the wellhead and manifold (HIPPS valve smart valve with IoT integration).

and a smaller manifold. Two valves connected to the HIPPS, HV1 and HV2, protect the low rating system which is the piping and valves inside the downstream manifold from the overpressure created in the wellhead and Christmas trees. Using two valves instead of one is a safety concept called redundancy, as explained in Chapter 10. If one valve fails, the other continues to function, ensuring system reliability. This setup enhances the reliability of the HIPPS.

Hydraulic Pressure Relief Valves

1. Introduction

Hydraulic pressure relief valves are essential in braking and pitch control systems, especially in wind turbines. Braking and pitch control systems are explained further in this chapter. Their main job is to prevent overpressure in hydraulic fluid to ensure safety and efficiency during braking. They relief excess hydraulic pressure back into the system safely, preventing fluid loss.

Excess pressure can damage brake cylinders, hoses, and actuators. These valves dynamically relieve pressure, protecting the system from overpressure while ensuring smooth and reliable braking.

2. Application

In braking systems, hydraulic pressure relief valves control pressure release in both normal and emergency braking. When the brake is engaged, hydraulic fluid pressurizes the braking mechanism. If pressure exceeds safe limits, the relief valve opens. It redirects excess fluid back to the reservoir, preventing overpressure. This protects the system from damage (e.g., blades and hub damage) and ensures consistent braking force. In wind turbines, these valves help stop the rotor safely during maintenance or extreme weather. This prevents turbine damage and improves operational safety.

3. Design and Working Principle

Hydraulic pressure relief valves are typically compact and robust, designed to handle dynamic and fluctuating pressures in real time. They consist of a spring-loaded mechanism or a pilot-operated system. When the hydraulic pressure exceeds a present limit, the valve opens gradually, allowing fluid to bypass or return to the reservoir. The spring or pilot system ensures the valve closes once the pressure normalizes. These valves are often integrated into the hydraulic circuit and are designed for frequent operation, offering reliable performance under demanding conditions. Materials like stainless steel or corrosion-resistant alloys are used to withstand the high pressures and harsh environments typical of braking systems.

Standard safety relief valves are not suitable for hydraulic braking systems because they are designed for emergency overpressure protection. They open fully at a set pressure and close once the pressure drops below a threshold. This on/off operation works well for static systems like boilers

TABLE 12.1
Comparing Hydraulic Relief Valve in Wind Turbine with Standard Safety Relief Valve in Oil and Gas

Feature	Hydraulic Relief Valve	Standard Safety Relief Valve
Primary function	Continuous pressure regulation and dynamic control	Emergency pressure release to prevent overpressure
Operation mode	Opens gradually and can remain in intermediate positions for precise control	Operates in an on/off manner (fully open or fully closed)
Response to pressure	Adjusts proportionally to maintain system balance	Opens fully when pressure exceeds a present threshold and closes after pressure drops
Application	Dynamic systems like hydraulic braking, and pitch control in wind turbine	Static systems like boilers, pipelines, and storage tanks
Frequency of use	Designed for frequent and continuous operation	Rarely activated; only during overpressure events
Integration	Integrates into hydraulic circuits with dynamic feedback systems	Acts as a stand-alone safety device; minimal interaction with system control
Durability for cycling	Handles frequent pressure changes without degradation	Not suitable for frequent cycling; designed for occasional use

and pipelines but is not ideal for dynamic systems like wind turbine brakes. Braking systems need continuous and precise pressure control for smooth operation. Hydraulic pressure relief valves are designed to open and close gradually, allowing controlled fluid release. They can handle frequent pressure changes and provide the responsiveness and durability needed for high-pressure, high-frequency applications. This makes them essential for wind turbine braking systems. Table 12.1 compares standard safety relief valves with hydraulic relief valves.

4. Advantages and Challenges

Hydraulic pressure relief valves have several advantages. They provide precise and continuous pressure regulation. This ensures smooth and reliable braking. They protect hydraulic components from overpressure damage. They also increase the safety and lifespan of the braking system. However, these valves have some challenges. They need regular maintenance to work properly. Debris or contamination in the hydraulic fluid can affect performance. Improper calibration or spring wear can cause malfunctions. This may compromise system safety. Despite these challenges, hydraulic pressure relief valves are essential in modern braking systems. They offer a strong and effective solution for managing hydraulic pressure.

Hydrogen Embrittlement

1. Introduction

Hydrogen embrittlement is a phenomenon where metals and alloys become brittle and prone to cracking due to the absorption and diffusion of hydrogen into the material. It is a critical challenge in valves and other components used in hydrogen-rich environments, especially as hydrogen energy gains traction as a clean fuel. This issue arises because hydrogen atoms infiltrate the metal lattice, causing dislocations and weakening the material's structural integrity. In valve technology, hydrogen embrittlement can compromise the mechanical performance and safety of components, leading to cracking or fracture in the valve body, seats, or stems, reduced fatigue life under cyclic loading as well as sealing failure.

2. Mitigation

New solutions are being developed to prevent hydrogen embrittlement in valves. These focus on better materials, coatings, and designs for safer use in hydrogen-rich environments. Hydrogen-resistant materials like nickel-based alloys and austenitic stainless steels (e.g., SS316L) help reduce cracking. Surface treatments, such as plasma nitriding and chemical passivation, create protective layers. These limit hydrogen penetration into the metal. Improved valve designs help reduce stress concentrations and prevent embrittlement. Hydrogen-specific standards like ISO 19880-3 ensure valves meet strict safety and performance requirements. These innovations help valves perform safely and reliably in hydrogen production, storage, and transport. They support the transition to clean hydrogen energy.

Hydrogen Valves

1. Hydrogen Plant Process

Hydrogen is produced using electrolysis or steam methane reforming (SMR). Electrolysis splits water into hydrogen and oxygen using electricity, often from solar or wind energy. The hydrogen gas is then collected, compressed, and purified for use in fuel cells, industry, or storage. Electrolysis is green because it produces zero emissions, uses renewable energy, and relies on water, a sustainable resource. In SMR, methane reacts with steam at high pressure and temperature to produce hydrogen and carbon monoxide. Carbon capture technology is often used in SMR to reduce CO_2 emissions. This modified process with carbon capture included is known as "blue hydrogen" production because it captures and stores the carbon dioxide (CO_2). Both methods need precise flow control, pressure regulation, and gas separation. Valves, compressors, and storage tanks are essential for handling hydrogen safely. It is stored in high-pressure tanks, transported through pipelines, or liquefied for long-distance pipelines.

Figure 12.11 illustrates a modern hydrogen production and transportation facility. The plant integrates renewable energy fuel production sources, including wind turbines with a focus on zero-emission hydrogen energy. The hydrogen is transported via a large white pipeline labeled "H_2 ZERO EMISSION." The large

FIGURE 12.11 A modern hydrogen plant showing a large pipeline for hydrogen transportation and a valve for flow control located on the pipeline. The background features storage tanks used for hydrogen containment before transport. (With permission: Shutterstock.)

valve on the pipeline is designed to regulate the flow of hydrogen and facilitate its transportation. This setup represents a hydrogen transportation and distribution system, where hydrogen is moved from production facilities to storage tanks or end users. In the background, two vessels or storage tanks are visible—one vertical and one horizontal. These vessels may serve as storage units for hydrogen before it is transported through pipelines. Additionally, the pressure vessels can be used for high-pressure hydrogen storage. Hydrogen can also be stored in liquid form for various advantages, including reduced volume for easier transport, lower fire risk, and minimized corrosion concerns. However, hydrogen is most commonly transported in gaseous form through pipelines.

2. Role of Valves in Hydrogen Plant

Valves are essential in hydrogen plants for control, safety, and efficiency. In electrolysis, control valves regulate water flow into the electrolyzer and manage the separation of hydrogen and oxygen gases. Isolation valves, like ball valves and gate valves, help shutoff sections of the plant during maintenance or emergencies. In SMR, control valves regulate methane and steam flow into the reformer. Pressure relief valves protect the system from overpressure in high-temperature reactions. Check valves prevent backflow in hydrogen pipelines. Cryogenic valves control liquefied hydrogen in storage and transport systems. Multi-port valves redirect gases between purification units, pipelines, and storage tanks. By managing the flow and ensuring system safety, valves are critical for hydrogen production, storage, and transport.

3. Materials and Design Requirements and Challenges

Valves in hydrogen applications must meet strict material and design standards for safety and performance. Corrosion-resistant materials like stainless steel, Inconel, and Hastelloy prevent wear and chemical damage. Valves in high-pressure systems, such as supercritical CO_2 or hydrogen, need strong construction to handle over 100 bar pressure. Seals and gaskets must provide tight shutoff to prevent leaks and meet ISO 15848 fugitive emission standards. In cryogenic applications, valve materials must stay functional at extremely low temperatures. In hydrogen applications, some valves require high-cycle durability such as control valves. High-cycle operation requires durable components to withstand frequent opening and closing. Modern valves should have easy maintenance, modular design for quick part replacement without a need to change the whole valve, and automation compatibility.

4. Example

Control Valve in a Hydrogen Plant

In a hydrogen plant using electrolysis, a control valve is crucial for regulating the flow of demineralized water into the electrolyzer, where water is split into hydrogen and oxygen. For instance, a pneumatically actuated globe valve might be used because of its precision in modulating flow rates and its ability to handle the low-pressure water supply with high accuracy.

In a hydrogen plant using SMR, a high-pressure control valve is used to regulate the flow of methane and steam into the reformer. These valves are typically constructed from corrosion-resistant materials like stainless steel or Inconel to withstand the high temperatures (600–1,000°C) and pressures (up to 50 bar) in the reforming process. The control valve plays a vital role in maintaining the correct gas mixture and ensuring optimal reaction conditions, which directly impacts hydrogen yield.

Hydropower Valves

1. Hydropower Process

The process in hydropower involves converting the energy of flowing or falling water into electricity. Figure 12.12 shows a hydropower plant, including key sections and parts. Water stored in a reservoir or naturally flowing in a river is directed through intake structures into penstocks, which are large pipes that channel the water toward turbines. An intake structure a type of gate or butterfly valve is the part of a hydropower plant that channels water from a reservoir, river, or other water sources into the hydropower system. It is typically located near the dam or riverbank and serves as the entry point for water to flow toward the turbines. The intake structure is designed to control the flow of water and prevent debris, fish, and other unwanted materials from entering the system. Penstocks are large, pressurized pipes or conduits that transport water from the intake structure to the turbines in a hydropower plant. These pipes are designed to withstand high water

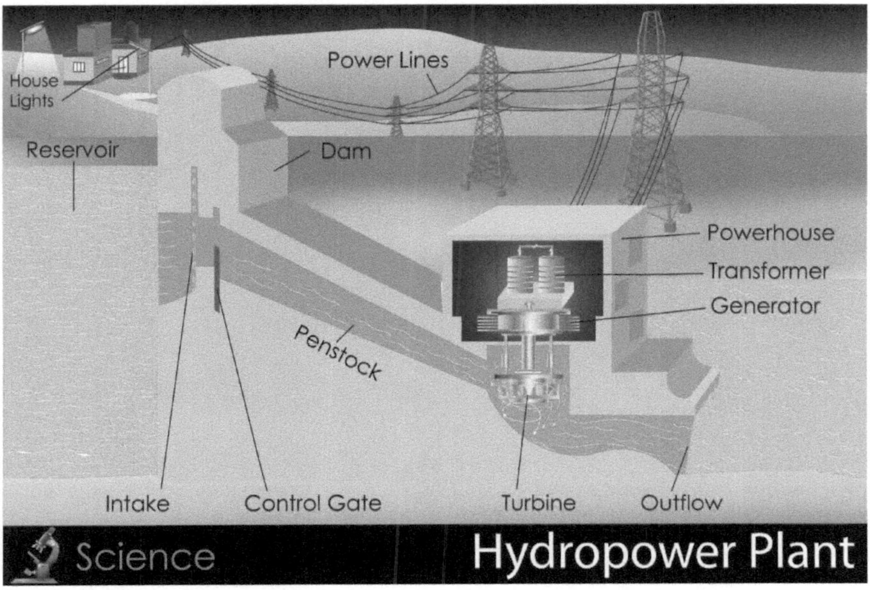

FIGURE 12.12 Hydropower plant schematic including the key components. (With permission: Shutterstock.)

pressure created by the height (or head) of the dam. The water's potential energy, due to its height or flow, is transformed into kinetic energy as it moves through the penstocks. This kinetic energy spins the turbines, which are connected to generators. The generators convert the mechanical energy of the turbines into electrical energy through electromagnetic induction. After passing through the turbines, the water is released back into the river via tailraces, ensuring a continuous flow.

A tailrace is the channel or conduit in a hydropower plant that carries water away from the turbines after it has been used to generate electricity. The water exits the turbines and flows into the tailrace, which directs it back into the river, reservoir, or other downstream water bodies.

Figure 12.13 shows a large water pipeline (penstock) with a big valve inside a hydroelectric power plant. The valve is used to control the flow of water before it reaches the turbines that generate electricity. The large pipe is part of a water intake system, which brings water into the plant. The big valve which looks like a butterfly helps open and close the water flow as needed. An actuator is installed on the valve, which is a device that automatically opens and closes the valve to control the flow of water. This helps regulate water movement without manual operation.

If the valve malfunctions, causing too much or too little water to flow to the turbine, it can create serious problems in the hydroelectric power plant. Excessive water flow increases turbine speed, which can lead to turbine mechanical damage as well as erosion, and cavitation. If not enough water reaches the turbine, it spins too slowly, reducing electricity output.

FIGURE 12.13 A large industrial valve with an actuator installed on a water intake pipeline in a hydroelectric power plant. The actuator automatically controls the valve, regulating water flow to the turbine. (With permission: Shutterstock.)

2. Role of Valves in Hydropower Plants

Hydropower plants use different valves to regulate water flow, ensure safety, and optimize performance. Butterfly valves are installed in the penstock near turbines to act as isolation valves. They allow or block water flow and are ideal for quick actuation and compact design. Operators use them to shut off water during maintenance or emergencies. Gate valves are used in large-diameter pipelines like intake structures. They provide a reliable seal and control high-pressure flows. Older hydropower plants used gate valves for turbine isolation because they were one of the most reliable options for large-diameter, high-pressure applications. As technology advanced, butterfly valves replaced gate valves for turbine isolation due to their faster actuation, smaller size, and easier operation. Some designs still use gate valves alongside butterfly valves for extra safety and flexibility during maintenance or operational changes. Both gate valves and butterfly valves can be used on a water outlet piping system where water is discharged from a dam. However, the choice between these two depends on operational requirements, flow control needs, and pressure conditions.

In bypass systems, needle valves help equalize pressure across the system. They precisely regulate water flow around turbines during start-up, shutdown, or maintenance. This ensures smooth transitions without damaging equipment. In addition, needle valves are installed in the bypass of butterfly or gate valves.

Ball and globe valves are used in auxiliary systems, such as the cooling water supply for turbines or lubrication systems for generators. These valves provide precise flow control or on/off to protect equipment and improve efficiency. To ensure system safety, pressure relief valves release excess pressure in pipelines or turbines. This prevents damage from overpressure conditions. Valves are essential for the reliable and efficient operation of hydropower plants.

3. Materials and Design Requirements and Challenges

Valve design in hydropower plants must handle high water pressures. They typically operate at 10–25 bar (145–360 psi) but can exceed 50 bar (725 psi) in high-head plants. Valves must also work reliably under changing flow conditions. High-velocity water can cause cavitation and erosion, especially in penstocks and turbines. To prevent this, valves are made from stainless steel, bronze, and duplex steel. These materials provide corrosion resistance and durability in wet environments. Butterfly, gate, and needle valves need strong sealing systems for tight shutoff and precise flow control under high pressure. Valves are frequently used during turbine start-up, shutdown, and maintenance, which increases wear over time. Cooling and lubrication system valves can cycle 10,000–50,000 times per year, depending on system demand. This requires designs that focus on long-term reliability and easy maintenance. A key challenge is balancing durability, precision, and cost-efficiency. Modern automation systems must also be integrated for efficient operation.

4. Example

Pressure Relief Valve in a Hydropower Plant

In a hydropower plant, a pressure relief valve is often installed on the penstock near the turbine to protect the system from overpressure conditions. For instance, during an emergency shutdown or sudden load rejection, the flow of water to the turbine may be abruptly stopped by the turbine's isolation valve (such as a butterfly valve). This sudden stoppage can cause a water hammer effect, generating a significant pressure surge in the penstock.

To prevent damage to the penstock or other components, a spring-loaded pressure relief valve or a hydraulic-operated relief valve opens automatically when the pressure exceeds a preset limit. The excess water is diverted to a safe discharge area, such as a spillway or tailrace, to relieve the pressure.

Hyperbaric Testing

A specialized term describing the testing of subsea valves under simulated subsea conditions, including extreme pressures and temperatures, to ensure functionality and durability. This test is done in a hyperbaric chamber as a part of valve qualification tests. Qualification test is explained in more detail in Chapter 6.

I

Industry 4.0

Industry 4.0, also known as the Fourth Industrial Revolution, refers to the ongoing transformation of traditional manufacturing and industrial practices through the integration of modern technologies like digitalization, the IoT, artificial intelligence (AI), and machine learning,

IoT

In the valve industry, IoT is used in smart valves. These valves have sensors and connectivity. They provide real-time monitoring of process data, valve position, etc., and remote control in industrial settings. IoT works through sensors, a control system, and internet connectivity. Sensors collect data such as pressure, temperature, or flow. The data is then processed and transmitted via the internet to a control system. The control system analyzes the data in real time and, if necessary, sends a command to adjust the valve.

IoT is a network of connected devices, sensors, and systems. These devices share data over the internet. They collect, send, and analyze data (e.g. process conditions like pressure and temperature) in real time. IoT allows automation, monitoring, and control without human input. IoT improves efficiency and productivity. It automates tasks, optimizes processes, and reduces errors. It enables real-time monitoring and control. Businesses can track equipment, predict maintenance, and prevent downtime.

K

Knife Gate Valves

A knife gate valve is a type of gate valve designed primarily for on/off control of slurries, viscous fluids, and materials containing solids, such as biomass, pulp, or wastewater. Its unique feature is a sharp-edged blade (or knife) that cuts through tough or fibrous materials. Unlike standard gate valves, the knife gate valve's special design minimizes material buildup and allows for smooth sealing, even in systems handling abrasive or solid-laden flows. It is commonly used in low-pressure applications across industries such as biomass processing, and pulp and paper. Made of durable materials like stainless steel, knife gate valves are highly resistant to wear and corrosion. The valve operates by sliding the blade to open or close the flow path, making it ideal for systems where precise shutoff is critical. Standards such as API 600, ASME B16.34, and MSS SP-81 govern its design, dimensions, and performance. MSS SP-81 is a standard published by the Manufacturers Standardization Society (MSS) specifically for knife gate valves designed for slurries and industrial applications. Knife gate valves are usually designed as wafer-style valves (see Figure 12.14), which means they have a compact, lightweight body that fits between two flanges in a pipeline. The wafer design is especially suitable for low-pressure applications and for handling slurries, solids, or viscous fluids, as it results in less resistance to flow. However, for higher pressures or specific applications, flanged-end designs of knife gate valves may also be used.

M

Machine Learning

Machine learning in valve applications involves using algorithms to analyze past and current data to identify patterns, make predictions, and optimize performance.

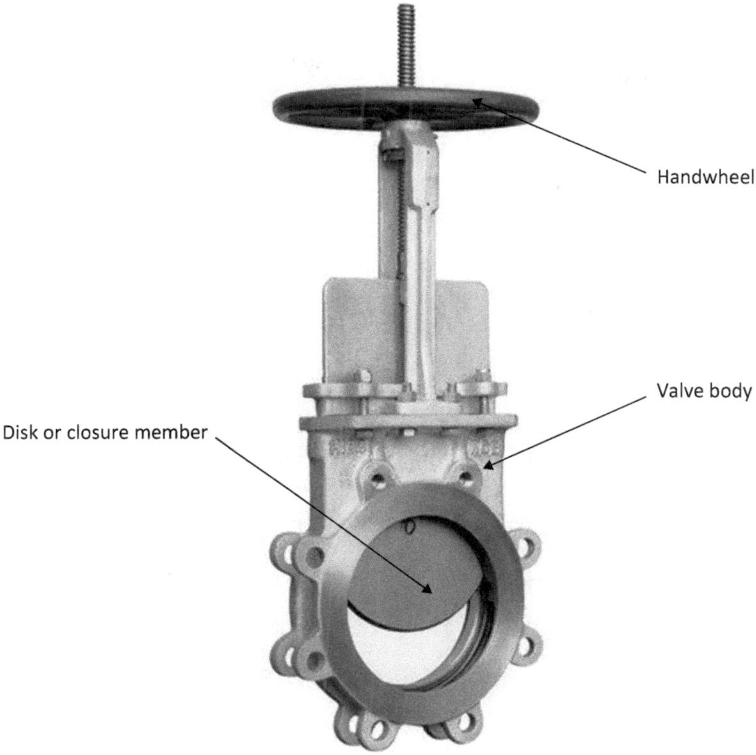

FIGURE 12.14 A wafer knife gate valve.

For example, a valve in an offshore pipeline is equipped with sensors that record data such as flow rates, pressures, and actuator torque. Over time, the machine learning system learns from this data and identifies patterns, such as how increasing torque often precedes actuator failure. Based on these patterns, it can predict that the valve might fail in 30 days if the torque trend continues. This allows operators to plan maintenance proactively, avoiding costly downtime. Unlike traditional methods, machine learning continuously improves its predictions by learning from new data, making it a powerful tool for predictive maintenance and operational efficiency in valve management. For more details on machine learning and its comparison with digital twin, refer to the "Digital Twin" section in this chapter

Multi-Port Valve

1. Role of Multi-Port Valve in Carbon Capture

A multi-port valve in carbon capture is a valve that features multiple ports for directing the flow of CO_2 or other process fluids to different pathways or process units within the system. These valves are particularly useful in carbon capture systems where flexible flow management is required, such as directing captured CO_2 between pipelines, compression units, or storage reservoirs. By allowing multiple flow paths through a single valve, multi-port valves reduce the need for multiple individual valves, simplifying piping configurations and saving space.

2. Application of Multi-Port Valve in Renewable Energy

Multi-port valves play a key role in renewable energy systems. They help manage and redirect fluid flow efficiently. In hydrogen production, they control the movement of hydrogen and oxygen gases. These valves switch between different ports to direct hydrogen and oxygen gases from the electrolysis unit to storage or distribution lines. In biogas plants, where bacteria break down organic matter without oxygen to produce biogas (a mixture of methane and carbon dioxide), multi-port valves regulate the flow of methane and CO_2. They control the movement of gases between the digester, storage units, and transportation systems, In geothermal energy, they redirect steam or working fluids. They control flow between turbines, condensers, and reinjection wells. Condenser in a geothermal plant is where steam is cooled and turned into water before reinjection into the well. For example, in the case of turbine maintenance, multi-port valve directs the steam flow to the condenser or vent it into atmosphere. In addition, these valves can switch the water injection flow between different wells. Chapter 2 provides more details about multi-port valves.

N

Nanolinings and Coatings

1. Introduction and Advantages

Nanolining and coating refers to the application of a thin layer of material, often in the range of nanometers, to improve the surface properties of a substrate. Nanolinings are predominantly applied to the internal surfaces of valves, targeting areas that directly interact with fluids. In contrast, nanocoatings can be applied to both internal and external surfaces of valves, with internal coatings protecting against fluid-related wear and corrosion, while external coatings shield against environmental factors. Both are used to increase performance, reliability, and durability of materials in industrial valves. Nanocoating and nanolining offer exceptional corrosion resistance by creating a protective barrier against corrosive environments and compounds such as moisture, chemicals, and saltwater. In addition, nanocoatings protect against environmental factors like UV exposure and abrasion. Nanolinings offer superior resistance to aggressive chemicals, and abrasive slurries flowing inside the piping system. Nanolinings reduce friction, allowing smoother operation and lower energy use. With high thermal stability, they perform well in extreme temperatures. These coatings provide strong protection in oil and gas, chemical processing, power plants, and renewable energy.

2. Applications

Lining

Oil and Gas Industry Diamond-like carbon (DLC) coating is a widely used nanocoating for valves in the oil and gas sector due to its superior properties such as high corrosion resistance, low friction, and exceptional wear resistance. DLC coating is primarily used on the inside of components, such as valves and pipes, to reduce wear, friction, and corrosion. However, it can also be applied to external surfaces. DLC coatings are particularly effective in subsea or offshore platforms, where valves are exposed to saltwater, abrasive slurries, and high-pressure fluid services during crude oil and gas processing. These coatings not only extend the lifespan of the valves but also ensure reliable sealing and reduced maintenance. Therefore, they are essential choice for harsh environments like deep-sea pipelines and drilling rigs.

Renewable Energy Ceramic-based nanocoatings, such as aluminum oxide (Al_2O_3) or titanium oxide (TiO_2), are commonly applied to valves in renewable energy systems, particularly in geothermal plants and biomass processing units. These coatings protect valves from high temperatures, abrasion, and chemical attack by geothermal brine or steam.

Coating

Oil and Gas Industry In the offshore oil and gas industry, external valve surfaces are exposed to harsh environmental conditions such as saltwater spray and high humidity. Nanocoatings, such as ceramic-based coatings, are applied to protect against corrosion, abrasion, and UV damage. These coatings form a robust barrier that prevents moisture, salt, and chemicals from interacting with the valve's surface, significantly extending its service life. For offshore installations, antifouling nanocoatings are often used to prevent the accumulation of biological growth, such as barnacles or algae. The durability and self-cleaning properties of nanocoatings also help reduce the operational costs associated with frequent surface treatments or replacements. Self-cleaning properties of nanocoatings mean that the surface can clean itself. These coatings create a very smooth and water-repellent layer (water does not stick to the coating surface), preventing dirt, oil, and other contaminants from sticking.

Renewable Energy In the renewable energy sector, valves in geothermal plants, wind farms, and solar installations face tough conditions like high temperatures, moisture, and pollution. Nanocoatings help protect these valves by preventing damage from heat, rust, and chemicals.

In *wind turbines*, while most valves are inside the turbine and protected, some external components, like hydraulic or cooling system valves, may benefit from UV-resistant or anticorrosion coatings to withstand environmental exposure. In solar power systems, coatings protect outdoor valves from heat, dust, and environmental pollutants. In geothermal plants, coatings shield valves from hot steam and salty water (after stem condensation), which can cause corrosion. These protective coatings help valves last longer, reduce maintenance, and keep renewable energy systems running smoothly.

P

Proportional Control Valves

1. Application

A proportional control valve in a wind turbine's hydraulic pitch control system regulates hydraulic fluid to adjust the blade angle in real time. This adjustment, called blade pitch, helps the turbine capture more energy in normal conditions and protects turbine blades from damage in extreme wind conditions by turning the blades parallel to the wind. This valve is not only used in wind turbines. It is also found in other industries like aerospace, where precise and dynamic (continuous) control of hydraulic fluid flow and pressure is needed.

2. Operation and Working Principle

Proportional control valves are used in turbines, particularly in hydraulic pitch control systems, to precisely regulate the flow and pressure of hydraulic fluid. Their operation is based on the principle of adjusting the valve's opening proportionally to an electrical signal. When the turbine's control system sends a signal to the valve, it moves a spool inside the valve body, altering the size of the flow passages. This adjustment allows hydraulic fluid to flow at controlled rates to actuators that adjust the angle (pitch) of the turbine blades. As a result of their continuous operation and smooth pressure regulation, proportional valves ensure optimal energy capture at all times, regardless of the wind condition. They also protect the turbine by turning the blades parallel to the wind in extreme weather, preventing damage. It is fascinating that some of proportional control valves do not actuate, yet they respond to control signals from the wind turbine's control system. Unlike actuators, which physically move components, proportional control valves regulate hydraulic flow. They operate by adjusting fluid pressure and flow rate, ensuring precise control. These valves are electrically controlled but hydraulically operated, allowing smooth and continuous regulation of actuator movement.

3. Design and Components

The core design of a proportional control valve consists of three main parts: a spool mechanism, an electromagnetic actuator, and a feedback system.

- The *spool mechanism* is a cylindrical component inside the valve body that slides back and forth to control the flow of hydraulic fluid. The position of the spool determines how much fluid passes through the valve, allowing precise adjustments.

- The *electromagnetic actuator*, often a solenoid, receives electrical signals from the turbine's control system. It converts these signals into mechanical movement, pushing or pulling the spool to the desired position.
- The *feedback system* ensures accuracy by monitoring the spool's position and sending data back to the control system. If the spool doesn't move as intended, the system adjusts the signal to correct it. This closed-loop mechanism guarantees precise and reliable operation.

Globe valves, while effective for throttling and on/off control, are not suitable for turbine applications like blade pitch adjustment. Globe valves operate using a disk that moves perpendicular to the flow to regulate it, which makes them slower and less responsive to rapid changes. Additionally, globe valves are typically designed for manual or basic automated control, lacking the real-time, dynamic adjustment capabilities needed for turbine systems. Thus, they also cannot handle high-frequency operations efficiently, which are essential in wind turbines where the blade pitch is adjusted continuously.

4. Advantages

Proportional control valves are indispensable in turbines because they provide dynamic (continuous), smooth, and precise control of hydraulic systems. In wind turbines, they regulate blade angles to maximize efficiency during normal operation and enhance safety by reducing stress on the turbine during extreme winds. Their spool mechanism ensures rapid and accurate flow adjustments, while the electromagnetic actuator offers a direct and reliable response to control signals. The feedback system adds an extra layer of precision, making these valves ideal for the highly dynamic environment of turbines. Their ability to handle frequent adjustments and respond quickly to system demands makes them superior to traditional valve types such as globe valves.

Q

Qualification Testing (Subsea)

Subsea qualification testing is a rigorous process to ensure that subsea equipment, including valves, can withstand the extreme environmental and operational conditions encountered in deepwater applications. This testing typically includes hyperbaric testing to simulate the high external pressures at depth, thermal cycling tests to assess performance under varying temperatures, and endurance testing to evaluate long-term reliability under repeated operations. Material testing for corrosion resistance, particularly against seawater and production fluids, is also critical. Additional tests, such as sealing integrity checks, pressure containment tests, and functionality verification under combined internal and external pressures, are conducted to confirm the equipment's robustness. Subsea qualification is guided by industry standards like API, ISO, and ASME to ensure safety, reliability, and compliance for extended service in remote, harsh underwater environments. More information about valve testing, including qualification testing, is available in Chapter 6.

R

Renewable Energy Valves

Valves play a crucial role in renewable energy systems to maintain safe and reliable operation. These valves control the flow of fluids and gases in various renewable energy processes, including solar, wind, hydropower, hydrogen, geothermal, and bioenergy. They are designed to withstand the unique challenges posed by renewable systems, such as high pressure, temperature variations, and exposure to corrosive or abrasive materials. As renewable energy technologies advance, the demand for specialized valves designed for these applications continues to grow.

ROV Compatibility

ROV valves are specialized subsea valves designed for operation and maintenance using ROVs in deep water environments. Manual subsea valves often include features designed for operation by ROVs, such as torque receptacles or T-handles, enabling maintenance and control without human intervention. Actuated subsea valves can have ROV operation to provide a backup or alternative method of operation in subsea environments where the actuator may become unavailable or inoperable.

Figure 12.15 illustrates an ROV-operated through conduit gate valve designed for subsea manifolds. The ROV interface is located at the top of the valve, with "S" and "O" markings indicating the valve's shut and open positions, respectively.

S

Slide Valves

A slide valve is a type of gate valve designed for on/off control of dry bulk materials, such as powders, granules, and biomass pellets, making it ideal for gravity-fed systems. Its flat gate design allows precise flow control while preventing leakage of fine materials, and its compact structure ensures easy integration into hoppers, silos, and conveyors. Slide valves are commonly used in industries like agriculture, biomass processing, and other bulk material–handling applications. They are typically constructed from stainless steel, carbon steel, or aluminum, with sealing components made of durable materials like rubber or PTFE, depending on the application. Standards such as DIN EN 15089 and CEMA guide their design, safety, and performance. Slide valves are cost-effective and efficient for dry material flow control but are not suitable for handling liquids or slurries. Figure 12.16 illustrates a slide gate valve.

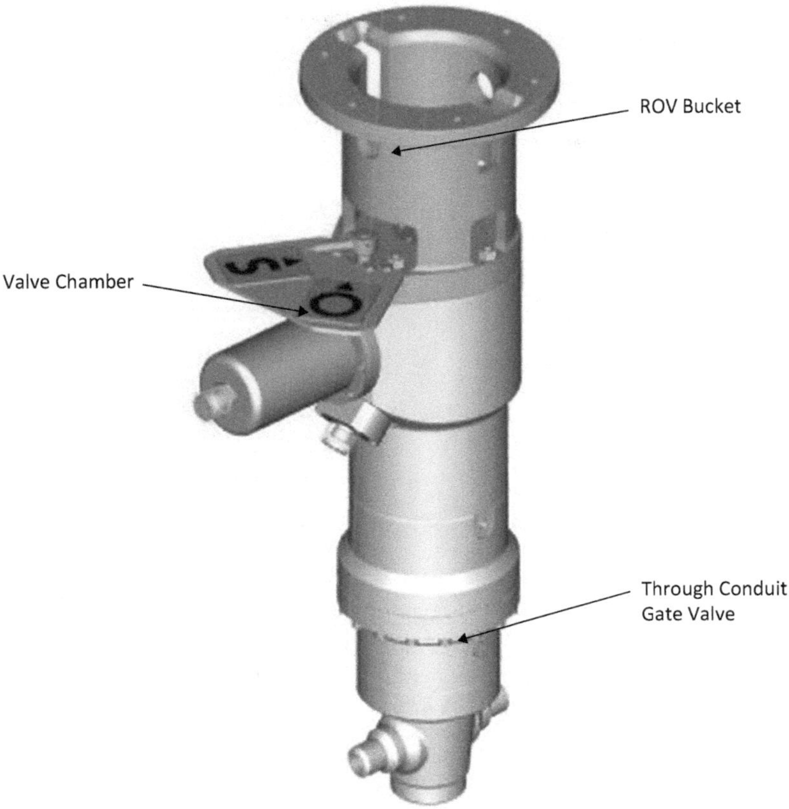

FIGURE 12.15 ROV-operated through conduit gate valve for subsea manifolds, featuring an interface at the top for ROV control and markings for "S" (shut) and "O" (open) to indicate valve positions.

Smart Materials

1. Definition

Smart materials are advanced materials that have the ability to respond dynamically to changes in their environment, such as temperature, pressure, light, pH, or magnetic fields. These materials can change their properties—such as shape, stiffness, conductivity, or color—in a controlled manner to perform specific functions.

2. Application

1. *Shape Memory Alloys (SMAs):* Smart materials are transforming valve technology in the oil and gas, carbon capture, and renewable energy sectors. In oil and gas, SMAs like nitinol (nitinol is an alloy of nickel [Ni] and titanium [Ti]) are used. Nitinol is primarily used in thermal

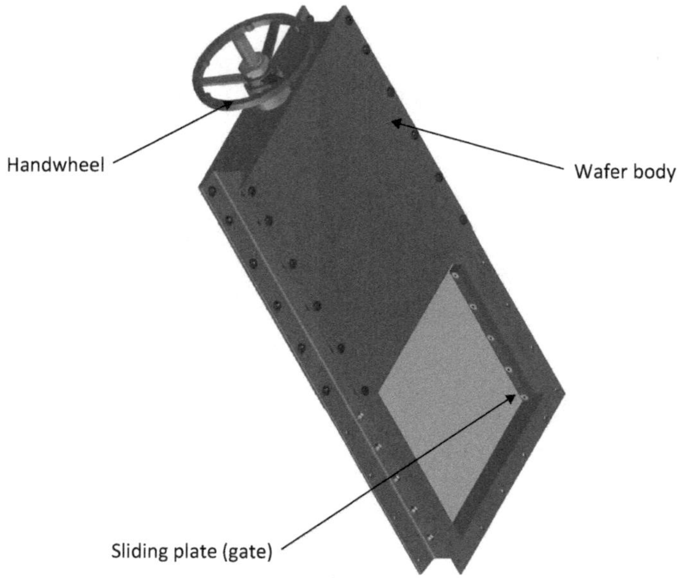

FIGURE 12.16 Slide gate valve. (With permission: Shutterstock.) A manual slide gate valve designed for controlling the flow of bulk materials or fluids in a biomass plant. The valve features a flat sliding plate (gate) that moves horizontally to open or close the flow path. It includes a manual handwheel or lever on top for operation, allowing precise control of biomass materials such as wood chips, pellets, or agricultural residues. The body is typically metallic, with flanged ends or wafer for secure connection to pipelines or chutes, making it ideal for biomass-handling and processing applications.

actuators, a specific type of actuator that relies on temperature changes to generate motion or force. Nitinol is predominantly used in thermal and linear actuators where temperature-driven, reliable, and precise actuation is essential. The function of a valve connected to a nitinol-based actuator is to control the flow of fluids (gases or liquids) by opening, closing, or modulating the flow based on temperature-driven actuation. The nitinol actuator provides the motion or force required to operate the valve without external power, enabling reliable and precise flow control in response to temperature changes.

2. *Piezoelectric materials* are materials that generate an electric charge when subjected to mechanical stress and, conversely, deform mechanically when exposed to an electric field. This dual property of mechanical–electrical energy conversion makes them highly versatile for applications in sensors, actuators, and energy harvesting systems. Piezoelectric materials enable precise actuation and flow control in high-frequency valves, enhancing performance in subsea or remote drilling operations.

3. *Magnetostrictive materials* change shape or dimensions when exposed to a magnetic field, enabling precise and rapid mechanical responses. Magnetostrictive materials offer rapid and accurate actuation in valves for critical applications, such as gas injection or pressure relief systems,
4. *Self-healing materials* can repair cracks and abrasions on their own making components last longer. In valves, they are used in coatings and seals to resist corrosion and wear, reducing maintenance and improving reliability. Self-healing coatings protect valve surfaces and seals, especially in corrosive environments. In carbon capture systems, CO_2 pipelines can accumulate water, leading to carbonic acid formation and CO_2 corrosion. The presence of moisture lowers the pH, making the environment more acidic. Hydrogel is a smart, self-healing polymeric material that responds to pH and moisture changes by absorbing excess water, preventing acid formation, and reducing the risk of corrosion. Electroactive polymers (EAPs) are smart materials that can be self-healing. They change shape when exposed to an electric field, making them ideal for flow control in renewable energy systems such as geothermal plants and hydrogen pipelines. EAPs are lightweight, flexible, and energy-efficient.

Smart Valves

Smart valves are advanced valves equipped with sensors, microprocessors (control systems), and communication modules that enable real-time monitoring and control. These valves collect data on parameters like pressure, flow rate, temperature, and valve position, transmitting this information to control systems through digital protocols such as Profibus or Wireless HART (highway addressable remote transducer). Profibus is an industrial communication network that connects actuators, sensors, and controllers in an automation system. Actuated valves integration with IoT platforms allows remote operation and predictive maintenance. This advanced data processing, monitoring, and control system is a clear example of digitalization in the valve industry. By enabling data-driven decision-making, reducing manual interventions, and improving operational efficiency, smart valves align with the principles of Industry 4.0. This advanced setup helps industries like oil and gas, chemical processing, and power generation achieve greater automation, reliability, and sustainability.

Figure 12.17 represents smart valve technology and digitalization in industrial settings. It combines traditional manual valves with a digital interface. The valve represents the integration of real-time monitoring and IoT. The holographic dashboard over the pipeline displays real-time data such as pressure equal to 8.15 bar and temperature of 27°C. The holographic dashboard indicates that the smart valve communicates with the control room and the operator can adjust the process parameters remotely. This concept is part of Industry 4.0, where industrial processes are enhanced with digital twin technology and IoT solutions.

FIGURE 12.17 A smart valve integrated into a digitally enhanced industrial system with a virtual screen. The image represents digitalization and user interface control. This technology allows remote monitoring and adjustment of process parameters such as pressure, temperature, and flow in modern industrial processes. (With permission: Shutterstock.)

Solar Energy Valves

1. Solar Energy Process

Solar energy is the process of capturing sunlight and converting it into usable energy, such as electricity or heat. Solar panels, made up of *photovoltaic* (PV) cells, absorb sunlight and convert it directly into electricity using a chemical process. In solar thermal systems, sunlight is used to heat a fluid, such as water or oil, which then generates steam to produce electricity or provide heating for homes and industries. *Concentrated solar power* (CSP) systems use mirrors or lenses to focus sunlight onto a small area, creating intense heat that drives turbines to generate electricity. Solar energy is a clean, renewable source that helps reduce dependence on fossil fuels and lowers greenhouse gas emissions.

Figure 12.18 shows a solar-powered water pumping system. Solar panels collect sunlight and turn it into electricity. The electricity powers a water pump, which pulls water from underground. The water is then sent up through inlet piping to a storage tank. From the tank, water flows through pipes and valves to be used for farming, drinking, or irrigation.

Various types of valves are used in this system to ensure efficient and safe operation. A check valve is installed after the underwater pump to prevent backflow, which could otherwise damage the pump and lead to operational issues like cavitation. Additionally, shutoff valves such as gate, ball, or butterfly valves must be

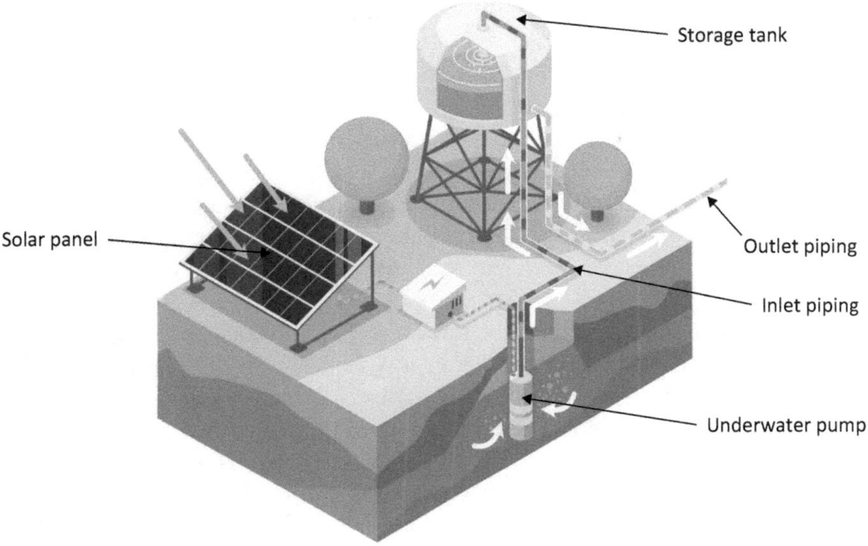

FIGURE 12.18 A solar-powered water pumping system using solar energy to produce electricity for pump operation and water transportation for drinking or irrigation. There are several valves required for this system. (With permission: Shutterstock.)

installed on the inlet and outlet of the piping system to regulate the start and stop of fluid flow. To protect the system from excessive pressure, pressure relief valves are essential, ensuring that overpressure does not damage the piping or equipment.

2. Solar Energy Valves Introduction

In solar energy systems, a heat transfer fluid (HTF) is a medium used to absorb, transport, and store heat generated from sunlight. In solar thermal systems, sunlight is stored as heat in receivers, collectors, or thermal storage systems, and HTFs like synthetic oils or water–glycol mixtures absorb this heat. HTFs circulate through solar collectors or receivers, absorbing heat from sunlight and transferring it to a heat exchanger where it generates steam from water to drive turbines. In solar thermal systems, water often acts as the HTF, absorbing heat from solar collectors and storing it in tanks for later use in heating or domestic hot water supply.

In solar energy systems, valves play a critical role in managing and controlling the flow of fluids, such as HTFs in CSP systems or water in solar thermal systems. It should be noted that there is no valve in PV cells as they convert sunlight directly into electricity. These valves ensure the efficient transfer of thermal energy from the solar collector to the storage or distribution systems, optimizing energy capture and utilization. The operational demands of solar energy systems, such as high temperatures, corrosive fluids, and precise flow control, necessitate the use of specialized valves designed to handle such conditions.

3. Type of Valves and Applications

Several types of valves are employed in solar energy systems, depending on the application. Ball valves and butterfly valves are commonly used for isolating flow in high-temperature piping. Globe valves provide precise control of fluid flow in systems requiring variable heat transfer rates. Pressure relief valves are essential for safety, releasing excess pressure that could damage the system during high-temperature operations. Additionally, check valves prevent backflow of fluids. Advanced valve materials like stainless steel and specialized coatings are used to withstand the corrosive nature of some HTFs and maintain long-term performance.

In CSP plants, valves regulate the flow of molten salts or synthetic oils, which act as HTFs, to capture and store solar energy for later use. In solar thermal systems for residential or industrial heating, valves control the movement of water or other fluids between the collectors, storage tanks, and distribution systems. The precise control offered by these valves enhances energy efficiency and safety. With the growing adoption of solar energy, innovative valve designs—such as automated and smart valves—are increasingly being used to optimize system performance, monitor conditions in real time, and reduce manual intervention.

Solenoid Valves

1. Introduction

Solenoid valves in renewable energy systems are critical components that regulate the flow of fluids or gases, enabling automated and precise control in processes like wind turbine lubrication, geothermal steam handling, or hydrogen storage. These valves operate using an electromagnetic coil that moves a plunger or spool to open or close the valve, either fully or partially, based on electrical signals. Their design is tailored to the specific needs of renewable energy applications—they are made from durable, corrosion-resistant materials like stainless steel or nickel alloys to withstand harsh environments, such as offshore wind farms or geothermal plants. Solenoid valves in these systems are often compact and lightweight to fit within space-constrained setups. They are robust enough to handle high pressures, temperatures, or specialized fluids like hydrogen.

2. Working Principle

A *solenoid valve* works by using an electromagnetic coil to control the movement of a plunger or spool within the valve body, which opens or closes the flow path for a fluid or gas. When an electrical current passes through the coil, it generates a magnetic field that changes the plunger position. The coil is de-energized to return the plunger to its default position, often assisted by a spring. In more advanced designs, proportional solenoid valves adjust the valve opening gradually based on the strength of the electrical signal, enabling precise flow control.

The solenoid valve contains a plunger and a spring within its assembly. In the *normally open position* (depicted on the right in Figure 12.19), as electrical

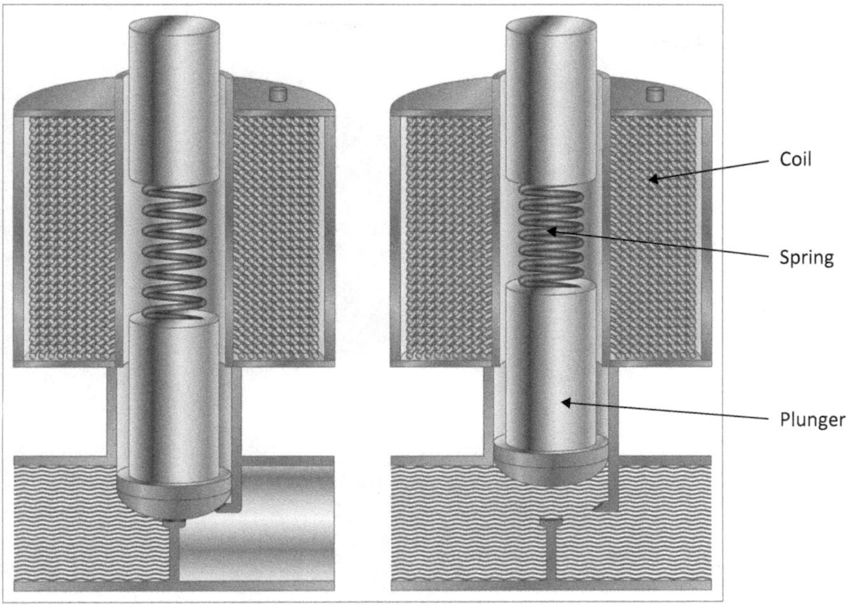

FIGURE 12.19 A solenoid valve operation in open and closed positions.

signal or current is sent from the logic solver to the solenoid, the coil generates a magnetic field. This magnetic field overcomes the spring's force, pushing the plunger upward. In this state, fluid flows freely through the valve under normal process conditions, providing hydraulic supply to the actuator of the valve. The spring exerts a downward force on the plunger, as illustrated in Figure 12.19 left side, in the absence of magnetic field. Thus, the spring force is closing the valve and stopping fluid flow.

Subsea Valves

1. Subsea Oil and Gas Process

The subsea oil and gas production process extracts hydrocarbons from reservoirs beneath the seabed. It uses specialized equipment designed for harsh underwater conditions. At the center of this system are subsea Christmas trees. These are assemblies of valves, chokes, and controls installed on wellheads. They regulate flow, prevent blowouts, and manage production. Subsea manifolds are assemblies of pipes, fittings, valves, and steel structures that collect and direct flow from multiple wells into pipelines. These pipelines connect to PLEMs (pipeline end manifolds), which link pipelines to floating production units or shore terminals. They also connect to PLETs (pipeline end terminations), which ensure smooth flow and pressure control. This integrated subsea system is essential for safe, efficient, and reliable oil and gas extraction and transportation.

FIGURE 12.20 A subsea manifold with subsea valves.

Figure 12.20 depicts a subsea manifold installed on the seafloor. It is an essential component of subsea production systems. The structure is designed to gather, control, and distribute the flow of oil and gas from multiple wells before transporting the fluids to a production facility. The large yellow structure protects critical components like subsea valves, piping, and control modules from external damage caused by water currents, debris. Subsea valves are located within the manifold, these hydraulically or electrically actuated valves regulate the flow of oil and gas from different wells. This subsea manifold contains several ball and through conduit gate valves.

2. Valves for Subsea Oil and Gas

Valves play a critical role in subsea oil and gas systems, ensuring safe and efficient flow control and isolation under extreme underwater conditions. Choke valves are used to control the flow rate of oil and gas from the wellhead, often integrated into subsea Christmas trees to manage production rates. Through conduit gate valves provide full-bore, flow isolation, critical in Christmas trees and manifolds to reduce pressure loss. Through conduit gate valves are also suited for piping applications in PLEMs, and PLETs, where their full-bore design supports efficient fluid transportation. Ball valves are employed for shutoff in various subsea applications, especially in the manifolds due to their durability and reliability under high pressure. Needle valves, along with other small-size valves, are used in subsea control systems for precise regulation of hydraulic fluids or chemical injection lines. Check valves prevent backflow in flowlines, protecting equipment from damage, while pressure relief valves safeguard the system by releasing excess pressure.

Hydraulic and small line valves in subsea systems are critical for controlling various auxiliary functions, such as actuating valves, injecting chemicals, or operating control systems. Hydraulic control fluids are transported in umbilicals to operate valves and actuators. Hydraulic fluids are typically synthetic or water-based, with properties like corrosion resistance, anti-freezing, and lubrication to handle high pressures and harsh underwater conditions. Chemical injection fluids include methanol or glycol to prevent hydrate formation. Paraffin inhibitors are needed to mitigate wax deposits, whereas corrosion inhibitors are used to protect pipelines and equipment. Finally, scale inhibitors prevent mineral buildup in subsea equipment and components. In subsea hydraulic and small line systems, check valves and small ball valves are widely used alongside other specialized valves such as needle valves due to their compact size and suitability. These valves are typically compact and designed for precise flow regulation, isolation, and pressure control. These valves are designed with corrosion-resistant materials like duplex, super-duplex stainless steel, or Inconel 625 to withstand harsh subsea environments, high pressures, and extreme temperatures.

3. Valve Challenges in Subsea Oil and Gas Industry

Subsea oil and gas valves work in harsh underwater conditions. They face high pressures, low temperatures, and corrosive seawater. Strong materials like duplex stainless steel or super alloys help prevent corrosion. Reliability is crucial because failures cause expensive downtime and difficult repairs. Repairs are challenging since valves are installed in remote subsea areas. These valves need advanced sealing to handle pressure and temperature changes. They must also work with different fluids. Automation and remote operation make design and maintenance more complex. Precise engineering ensures they function well for a long time.

4. Essential Subsea Valve Features

Subsea valves utilize several important and unique expressions to describe their features, performance, and operational requirements:

Compensation System

Pressure compensation is essential for subsea valves and actuators. It helps maintain performance and integrity under extreme underwater conditions. High hydrostatic pressure can deform structures, damage seals, and allow seawater ingress. This can cause corrosion, contamination, and failure. Pressure compensation balances internal and external pressures using open or closed systems.

Open systems let seawater enter a flexible bladder or diaphragm, equalizing pressure while protecting components. Closed systems are sealed and use noncompressible fluids like hydraulic oil to maintain balance. This prevents contamination and reduces environmental risks. Without pressure compensation, actuators can collapse, seals may fail, and leaks can cause hazards. By ensuring pressure balance, this system improves reliability, longevity, and environmental safety in subsea operations.

Depth Rating

Refer to the depth rating section in this chapter for additional information.

HIPPS Valves

Refer to the "High-Integrity Pressure Protection System (HIPPS) Valves" section in this chapter for additional information.

Hyperbaric Testing

Refer to the "ROV Hyperbaric Testing" section in this chapter for additional information.

Qualification Testing

Refer to the "Qualification Testing" section in this chapter for additional information.

ROV Compatibility

Refer to the "ROV Compatibility" section in this chapter for additional information.

T

Thermostatic Valve

Thermostatic valves are important in renewable energy systems like solar thermal, wind turbines, and biomass heating. They regulate temperature to improve efficiency and safety. Thermostatic valves can be designed to either open or close when the temperature rises.

Since the renewable energy valves explained in this section open due to high temperature, this paragraph focuses on thermostatic valves that are normally closed but open when the temperature rises. Inside the valve, there is a temperature-sensitive element, such as wax or fluid, that expands or contracts with temperature changes. When the temperature rises, this element expands, pushing a small piston or diaphragm inside the valve. The piston moves the closure part (valve plug or disc) away from the valve seat, opening the valve. When the temperature drops, the thermal element contracts. This removes pressure from the piston or diaphragm inside the valve. Then closure member returns back to the seat and closes the valve.

In solar thermal systems, thermostatic valves control the flow of HTF, which is usually a mixture of water and glycol. The solar collector absorbs sunlight and heats the HTF inside it. If the fluid is hot enough, the thermostatic valve opens, allowing the hot water to flow to the storage tank. If the fluid is not hot enough, the valve keeps it inside the solar collector to absorb more heat before it is sent to the storage tank

In wind turbines, they help control hydraulic and cooling system temperatures, preventing overheating or freezing. Wind turbines have cooling circuits to manage the temperature of the gearbox and generator. As explained further in this chapter, heat is removed by air or liquid coolers. When the temperature rises beyond the safe limit, the thermostatic valve senses the change and opens, allowing coolant (air or liquid) to flow and cool down the gearbox or generator.

W

Waste-to Energy Valves

1. Waste-to-Energy (WtE) Process Plant

A WtE plant turns trash into energy. It produces electricity, heat, or steam. The main methods are incineration, gasification, and anaerobic digestion.

Incineration burns waste at very high temperatures, usually above 850°C. The heat turns water into steam, which powers turbines to generate electricity. This method also reduces waste volume by up to 90%. To protect the environment, flue gas cleaning systems remove harmful pollutants like sulfur dioxide, dioxins, and heavy metals.

Gasification heats waste in a low-oxygen environment. Instead of burning completely, the waste breaks down into syngas (synthetic gas). This gas mainly contains carbon monoxide, hydrogen, and methane. Syngas can be used as a fuel or electricity generation in turbine.

Anaerobic digestion is used mainly for organic waste like food scraps, sewage, and agricultural waste. It happens in a sealed tank without oxygen, where bacteria break down the waste. This process produces biogas and fertilizers.

2. Valves

Valves in WtE plants must handle high temperatures, pressures, corrosive gases, and abrasive media such as ash and sludge. Common valves include gate valves and knife gate valves for isolating solid-laden fluids in ash-handling systems. Globe valves are used for precise steam and water flow control, and butterfly valves for regulating flue gases and air in combustion chambers. Safety valves protect boilers and pressure vessels by releasing excess pressure, while blowdown valves remove impurities from boiler systems to maintain efficiency. These valves are often made of durable, heat-resistant, and corrosion-resistant materials like stainless steel or high-alloy metals to ensure long-term reliability.

Wind Energy Valves

1. Wind Turbine and Energy Process

Wind turbines are mechanical devices that convert the kinetic energy of wind into electrical energy. The primary components of a wind turbine include the rotor blades, hub, nacelle, tower, and control systems. The tower, which supports the structure; the nacelle, housing key mechanical and electrical systems (rotor shaft, gearbox, generator, break system, yaw system, pitch control system, cooling system, and control system); and the rotor blades, which capture wind energy, causing them to spin. The hub serves as the central part of the rotor assembly and blades, connecting the rotor blades to the main shaft and other mechanical systems within the turbine, and drive a shaft connected to a gearbox. The gearbox increases the rotational speed to drive the generator, which converts mechanical energy into electricity.

Within a turbine, several systems—such as hydraulic pitch control, braking systems, lubrication mechanisms, and cooling circuits—ensure efficient operation. Wind turbines rely on several critical subsystems to operate efficiently and safely. The hydraulic pitch control system adjusts the blade angles to optimize energy capture during normal operation and protect the turbine from high wind speeds. This system uses hydraulic fluid regulated by valves to precisely control the blade pitch. Blade pitch refers to the angle or orientation of a wind turbine's blades relative to the direction of the wind. It is a critical mechanism that allows the turbine to optimize energy capture and ensure safe operation. The braking system ensures the turbine can stop safely during emergencies or maintenance. It utilizes hydraulic pressure controlled by valves to activate the brakes, halting the rotor's rotation. The lubrication system distributes oil or grease to moving components such as the gearbox and bearings, reducing friction and wear to ensure smooth operation. Additionally, the cooling system manages heat generated by the generator and gearbox by circulating coolant through these components to maintain safe operating temperatures. The yaw system adjusts the nacelle to align with the wind direction. These subsystems, supported by advanced valve technology, are integral to the turbine's reliability and efficiency.

Figure 12.21 illustrates inside view of a wind turbine nacelle, showing key systems that ensure efficient power generation. The nacelle is the main body of the wind turbine that sits on top of the tower. It houses all the key components responsible for converting wind into electricity.

Rotor and Blade System: Front Section

Rotor and blade system is located in front. This system is designed to capture the wind energy and turn it to motion. Blades are designed to move when wind flows over them. Hub is the circular section in front of the turbine which connects all blades and transfer the rotational movements to the shaft.

Pitch Control System

Inside the hub, there is pitch control system which is not visible in the picture. This system adjusts the blade angle for optimal capture and safety in high winds.

Main Shaft and Gearbox

Shaft transfers the rotational motion of the blades to the gearbox, which increases speed for efficient power generation. The slow rotation from the blades is

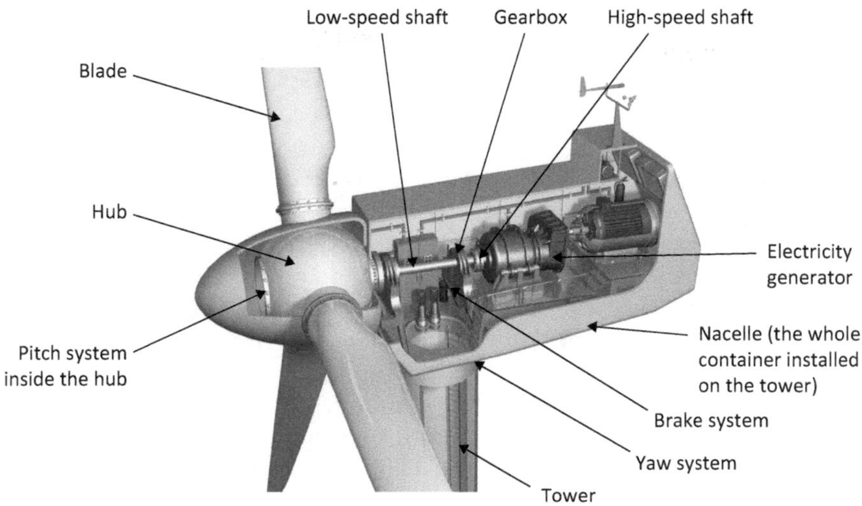

FIGURE 12.21 An illustration of a wind turbine nacelle inside view, showing key systems that ensure efficient power generation. (With permission: Shutterstock.)

transferred to the gearbox via the main shaft. The gearbox increases speed, converting low RPM (rotations per minute) into higher RPM needed for the generator. The shaft before the gearbox is low speed and the one after gearbox is high speed.

Electricity Generator

It converts mechanical energy from the gearbox output shaft into electricity.

Yaw System: Turbine Directional Control

It ensures that the turbine faces the wind for maximum efficiency. It is connected to sensors that detect the wind direction. Yaw motors rotate the nacelle to align the blades aligned to the wind. The yaw system is exactly where the nacelle connects to the tower.

Breaking System

It stops turbine rotation in high winds, emergencies, or maintenance.

Cooling and Lubrication System

It keeps the gearbox, generator, and hydraulic components at safe operating temperatures. The cooling and lubrication systems are inside the nacelle, mainly around the gearbox.

2. Valves in Wind Turbine

Wind turbines and wind energy systems rely on various types of valves to manage fluid flow, pressure, and safety across their critical subsystems.

In a hydraulic pitch control system, proportional control valves regulate hydraulic fluid. They adjust the blade angles to optimize energy capture. They also protect the turbine during high winds. (See the "Proportional Control Valves" section for more details.) The wind turbine control system continuously monitors wind speed and direction. Based on this data, the control system sends an electrical signal to the proportional control valve. The proportional control valve responds to the electrical signal by adjusting its opening size. A larger opening increases hydraulic flow, causing faster blade movement. Hydraulic safety valves are also used in the system. They work alongside proportional control valves but have a different role. These valves protect the hydraulic circuit from excessive pressure. They keep the system within safe limits. (See the "Hydraulic Pressure Relief Valves" section for more details.)

The braking system uses hydraulic pressure relief valves to control braking force. These valves help stop the rotor during emergencies or maintenance. (See the "Hydraulic Pressure Relief Valves" section for more details.) When the brake is engaged in turbine operation, hydraulic fluid pressurizes the braking mechanism. If pressure exceeds safe limits, the relief valve opens. It redirects excess fluid back to the reservoir, preventing overpressure and damage to turbine components like blades. Hydraulic safety valves in the pitch control and braking systems have a similar purpose. Both protect the hydraulic circuit from overpressure. However, they differ in design, operation, and application based on each system's needs.

Lubrication systems use solenoid valves, check valves, and thermostatic valves to control oil flow. These valves ensure that critical parts like the gearbox and bearings receive proper lubrication. This helps reduce friction and wear. Check valves maintain a one-way flow of oil or grease. They prevent backflow, which could lower lubrication pressure. Without proper pressure, components may suffer from wear or overheating. In wind turbines, lubrication is vital for high-speed rotating parts that experience heavy stress. A reliable check valve ensures continuous lubrication, even when pressure fluctuates. Most check valves use a spring-loaded mechanism with a piston or ball. This design allows flow in one direction while blocking it in the opposite direction. These valves are compact and made from stainless steel or brass for durability and oil compatibility. Solenoid valves provide precise control of lubrication flow. They are electrically operated and respond to signals from the turbine's control system. (See the "Solenoid Valves" section for more details.) Thermostatic valves help regulate oil or grease in lubrication systems. Thermostatic valves work differently in oil- or grease-based lubrication systems. In grease-based lubrication systems, if the grease gets too hot, the thermostatic valve opens to allow air or cooling fluid to reduce the temperature of the gearbox and generator.

The cooling system in a wind turbine prevents overheating of critical components such as the generator and gearbox. It relies on control valves to regulate coolant flow and thermostatic valves to maintain optimal operating temperatures. (See the "Thermostatic Valves" section for more details.) Control valves manage the flow of coolants like water, oil, or refrigerant, ensuring that the generator, gearbox, and power electronics stay within safe temperature limits. These valves adjust the flow rate dynamically based on real-time signals from temperature sensors, providing precise cooling under different load conditions and environmental changes. By controlling the coolant flow, they help dissipate excess heat from the turbine's high-speed operations, preventing overheating and ensuring efficient energy production. To withstand harsh conditions, control valves are made from corrosion-resistant materials and designed to handle high pressures and temperature fluctuations. In contrast, thermostatic valves operate automatically based on temperature changes, without needing signals from external control systems. These valves contain a temperature-sensitive element, such as wax or a bimetallic strip, that expands or contracts with temperature fluctuations. As the temperature rises or falls, the valve opens or closes, adjusting the coolant flow accordingly. This ensures steady and efficient cooling without requiring manual intervention. Together, control valves and thermostatic valves work to maintain the right temperature for wind turbine components, preventing damage and ensuring the system operates efficiently.

In yaw systems, directional control valves help adjust the nacelle's orientation to face the wind. (See the "Directional Control Valves" for more details.) These valves control hydraulic or electric motors that turn the nacelle. They direct hydraulic fluid or electrical signals to ensure smooth movement. By regulating flow and pressure, they help reduce wear and tear on components. This prevents stress on the system and improves reliability. Directional control valves also enhance safety by keeping the turbine aligned with the wind. This ensures efficient energy production and protects the blades and tower from excessive strain.

QUESTIONS AND ANSWERS

1. What is the primary purpose of a high-integrity pressure protection system (HIPPS) in subsea applications?
   A. To enhance the flow rate of oil and gas through pipelines
   B. To protect downstream equipment from overpressure conditions
   C. To ensure the structural integrity of the wellhead
   D. To replace traditional pressure relief valves in all applications

   Answer: Correct answer is B.

2. Which type of valve is typically used to control the pitch of wind turbine blades?
   A. Proportional control valve
   B. Ball valve
   C. Needle valve
   D. Check valve

   Answer: Correct answer is A.

3. What characteristic of proportional valves makes them suitable for dynamic applications in wind turbines?
   A. Their ability to fully open or close fluid pathways
   B. Their ability to operate at high temperatures
   C. Their ability to provide variable flow control based on input signals
   D. Their low manufacturing cost

   Answer: Correct answer is C.

4. What is the primary function of valves in a wind turbine's lubrication system?
   A. To control the direction of hydraulic fluid flow
   B. To regulate and distribute lubricating oil to critical components
   C. To monitor the temperature of lubricating oil
   D. To measure the speed of turbine rotation

   Answer: Correct answer is B.

5. Which of the following valves is best suited for handling abrasive media, such as ash or slurry, in a biomass plant?
   A. Gate valve
   B. Ball valve
   C. Knife gate valve
   D. Butterfly valve

   Answer: Correct answer is C

6. Why is valve material selection critical in biomass systems?
   A. To ensure compatibility with high temperatures and corrosive media
   B. To reduce energy consumption in the system
   C. To improve the combustion efficiency of biomass
   D. To enhance the automation of the system

   Answer: Correct answer is A. Biomass combustion and gasification processes often reach extremely high temperatures (above 800–1,000°C in some cases). Therefore, valves must be made of heat-resistant materials. Biomass fuels (wood, agricultural waste, municipal solid waste) release aggressive gases like sulfur compounds, chlorine, and acidic vapors during combustion.

7. Which type of valve is commonly used as the main inlet valve in the intake structure of a hydropower plant?
   A. Butterfly valve
   B. Needle valve
   C. Gate valve
   D. Ball valve

   Answer: Correct answer is C.

8. What is the purpose of a bypass valve in a hydropower plant?
   A. To reduce sediment in the water flow
   B. To measure the flow rate of water
   C. To control the speed of the turbine
   D. To divert water around the turbine for maintenance or emergencies

   Answer: Correct answer is D.

9. What is a critical design feature of valves used in hydrogen systems?
   A. Resistance to high-temperature conditions
   B. Compatibility with low-pressure applications
   C. High resistance to hydrogen embrittlement and leakage
   D. Ability to operate in noncorrosive environments

   Answer: Correct answer is C.

10. What material is commonly used for valves in LNG and hydrogen applications to handle cryogenic temperatures?
    A. Austenitic stainless steel
    B. Carbon steel
    C. Cast iron
    D. Aluminum

    Answer: Correct answer is A.

11. Which type of valve is commonly used in subsea systems for flow isolation?
    A. Through conduit gate valve
    B. Check valve
    C. Ball valve
    D. Both options A and C are correct

    Answer: Correct answer is D.

12. Why are materials like duplex stainless steel commonly used for subsea valves?
    A. They are cost-effective.
    B. They are lightweight and easy to handle.
    C. They resist corrosion and high pressure in seawater environments.
    D. They are flexible and allow for easy installation.

    Answer: Correct answer is C.

13. What is the main function of a subsea choke valve?
    A. To isolate flow during maintenance
    B. To control the flow rate of fluids
    C. To prevent gas leakage
    D. To connect subsea pipelines

    Answer: Correct answer is B.

14. What is a significant challenge for subsea valve operation?
    A. Difficulty in accessing valves for maintenance
    B. Leakage due to high viscosity of fluids
    C. Limited flow control options
    D. High cost of materials

    Answer: Correct answer is A.

15. Which type of valve is commonly used to manage high-temperature heat transfer fluids in concentrated solar power systems?
    A. Butterfly valve
    B. Globe valve
    C. Ball valve
    D. Gate valve

    Answer: Correct answer is C. In concentrated solar power (CSP) systems, high-temperature heat transfer fluids (HTFs) like molten salt, synthetic oils, or steam require valves that can withstand extreme heat and provide tight shutoff

16. Which emerging technology is expected to significantly impact valve performance in modern industrial applications?
    A. Manual valve operation
    B. Advanced materials for extreme environments, such as smart alloys
    C. Steam-powered actuation systems
    D. Mechanical lubrication systems

    Answer: Correct answer is B.

17. What role does the Internet of Things (IoT) play in valve technology?
    A. Allows manual monitoring of valve conditions
    B. Enables remote monitoring and control of valve performance
    C. Reduces the size of valves for compact systems
    D. Eliminates the need for actuators

    Answer: Correct answer is B.

18. Which emerging technology can help predict valve failure before it occurs?
    A. Machine learning and predictive analytics
    B. Enhanced manual inspection procedures
    C. Traditional mechanical systems
    D. Automated cleaning mechanisms

    Answer: Correct answer is A.

19. What is the purpose of 3D printing in valve manufacturing?
    A. To create traditional valve components at lower costs
    B. To allow customized designs and rapid prototyping of valves
    C. To replace all existing valve manufacturing processes
    D. To increase the weight of valves for high-pressure systems

    Answer: Correct answer is B.

20. Which emerging valve technology offers the highest precision in controlling flowrates?
    A. Proportional solenoid valves
    B. Manual butterfly valves
    C. Pneumatic control valves
    D. Fixed-orifice valves

    Answer: Correct answer is A.

BIBLIOGRAPHY

Ali, F., Dawood, A., Hussain, A., Alnasir, M. H., Khan, M. A., Butt, T. M., Janjua, N. K., & Hamid, A. (2024). Fueling the future: Biomass applications for green and sustainable energy. Discover Sustainability, 5(1), 156.

American Petroleum Institute. (2015). API 6DSS: Specification for Subsea Pipeline Valves. 2nd edition. API Publishing Services: Washington, DC.

American Petroleum Institute. (2018). API 17D: Specification for Subsea Wellhead and Tree Equipment. 4th edition. API Publishing Services: Washington, DC.

American Petroleum Institute. (2021). API 600: Steel Gate Valves for Petroleum and Natural Gas Industries. 13th edition. API Publishing Services: Washington, DC.

American Petroleum Institute. (2021). API 602: Compact Steel Gate Valves for Refinery Use. 9th edition. API Publishing Services: Washington, DC.

American Petroleum Institute. (2022). API 607: Fire Test for Quarter-Turn Valves and Valves Equipped with Non-Metallic Seats. 8th edition. API Publishing Services: Washington, DC.

American Society of Mechanical Engineers. (2012). Design and Fabrication of Pressure Vessels: Boiler and Pressure Vessel Code (ASME Section VIII Div.02). ASME Publishing: New York, NY.

American Society of Mechanical Engineers. (2021). ASME B16.34: Valves—Flanged, Threaded, and Welding End. ASME Publishing: New York, NY.

British Standards Institution. (2021). BS 6364: Valves for Cryogenic Service. 3rd edition. BSI: London, UK.

Crosby Valve Inc. (1997). Crosby Pressure Relief Valve Engineering Handbook. Technical document number TP-V300.

Dahiya, A. (2020). Bioenergy: Biomass to Biofuels and Waste to Energy. 2nd edition. Academic Press: New Delhi, India.

Fennell, M. (2011). Hydropower: Basic Concepts. 1st edition. CreateSpace Independent Publishing Platform: Scotts Valley, CA.

Gipe, P. (2009). Wind Energy Basics: A Guide to Home and Community Scale Wind-Energy Systems. 2nd edition. Chelsea Green Publishing: White River Junction, VT.

Gipe, P. (2016). Wind Energy for the Rest of Us: A Comprehensive Guide to Wind Power and How to Use It. 1st edition. Wind-Works.org: Bakersfield, CA.

Herzog, H. J. (2018). Carbon Capture. 1st edition. The MIT Press: Cambridge, MA.

International Organization for Standardization. (2013). ISO 28921-1: Industrial Valves—Isolating Valves for Low-Temperature Applications: Part 1: Design, Manufacturing, and Production Testing Requirements. 1st edition. ISO: Geneva, Switzerland.

International Organization for Standardization. (2018). ISO 19880-3: Gaseous Hydrogen—Fuelling Stations: Part 3: Valves. ISO: Geneva, Switzerland.

International Organization for Standardization. (2022). ISO 15488-1: Industrial Valves—Pressure Testing of Metallic Valves: Part 1: Test Requirements. 2nd edition. ISO: Geneva, Switzerland.

Manufacturers Standardization Society of the Valve and Fittings Industry. (2021). MSS SP-80: Bronze Gate, Globe, Angle, and Check Valves. MSS: Alexandria, VA.

Nesbitt, B. (2007). Handbook of Valves and Actuators: Valves Manual International. 1st edition. Elsevier: Oxford, UK.

Smit, B., Reimer, J. R., Oldenburg, C. M., & Bourg, I. C. (2014). Introduction to Carbon Capture and Sequestration. Imperial College Press: London, UK.

Sotoodeh, K. (2019). A review on subsea process and valve technology. Marine Systems & Ocean Technology, 14, 210–219.

Sotoodeh, K. (2021). Subsea Valves and Actuators for the Oil and Gas Industry. 1st edition. Gulf Professional Publishing: Houston, TX.

Sotoodeh, K. (2022). Cryogenic Valves for Liquefied Natural Gas Plants. 1st edition. Gulf Professional Publishing: Houston, TX.

13 Valve Life Cycle and Cost of Ownership

INTRODUCTION TO VALVE LIFE CYCLE AND COST OF OWNERSHIP

This chapter provides a comprehensive overview of the valve life cycle and the associated cost of ownership, structured to enhance understanding and ease of reference. The chapter is divided into five main sections: Valve Life Cycle Stages, Costs and Analysis, Maintenance and Reliability, Materials and Technology, and Energy-Related Terms. Each section groups related terms alphabetically to facilitate quick access while maintaining a logical flow.

- The Valve Life Cycle Stages section explores the various phases a valve undergoes, from design and manufacturing to installation, operation, maintenance, and eventual decommissioning. Understanding these stages is essential for optimizing valve performance, minimizing downtime, and ensuring safety and reliability throughout the valve's lifespan.
- The Costs and Analysis section focuses on evaluating the total cost of ownership, including initial purchase costs, operational and maintenance expenses, and replacement costs. This section helps engineers make informed decisions based on long-term value rather than just upfront costs.
- The Maintenance and Reliability section discusses strategies to extend valve lifespan, reduce downtime and associated costs, and improve reliability.

The remaining sections delve deeper into specific aspects that influence valve selection and usage. The Materials and Technology section examines the materials and advanced technologies used in valve design to enhance durability, corrosion resistance, and overall efficiency. Finally, the Energy-Related Terms section addresses the role of valves in optimizing energy usage, ensuring hydraulic and thermal efficiency, and supporting sustainability goals. Each section is thoughtfully organized to offer readers a clear, concise understanding of the key terms and concepts that define valve life cycle and cost of ownership.

PART 1: VALVE LIFE CYCLE STAGES

Commissioning

The *commissioning stage* of a valve occurs after installation and involves testing, adjusting, and validating the valve to ensure it operates correctly and meets

DOI: 10.1201/9781003640967-13

design specifications. During this stage, the valve is integrated into the system. Pre-commissioning activities include pressure testing to verify the valve's integrity, while commissioning involves functional tests such as leak testing and stroke testing (for control valves) to ensure proper operation under real operating conditions. Adjustments may be made to ensure the valve responds accurately to control signals and system requirements. Documentation of the test results and any final calibrations is completed during commissioning. This stage ensures the valve is ready for reliable and safe operation within the system. In short, *post-installation checks and function tests* are part of the *commissioning stage.*

Note: When leak testing is done under pressure conditions (e.g., hydrostatic or pneumatic pressure test), it is part of the pressure test. When leak testing is performed at operational conditions, it is considered a functional test.

Decommissioning

The decommissioning stage of a valve marks the end of its life cycle, involving the safe removal, disposal, or replacement of the valve. This stage typically begins when the valve is no longer functional, economically repairable, or required for the system. Key activities include isolating the valve from the system, depressurizing and draining the line, and dismantling the valve. If the valve is being replaced, the process may also involve preparing the site for installation of a new valve. Proper decommissioning involves handling the valve in a way that adheres to environmental regulations, especially if the valve has been in contact with hazardous fluids such as toxic chemicals or corrosive substances. Recyclable materials, such as metal components, are recycled whenever possible, while non-recyclable or contaminated parts are disposed of responsibly according to environmental regulations. This careful process not only ensures compliance with safety and legal standards but also prepares the system for seamless upgrades or replacement.

Design Stage

The *design stage* of a valve is the initial phase in its life cycle, where the functional and performance requirements are defined for valves according to specifications. During this stage, engineers determine the valve's type, size, materials, and pressure–temperature ratings based on the process data such as pressure, temperature, and fluid characteristics. Advanced computational tools, such as finite element analysis (FEA) and computational fluid dynamics (CFD), are often used to optimize the valve's design for stress resistance and flow efficiency. Standards like ASME B16.34 or API 600 are applicable to valve design to comply with industrial regulations. This stage is the foundation for the valve's functionality and reliability. Many parameters are addressed during valve design like cost, ease of maintenance, and long-term performance.

Installation

The *valve installation stage* after procurement stage involves placing the valve into the system in the yard and ensuring it is properly connected, aligned, and secured to function as intended. This stage includes preparing the piping system, verifying compatibility with flanges, gaskets, and fasteners, and ensuring proper torque and alignment to prevent leaks or operational issues. Technicians follow installation guidelines and standards, such as ASME and manufacturer instructions, to ensure the valve is installed correctly. Post-installation checks, including pressure and leak tests, are performed to confirm the valve is correctly installed and ready for operation. Post-installation checks and tests are part of the pre-commissioning and commissioning processes.

Maintenance

Valve maintenance, conducted during the operation stage, involves regular inspection, cleaning, and servicing to maintain optimal performance, safety, and durability. Key activities include checking for leaks, wear, or corrosion, lubricating moving parts, and replacing worn-out seals, gaskets, or components. Advanced strategies like predictive maintenance utilize IoT sensors to monitor valve conditions in real time. However, predictive maintenance requires a processing system that analyzes the sensor data, often incorporating machine learning techniques, to identify trends and predict potential failures before they occur. Periodic functional tests, such as stroke testing or pressure testing, are essential to verify operational reliability. Proper maintenance minimizes downtime, reduces repair costs, and ensures efficient operation within designed specifications.

Mechanical Completion

The mechanical completion stage of a valve refers to the phase where all physical installation and assembly tasks are finalized, and the valve is prepared for testing and commissioning. This stage ensures that the valve and associated components are installed according to design specifications, and manufacturer guideline, and are mechanically ready for operation. Mechanical completion is done before pre-commissioning. During pre-commissioning, activities such as pressure testing and nondestructive testing (NDT) are performed. Both mechanical completion and pre-commissioning are carried out by the pre-commissioning team

Key activities in the mechanical completion stage of a valve include final inspections to ensure proper installation, alignment, and secure connections to the piping system. Bolts, gaskets, and flanges are checked for correct torque, and preliminary leak or pressure tests are conducted to identify potential issues. Documentation, including completion checklists and installation records, is finalized to confirm adherence to design specifications and industry standards. Additionally, actuators or control mechanisms, if applicable, are mounted and verified for functionality. Once all tasks are complete and approved, the valve is handed over to the commissioning team for further testing and integration into the system.

Operation

The *operation stage* of a valve begins after the installation stage and involves the active use of the valve within a system to regulate, control, or isolate fluid flow. During this stage, the valve performs its designed function under specific operating conditions, such as pressure, temperature, and flow requirements. Proper operation ensures the system runs efficiently and safely, with the valve responding accurately to manual, automated, or remote controls. Regular monitoring during this phase is essential to detect any signs of wear, leaks, or performance issues, helping to maintain system reliability and prevent unexpected downtime.

Pre-commissioning

Pre-commissioning is the phase that occurs after mechanical completion and before commissioning, ensuring that all systems are installed correctly and ready for functional testing. It includes activities such as pressure testing, NDT, flushing, cleaning, and instrument calibration to verify system integrity and operational readiness. Once pre-commissioning is successfully completed, the system is handed over to the commissioning team for final functional tests.

Procurement

The *procurement stage* of a valve involves sourcing and acquiring the valve based on the specifications defined during the design phase. This stage includes evaluating suppliers, ensuring valve manufacturers (suppliers) compliance with industry standards (e.g., ASME, API), and follow up during manufacturing, testing, and final inspection. Procurement also requires careful consideration of cost, time, and quality to ensure the valve meets operational and safety requirements. Inspections and tests, such as factory acceptance testing (FAT), are often conducted during this stage to confirm the valve's suitability before it is delivered for installation.

Valve Life Cycle

The *valve life cycle* encompasses all stages a valve undergoes, from initial design and manufacturing to installation, operation, maintenance, and eventual decommissioning or replacement. Each stage involves specific activities aimed at ensuring optimal performance, reliability, and cost-effectiveness throughout the valve's service life.

Valve Optimization

Valve optimization involves improving a valve's design, selection, and operation to achieve maximum efficiency, reliability, and cost-effectiveness within a system. This process includes selecting the right valve type and materials for the application, minimizing energy losses (such as pressure drops), and implementing effective maintenance strategies like predictive maintenance to extend the valve's lifespan. Optimization ensures the valve performs effectively under its operating conditions while reducing downtime and overall life cycle costs.

PART 2: COST ANALYSIS

Cost-Effectiveness

Cost-effectiveness in the valve cost of ownership refers to achieving the best balance between the initial, operational, maintenance, and replacement costs while ensuring optimal performance and reliability over the valve's life cycle. A cost-effective valve minimizes total expenses without compromising safety, functionality, or process efficiency. For instance, investing in a high-quality valve with superior materials might have a higher initial cost but can reduce long-term expenses by requiring less maintenance, consuming less energy, and extending the replacement interval. Proper selection, routine maintenance, and energy-efficient operation are critical for enhancing cost-effectiveness, as they reduce unexpected repairs, downtime, and operational inefficiencies, leading to better overall financial performance.

1. Case Study: Cost-Effectiveness of a High-Performance Butterfly Valve

Background

A petrochemical plant needed to replace a standard metal-seated gate valve in its cooling water system due to frequent failures caused by corrosion and wear. The existing valve had an initial cost of $800 but required yearly maintenance costing $300, frequent repairs averaging $1,000 every three years, and replacement every six years due to severe corrosion. Over 12 years, the total cost of ownership (TCO) for this valve amounted to $8,400. The TCO over 12 years is calculated as follows—the initial cost of $800, plus yearly maintenance of $300 for 12 years ($3,600), plus repair and replacement costs of $1,000 every three years, occurring four times ($4,000), resulting in a total of $8,400.

2. Approach

To improve cost-effectiveness, the plant replaced the gate valve with a high-performance butterfly valve made of corrosion-resistant stainless steel, costing $2,000 initially. This valve had better durability, requiring maintenance every five years at $500 and no major repairs due to its enhanced materials and design. Its life cycle was projected to be 15 years, far exceeding the gate valve.

3. Results

Over 12 years, the butterfly valve incurred:

- *Initial Cost:* $2,000
- *Maintenance Costs:* $1,000 (two maintenance cycles)

- *No Repair Costs*
- *TCO:* $3,000

In contrast, the gate valve cost $8,400 over the same period. The butterfly valve's higher initial cost was offset by reduced maintenance and eliminated repair expenses, proving to be a cost-effective solution with enhanced reliability and reduced downtime.

4. Conclusion

This case study highlights the importance of considering long-term costs when selecting a valve. Although the butterfly valve had a higher upfront cost, its superior performance and durability resulted in significant savings, improved system reliability, and reduced operational disruptions.

Downtime

Downtime in valves refers to the period during which a valve is out of service and unable to perform its intended function due to maintenance, repairs, or failures. This can result from various issues, such as mechanical wear, corrosion, leakage, or actuator malfunctions. Downtime is a critical factor in valve operations because it can lead to production halts, reduced efficiency, and financial losses in industrial processes.

For example, in a refinery, if a critical control valve in a process line fails and takes 10 h to repair, it can result in significant production losses valued at thousands of dollars per hour. Downtime costs include not only the repair or replacement expenses but also the indirect costs, such as lost production, delayed operations, and potential penalties for unmet deadlines.

Minimizing downtime involves selecting durable valves during design phase, implementing preventive maintenance programs, and using real-time monitoring systems to detect issues early. By reducing downtime, companies can improve reliability, maintain steady operations, and optimize overall cost-effectiveness.

Initial Costs

The initial cost of a valve includes the upfront expense of purchasing the valve, along with costs associated with manufacturing and testing, customization, coatings, actuators, and accessories. This cost excludes installation and operational expenses, focusing solely on procurement costs.

1. Example of the Initial Cost of a Valve

The initial cost of a valve includes the base price, additional features, and testing expenses. For instance, a 6-inch carbon steel ball valve (Class 150) might have a base price of $1,200. Adding epoxy coating for corrosion resistance costs $150, while a pneumatic actuator adds $800. Testing expenses, such as hydrostatic and functional testing, contribute another $200. This brings the total initial cost to

$2,350, encompassing all procurement-related costs to make the valve operational, excluding installation or maintenance expenses.

Investment

Investment in the cost of ownership for valves refers to the upfront and ongoing expenses required to ensure the valve's long-term reliability, efficiency, and functionality. This includes the initial purchase cost, which may involve higher expenses for high-quality materials, coatings, testing, and customization to meet specific operational demands (e.g., extended bonnet and stem to facilitate wrapping and insulation around the valve). It also covers preventive measures, such as installing monitoring systems or smart valves with predictive maintenance capabilities, which help reduce downtime and unplanned failures. Additionally, investing in specialized designs and durable materials suited for harsh environments can minimize repair and replacement costs over the valve's life cycle. Training operators and maintenance personnel further enhances valve performance by providing better valve handling and more effective maintenance. Strategic investments in these areas lead to lower overall costs and improved cost-effectiveness throughout the valve's life cycle.

Life Cycle Cost Analysis (LCCA)

LCCA is a method used to evaluate the total cost of owning, operating, maintaining, and ultimately disposing of a product such as valve or a system such as piping over its entire life cycle. It considers all costs incurred from the initial investment to the end of the product's useful life, helping decision-makers assess long-term value rather than focusing solely on upfront costs.

The life cycle cost of a product or system encompasses all expenses incurred throughout its lifespan, from acquisition to disposal. This includes initial costs such as purchase, transportation, installation, and commissioning; operational costs like energy consumption and labor; and maintenance costs for routine servicing, repairs, and part replacements. It also accounts for downtime costs caused by system failures or inefficiencies and end-of-life costs such as decommissioning, recycling, or disposal. Additionally, indirect costs, such as insurance, training, and obtaining certificates regarding regulatory or environmental compliance, may also be considered. Life cycle cost analysis helps evaluate the TCO to ensure cost-effective decisions and long-term value.

Long-Term Value

Long-term value in the cost of valves refers to the overall benefits and cost-savings realized over the entire life cycle of a valve, taking into account its durability, reliability, efficiency, and reduced TCO. A valve with high long-term value minimizes expenses related to maintenance, operational inefficiencies, and downtime while delivering consistent performance in demanding applications.

For example, a valve made from corrosion-resistant materials may have a higher initial cost but offers extended service life, fewer repairs, and lower downtime, resulting in significant cost-savings over time. Long-term value emphasizes not just the purchase price but also the valve's ability to provide sustained performance and cost-effectiveness throughout its operational life.

Maintenance Cost

The maintenance cost of a valve includes all expenses related to ensuring the valve remains functional and efficient throughout its life cycle. This involves routine inspections, lubrication, replacement of worn-out parts, and repairs. Additionally, downtime cost—caused when the valve is offline for maintenance—adds to the overall maintenance expense.

Example: Total Maintenance Cost of a Valve

A control valve may require annual maintenance costing $300 for inspections and part replacements. Every five years, it might need a major overhaul costing $2,000. If the valve causes 10 h of downtime during repairs, and the production loss is valued at $500/h, the downtime cost for a single maintenance event would be $5,000. Over 20 years, these cumulative maintenance costs can significantly impact operational budgets, making proper planning and regular upkeep essential to minimize unexpected downtime and expensive repairs.

The total maintenance cost over 20 years for a valve includes the following:

- *Routine Maintenance (20 years):* $6,000 (annual cost of $300 × 20 years)
- *Major Overhauls (Every 5 years):* $8,000 ($2,000 × 4 overhauls)
- *Downtime Cost (during Major Overhauls):* $20,000 ($5,000 × 4 overhauls)

Total Maintenance Cost (20 Years): $34,000

This calculation highlights the importance of considering both direct maintenance expenses and downtime costs when evaluating the long-term financial impact of valve ownership.

Note: A valve overhaul is a detailed maintenance procedure performed to restore a valve to its optimal working condition. It involves disassembling the valve, inspecting all components for wear, corrosion, or damage, and replacing worn parts like seals and gaskets. The process also includes thorough cleaning to remove deposits and reassembling the valve to ensure proper functionality. Once reassembled, the valve undergoes rigorous pressure and functional testing to confirm it meets performance standards. Overhauls are typically scheduled every three–five years, depending on the valve's type, operating conditions, and criticality, to prevent failures.

Operational Cost

The operational cost of a valve includes energy consumption and operator costs, calculated over the valve's lifespan.

Example: Operation Cost Calculation of a Ball Valve

If a motorized ball valve consumes electricity costing $200 per year and the operator costs are $1,000 annually, the total operational cost over 20 years would be as follows:

- *Energy Cost (20 years):* $4,000
- *Operator Cost (20 years):* $20,000
- *Total Operational Cost (20 years):* $24,000

These costs reflect the ongoing expenses tied to the valve's operation, excluding maintenance, downtime, or repair costs.

Repair Cost

The repair cost of a valve refers to the expenses incurred to restore a malfunctioning or damaged valve to working condition. This can include costs for spare parts, labor, and testing required after repairs. Repairs are typically needed due to wear and tear, corrosion, leakage, or mechanical failures.

Example: Repair Cost Calculation for a Control Valve

Repairing a control valve might involve replacing seals and seats, which could cost $500 in parts and $1,000 in labor. Additional costs may include transport or handling fees if the valve needs to be serviced off-site. In some cases, specialized repairs like actuator reconditioning or body welding for high-pressure valves can significantly increase expenses. Repair costs often depend on the valve's type, size, material, and the complexity of the issue. Thus, it is essential to make regular maintenance to minimize expensive repairs.

For example, repairing a 10-inch globe valve used in a high-pressure system may involve replacing the damaged seat and stem, costing $800 for parts. The labor required to disassemble, repair, and reassemble the valve could add $1,200. Additionally, hydrostatic testing after the repair to ensure the valve meets safety and performance standards may cost $300. If the valve is sent off-site for repair, transportation costs might be another $200. In total, the repair cost for this valve would amount to $2,500, emphasizing the importance of preventive maintenance to reduce the frequency and expense of major repairs.

Replacement Cost

The replacement cost of a valve refers to the total expense required to procure and install a new valve when the existing one becomes irreparable. This cost includes the purchase price of the new valve, labor for removing the old valve and installing the new one, and any additional expenses for testing, commissioning, or transportation. For example, replacing a 6-inch stainless steel ball valve may cost $2,000 for the valve itself, $500 for labor, and $300 for pressure testing after installation, totaling $2,800. Replacement costs can vary widely depending on the valve type, material, size, and the complexity of installation. Thus, replacement costs make them significant considerations in the life cycle cost of a valve.

Return on Investment (ROI)

ROI in valve cost of ownership measures the financial benefits gained over the valve's life cycle compared to its total costs, including initial purchase, operational expenses, maintenance, and eventual replacement. A high ROI indicates that the valve provides significant value relative to its cost, contributing to efficiency, reliability, and reduced downtime in operations. ROI is calculated to measure the profitability of an investment by comparing the net benefit (or profit) to the total cost of the investment. It is expressed as a percentage and indicates how much return you gain (or lose) relative to what you spent.

Example: ROI Calculation for a High-Performance Valve

A high-performance control valve made of durable stainless steel may cost $5,000 upfront, while a lower quality valve costs $2,000. Over ten years, the high-quality valve only requires $500 for maintenance every five years (totaling $1,000), compared to $1,000 annually for the cheaper valve (totaling $10,000). Additionally, the high-quality valve reduces downtime, saving $10,000 in lost production.

This results in a total savings of $19,000 ($10,000 from reduced downtime + $9,000 from lower maintenance costs). With a total life cycle cost of $6,000 ($5,000 upfront plus $1,000 maintenance costs) for the high-quality valve, its ROI is as follows:

$$\text{ROI} = \frac{\text{Total savings} - \text{Total life cycle cost}}{\text{Total saving}} \times 100 = \frac{19{,}000 - 6{,}000}{6{,}000} \times 100 = 216.67\%$$

This means that the high-quality valve returns 216.67% of its cost in savings over ten years.

Total Cost of Ownership

TCO for a valve includes all costs associated with purchasing, operating, maintaining, and eventually replacing the valve throughout its lifespan. It begins with the *initial costs*, such as the purchase price, installation, and commissioning expenses. During operation, energy consumption and associated labor costs contribute to the TCO, along with maintenance *costs* for inspections, repairs, and part replacements. Periods of downtime or inefficiency, leading to lost productivity, also add to the cost. Finally, end-of-life *costs*, including decommissioning, disposal, or recycling, are factored in. TCO emphasizes evaluating a valve's long-term economic impact rather than focusing solely on its initial price. This helps engineers and decision-makers choose options that balance performance, durability, and cost-effectiveness.

Example: Gate Valve for a Chemical Processing Plant (Ten Years Operation)

1. **Initial Costs**
   - *Purchase Price:* $1,500
   - *Installation and Commissioning*: $500
   - *Initial Total*: $2,000
2. **Operational Costs (over ten years)**
   - *Energy Consumption* (e.g., additional pump power due to pressure drop): $100 per year × 10 years = $1,000
   - *Labor Costs* for operation adjustments: $50 per year × 10 years = $500
3. **Maintenance Costs**
   - *Routine Maintenance* (e.g., lubrication, cleaning): $200 per year × 10 years = $2,000
   - *Repair and Part Replacement* (e.g., replacing seals): $300 every 3 years × 3 cycles = $900
4. **Downtime Costs**
   - Valve failure causes 2 h of downtime at $500/h, once during its lifetime: $1,000
5. **End-of-Life Costs**
   - *Decommissioning and Disposal:* $300

Total Cost of Ownership:

- *Initial Costs:* $2,000
- *Operational Costs:* $1,500
- *Maintenance Costs:* $2,900
- *Downtime Costs:* $1,000
- *End-of-Life Costs:* $300

Total cost = $7,700

PART 3: MAINTENANCE AND RELIABILITY

Condition-Based Maintenance (CBM)

CBM is a maintenance strategy where equipment is monitored in real time using sensors to assess its current condition. Maintenance actions are performed only when specific parameters (e.g., temperature, pressure, vibration) exceed predefined thresholds, indicating potential issues.

Corrective Maintenance

Corrective maintenance refers to the repair or replacement of valves when a fault or malfunction is detected. The purpose of this procedure is to restore a valve to its original functional state to ensure reliable operation. While corrective maintenance is often performed after a failure has occurred, it can also be carried out when a defect or degradation is identified before total failure. In contrast to preventive maintenance, which is scheduled to prevent failures from occurring, corrective maintenance can be performed both after a failure and when early signs of failure are detected. A valve that leaks (for example, due to worn-out seals or gaskets) would require corrective maintenance that involves replacing these components, which means performing repairs necessary to stop the leak and restore tight sealing. Additionally, the valve stem may wear out over time due to frequent operation, corrosion, or misalignment. The replacement of a damaged or corroded stem is a necessary corrective action to restore the smooth operation of the valve.

Durability

Durability of a valve refers to its ability to withstand physical wear, operational and environmental conditions, over time while maintaining its performance and functionality. A durable valve is designed to endure extended usage (e.g., frequent cycling) and harsh conditions, such as high pressure, extreme temperatures, corrosive media, or abrasive flow, with minimal degradation or failure.

Fault Detection

Fault detection refers to the process of identifying abnormalities, malfunctions, or deviations in the performance of a system, component, or device, such as a valve. It involves using sensors, monitoring tools, and analytical techniques to recognize issues early, enabling timely intervention to prevent failures, downtime, or safety risks.

IoT Sensors

IoT sensors are devices that detect and measure physical or environmental conditions (such as temperature, pressure, vibration, or motion) and transmit the collected data over a network (e.g., internet, 4G, 5G) to control systems or control rooms for analysis, monitoring, and automation. These sensors are key components of IoT systems, enabling devices, machines, and systems to communicate and function intelligently.

Performance Efficiency

Performance efficiency in valves refers to how effectively a valve performs its intended function, such as controlling flow, avoiding overpressure, or ensuring tight shutoff, relative to its design specifications. It is influenced by various factors, including the valve's design, material, operating conditions, and maintenance practices.

Performance efficiency in valves encompasses several key aspects that determine how effectively a valve performs its intended function. *Flow efficiency* focuses on minimizing pressure drops and optimizing the flow coefficient (C_v or K_v) through streamlined design. *Sealing efficiency* ensures tight shutoff with low leakage rates, often achieving bubble-tight performance in critical applications. *Operational efficiency* includes fast and accurate actuation, low energy consumption, and reliable durability with minimal downtime. *Thermal efficiency* addresses heat transfer and insulation to maintain performance in high-temperature systems. Efficient valves handle their rated *pressure and temperature* ranges without failure and maintain *corrosion and wear resistance* through suitable materials and coatings. *Automation and control efficiency* involves precise positioning, real-time feedback, and integration with control systems. *Environmental efficiency* ensures low fugitive emissions and minimal noise and vibration for compliance and safety. Finally, *maintenance efficiency* is achieved with modular designs, ease of repair, and predictive maintenance capabilities, ensuring sustained performance and reduced downtime. Together, these factors contribute to the overall performance efficiency of valves in diverse applications.

Predictive Maintenance

Predictive maintenance is a proactive approach to maintaining equipment, including valves, by using advanced monitoring technologies and data analysis to predict when a component is likely to fail. This allows maintenance to be performed at the optimal time—before failure occurs—minimizing unplanned downtime, reducing repair costs, and extending the equipment's life cycle. For example, IoT enables predictive maintenance when combined with machine learning (ML). Otherwise, without ML, IoT mainly supports CBM.

Preventive Maintenance

Valve preventive maintenance refers to routine, planned activities designed to prevent valve failures before they occur. The purpose of this type of maintenance is to extend the valve's lifespan, to ensure efficient operation, and to minimize unexpected breakdowns. Inspection, cleaning, lubrication, and minor adjustments all are components of preventive maintenance. Checking valves periodically for signs of wear, corrosion, and damage can help detect potential issues early. It may be necessary to inspect the valve bodies for rust or to check for leaks in order to determine whether maintenance is required before a serious failure occurs. Lubrication of valve stems and other moving parts ensures smooth operation and prevents wear on valves with moving parts (such as ball, gate, and butterfly valves). As a result of regular lubrication, friction is reduced, which helps prevent stem seizure and prolongs the life of valves.

Reliability

Reliability in the context of a valve refers to its ability to consistently perform its intended function under specified operating conditions without failure over a given period. A reliable valve ensures uninterrupted flow control, sealing, and operational integrity, even in demanding environments, such as high pressure, extreme temperatures, or corrosive media.

Valve Life Extension Techniques

Valve life extension techniques are methods and practices used to maximize the operational lifespan of valves, reduce maintenance costs, and enhance their performance and reliability. These techniques are crucial for ensuring safety, efficiency, and cost-effectiveness in industrial systems. Valve life extension techniques involve a combination of proper material selection, regular maintenance, and optimized operating practices to maximize performance and durability. Selecting corrosion- and wear-resistant materials and applying protective coatings or linings like PTFE or ceramic help prevent damage from aggressive environments. Preventive and predictive maintenance, including regular inspections, lubrication, cleaning, and condition monitoring, ensures early detection and resolution of potential issues. Proper installation, including alignment and following manufacturer guidelines, reduces stress on valves, while replacing seals, packing, and wear-prone parts like seats and stems maintains sealing integrity and functionality.

Example: Poor Valve Material Selection Results in Shortening Valve Life Cycle

One example of poor valve selection reducing valve life cycle can be seen in *corrosive chemical processing plants,* where valves must handle highly aggressive fluids. Suppose a valve made of standard carbon steel is selected instead of a corrosion-resistant material like stainless steel or Hastelloy. In such cases:

- The carbon steel valve may rapidly corrode due to the chemical nature of the fluid (e.g., acids or chlorides).
- This corrosion can lead to thinning of the valve body, leakage, and even catastrophic failure.
- Additionally, the valve's sealing components, if not chosen to withstand the chemicals (e.g., standard elastomers instead of PTFE or Viton), can degrade quickly, causing frequent leaks and operational inefficiencies.

This poor material selection shortens the valve's life cycle, increases maintenance costs, and poses safety risks, ultimately impacting the system's reliability and efficiency. Properly selecting a valve suited to the chemical composition, pressure, and temperature of the fluid would significantly extend its service life and reduce operational disruptions.

Valve Monitoring

Valve monitoring refers to the process of continuously tracking and analyzing the performance and condition of valves using sensors and diagnostic tools. It is used to ensure valves operate efficiently, detect potential issues early, and prevent unexpected failures. This monitoring can be performed in real time, often as part of an *IoT* system, to provide process data such as pressure and temperature during valve operations. In modern valve monitoring systems, sensors installed on valves continuously measure critical parameters such as position, pressure, flow rate, temperature, and vibration. These sensors are connected to a *central control room* or a control system through wired or wireless networks, enabling real-time monitoring and management of valve performance across an entire facility.

Example: In a chemical plant, sensors on a high-pressure valve monitor flow rate and upstream pressure. If the pressure exceeds safe limits, the sensors relay this information to the control room in real time. The operators are alerted and can adjust the valve remotely to restore safe conditions, or the system can automatically regulate the valve if pre-programmed thresholds are exceeded. Connecting sensors to a control room for valve monitoring enhances operational reliability and improves safety.

Valve Testing

Valve testing is the process of evaluating a valve's performance, reliability, and compliance with specified standards to ensure it functions correctly in its intended application. Testing is conducted at various stages, including manufacturing, installation, operation, and maintenance, to validate factors such as structural integrity against pressure, flow control, sealing performance, frequent cycling, and durability. Valve testing includes hydrostatic, seat leakage, operational/functional, pressure relief, air/gas, cryogenic, fire, fugitive emission, flow, and endurance testing. More information about valve testing is provided in Chapter 6.

PART 4: MATERIALS TECHNOLOGY AND VALVE LIFE CYCLE

Advanced Coating Technology

Advanced coating technology in valves involves applying thin, specialized surface layers to enhance the valve's resistance to corrosion, wear, and extreme conditions without altering the base material's structural integrity. Unlike linings, coatings are typically much thinner and adhere directly to the substrate through processes like thermal spraying, electroplating, or chemical vapor deposition (CVD).

Thermal spray aluminum (TSA), *phenolic epoxy*, and *zinc-rich epoxy* are advanced coating technologies widely used in the oil and gas industry to protect valves from corrosion. *TSA* involves applying molten aluminum onto valve surfaces, forming a robust and corrosion-resistant layer ideal for high-temperature and offshore environments. It offers exceptional protection against atmospheric and marine environment corrosion, particularly in high-chloride conditions.

Phenolic epoxy coatings are highly resistant to chemicals and heat, making them ideal for valves in processing plants exposed to aggressive chemicals and elevated temperatures. These coatings excel in protecting against acidic and caustic environments. *Zinc-rich epoxy*, composed of a high concentration of zinc dust, provides cathodic protection by acting as a sacrificial layer to safeguard the underlying steel. This coating is commonly used on valves in moderately corrosive environments and often serves as a primer for additional protective layers.

Corrosion Resistance

Corrosion resistance refers to a material's ability to withstand deterioration caused by its reaction with environmental elements such as moisture, chemicals, or gases. Materials with high corrosion resistance, such as stainless steel, nickel alloys (e.g., Inconel, Hastelloy), and nonmetallics like PTFE, are often selected for applications in harsh environments to extend equipment life and ensure safety. The choice of material depends on the type of corrosion expected, such as uniform corrosion, pitting, crevice corrosion, stress corrosion cracking, or galvanic corrosion.

Corrosion-resistant technologies enhance the durability of materials through advanced treatments and coatings. Common methods include the following:

- *Protective Coatings:* Epoxy, phenolic, or zinc-rich coatings create a physical barrier against corrosive agents.
- *Protective Lining:* Lining carbon steel with materials like Inconel 625 provides excellent protection against corrosion in extreme conditions. The lining combines the structural strength and cost-effectiveness of carbon steel with the superior corrosion resistance of Inconel, making it ideal for applications like chemical processing and offshore equipment.
- *Metal Plating:* Techniques like chromium plating or galvanization deposit a thin layer of a more corrosion-resistant metal onto the surface.
- *Cathodic Protection:* Involves using sacrificial anodes or impressed current systems to protect metal surfaces.
- *Alloy Development:* Engineering high-performance alloys, such as duplex stainless steel or titanium, improves inherent corrosion resistance.

The combination of material selection and technological enhancement ensures optimal performance, especially in critical industries like oil and gas, chemical processing, and marine applications, where corrosion can cause significant damage and safety concerns.

Fluid Dynamics

Fluid dynamics in material technology involves studying how fluids (liquids and gases) interact with materials, focusing on the effects of fluid flow on the material's performance, durability, and structural integrity. This field is crucial in applications where materials are exposed to moving fluids, such as in pipelines, valves, turbines, and heat exchangers.

In material technology, fluid dynamics focuses on key aspects such as *erosion and wear resistance*, analyzing how high-velocity or particulate-laden fluids cause material degradation, and selecting hard or coated materials to reduce damage. *Corrosion analysis* examines the interaction between fluid chemistry and material surfaces, predicting and mitigating corrosion in aggressive environments like chemical processing or marine systems. *Flow-induced vibrations* are studied to ensure materials can withstand stresses caused by turbulent or high-velocity flows without compromising structural integrity. Additionally, *heat transfer and thermodynamics* explore how materials handle thermal stress and enable efficient heat exchange, while *fouling and deposit resistance* address the prevention of efficiency-reducing buildups on material surfaces. These aspects collectively guide material selection and design for durability and optimal performance in fluid-exposed systems.

Material Selection

Material selection for valves is a critical aspect of valve design and performance, as it directly influences the valve's durability, functionality, and ability to withstand operating conditions. The right material ensures compatibility with the fluid being handled, resists environmental stresses, and meets safety and regulatory requirements. The selection process requires careful consideration of several factors. More information is provided in Chapter 4 under the "Material Selection" section.

Wear Resistance

Wear resistance in materials refers to their ability to withstand gradual material loss caused by mechanical interactions such as friction, abrasion, or erosion. Materials like tungsten carbide, ceramics (e.g., silicon carbide), and hardened steels are commonly used for their superior resistance to wear. Technologies like hard-facing, surface hardening, and wear-resistant linings further enhance the durability of materials, making them ideal for applications in high-friction or abrasive environments, such as oil and gas.

Wear-resistant materials and technologies are critical for enhancing the durability of industrial components, minimizing operational disruptions, and ensuring cost-effectiveness in challenging environments. More information about wear resistance is available in Chapter 4 under the "Wear Resistance" section.

PART 5: ENERGY-RELATED TERMS AND VALVE COST

Energy Consumption

1. Definition

Energy consumption for valves refers to the amount of energy required to operate a valve and maintain its functionality within a system.

2. How Valves Consume Energy?

One significant aspect of valve energy consumption is *actuation energy*. Valves operated by actuators—such as electric, pneumatic, or hydraulic types—require energy to open, close, or modulate. For instance, motorized ball valves consume electrical power to rotate the ball, while solenoid valves rely on electromagnetic energy for position changes. The energy requirement varies based on the actuator type, valve size, and operating conditions. In large industrial setups, frequent actuator operation can contribute significantly to overall energy costs, particularly under high-pressure conditions.

Another key factor is *pressure loss*, also known as hydraulic energy loss. As fluids pass through a valve, resistance caused by the valve design results in a drop in pressure. This pressure loss necessitates additional energy from pumps or compressors to maintain the desired flow rate and pressure in the system. For example, globe valves, with their non-straight flow path, typically cause higher pressure losses compared to streamlined designs like ball or gate valves. This makes globe valves more energy-intensive in systems requiring consistent flow rates.

Frictional losses within the valve also contribute to energy consumption. These losses occur due to friction between moving parts such as the valve stem, disk, or seats. Over time, wear and tear can increase friction, requiring more energy to operate the valve. Proper lubrication and maintenance are essential to minimize these losses and ensure efficient operation.

3. Factors Affecting Valve Energy Consumption

Several factors influence the energy consumption of valves. The *valve type and design* play a crucial role; for instance, butterfly valves are generally more energy-efficient due to their low-pressure drop design, while throttling valves like globe valves consume more energy. *Flow conditions*, including high flow rates, fluid viscosity, or turbulent flow, also increase the energy needed for valve operation. Additionally, the *frequency of valve operation* affects energy use, with frequently cycling valves consuming more energy, especially if they are power-actuated. Finally, *system integration* can impact energy efficiency; poorly designed systems with mismatched valve types or sizes may cause unnecessary losses.

4. Why Energy Consumption Matters?

High energy consumption in valves has direct implications for cost, sustainability, and system performance. From a financial perspective, increased energy use translates to higher operational expenses, which can be significant in industries with extensive valve networks. From an environmental standpoint, reducing valve energy consumption supports sustainability goals by lowering carbon emissions. In terms of system performance, excessive energy use can indicate inefficiencies or mismatched components, potentially leading to reduced reliability or failures.

5. Strategies to Minimize Energy Consumption

To minimize energy consumption, several strategies can be implemented. One key approach is *optimized valve selection*, ensuring that the chosen valve type minimizes pressure loss and matches the application requirements. *Regular maintenance* is also vital to ensure smooth operation and address wear that could increase friction or lead to leaks. Utilizing *energy-efficient actuators* and optimizing their operation with advanced control systems can further reduce energy consumption. Additionally, proper *system design* is essential to avoid unnecessary pressure losses or turbulence that could elevate energy demands.

Energy Efficiency

Energy efficiency for a valve refers to the valve's ability to minimize energy losses during operation, ensuring optimal performance with the least possible energy consumption. This concept is particularly important in systems where fluid flow is regulated, as inefficient valves can lead to significant energy waste and increased operational costs.

Valve energy efficiency is influenced by several key factors, including the valve's *flow characteristics*, where streamlined designs and high flow coefficients (Cv or K_v) reduce turbulence and minimize energy losses. A *low-pressure drop* across the valve ensures that less energy is required to maintain fluid flow, with the right valve type (e.g., ball, butterfly, or globe) selected based on the application. Actuation energy efficiency refers to the energy required by a valve actuator to open, close, or regulate a valve. An efficient actuator minimizes energy use while maintaining precise and reliable operation. *Actuation energy* also plays a role, as efficient motorized or pneumatic actuators optimize energy consumption during operation. Proper *seal and leakage performance* is essential to prevent energy loss, especially in high-pressure systems or those involving compressed air or steam. Additionally, if a valve is not suited to the operating conditions, it may wear out faster, leak, or cause *energy losses* due to inefficiencies in controlling flow. Together, these factors contribute to the overall energy efficiency and performance of a valve in its operating system.

The benefits of valve energy efficiency are significant for both operational and environmental aspects. Energy-efficient valves help *reduce operating costs* by minimizing energy consumption in pumping, compressing, or regulating fluid flow, which is particularly important in large-scale industrial systems. They also contribute to *environmental sustainability* by lowering the system's overall energy usage and carbon footprint. Enhanced *system reliability* is another key benefit, as efficient valves reduce wear and strain on other components, leading to fewer failures and longer equipment lifespans. For example, high-pressure drop in a valve can lead to wear and erosion. Furthermore, improved energy efficiency aligns with *regulatory compliance* and sustainability goals, offering economic and reputational advantages to organizations. By optimizing energy use, these valves ensure cost-effective, sustainable, and reliable operations.

Example: Using Smart Actuators for Energy Efficiency

Consider a *pneumatic actuator* used to control a valve in a pipeline transporting natural gas. If the actuator is oversized or poorly calibrated, it may consume excessive compressed air, leading to significant energy waste. By switching to a *smart actuator* with optimized sizing and control features, the system can achieve energy savings. The smart actuator adjusts the air pressure and volume needed based on the actual load, reducing compressed air consumption by up to 30%.

Flow Coefficient (C_v)

Valve coefficients (often represented as C_v for imperial units or K_v for metric units) measure the capacity of a valve to allow fluid to flow. It indicates how much fluid (usually water) can pass through a valve at a specific pressure drop. Using this coefficient, engineers are able to determine the ideal valve size and type for a particular application, ensuring optimal performance and control. In the case of a valve, C_v is the flow rate measured in gallons per minute (GPM) at a pressure drop of 1 psi across the valve at a temperature of 60°F. For instance, a valve with a C_v of 10 will allow 10 GPM of water to flow through with a 1 psi drop. K_v is a similar quantity, but uses metric units. In m^3/h, it indicates the flow rate of water at a pressure drop of 1 bar across the valve.

The flow coefficient (C_v) for liquids is calculated from Equation 13.1.

Equation 13.1: Flow Coefficient (C_v) Calculation for Liquids

$$C_v = Q\sqrt{\frac{\mathrm{SG}}{\Delta P}}$$

where

C_v: flow coefficient $\left(\frac{\text{US gal per minute (gpm)}}{\text{psi}}\right)$;

Q: Flow rate $\left(\frac{\text{US gal}}{\text{min}}\right)$;

SG: Fluid specific gravity (dimensionless) calculated by Equation 13.2;

$\varnothing P$: Differential pressure across the valve (pound per square inch [psi]);

Note: Water specific gravity is equal to 1

Equation 13.2: Specific Gravity (SG) Calculation

$$\mathrm{SG} = \frac{\rho_{\text{fluid}}}{\rho_{\text{water}}}$$

where

ρ_{fluid}: Density of the fluid passing through the valve;

ρ_{water}: Water density equal to $1\frac{\text{g}}{\text{cm}^3}$ or $1{,}000\frac{\text{kg}}{\text{m}^3}$.

Example: Suppose you have a valve with a C_v of 20, and you want to know how much water can pass through it with a 4 psi pressure drop. Plugging into the C_v formula:

$$C_v = Q\sqrt{\frac{SG}{\Delta P}} \rightarrow 20 = Q\sqrt{\frac{1}{4}} \rightarrow Q = 40\,\text{GPM}$$

Hydraulic Efficiency

Hydraulic efficiency in valves refers to their ability to facilitate fluid flow with minimal energy loss due to resistance or turbulence. A hydraulically efficient valve design reduces pressure drop across the valve, ensuring that the system requires less energy to maintain fluid flow. Factors influencing hydraulic efficiency include the valve's internal geometry that affects flow path and surface smoothness. For example, ball and butterfly valves typically offer higher hydraulic efficiency compared to globe valves due to their streamlined flow paths. High hydraulic efficiency minimizes pumping energy, and reduces operational costs.

Pressure Loss

Pressure loss in a valve refers to the reduction in pressure as a fluid flows through the valve. This loss occurs because the fluid encounters resistance due to the valve's internal components, changes in flow direction, turbulence, and friction with the valve body and internal surfaces.

Pressure loss in a valve is influenced by several key factors, including the *valve design*, where more tortuous flow paths in valves like globe valves result in higher losses compared to streamlined designs like ball or gate valves. The *valve size* and its compatibility with the system flow capacity also play a significant role, as undersized valves create excessive resistance while oversized valves may reduce efficiency. The *flow rate* impacts pressure loss, with higher velocities increasing turbulence and friction. As the flow rate increases, the velocity of the fluid through the valve also increases, leading to greater turbulence and friction. Additionally, the *fluid properties*, such as density and viscosity, affect the level of resistance encountered. The *valve opening position* is critical, as partially open valves significantly restrict flow and amplify turbulence, causing greater losses. Finally, *internal features* like sharp edges, flow rotations, and obstruction by the valve mechanism further contribute to resistance and pressure drop.

Sustainability

Sustainability in the context of valves refers to the design, manufacturing, operation, and end-of-life disposal practices that minimize environmental impact while ensuring high performance and reliability. Sustainable valve practices focus on reducing emissions, conserving energy, and using environment-friendly materials throughout the valve's life cycle. Chapter 14 provides more information about sustainability and environment impact.

1. Materials and Manufacturing

The materials used in valve production significantly influence sustainability. Opting for recyclable and durable materials such as stainless steel, or high-grade alloys extends the valve's lifespan and reduces the need for frequent replacements. Manufacturing processes should aim to minimize waste, energy consumption, and the use of harmful chemicals. Advanced manufacturing techniques, like additive manufacturing, can contribute to sustainability by reducing material waste and energy use during production.

2. Reducing Fugitive Emissions

Valves are potential sources of fugitive emissions, particularly in oil and gas applications. Sustainable valves are designed with advanced sealing technologies to prevent leaks of greenhouse gases like methane. Standards such as ISO 15848 and API 624 address the need for low-emission valves, pushing manufacturers to adopt more sustainable designs. Valves with bellows seals or low-emission packing systems are examples of solutions that meet stringent environmental requirements.

3. Digitalization and Smart Technologies

The integration of digitalization and smart valve technologies enhances sustainability by enabling real-time monitoring and CBM. When combined with machine learning, these technologies also support predictive maintenance, helping to anticipate failures before they occur. This reduces unplanned downtime and extends the valve's operational life. For example, smart valves can detect early signs of wear and tear, allowing timely interventions that prevent catastrophic failures and associated environmental impacts.

4. Valve Life Cycle Management

Extending the life cycle of valves through proper maintenance, refurbishment, and repurposing contributes to sustainability. Instead of discarding old valves, industries are increasingly adopting practices to repair and reuse components. This not only reduces waste but also supports a circular economy by keeping materials in use for as long as possible.

System Optimization

1. Definition

System optimization in relation to valves refers to the process of designing, selecting, and maintaining valves to maximize the performance, efficiency, and reliability of the entire fluid control system. Valves play a critical role in managing the flow, pressure, and direction of fluids within industrial and commercial systems. As a result, their optimization is essential for reducing energy consumption, minimizing costs, and improving overall system functionality.

2. Key Aspects

Valve selection is a foundational aspect of system optimization. It involves choosing the right type of valve (e.g., ball, gate, globe, butterfly) based on the specific application's requirements. This includes considering factors such as the valve's size, material, and pressure/temperature ratings to ensure compatibility with the system's operating conditions. Additionally, selecting valves with appropriate flow coefficients (C_v) helps minimize pressure losses and optimize energy use. *Flow management* is another crucial element of optimization. It ensures consistent and reliable operation by reducing turbulence and pressure drops through streamlined valve designs and proper placement. Valves must operate within their designed flow range to prevent issues such as cavitation, erosion, or system instability. Efficient flow control improves overall system functionality and longevity. *Energy efficiency* is a key focus area in valve system optimization. Energy consumption can be minimized by reducing pressure losses across valves and using energy-efficient actuators in automated systems. *Leak prevention* is essential for maintaining system efficiency and safety. Ensuring tight sealing through advanced sealing technologies or optimized packing systems prevents fluid leakage, which can lead to energy losses, environmental harm, and safety hazards. *Integration with smart systems* is becoming increasingly important in modern valve optimization such as smart valves equipped with sensors. *Maintenance and reliability* are critical for ensuring valves operate efficiently over their lifespan.

3. Benefits

Optimizing valve systems offers numerous *benefits*, including cost-savings from reduced energy consumption and maintenance expenses. Enhanced performance and increased duration of valves and associated equipment improve system stability and reliability.

THERMAL EFFICIENCY

Thermal efficiency in valves refers to their ability to perform reliably under varying temperature conditions while minimizing heat loss or gain in the system. It is influenced by factors such as *material selection*, with high-temperature alloys like stainless steel or Inconel used for elevated temperatures and austenitic stainless steel for cryogenic conditions. *Insulation* helps reduce heat loss in high-temperature systems, while designs that account for *thermal expansion* prevent leaks or mechanical failures. *Sealing components*, such as graphite or PTFE gaskets, ensure tight shutoff across a range of temperatures, and features like *thermal barriers* or specialized coatings control heat transfer to protect sensitive parts. Valves with good thermal efficiency improve energy conservation, enhance reliability, and ensure safety in temperature-critical applications like power plants, LNG systems, and chemical processing plants.

Examples: Material Selection for Thermal Efficiency

Example 1: An example of using *graphite or PTFE* to improve the thermal efficiency of a valve can be found in *steam systems* operating at high temperatures and pressures. In such systems, the valve's sealing components, such as gaskets and packing, must maintain a tight shutoff to prevent leakage, which could lead to energy loss and reduced thermal efficiency.

Example 2: An example of a *thermal barrier in valves* for thermal efficiency can be found in *cryogenic applications,* such as in LNG (liquefied natural gas) processing and transportation. Cryogenic valves are exposed to extremely low temperatures, and without proper thermal barriers, the cold from the fluid could transfer through the valve body to external components, causing freezing and operational issues. To address this, manufacturers incorporate a *long stem extension* as a thermal barrier. The extended stem design increases the distance between the cryogenic fluid and the actuator or manual handle, reducing heat transfer and preventing freezing of external components.

QUESTIONS AND ANSWERS

1. What does the cost of ownership of a valve include?
   A. Only the purchase price of the valve
   B. Initial cost, operational cost, maintenance cost, and replacement cost
   C. Initial cost and testing cost
   D. Operational and downtime costs only

   Answer: The best and most correct answer is option B.

2. What is the primary impact of valve downtime?
   A. Increased flow rates
   B. Reduced valve lifespan
   C. Production halts and financial losses
   D. Decreased energy consumption

   Answer: Option C is the correct answer.

3. How can cost-effectiveness be achieved in the valve life cycle?
   A. By investing in durable valves and conducting preventive maintenance
   B. By reducing the valve's operational efficiency
   C. By ignoring maintenance schedules to save costs
   D. By reducing the valve's operational efficiency

   Answer: Option A is the correct answer.

4. You invest $7,000 in a high-quality valve, which reduces downtime and maintenance costs, saving $20,000 over 10 years. What is the return on investment?
   A. 171%
   B. 180%
   C. 186%
   D. 195%

Answer: $ROI = \frac{\text{Savings} - \text{Total cost}}{\text{Total cost}} \times 100 = \frac{20{,}000 - 7{,}000}{7{,}000} \times 100 = 186\%$

Option C is the correct answer.

5. Which of the following statements correctly describe the interplay of energy efficiency, hydraulic efficiency, and thermal efficiency in valve systems? Select all that apply:
   A. *Energy efficiency* in valves is maximized by reducing pressure drop, optimizing flow path design, and ensuring leak-tight sealing to minimize energy losses in pumping or compressing fluids.
   B. *Hydraulic efficiency* refers to the valve's ability to handle high-pressure fluid flow without turbulence, ensuring minimal friction losses and a reduced flow coefficient (C_v).
   C. *Thermal efficiency* in valves can be improved by using materials like stainless steel for high temperatures and lip seals for cryogenic conditions to minimize heat transfer and energy loss.
   D. A valve with poor hydraulic efficiency will inherently have low energy efficiency, as increased turbulence or pressure drop leads to higher energy consumption.
   E. Thermal barriers, such as long stem extensions in cryogenic valves, contribute to energy efficiency by reducing heat transfer and preventing external freezing or energy losses.
   F. Energy-efficient valves are inherently thermally efficient and hydraulically efficient without requiring specific designs or material considerations.

   Answer: Options A, C, D, and E are correct. Option B is not entirely correct as hydraulic efficiency is not only related to high-pressure fluid. Option F is not correct, because energy efficiency in valves is not inherently linked to thermal or hydraulic efficiency without specific design or material considerations

6. Which of the following factors is most critical in determining the energy efficiency of a valve in a high-pressure pipeline system?
   A. The valve's flow coefficient (C_v) and associated pressure drop.
   B. The material of the valve body (e.g., steel or alloy).
   C. The actuator's power consumption during standby mode.
   D. The installation location of the valve in the pipeline.

   Answer: Option A is the correct answer.

   Explanation: The flow coefficient (C_v) is a measure of a valve's capacity to allow fluid flow and directly impacts the pressure drop across the valve. A higher pressure drop increases the energy required by pumps or compressors to maintain the desired flow rate, making it the most critical factor in energy efficiency. While actuator power consumption and material choice are important, they are secondary to the effects of pressure loss caused by valve design and sizing.

7. A high-temperature valve is installed in a steam pipeline system operating at 250°C. The valve's thermal efficiency is influenced by several factors. Which combination of the following factors is most critical for maintaining optimal thermal efficiency in this scenario?
   1. The valve's insulation effectiveness.
   2. The type of packing material used in the valve stem.
   3. The valve's pressure class and allowable stress levels.
   4. The ambient temperature surrounding the pipeline.
   5. The valve's ability to minimize steam leakage under high-pressure conditions.

   Select the correct combination:
   A. 1,2,5
   B. 2,3,4
   C. 1,4,5
   D. 3,4,5

   Answer: Option A is the correct answer. *Insulation effectiveness* minimizes heat loss through the valve body, maintaining the system's thermal efficiency. *Packing material* must withstand high temperatures to prevent thermal degradation and leakage, which could result in heat loss. Minimizing *steam leakage* under high pressure is essential for retaining energy within the system and preventing efficiency losses. Factors like pressure class (3) and ambient temperature (4) have an indirect impact but are less critical for directly optimizing thermal efficiency in this context.

8. Which of the following strategies best contributes to improving the sustainability of industrial valves throughout their life cycle?
   A. Increasing the frequency of valve replacements to ensure optimal performance
   B. Designing valves exclusively with nonrecyclable, high-performance materials
   C. Utilizing advanced sealing technologies to reduce fugitive emissions
   D. Focusing solely on operational efficiency without considering end-of-life disposal

   Answer: Option C is the correct answer.

9. Hydraulic efficiency in valve systems is maximized when:
   A. Pressure loss across the valve is minimized.
   B. The valve material has high thermal conductivity.
   C. Turbulence within the valve is increased to promote mixing.
   D. The valve is operated at a pressure higher than its rated capacity.

   Answer: Option A is the correct answer.

10. Which valve type typically exhibits the least pressure loss in a fully open position?
    A. Gate valve
    B. Needle valve
    C. Check valve
    D. Globe valve

    Answer: Option A is the correct answer.

BIBLIOGRAPHY

Al Ruqaishi, M. A., & Al Kalbani, F. K. (2024, November). Revolutionizing Leak Detection and Cost Savings in Valve Predictive Maintenance. Paper presented at the Abu Dhabi International Petroleum Exhibition & Conference (ADIPEC), Abu Dhabi, United Arab Emirates. https://doi.org/10.2118/222620-MS

American Petroleum Institute. (2015). API 600: Steel Gate Valves—Flanged and Butt-Welding Ends, Bolted Bonnets. 13th edition. API Publishing Services: Washington, DC.

American Petroleum Institute. (2021). API 624: Type Testing of Rising Stem Valves Equipped with Flexible Graphite Packing for Fugitive Emissions. 3rd edition. API Publishing Services: Washington, DC.

American Society of Mechanical Engineers. (2010). ASME B31.3: Process Piping. ASME Publishing: New York, NY.

American Society of Mechanical Engineers. (2012). Design and Fabrication of Pressure Vessels: Boiler and Pressure Vessel Code (Section VIII Div. 02). ASME Publishing: New York, NY.

American Society of Mechanical Engineers. (2012). Rules for Construction of Pressure Vessels: Boiler and Pressure Vessel Code (Section VIII Div. 01). ASME Publishing: New York, NY.

American Society of Mechanical Engineers. (2020). ASME B16.34: Valves—Flanged, Threaded, and Welding End. 2020 edition. ASME Publishing: New York, NY.

Centre Technique des Industries Mécaniques. (2008). Valve Life Cycle Costs: A Guide to LCC Analysis for Valve Systems Performances. CETIM: France.

Erhan, O. (2024). Increasing energy efficiency and flow coefficient by determining fluid behaviours with computer aided design and CFD analysis in isolation valves. Advances in Mechanical Engineering, 16(10), 123–145.

International Organization for Standardization. (2015). Industrial Valves—Measurement, Test, and Qualification Procedures for Fugitive Emissions: Part 1: Classification System and Qualification Procedures for Type Testing of Valves (ISO 15848-1:2015). International Organization for Standardization: Geneva, Switzerland.

Lejeune, H. (2024, October). Valve Packing Calculation for Tightening Optimization Regarding Emissions and Energy Consumption. Paper presented at the ASME BJRS 2024 Conference, Houston, TX.

Mottahedin, A., Giudicianni, C., Di Nunno, F., Granata, F., Brentan, B. M., & Creaco, E. (2024). Analysis, design, and maintenance of isolation valves in water distribution networks: State of the art review, insights from field experiences and future directions. Water Research, 262, 122088. https://doi.org/10.1016/j.watres.2024.122088

Nesbitt, B. (2007). Handbook of Valves and Actuators: Valves Manual International. 1st edition. Elsevier: Oxford, UK.

NORSOK Z-007. (2015). Mechanical Completion and Commissioning. NORSOK: Lysaker, Norway.

Storm, K. (2017). Industrial Piping and Equipment Estimating Manual. 1st edition. Gulf Publishing Company: Houston, TX.

Wolski, G. (2011). Total cost of ownership and nuclear valve projects. Power Engineering, 115(2), 46–48.

Yan, J., Heng-hu, Y., Hong, Y., Feng, Z., Zhen, L., Ping, W., & Yan, Y. (2015). Non-destructive detection of valves using acoustic emission technique. Advances in Materials Science and Engineering, 2015, 749371. https://doi.org/10.1155/2015/749371

14 Environmental Impact and Sustainability of Valves

A

Advanced Coating for Longevity

Advanced coatings are special protective layers applied to the inside or outside of valve parts. These coatings make valves stronger, reduce wear, and protect against corrosion. Ceramic coatings resist abrasion and high heat, while polymer-based coatings like PTFE provide chemical resistance and low friction. Thermal spray coatings such as tungsten carbide and chrome carbide help valves last longer in harsh environments, especially offshore. These coatings form a barrier that prevents corrosion, erosion, and fouling, helping valves last longer and need less maintenance, which reduces costs and improves performance.

Air Quality

Air quality refers to the condition of the air in a specific area, particularly with regard to how clean or polluted it is and how safe it is for humans, animals, and the environment. It is determined by measuring the concentration of pollutants in the air, such as particulate matter (PM), carbon monoxide, nitrogen dioxide, sulfur dioxide, ground-level ozone (O_3), and volatile organic compounds (VOCs). PM refers to tiny solid and liquid particles suspended in the air. These particles come from various sources like vehicle emissions, industrial processes, and burning fuels.

B

Biodegradable Sealing Materials

Biodegradable sealing materials are environment-friendly materials used in valve seals and gaskets that can break down naturally into nontoxic components when exposed to environmental conditions such as microorganisms, moisture, or sunlight. Unlike conventional synthetic materials, such as PTFE or rubber, which can persist in the environment for decades, biodegradable sealing materials decompose more quickly, reducing waste and pollution. These materials are typically made from renewable sources, such as natural rubber, plant-based polymers,

DOI: 10.1201/9781003640967-14

or biocomposites. They are designed to maintain the required performance characteristics, such as elasticity and durability, during their service life.

C

Carbon Footprint

Carbon footprint regarding valves refers to the total amount of greenhouse gas (GHG) emissions, expressed as carbon dioxide equivalent (CO_2e), generated during the entire life cycle of a valve. This includes emissions from raw material extraction, manufacturing, transportation, installation, operation, maintenance, and end-of-life disposal or recycling.

Carbon Neutrality

Carbon neutrality refers to achieving a balance between the amount of carbon dioxide emitted into the atmosphere and the amount removed or offset, resulting in a net-zero carbon footprint. It is a key strategy in combating climate change and aligns with global efforts to limit global warming to 1.5°C above pre-industrial period levels, as outlined in the Paris Agreement. The pre-industrial period is generally considered to be the time before the mid-19th century, around 1850–1900. The global average surface temperature during this time is estimated to have been approximately 13.7°C (56.66°F). Adding 1.5°C to this baseline, the target temperature would be approximately 15.2°C (59.36°F).

Circular Economy

The circular economy for valves aims to extend their lifespan, reduce waste, and use resources more efficiently. Instead of the traditional model of making, using, and throwing away valves, this approach focuses on reuse, repair, refurbishment, and recycling. Valves designed with modular parts and durable materials allow industries to replace or fix specific components instead of discarding the whole valve. This helps reduce waste and saves resources.

The circular economy for valves includes refurbishing, recycling, and sustainable manufacturing. Refurbishing restores old or damaged valves to working condition, like repairing worn seals or replacing corroded parts. Recycling uses materials like metals and alloys to make new valves, reducing waste. For example, used valve materials can be repurposed for valves in less demanding applications, such as those with lower corrosion rates. This helps extend material life and minimizes resource consumption. Sustainable manufacturing, such as 3D printing, cuts material use and lowers environmental impact. It reduces waste by using only the necessary amount of material, unlike traditional methods that generate excess scrap. Three-dimensional printing also allows for lightweight designs, reducing energy consumption in transportation and operation. Industries also use predictive maintenance and IoT monitoring to prevent failures, like sensors detecting leaks early. These practices protect the environment by reducing waste and conserving resources, while also lowering costs by extending valve life.

Example: Recycling Materials from Decommissioned Valves

When valves reach the end of their operational life and cannot be repaired, the materials can be recycled. For instance, a butterfly valve with a stainless steel body and a PTFE lining used in a water treatment plant can be dismantled, and the metal components can be melted down and reused in the manufacturing of new valves or other industrial products. Nonmetallic parts, such as the PTFE lining, can be processed for reuse where feasible. This reduces landfill waste and supports resource conservation, contributing to a circular economy while lowering manufacturing costs for new products.

Climate Change Mitigation

Climate change mitigation refers to the efforts and strategies aimed at reducing or preventing the emission of GHGs into the atmosphere to slow down global warming and minimize its long-term undesirable impacts on the planet. It focuses on addressing the root causes of climate change by transitioning to low-carbon, sustainable practices across all sectors of society, including energy, transportation, agriculture, and industry.

Condition-Based Monitoring (CBM) and Sustainability

CBM uses sensors and data to check valve performance in real time, helping industries maintain valves based on actual wear instead of fixed schedules. It tracks key parameters like flow, pressure, vibration, and temperature to detect inefficiencies and leaks before they cause major failures. For example, sudden pressure drops can indicate internal damage, while unusual vibrations may signal mechanical issues. By identifying problems early, CBM reduces waste by preventing fluid leakage, and lowers unplanned downtime, improving overall system reliability. This protects the environment by stopping leaks that could cause contamination, pollution, or harmful emissions. It also improves efficiency by ensuring valves operate at optimal performance, reducing excess energy loss in the valve and actuators. Additionally, CBM helps industries save costs by avoiding unnecessary maintenance and extending valve lifespan, making operations more sustainable and economical.

Condition monitoring (CM) focuses on assessing the long-term health of equipment by analyzing periodic or continuous data trends, such as vibration, temperature, or pressure, to predict potential failures and schedule maintenance proactively. In contrast, real-time monitoring (RTM) provides instant feedback on operational parameters like flow rate or valve position, allowing immediate responses to anomalies or deviations during operation. For example, CM might analyze vibration data over weeks to predict a valve seal's degradation, while

RTM would detect a sudden drop in pressure as it happens and alert operators to take immediate action. Thus, CM is generally stronger than RTM for future valve and actuator prediction. However, both CM and RTM can be integrated with machine learning (ML) to enhance predictive capabilities and improve maintenance strategies. Together, they complement each other by ensuring both future reliability and present operational efficiency.

Corrosion-Resistant Alloys (CRAs)

CRAs are specialized materials used in valves to withstand corrosive environments and prevent degradation caused by corrosion types, such as pitting, crevice corrosion, or stress corrosion cracking. Common CRAs include stainless steels (e.g., 316L), duplex stainless steels, and nickel-based alloys (e.g., Inconel, Hastelloy), which offer excellent resistance to corrosive fluids, high temperatures, and aggressive chemical environments. These alloys are essential in industries like oil and gas, chemical processing, and seawater systems, where durability and reliability are critical. For a more in-depth discussion on the selection, properties, and applications of CRAs, refer to Chapter 4 on materials, which provides detailed insights into material selection and their impact on valve performance.

D

Digital Twin and Sustainability

Digital twins play a pivotal role in promoting sustainability in valves by creating virtual replicas of physical valves and their operating environments, enabling RTM, simulation, and optimization. Through the integration of IoT sensors, a digital twin collects and analyzes data on valve performance, such as flow, pressure, temperature, and wear. This helps identify inefficiencies, predict maintenance needs, and prevent failures. All these capabilities result in reducing resource waste and environmental risks like leaks or spills.

E

Eco-Efficiency

Eco-efficiency regarding valves refers to the practice of optimizing valve performance to achieve maximum functionality while minimizing environmental impact and resource consumption.

Eco-efficiency in valves focuses on better performance with less environmental impact. Energy-efficient designs reduce pressure drops and optimize flow, lowering energy use in fluid systems. Using durable and recyclable materials, like stainless steel and alloys, extends valve life and reduces waste. Low-emission technologies, following standards like ISO 15848, help prevent harmful leaks, especially in oil and gas industries. Predictive maintenance, modular designs, and recycling make valve use more sustainable. Sustainable manufacturing cuts waste, saves energy, and uses advanced methods like 3D printing to reduce material use. These practices help valves work efficiently while lowering their environmental footprint.

Eco-Friendly Design

Eco-friendly valve design reduces environmental impact from manufacturing to disposal. It uses sustainable and recyclable materials like lead-free alloys and biodegradable polymers. Eco-friendly coatings, such as water-based paints and powder coatings, avoid toxic substances like heavy metals and VOCs. Valves are built to prevent leaks and emissions, following ISO 15848 standards with advanced sealing technologies. Durable designs ensure long service life and easy maintenance, reducing waste. Resource-efficient designs use smaller, lighter components without losing functionality. At the end of life, valves can be recycled or safely disposed of. These practices help lower carbon footprints, meet environmental rules, and maintain high performance.

Energy Efficiency

Energy efficiency in valves reduces energy use while ensuring optimal performance. Advanced actuators, like electric ones with VFDs, adjust motor speed to match demand. Pneumatic actuators with air recirculation systems cut compressed air waste. Energy-efficient designs, such as low-friction seals and streamlined flow paths, reduce pressure drops and energy loss. IoT-enabled smart valves improve efficiency by monitoring real-time conditions and making precise adjustments. These strategies lower energy costs, reduce carbon footprints, and support sustainability goals while meeting regulatory standards.

Energy-Efficient Actuation

Energy-efficient actuation for valves involves technologies and mechanisms designed to minimize energy consumption while controlling valve movement, improving operational sustainability and cost-effectiveness.

Electric actuators improve energy efficiency with optimized motor designs, variable-speed drives, and advanced control systems. Variable-speed drives adjust motor speed to match demand, saving energy. Advanced control systems reduce power use during idle or partial operation by adjusting performance based on real-time needs. In older systems, motors may keep running even when actuation is not needed, wasting energy. For example, a basic actuator might keep running to hold a valve open. In contrast, systems with advanced controls stop the motor once the position is reached, using a mechanical lock to hold the valve in place. These features lower power consumption and improve efficiency.

Pneumatic actuators improve efficiency by using air recovery systems to recycle exhaust air, reducing the need for constant air compression, which consumes a lot of energy. A quick exhaust system normally releases air quickly to speed up actuator response. However, with modifications, this air can be recaptured and reused instead of being wasted. This setup helps lower energy use and enhances overall system efficiency.

Hydraulic actuators improve efficiency with low-friction seals, advanced fluid dynamics, and energy-saving pumps. Low-friction seals reduce resistance, lowering the energy needed for operation while maintaining pressure and preventing leaks. Sealing friction affects energy use, as well-designed seals ensure smooth movement with less resistance. Advanced fluid dynamics optimize flow paths, minimizing power loss during actuation. For example, engineers use CFD simulations to analyze and improve fluid flow inside the actuator, identifying areas of inefficiency and making necessary design adjustments.

These technologies collectively reduce energy demands, extend equipment life, and support sustainability initiatives across industries such as oil and gas, water management, and manufacturing.

Energy Recovery

Energy recovery in valves captures and reuses energy from fluid systems to improve efficiency. It harnesses pressure, hydraulic, and heat energy during valve operation. Pressure recovery uses mini turbines inside or alongside of valves to convert excess pressure into electricity. Hydraulic energy recovery captures energy from high-pressure fluids for mechanical or electrical use. Heat recovery systems, like steam traps, reclaim energy from condensate or flash steam in industries. Regenerative energy valves recover energy during operation, reducing waste. Smart valves with sensors optimize flow and prevent unnecessary energy loss. These methods lower energy use, improve system efficiency, and support sustainability in industries like water distribution, oil and gas, and power generation.

1. Example of Pressure Use in Valves

Pressure-Reducing Valves (PRVs) with Turbines in Water Distribution Systems: In municipal water supply systems, *PRVs* are often installed to manage excess pressure in pipelines, especially in hilly or mountainous areas. Modern PRVs can be equipped with turbines to capture the excess pressure energy and convert it into electricity. For example, a PRV installed in a water pipeline in a mountainous region may throttle water flow to reduce pressure downstream while driving a small turbine attached to the valve. This turbine generates electricity that can power nearby infrastructure or feed back into the grid.

2. Hydraulic Use in an Oil and Gas Plant Valve

Energy Recovery from High-Pressure Gas in Natural Gas Processing Plants: In natural gas processing plants, *control valves* or *PRVs* are used to manage the pressure of natural gas as it moves through pipelines or into storage. A common application involves reducing high-pressure gas to meet pipeline specifications. Instead of dissipating the pressure energy as heat, this energy can be harnessed using a turboexpander system. For example, turboexpanders are integrated with the valve system. When gas passes through the valve and expands, the turboexpander recovers the hydraulic energy from the high-pressure gas. This energy is converted into mechanical energy, which can drive compressors or generators, reducing the plant's overall energy consumption.

Energy Star Compliance

Energy Star compliance in valves means that valve products or systems help meet the energy efficiency and environmental standards set by the US Environmental Protection Agency (EPA) and the Department of Energy (DOE) under the Energy Star program. Although Energy Star mainly applies to appliances and equipment, valves that improve energy efficiency in fluid systems can support Energy Star goals. These valves help reduce energy loss, improve system performance, and lower emissions by optimizing flow control, minimizing pressure drops, and preventing leaks.

Example: In a wind turbine, an example of an Energy Star–compliant valve is a hydraulic control valve used in the pitch control system, which adjusts the angle of the blades to optimize wind energy capture. These valves regulate the flow and pressure of hydraulic fluid efficiently, ensuring precise blade movement with minimal energy loss. Their low-friction design reduces power consumption, while tight sealing prevents leaks and fluid waste, aligning with Energy Star's focus on energy efficiency and sustainability.

Environmental Compliance

Environmental compliance refers to the adherence to environmental laws, regulations, standards, and other requirements set by governments or regulatory bodies to minimize environmental impact and ensure sustainable practices. For industries, including those utilizing valves, environmental compliance involves taking steps to reduce pollution, conserve resources, and operate responsibly while meeting both legal and ethical obligations. Figure 14.1 illustrates an industrial piping system featuring multiple medium-sized butterfly valves and pipelines within a well-maintained facility. An engineer, equipped with safety gear, is inspecting the system while reviewing technical drawings. The visible piping network is part of an HVAC (heating, ventilation, and air-conditioning) system and is designed to transport water. While water leakage may not directly harm the environment, it leads to resource waste, reduced energy efficiency, and potential corrosion risks. This image highlights environmental compliance in piping and valve systems, emphasizing the importance of regular inspections, a clean and organized workspace, and adherence to safety standards.

Environment Protection Agency

EPA is a federal agency in the United States responsible for protecting human health and the environment. Established in 1970, the EPA develops and enforces

FIGURE 14.1 Industrial piping system with medium-sized butterfly valves in an HVAC water distribution network. An engineer is inspecting the system while reviewing technical drawings. The clean and organized facility and regular inspection demonstrate environmental compliance through regular maintenance. (With permission: Shutterstock.)

regulations based on laws passed by Congress to address issues such as air and water pollution, hazardous waste management, chemical safety, and climate change.

Environmental Regulations

Environmental regulations play a crucial role in governing valve design, manufacturing, and operation to minimize environmental impact and ensure public health and safety.

Key environmental regulations for valves focus on reducing emissions, managing hazardous materials, and minimizing pollution. In the United States, the Clean Air Act (CAA) requires controlling fugitive emissions like methane and VOCs through Leak Detection and Repair (LDAR) programs. The Resource Conservation and Recovery Act (RCRA) regulates the handling and disposal of hazardous materials used in valve manufacturing and maintenance. In the European Union, the Industrial Emissions Directive (IED) enforces Best Available Techniques (BAT) to limit industrial pollution. Globally, ISO 15848 sets standards for low-emission valve design, ensuring compliance with fugitive emission limits. These regulations promote sustainability by reducing leaks, improving waste management, and lowering the environmental impact of valves.

Figure 14.2 illustrates a smart leak detection system integrated into a piping network. The leak detection sensor is positioned downstream of the valve,

FIGURE 14.2 A smart leak detection system integrated into a piping network. (With permission: Shutterstock.)

continuously monitoring for leaks. Upon detecting a leakage in the piping system, the sensor sends a signal to the valve, triggering its automatic closure to prevent further fluid or gas loss. This sensor-to-valve connection represents a digitalized system (IoT [Internet of Things]) technology for RTM, and automated response

F

Flow Optimization

Flow optimization in valves ensures smooth and efficient fluid movement, reducing energy losses and improving performance. This is achieved through streamlined flow paths, full-bore designs, and low-friction coatings and lining that minimize turbulence and pressure drops. Smart actuators and IoT sensors enable real-time adjustments, keeping valves in ideal positions based on system demands. Choosing the right valve type is also important—butterfly valves handle large flow volumes efficiently with less pressure drop than globe valves. However, globe valves provide precise flow control. Optimized valves help save energy and improve reliability for long-term performance. Figure 14.3 shows an engineer is using an ultrasonic flow meter, which is a nonintrusive method to measure flow rate inside the firewater valve and piping system. The red piping indicates a fire water or emergency water distribution system, where efficient flow is critical. The measurement helps in detecting pressure drops and flow inefficiencies inside the valve. This is a clear example of flow optimization technique in the piping systems, including valves.

FIGURE 14.3 An engineer using an ultrasonic flow meter to evaluate the valve flow efficiency and optimization. (With permission: Shutterstock.)

Fugitive Emission

Fugitive emissions in valves are leaks of gases or liquids into the air. These leaks happen in valve parts like stem packing, body joints, or body and bonnet joints. They are often invisible. Fugitive emissions affect the environment and sustainability. They include GHGs like methane and VOCs, which cause air pollution and climate change. Sustainable valve design reduces these leaks. It uses better sealing and packing, follows ISO 15848, and includes RTM. For example, Bellows stem seals provide a leak-free seal by using a flexible metal bellow to prevent fugitive emissions. Live-loaded packing uses springs or Belleville washers to maintain constant compression on the packing, reducing wear and ensuring a tighter, long-lasting seal. Chapter 5 provides more information about bellow stem seals, live loaded packings, and valve design for fugitive emission prevention. Reducing fugitive emissions helps industries meet regulations. It also lowers carbon footprints, saves resources, and improves energy efficiency. Figure 14.4 shows a pipeline valve leaking gas in an industrial plant. A faint cloud of vapor is visible.

G

Global Warming

Global warming refers to the long-term increase in Earth's average surface temperature as a result of human activities, particularly the emission of GHGs, such

FIGURE 14.4 Fugitive emissions from a valve in an industrial plant, illustrating gas or vapor leaks. (With permission: Shutterstock.)

as carbon dioxide, methane, and nitrous oxide. As a result of these gases, the Earth's climate warms, resulting in a number of environmental changes, including rising sea levels, more extreme weather events, and disruptions to ecosystems. In industrial settings, especially in the oil, gas, and chemical industries, there are several key considerations regarding global warming and valves. In the oil and gas industry, valves can be significant sources of fugitive emissions. Unless these emissions are properly controlled or repaired, they may directly contribute to global warming. Fugitives are gas or liquid leaks occurring from valves, pipes, flanges, and other equipment, typically caused by poor material selection and corrosion, worn seals, loose connections, or inadequate maintenance. Figure 14.5 illustrates the effects of global warming on Earth. It shows Earth is partially transformed into an industrial area with factories emitting thick black smoke. The lower half of the planet is burning, and drying out, symbolizing climate change, extreme heat, and environmental destruction.

Greenhouse Gas Emission/Greenhouse Effect

GHG emissions related to valves are primarily associated with fugitive emissions, where valves unintentionally release gases like methane, carbon dioxide, or other VOCs into the atmosphere.

Valves control fluids that contribute to GHG emissions in industries. CH_4 is the main component of natural gas and a strong GHG. It is handled in pipelines,

FIGURE 14.5 Effect of global warming on Earth and ecosystems. (With permission: Shutterstock.)

compressor stations, and LNG facilities. CO_2 is common in power generation and carbon capture. It plays a major role in global warming, and valves control its flow and storage. VOCs are used in oil refining and petrochemicals. They form secondary pollutants like ozone, which affect climate. Refrigerants like HFCs, CFCs, and HCFCs have high global warming potential (GWP). They are used in refrigeration and air-conditioning, requiring leak-proof valves. Ammonia is used in agriculture and chemical manufacturing. It leads to N_2O emissions, a GHG 300 times stronger than CO_2. Valves also regulate N_2O in chemical production and emissions control. Preventing leaks and ensuring valve efficiency are key to reducing emissions and protecting the environment.

Figure 14.6 shows a large industrial plant emitting thick clouds of smoke from multiple tall smokestacks, symbolizing the release of GHGs into the atmosphere. The sky appears hazy, reflecting the impact of pollution on air quality. The surrounding area is barren, with minimal vegetation, illustrating the adverse environmental effects of industrial emissions. The scene shows the significance of GHG emissions as a contributor to climate change and highlights the need for sustainable solutions to mitigate environmental damage.

Green Manufacturing

Green manufacturing is the practice of designing and implementing manufacturing processes that minimize environmental impact while conserving energy

FIGURE 14.6 Industrial greenhouse gas emissions from smokestacks, illustrating the impact of pollution on air quality and the environment. (With permission: Shutterstock.)

and natural resources. Green manufacturing for valves focuses on reducing environmental impact at every stage. It starts with using eco-friendly materials like recycled metals and low-carbon alloys. Energy-efficient processes, such as precision machining and additive manufacturing, reduce waste and energy use. Factories run on renewable energy and use water recycling systems to lower resource consumption. Advanced sealing technologies help prevent fugitive emissions, making valves more sustainable in industrial applications. Compliance with standards like ISO 14001 and ISO 15848 ensures low emissions and long-lasting performance. Digital monitoring and smart sensors improve efficiency and reduce maintenance needs. At the end of their life cycle, valves are designed for easy recycling or refurbishment, reducing waste and promoting sustainability.

H

Hazardous Air Pollutants (HAP)

HAP are toxic substances that are known or suspected to cause serious health effects, such as cancer, birth defects, reproductive issues, or other significant environmental harm. These pollutants are also referred to as "air toxics" and are regulated under the Clean Air Act by the US EPA. Mercury, lead, and asbestos are examples of HAP.

Hazardous Waste Management

Hazardous waste management for valves involves safe handling and disposing of valves exposed to harmful substances. Valves may contain toxic metals, coatings, or chemical residues that can harm people and the environment. For example, lead-based alloys, chromium coatings, or chemical buildup from process fluids must be carefully removed before valve disposal. During decommissioning, valves are checked for hazardous materials before disposal or recycling. Treatment methods include neutralization (using acids for chemical residues), encapsulation (sealing harmful substances), or incineration (burning waste safely). Some coatings, like cadmium plating, need special disposal to prevent contamination. Recycling programs separate nonhazardous metals like stainless steel or brass while ensuring hazardous parts are safely processed. Waste disposal follows laws from EPA, EU regulations, or local environmental agencies to prevent pollution. Proper management helps industries stay compliant and reduces environmental risks.

Hydraulic Efficiency

Hydraulic efficiency in valves refers to the ability of a valve to minimize energy losses caused by flow resistance, turbulence, and pressure drops, ensuring optimal fluid flow through the system; for more information, refer to Chapter 13 under the “Hydraulic Efficiency” section.

I

IoT for Sustainability

IoT for sustainability in valves involves integrating smart sensors, connectivity, and data analytics to optimize valve operations in industrial systems. IoT-enabled valves continuously monitor parameters like flow, pressure, and temperature, enabling real-time adjustments to reduce energy consumption and minimize resource waste. These valves can detect leaks or inefficiencies early, preventing environmental damage and improving system efficiency. Additionally, IoT systems collect data over time, allowing predictive maintenance to extend valve lifespan and reduce material waste.

L

Leak Detection and Repair

LDAR is a regulatory and operational program designed to identify and fix leaks of hazardous substances, particularly VOCs and GHGs, from industrial equipment such as valves, pumps, flanges, and connectors. LDAR programs are widely implemented in industries like oil and gas, petrochemicals, and chemical processing. LDAR main goals are to reduce emissions, improve safety, and comply with environmental regulations. Key components of LDAR are monitoring, detection, repair, and record keeping.

Example: Leak Detection and Repair for a Valve

In an industrial facility, LDAR for a valve typically involves routine monitoring to identify and mitigate fugitive emissions of VOCs or other harmful gases. For instance, a control valve in a chemical processing plant might be inspected using a portable VOC analyzer or an infrared camera to detect leaks. If a leak is identified through a flange connection or the stem seal, a repair team is notified to take corrective action. Possible actions are tightening bolts, replacing gaskets, or stem packing. The process is documented, including the leak rate, repair details, and follow-up inspections. This practice ensures compliance with environmental regulations like the US EPA's Method 21 or equivalent. In conclusion, this systematic approach minimizes emissions, maintains regulatory compliance, and reduces environmental impact.

LEED (Leadership in Energy and Environmental Design) Certification

LEED is a global certification that measures how eco-friendly and energy-efficient a building is. It was created by the US Green Building Council (USGBC) to help design and build sustainable buildings and communities. LEED encourages using green materials, saving energy and water, and reducing pollution to protect the environment.

LEED for Valves Installed on Offshore Oil and Gas Plant

In an offshore oil platform aiming for LEED certification, valves designed for sustainability can play a key role in minimizing environmental impact and optimizing resource use. For example, low-emission valves (e.g., valves meeting ISO 15848 standards) are used in process systems to reduce fugitive emissions of VOCs and GHGs. Additionally, pressure-regulating valves are installed in water injection systems to the reservoir for advanced oil recovery. These valves ensure precise control, reducing water waste and improving efficiency in reservoir management. If a pressure-PRV fails to regulate pressure effectively, water may be injected at excessively high pressures. Overpressure can force injected water or chemicals into freshwater aquifers, contaminating groundwater resources. This can fracture the reservoir rock, leading to uncontrolled water migration into unintended zones. Reservoir fractures may disrupt oil containment, causing oil seepage into the surrounding environment. Smart valves integrated with IoT sensors could be used for safety-critical valves to detect leaks and send alerts in real time, preventing resource wastage and environmental contamination. These measures not only enhance operational efficiency but also support the platform's efforts to achieve LEED credits in categories such as Energy and Atmosphere, Water Efficiency, and Indoor Environmental Quality.

Life Cycle Assessment (LCA)

Valve LCA is a systematic process of evaluating the environmental, economic, and functional impact of a valve throughout its entire life cycle. This assessment provides a comprehensive understanding of the valve's performance and sustainability by examining every stage, from raw material extraction to end-of-life disposal or recycling. The goal is to identify opportunities for improvement, reduce environmental footprints, and optimize resource efficiency.

Low-Emission Valve

A *low-emission valve* is designed to minimize the release of fugitive emissions, which are leaks of gases or vapors (such as methane or VOCs) into the atmosphere from valve components. These valves are engineered with advanced sealing systems, such as low-friction, high-performance stem seals, and gaskets that reduce leaks at critical points like the stem and body connections. They are rigorously tested for fugitive emissions to ensure compliance with stringent environmental standards, such as ISO 15848 or API 624. Low-emission valves are crucial in industries like oil and gas, chemical processing, and power generation. Reducing emissions is critical for environmental protection, regulatory compliance, and operational efficiency. By minimizing leaks, these valves also enhance safety and reduce product loss, making them a sustainable choice for industrial applications.

M

Methane Leakage

Methane leakage refers to the unintended release of methane gas into the atmosphere during the production, transportation, storage, or use of natural gas and other methane-rich fuels. Methane is a potent GHG, with a GWP approximately 25–30 times greater than carbon dioxide over a 100-year period, making its control critical for addressing climate change. The relationship between methane leakage and valves lies in the critical role valves play in controlling the flow of natural gas and other methane-rich substances in industrial systems. Methane leakage from valves is a significant contributor to GHG emissions and energy loss in industries like oil and gas. By designing and maintaining valves to minimize leakage, industries can enhance sustainability, comply with environmental regulations, and reduce their carbon footprint.

P

Pollution Control

Pollution control refers to the strategies, technologies, and actions implemented to reduce, manage, or eliminate the release of pollutants into the environment.

It aims to minimize the harmful effects of pollutants on human health, ecosystems, and natural resources by addressing their sources and mitigating their impact.

Example 1: Using Electrical or Pneumatic Actuators for the Valves: Some valves are integrated with pneumatic or electric actuators that eliminate the need for pressurized gas-powered actuators, which can vent emissions to the atmosphere. For example, in natural gas processing plants, replacing gas-operated actuators with electric actuators prevents methane leaks, a significant contributor to climate change.

Example 2: Valves equipped with advanced sealing technologies, such as graphite-based packing materials or live-loaded packing systems or bellows stem seals.

Predictive Maintenance and Sustainability

Predictive maintenance for valves plays a critical role in sustainability by leveraging advanced monitoring technologies to anticipate failures and optimize maintenance schedules. The integration of IoT sensors and data analytics in industrial valves has revolutionized predictive maintenance, improving both operational reliability and environmental safety. This approach minimizes unplanned downtime, prevents leaks, and reduces the risk of catastrophic failures that could lead to hazardous spills or emissions. By addressing potential issues before they escalate, it extends the operational lifespan of valves, conserving materials and reducing the need for frequent replacements.

Example 1: IoT Enables Predictive Maintenance for Leakage Prevention in High-Pressure Scenario

By continuously monitoring real-time performance metrics such as pressure, temperature, vibration, and flow rate, IoT-enabled systems help prevent catastrophic failures before they occur. For example, in an oil and gas processing plant, if a valve experiences an unexpected pressure surge that exceeds safe limits, this could lead to a valve rupture, releasing hazardous substances into the environment. However, an IoT-based predictive maintenance system can detect the anomaly before failure occurs. When a high-pressure condition is detected, the IoT system can automatically trigger a safety response, such as closing another valve upstream to isolate the affected area and prevent further pressure buildup. This real-time corrective action not only prevents valve damage but also reduces the risk of environmental contamination, such as methane leaks or oil spills.

Example 2: IoT Enables Predictive Maintenance for Energy Efficiency for a Valve

This example shows how IoT leads to energy efficiency. When a valve is not fully open or properly aligned, it restricts the flow of liquid or gas, creating a pressure drop. IoT sensors continuously monitor pressure, flow rate, and valve position. If they detect higher-than-normal pressure before the valve and lower pressure after the valve, it signals a partial blockage or misalignment. The system sends an alert to operators or automatically adjusts the valve to fix the issue.

Pressure Drop Reduction

Pressure drop reduction in valves focuses on minimizing the loss of fluid pressure as it flows through the valve, ensuring energy efficiency and optimal system performance. This is achieved by designing valves with smooth internal surfaces, streamlined flow paths, and larger port areas to reduce turbulence and friction. Features like full-bore designs, low-friction lining, and optimized geometries prevent unnecessary restrictions in the flow, allowing the fluid to pass through with minimal resistance.

R

Real-Time Monitoring and Sustainability

RTM in valves involves the use of sensors and IoT technology to continuously track parameters like pressure, flow rate, temperature, and valve position. This enhances sustainability by enabling precise control and efficient operation of fluid systems, minimizing energy and resource waste. By detecting leaks, inefficiencies, or anomalies instantly, RTM prevents fluid loss and mitigates environmental risks such as spills or emissions. It also optimizes energy usage in processes like water distribution or steam systems by ensuring valves function only as needed. Furthermore, the data collected supports predictive maintenance, reducing unnecessary repairs and extending valve lifespan, which conserves materials and reduces waste. More information about the difference between RTM and condition-CBM is provided under condition-monitoring section.

Figure 14.7 illustrates an engineer using a laptop with a holographic interface, likely displaying a process flow diagram (PFD) or piping and instrumentation diagram (P&ID). This digitalized monitoring system helps track valve and equipment operations, as well as process parameters such as pressure and flow rates in real time. The system is connected to control systems and sensors via the IoT or other communication networks, allowing for remote monitoring and automation. By visualizing the real-time status of the process system, operators can detect anomalies, optimize performance, and take corrective actions before failures occur. Using online monitoring improves safety and reliability of the systems.

Renewable Energy Valves

Renewable energy valve refers to the use of valves as critical components in systems that control and optimize renewable energy sources, such as wind, solar, hydropower, and geothermal energy. Valves play an essential role in managing

FIGURE 14.7 An engineer using a laptop with a holographic interface to monitor a process system, including industrial valves. (With permission: Shutterstock.)

the flow, pressure, and temperature of fluids or gases in renewable energy systems. Chapter 12 explains renewable energy valves in more detail.

Renewable Resources: Valve Manufacturing

Renewable resources in valve manufacturing refer to materials and energy sources that are naturally replenished and used to reduce the environmental impact of production and operation. For example, renewable energy sources like solar, wind, or hydropower can be utilized to power factories during valve manufacturing, minimizing carbon emissions compared to fossil fuels.

S

Smart Valves and Sustainability

Smart valves are advanced valves equipped with sensors, actuators, and IoT connectivity to monitor and control fluid flow, pressure, and temperature in real time. These valves enhance sustainability by optimizing resource usage, reducing energy consumption, and minimizing waste. By providing instant feedback and remote control capabilities, smart valves improve operational efficiency, detect leaks early, and prevent environmental hazards such as fluid spills or excess emissions. They also support predictive maintenance, reducing downtime and extending valve lifespan, which conserves materials and lowers operational costs.

Spill Containment

Spill containment in valves includes systems that prevent and manage fluid leaks, especially hazardous substances. Valves control liquids and gases that can be dangerous if spills occur. Drip trays are placed under valves to catch small leaks, making disposal or recycling easier. Secondary containment systems, like berms, bunds, or trenches, provide extra protection by capturing larger spills.

Other safety measures include absorbent materials like pads or granules that quickly soak up spills in emergencies. Double-seal mechanisms reduce leakage risk, while leak detection ports help identify leaks early. Flange guards protect against spray leaks, keeping fluids from harming people or the environment. These systems work together to enhance safety and prevent environmental damage.

Example: How Leak Detection Port Works in a Valve

A leak detection port is a feature in a valve that allows for early identification of leaks by monitoring the space between seals or in a specific cavity designed for this purpose. It helps in detecting leaks before they escape to the external environment, minimizing potential hazards and enabling timely maintenance. The leak detection port is typically located between the primary and secondary seals in a double-seal valve or within a cavity in the valve body. The port is connected to a monitoring device or left open for manual inspection. It can be used to detect leaks via pressure changes, fluid presence, or gas detectors. If a leak is detected, the port may trigger an alarm or visual indicator to alert operators for immediate action.

Standards

Standards help industries improve sustainability by providing guidelines for environmental management, efficiency, and safety. They ensure organizations minimize environmental impact, reduce waste, and comply with regulations.

For example, ISO 14001 focuses on environmental management systems, helping businesses manage resources efficiently and reduce waste. ISO 15848 sets strict leakage limits for industrial valves, reducing fugitive emissions and improving air quality. In the oil and gas industry, API 624 ensures that rising stem valves meet low-emission standards, supporting durability and environmental compliance.

By following these standards, industries can adopt sustainable practices, lower their environmental footprint, and contribute to greener technologies worldwide.

Sustainability

Sustainability in valves means making them eco-friendly, durable, and efficient. This involves using recyclable materials, reducing waste, and adopting clean manufacturing methods. Smart technologies like IoT sensors and predictive maintenance help cut energy use and prevent leaks. Sustainable valves last longer, lowering costs and supporting a circular economy. These practices help industries follow environmental laws, reduce carbon footprints, and save resources.

Sustainable Development Goals (SDGs)

SDGs are a set of 17 global objectives established by the United Nations in 2015 to promote peace, prosperity, and environmental protection for all by 2030. They are part of the 2030 Agenda for Sustainable Development and aim to address critical challenges such as climate change, environmental degradation, and sustainable economic growth.

Examples of Goals Relevant to Valves and Industries

- *Goal 7 (Affordable and Clean Energy):* Encourages the use of energy-efficient systems, such as low-pressure-drop valves, to reduce energy consumption.
- *Goal 9 (Industry, Innovation, and Infrastructure):* Promotes sustainable and resilient industrial practices, including the adoption of durable, recyclable materials in valve manufacturing.
- *Goal 12 (Responsible Consumption and Production):* Encourages circular economy practices, such as recycling and reusing valve components to minimize waste.
- *Goal 13 (Climate Action):* Advocates for reducing emissions, including fugitive emissions from valves, to combat climate change.

Sustainable Materials

Sustainable materials related to valves are eco-friendly and responsibly sourced materials designed to minimize environmental impact throughout the valve's life cycle. These materials include recyclable metals like stainless steel and bronze, which can be reused multiple times without degrading in quality. Sustainable materials also encompass lead-free alloys and coatings free from toxic substances, which comply with environmental regulations like RoHS (Restriction of Hazardous Substances). Additionally, composite materials made from renewable resources, such as bio-based plastics or natural fibers, are being used to replace traditional synthetics. By incorporating these materials, valve manufacturers can reduce carbon footprints, promote a circular economy, and comply with green engineering standards.

T

Thermal Efficiency

For more information, refer to Chapter 13 under the "Thermal Efficiency" section.

Toxicity

In the context of sustainability and the environmental impact of valves, toxicity refers to the potential of the materials, coatings, or substances used in valves to harm living organisms or the environment. This can be due to the release of toxic substances during the valve's manufacturing, use, or disposal.

Toxicity in valves refers to the environmental and health risks associated with the materials, coatings, and fluids they involve throughout their life cycle. Key concerns include the use of toxic metals (e.g., lead, cadmium) or hazardous coatings (e.g., those containing VOCs or heavy metals) that can have harmful substances during degradation or disposal. Corrosion can release toxic ions and leaks may expose the environment to hazardous fluids. Improper recycling or disposal of valves can further contaminate soil, water, or air. Reducing toxicity means choosing safe materials, using nontoxic coatings, making valves durable to prevent leaks, following regulations, and recycling properly to protect the environment.

V

Valve Life Cycle Optimization

Valve life cycle optimization involves managing a valve's performance, efficiency, and reliability throughout its entire life cycle, from design and manufacturing to operation, maintenance, and disposal. The goal is to maximize the valve's operational lifespan while minimizing costs, environmental impact, and downtime. This includes selecting durable materials like CRAs, using advanced coatings and linings to reduce corrosion and wear, and integrating smart technologies such as IoT sensors for RTM and predictive maintenance. Regular maintenance schedules, proper installation, and efficient recycling or repurposing at the end of life further contribute to life cycle optimization. By implementing these strategies, industries can ensure consistent performance, reduce total cost of ownership, and align with sustainability goals.

Valve Performance Analysis

Valve performance analysis with regard to sustainability involves evaluating a valve's efficiency, durability, and environmental impact to optimize its role in fluid control systems. This analysis focuses on parameters such as flow capacity, pressure drop, leakage rate, energy efficiency, and material durability. By using advanced tools like IoT sensors, computational modeling, and RTM, valve performance analysis identifies inefficiencies, predicts failures, and suggests operational adjustments to enhance sustainability.

Example: In a city's water distribution system, valve performance analysis was conducted to address excessive energy consumption and water wastage due to leaks and inefficiencies. IoT sensors monitored real-time data on flow rates, pressure, and valve positioning, while historical data and acoustic monitoring tools identified leaks and undersized valves causing high-pressure drops and energy inefficiencies. The analysis revealed worn-out seals, corrosion in valves, and unnecessary throttling in high-flow areas. As a result, leaking valves were replaced with low-leakage models, undersized valves were upgraded, and pressure-regulating valves with integrated turbines were installed to recover energy. Corroded valves were replaced with durable, corrosion-resistant materials, extending their lifespan. These improvements reduced energy consumption by 20%, conserved thousands of liters of water monthly, cut operational costs by 25%, and significantly lowered the system's carbon footprint. This example highlights the critical role of valve performance analysis in enhancing sustainability.

Valve Recycling

Valve recycling is the process of recovering and reusing materials and components from decommissioned or discarded valves to minimize waste and reduce environmental impact. It involves disassembling the valve into its constituent parts, such as metals, plastics, and elastomers, which are then sorted for recycling or disposal. Metals like stainless steel, brass, and cast iron can be melted and reused, while nonmetallic components, such as gaskets or coatings, may be repurposed or safely disposed of. Hazardous materials, including residues from coatings or fluids, must be treated according to environmental regulations before recycling. Recycling valves not only conserves resources but also reduces landfill waste and the carbon footprint associated with manufacturing new valves. It aligns with sustainability practices by promoting a circular economy in the valve industry.

Figure 14.8 illustrates the recycling process of brass valves, which are commonly used in plumbing and HVAC systems. Brass, an alloy of copper and zinc, is a nonferrous metal, making it highly recyclable. The image shows a large collection of brass valve bodies, prepared for melting and repurposing. Recycling brass reduces the demand for newly mined raw materials, leading to lower energy consumption and decreased CO_2 emissions, contributing to environmental sustainability.

Valve Repurposing (Refurbishment)

Valve repurposing involves reusing old or decommissioned valves for applications other than their original purpose, extending their useful life and reducing waste. Instead of being discarded, valves are inspected, repaired, or modified to meet the requirements of a different system or function. For example, a valve from an industrial process may be refurbished and repurposed for use in a less demanding application, such as in water management or noncritical systems.

Volatile Organic Compounds

VOCs are a group of organic chemicals that readily evaporate into the atmosphere at room temperature. They can exist in coating or paint and air fresheners. In addition, emission from valves handling natural gas and methane can release

FIGURE 14.8 Recycling of brass valves commonly used in plumbing and HVAC (heating, ventilation, and air-conditioning) systems. (With permission: Shutterstock.)

VOCs. In industrial emissions, these compounds are commonly found and can have a negative impact on the environment and human health. VOCs contribute to air pollution by forming smog and ground-level ozone, which are harmful to health and the environment. They react with nitrogen oxides (NO_x) in sunlight, creating ozone (O_3) and smog. VOCs come from both natural sources and human activities. Methane is a kind of VOC, but it behaves differently from many other VOCs. Unlike most VOCs, which react quickly in the atmosphere to form ground-level ozone and smog, methane is less reactive in this process. However, methane is a potent GHG, trapping heat about 80 times more effectively than CO_2 over 20 years.

Z

Zero Waste

Zero waste in valve manufacturing is a sustainable approach that aims to eliminate waste throughout the production, use, and end-of-life stages of valves. This involves reusing or recycling materials like metals (e.g., stainless steel) from decommissioned valves. In addition, repurposing manufacturing by-products (finding alternative uses for the leftover materials, scraps, or waste generated during the manufacturing process), instead of discarding them. In addition, designing valves with modular components that can be repaired or replaced instead of discarded. Modular components are individual parts of a system or product that are designed to be easily interchangeable, replaceable, or reconfigurable such as an actuator or a renewable valve seat (a valve seat welded or screwed or pressed into the body) without affecting the rest of the system. Additionally, production processes are optimized to minimize material scraps.

QUESTIONS AND ANSWERS

1. Which of the following combinations of sustainable development goals (SDGs) most directly aligns with promoting sustainability in valve manufacturing and industrial operations?
   A. Goal 3 (good health and well-being), Goal 8 (decent work and economic growth), and Goal 15 (life on land)
   B. Goal 6 (clean water and sanitation), Goal 7 (affordable and clean energy), and Goal 12 (responsible consumption and production)
   C. Goal 9 (industry, innovation, and infrastructure), Goal 11 (sustainable cities and communities), and Goal 16 (peace, justice, and strong institutions)
   D. Goal 5 (gender equality), Goal 8 (decent work and economic growth), and Goal 13 (climate action)

   Answer: Option B is the correct answer.

2. Which of the following practices best aligns with the principles of zero waste in valve manufacturing?
   A. Using modular valve designs that allow for the replacement of individual components
   B. Disposing of all worn-out valve components in landfills
   C. Manufacturing valves with nonrecyclable materials to maximize durability
   D. Ignoring by-products during the production process to reduce operational complexity

   Answer: Option A is the correct answer.

3. Which of the following stages is not typically included in a comprehensive life cycle assessment (LCA) for industrial valves?
   A. Raw material extraction
   B. Manufacturing and assembly
   C. Marketing and advertisement
   D. End-of-life management

   Answer: Option C is the correct answer. Life cycle assessment focuses on the environmental and resource impacts of a product throughout its life cycle, including raw material extraction, manufacturing, transportation, operation, and end-of-life management. Marketing and advertisement are not part of the LCA framework.

4. Which material in valves is commonly recycled due to its high reusability?
   A. Plastics
   B. Stainless steel
   C. Rubber gasket
   D. Coating

   Answer: Option B is the correct answer.

5. What must be done with hazardous materials found in valves during recycling?
   A. Ignored
   B. Reused without treatment
   C. Disposed of in regular landfills
   D. Treated according to environmental regulations

   Answer: Option D is the correct answer.

6. What is a key feature of energy-efficient pneumatic actuators?
   A. They operate without compressed air.
   B. They use air recirculation systems to reduce energy loss.
   C. They eliminate the need for seals.
   D. They require more pressure to operate.

   Answer: Option B is the correct answer.

7. Which of the following reduces energy waste in electric actuators during idle times?
   A. Increasing motor power
   B. Standby or low-power mode
   C. Continuous full-speed operation
   D. Removing IoT sensors

   Answer: Option B is the correct answer.

8. What is the primary benefit of using biodegradable sealing materials in valves?
   A. They decompose naturally, reducing environmental impact.
   B. They are more resistant to high temperatures.
   C. They are stronger than all synthetic material.
   D. They eliminate the need for maintenance.

   Answer: Option A is the correct answer.

9. Which of the following is an example of a biodegradable sealing material?
   A. PTFE
   B. Stainless steel
   C. Silicone
   D. Natural rubber

   Answer: Option D is the correct answer.

10. Which combination of design features in a valve contributes most effectively to fluid flow optimization while minimizing pressure drop, turbulence, and energy loss in a high-flow system?
    A. Full-bore design, streamlined internal geometry, and low-friction lining.
    B. Narrow bore, sharp internal transitions, and increased actuator force.
    C. Full-bore design, increased seal friction, and irregular flow path.
    D. Narrow bore, low-friction coatings, and advanced actuator controls.

    Answer: Option A is the correct answer.

11. Which of the following accurately describes how IoT, digital twins, real-time monitoring, and condition-based monitoring collectively contribute to sustainability in valve systems?
    A. IoT ensures continuous data collection, digital twins simulate operational scenarios to optimize efficiency, real-time monitoring detects anomalies instantly, and condition-based monitoring predicts maintenance needs based on valve wear.
    B. IoT provides simulations, digital twins detect leaks, real-time monitoring schedules long-term maintenance, and condition-based monitoring focuses on energy savings.
    C. IoT replaces the need for sensors, digital twins provide 3D-printed valve models, real-time monitoring eliminates manual oversight, and condition-based monitoring schedules inspections only when failures occur.
    D. IoT ensures maintenance schedules, digital twins operate valves autonomously, real-time monitoring handles material selection, and condition-based monitoring eliminates the need for predictive analytics.

    Answer: Option A is the correct answer.

BIBLIOGRAPHY

Al Ruqaishi, M. A., & Al Kalbani, F. K. (2024, November). Revolutionizing Leak Detection and Cost Savings in Valve Predictive Maintenance. Paper presented at ADIPEC. https://doi.org/10.2118/222620-MS

American Petroleum Institute. (2014). Type Testing of Rising Stem Valves Equipped with Flexible Graphite Packing for Fugitive Emissions (API Standard 624). 1st edition. American Petroleum Institute: Washington, DC.

Babaremu, K., Adediji, A., Olumba, N., Okoya, S., Akinlabi, E., & Oyinlola, M. (2024). Technological advances in mechanical recycling innovations and corresponding impacts on the circular economy of plastics. Environments, 11(3), 38. https://doi.org/10.3390/environments11030038

International Organization for Standardization. (2015). Industrial Valves: Measurement, Test and Qualification Procedures for Fugitive Emissions (ISO 15848-1:2015). ISO: Geneva, Switzerland.

Nesbitt, B. (2007). Handbook of Valves and Actuators: Valves Manual International. 1st edition. Elsevier: Oxford, UK.

Rangappa, R., Murthy, B. R. N., Porali, A., Rajesha, S., & Natarajan, N. (2021). Computational design optimization of industrial single-piece ball valve to enhance the flow performance. Journal of Physics: Conference Series, 2070(1), 012197. https://doi.org/10.1088/1742-6596/2070/1/012197

Sinagra, M., Creaco, E., & Morreale, G. et al. (2023). Energy recovery optimization by means of a turbine in a pressure regulation node of a real water network through a data-driven digital twin. Water Resources Management, 37, 4733–4749. https://doi.org/10.1007/s11269-023-03575-0

Sotoodeh, K. (2021). Prevention of Valve Fugitive Emissions in the Oil and Gas Industry. Gulf Professional Publishing: Houston, TX. https://doi.org/10.1016/C2021-0-00122-X

Sotoodeh, K. (2020). Development of a numeric method to validate the reliability improvement of safety critical valves during operation through online monitoring implementation. Sensing and Imaging, 21(1), 57. https://doi.org/10.1007/s11220-020-00323-1

Teles, D. B., Strohaecker, T. R., & Mariano, J. F. (2016, October). Project and validation of industrial valve for low fugitive emission. In Proceedings of the Rio Oil & Gas Expo and Conference 2016. Brazilian Institute for Oil, Gas and Biofuels: Rio de Janeiro, Brazil.

Zhang, Z., Zhou, Q., Jin, H., & Dai, Y. (2024). A real-time intelligent valve monitoring approach through cameras based on computer vision methods. Sensors, 24(16), 5337. https://doi.org/10.3390/s24165337

Index

A

Actuation system, 314
Actuator, 421
Actuator operation, 314, 450, 530
Additive manufacturing, 456
Advanced materials, 144
Air release valve, 63
Ambient conditions, 246, 374
Angle valve, 47, 475
Anti-blowout stem design, 159
Anti-cavitation trim, 315
Anti-static device, 93
API, 2, 15
API 607, 162
API 6D, 162
API 6DSS, 162
API 6FA, 162
ASME, 2, 162–163
ASME B16.10, 163
ASME B16.34, 163
ASME Section VIII, 369, 392, 397, 399
ATEX, 368
Auxiliary valve connections, 71
Automated valve, 2
Automation, 294
Axial flow valve, 29–30

B

Backflow, 33
Backpressure, 368
Backseat, 71, 233
Ball, 73–74
Ball valve, 30
Ball valve design, 30
Ball valve evolution, 15
Ball valve seat, 30
Base metal, 101
Bearing, 74
Belleville washer, 74–75, 185, 550
Bellows seal assembly, 75
Bellows design, 165
Bellows seal, 75
Bellows seal design, 165
Bidirectional seat, 166
Bidirectional valve, 166
Bimetallic, 507
Bending moment test, 234
Biomass applications, 511
Bleed valve, 172
Block and bleed valve, 172
Blow-out proof stem, 166
Body, 76
Body cavity, 80
Body cavity relief, 195
Body connections, 556
Body design, 203
Body test, 234
Body seals, 76
Bolt tightening, 211
Bolting, 77
Bolting materials, 77
Bonnet, 77
Bonnet design, 77
Boring, 211
Bottom cover, 78
Brittleness, 102
Bronze valves, 15
Bubble tight, 234
Buna-N, 102
Burst test, 234
Bushing, 78
Butterfly valve, 32
Butterfly valve disc, 32
Butterfly valve seat, 32
Bypass valve, 296

C

Cage, 79
Caged valve, 79
Carbon capture valve, 464
Carbon dioxide, 102
Carbon dioxide corrosion, 102
Carbon equivalent, 103
Carbon steel valves, 136
Cast iron, 104
Casting, 212
Cathodic protection, 105
Cavitation, 296
Cavity, 79
Cavity free design, 167
CE Marking, 168
Chamfering, 212
Charpy impact test, 107
Chattering, 372
Check valve, 33
Check valve disc, 33

Check valve types (CHVT), 34
Chemical industry, 3
Chemical resistance, 541
Christmas tree valve, 467
Clamp connection (with hubs), 80
Cleaning, 212
CNC Machining, 213
Closed loop control, 432
Closure member, 80
Coating, 214
Code, 168
Cold Differential Set Pressure (CDSP), 373
Combination valve, 39
Commissioning tests, 262
Compact ball actuators, 430
Composite materials, 561
Condition-based monitoring, 356
Concentric butterfly valve, 197
Control valve, 4
Corrosion allowance, 111
Corrosion resistance application, 214
Crevice corrosion, 114
Cryogenic applications, 169
Cryogenic extension bonnets, 174
Cryogenic materials, 114
Cryogenic test, 235
Cryogenic valve design, 169

D

Dash pot, 81
Data sheet, 169
Deflection, 276
Depth rating, 472
Design life, 169
Diaphragm valve, 58
Digital twin, 472
Digitalization, 472
Dimensioning, 215
Directional control valve, 474
Disc, 81
Discharge capacity, 301
Documentation and Records, 262
Double block and bleed valve, 172
Double offset butterfly valve, 173
Double piston effect seat, 173
Drain, 82
Drainage, 301
Drilling, 215
Downstream, 6
Dual-mode operation, 301
Ductility, 116
Duplex stainless steel, 116
Dye penetration testing, 252
Dynamic loads, 302

E

Eccentric butterfly valve, 173
Economic consideration in material selection, 117
Elastic deformation, 117
Elasticity, 117
Elastomer, 118
Electrical actuator, 314, 430
Electrohydraulic actuator, 431
Electromechanical actuator, 431
Emergency shutdown valves, 437
Endurance testing, 237
Energy industry, 6
Erosion, 6
Explosive decompression, 120
Extended bonnet, 174
External coatings, 488
External leakage, 241
External pressure, 432, 491

F

Face-to-face dimensions, 284
Factory acceptance testing, 237
Factory internal testing, 237
Fail-close actuator, 382
Fail-open actuator, 382
Failure modes, 336
Fatigue resistance, 138
Feedback system, 438
Ferrite, 121
Ferritic, 121
Ferritic stainless steel, 121
Final control element, 383
Fire test, 383
Fire-safe design, 175
Flange bolting, 211
Flange connections, 83, 215
Flange drilling, 217
Flange face, 176
Flange installation, 217
Flange inspection, 263
Flanged ends, 494
Flashing, 304
Flexible wedge gate valve, 176
Floating ball, 177
Flow area, 29
Flow characteristics, 304
Flow coefficient (Cv), 178
Flow curve, 304
Flow control, 6, 306
Flow direction marking, 264
Flow dynamics, 92
Flow management, 488
Flow path, 6

Flow path alignment, 275
Flow testing, 326
Fluid, 6
Fluid power, 438
Fluid compatibility, 68
Fluid dynamics, 311
Fluoropolymer coatings, 149
Forging, 217
Forge master, 218
Foundry, 218
Fouling resistance, 529
Frictional losses, 530
Fugitive emission, 7
Fugitive emission prevention design, 180
Full bore, 179
Functional testing, 279

G

Gasket, 82
Gate, 84
Gate valve, 40–46
Gate valve designs, 40–46
Gate valve pressure classes, 164
Gate valve types (GTVT), 40–46
Gearbox, 84
Gear-box operated design, 180
Gear ratio, 180
Gland, 84
Gland flange, 84
Gland flange (live loaded), 185
Gland bolting, 345, 351, 355
Global warming, 550–551
Globe valve, 47–49
Globe valve angle type, 47
Globe valve body styles, 47–49
Globe valve designs, 47–49
Globe valve tee type, 48
Globe valve types (GLVT), 47–49
Globe valve Y-type, 49
Graphite, 124
Grease injection for plug valves, 186
Grease injection system, 186
Greenhouse gas emission, 551–552
Green manufacturing, 552–553
Guided disc globe valve, 181

H

Handwheel, 84–86
Hard facing, 124, 151
Hardness difference, 338
Hardware fault tolerance, 385
Hastelloy, 126
Hazardous air pollutant (HAP), 553
Hazardous waste management, 554
Heat affected zone (HAZ), 126
Heat treatment, 126
Helical threads, 440
Helium leak test, 241
High cycle fatigue, 337
High integrity pressure protection systems (HIPPS), 386
High pressure drop, 48
High pressure systems, 7
High strength bolting, 182
Hinge, 86
Hinge pin, 86
Hot dip galvanized, 128
Hub, 87
HVAC, 7
Hydraulic actuator, 429
Hydraulic control valve, 547
Hydraulic efficiency, 554
Hydraulic system valve, 7
Hydrogen embrittlement, 128
Hydrogen-induced cracking corrosion, 128
Hydrogen sulfide, 128
Hydropower valve, 482
Hydrostatic pressure, 432–433
Hydrostatic test, 240
Hyperbaric test, 241

I

Identification tags, 265
Impact testing, 107
Industrial Revelation, 8
Industry 4.0, 485
Inherently safe design, 386
Inlet pressure, 356
Inside screw and yoke, 182
Instrument connection, 275
Instrumentation consideration, 262
Integrated seat, 182
Integrity verification, 275
Internal coatings, 488
Internal leakage, 242, 246
Internal parts hardness, 76–77
Internal testing, 237
IoT (Internet of things) sensors, 524
IoT for sustainability, 554
ISO 15848-1 testing, 239
ISO standards, 182
Isolation valves, 28

J

Jacketed design, 184
Jamming, 338
Joint efficiency, 276
Joint integrity, 276

K

Knife gate valve, 486

L

Leakage class, 184
Lever operated valve, 185
Lever operation, 313
Life cycle cost analysis, 519
Lifting lever for safety valve, 185
Lifting lug, 87
Line boring, 211
Liner flow, 313
Linear flow characteristic, 313
Liner, 88
Lip seal, 88
Live load gland flange, 185
Live load packing, 185
Locking devices, 88
Locking mechanism, 160
Loctite, 221
Long-term value, 519
Low emission, 553
Low temperature applications, 116
Lubricated plug valve, 186
Lug type design for butterfly valve, 187

M

Machining, 221
Maintenance procedures, 362
Manual valve, 9
Marking of flow direction, 231
Material certificates, 249
Material compatibility, 266
Material hardness, 125
Material selection, 133, 187
Material test record, 251
Material verification, 248
Mechanical seal, 418
Metal seated ball valve, 219
Metal spray, 242
Milling, 223
Milling slots, 223
Modular valve, 50
Monitoring, 362
Mounting bolt, 271
Mounting flange, 430
Multiport valve, 58
Multi-turn valves, 190

N

NDE (Non-Destructive Examination), 138
Needle valve, 27, 39
Non-lubricated plug valve, 52–53
Non-return valve, 28, 295
Non-rising stem valve, 190
Non-slam check valve, 190
Nozzle design for safety valve, 191
NPS (Nominal Pipe Size), 171
NPT threaded connection, 190

O

On/off control, 315
Operating mechanism, 316
Operation procedures, 324
Operational torque, 95
Overpressure prevention, 316

P

Packing for fugitive emission prevention, 550
Packing adjustment, 345
Packing box, 90
Packing gland, 75, 95
Packing leakage, 239
Packing materials, 180
Packing replacement, 72
Packing sealing, 9
Painting, 214
Passivation, 225
Pneumatic actuator, 434
Pneumatic control systems, 434
Pneumatic piping valves, 9
Plug, 90
Plug valve, 52–53
Plug valve design, 52–53
Plunger, 91
Polishing, 225
Porting, 225
Position indicator, 91
Position indicator, 444
Position sensor, 444
Positioner, 91
Positive material identification, 225
Post-installation checking and testing, 514
Post-weld heat treatment, 226
Predictive maintenance, 345
Preparation for shipment, 226
Pressure rating (pressure class based on ASME), 163
Pressure relief valve, 398–399
Pressure-containing parts, 92
Pressure-controlling parts, 92
Preventive maintenance, 346
Proactive maintenance, 346
Procedure qualification record (PQR), 227
Process systems, 10

Process wetted parts, 92
Protective coating, 158
PTFE sealing, 396
Pup piece, 92
Purging, 318

Q

Quarter-turn valve, 195
Quality assurance, 210
Quality control, 227
Quick closing valve, 318–319
Quick exhaust valve, 434
Quick opening valve, 305

R

Rack and pinion actuator, 446
Radiographic testing, 138
Rangeability, 319
Reaming, 245
Redundancy in safety system, 447
Regulating valve, 10
Relief valve (pressure), 55
Remote control, 447
Resistance to corrosion (cladding), 108
Resistance to wear, 529
Response time, 319
Reverse flow prevention, 320
Rotary actuator, 425
Rotary motion valve, 453

S

Safety integrity level (SIL), 358
Safety relief, 320
Safety relief valve, 320
Safety valve backpressure, 368
Safety valve integration, 277
Scotch yoke actuator, 423
Seal pocket, 92
Seal replacement, 347
Seat, 93
Seat leakage testing, 246
Seat material selection, 149
Seat pocket, 347
Seat replacement, 347
Seating surface, 93
Self-relieving seat, 173, 195
Shaft, 93
Shell testing, 246
Shutdown valve, 303
Signal wiring, 277
Smart valve design, 199
Soft seated valve, 199
Solid wedge, 199
Special alloy materials, 67
Special valves, 55, 466
Special valve types (SPVT), 56
Spring, 93
Spring return actuator and mechanism, 451
SSV (Surface Safety Valve), 21
Stem backseat, 71–73
Stem extension, 16, 175
Stem guide, 340
Stem key, 94
Stem seal, 94
Stem thread, 94
Stuffing box, 94
Subsea actuator, 432–433
Subsea valves, 499–500
Surface preparation, 216
Sustainable valve solutions, 533
System flushing, 277
System integration, 275

T

Tack weld, 149
Tagout, 266
Temporary tag plate, 266
Testing and verification, 278
Thermal relief, 431
Throat, 94
Thrust washer, 94
Torque efficiency, 425
Trim, 95
Trimming, 230
Triple offset valve, 173
Trunnion, 177, 202
Trunnion mounted ball valve, 202
Turning, 230
Two-way valve, 58

U

Ultimet, 151
Ultrasonic cleaning, 230
Under-torque, 230
Underground valves, 203
Unified numbering system, 151
Uniform corrosion, 152
Utility services, 173

V

Valve accessibility, 267
Valve actuator alignment, 268
Valve actuator integrity 275
Valve actuator signal 428

Valve automation, 421–455
Valve body and bonnet, 8
Valve cleaning, 213
Valve closure member, 80
Valve commissioning, 262
Valve configuration, 203
Valve design considerations, 470
Valve end connections, 173
Valve engineering standards, 171, 561
Valve fabrication, 149
Valve general arrangement drawing, 251
Valve handle, 88, 391
Valve installation, 262
Valve installation documentations and records, 7
Valve life cycle, 513
Valve maintenance, 515
Valve mechanical completion, 515
Valve orientation, 267
Valve packing material, 345
Valve piggability, 203
Valve preparation for installation, 265–266
Valve pressure classification, 163
Valve pressure drop, 194
Valve pressure relief testing, 244
Valve seat design, 200
Valve sizing, 204
Valve standards, 14, 162
Valve stem sealing
Valve testing, 14, 204, 232
Valve torque, 427
Valve types, 14, 26–70
Vibration resistance, 259
Viton, 152
Volatile Corrosion Inhibitor (VCI), 232

W

Wafer design, 204
Washer, 74, 94
Wear resistance, 529
Wedge gate valve, 46
Wedging effect, 204
Welded bonnet, 77
Welded valve body, 142
Weld-end valve, 174, 203
Welding Procedure Specification (WPS), 232
Wetted parts, 95
Worm gear operator, 309
Wrapping, 232

Y

Y-pattern globe valve, 49
Yield strength, 154
Yoke, 96

Z

Zero leakage, 248
Zero pressure, 366

FSC
www.fsc.org
MIX
Papier | Fördert
gute Waldnutzung
FSC® C083411

Kolibri 360 GmbH
Ferdinand-Jühlke-Straße 7
99095 Erfurt, Deutschland
produktsicherheit@kolibri360.de